Scientific Writing and Communication

PAPERS, PROPOSALS, AND PRESENTATIONS

Fourth Edition

Angelika H. Hofmann, PhD
Yale University

New York   Oxford
OXFORD UNIVERSITY PRESS

Oxford University Press is a department of the University of Oxford.
It furthers the University's objective of excellence in research, scholarship,
and education by publishing worldwide. Oxford is a registered trade mark of
Oxford University Press in the UK and certain other countries.

Published in the United States of America by Oxford University Press
198 Madison Avenue, New York, NY 10016, United States of America.

For titles covered by Section 112 of the US Higher Education
Opportunity Act, please visit www.oup.com/us/he for the latest
information about pricing and alternate formats.

Library of Congress Cataloging-in-Publication Data

Names: Hofmann, Angelika H., author.
Title: Scientific writing and communication : papers, proposals, and
    presentations / Angelika H. Hofmann, PhD.
Description: Fourth edition. | New York : Oxford University Press, c2020. |
    Includes bibliographical references and index. |
Identifiers: LCCN 2019017358 (print) | LCCN 2019020460 (ebook) | ISBN
    9780190063290 (epub) | ISBN 9780190063283 (pbk.)
Subjects: LCSH: Communication in science. | Scientific literature. |
    Technical writing.
Classification: LCC Q223 (ebook) | LCC Q223 .H63 2020 (print) | DDC
    808.06/65—dc23
LC record available at https://lccn.loc.gov/2019017358

Printing number: 9 8 7 6 5 4 3 2 1
Printed by Sheridan Books, Inc., United States of America

BRIEF CONTENTS

Preface   xvii

Part I   SCIENTIFIC WRITING BASICS: STYLE AND COMPOSITION   1

CHAPTER 1   Science and Communication   3

CHAPTER 2   Individual Words   14

CHAPTER 3   Word Location   36

CHAPTER 4   Technical Sentences   48

CHAPTER 5   Common Grammar Concerns   82

CHAPTER 6   From Sentences to Paragraphs   96

Part II   PLANNING AND LAYING THE FOUNDATION   123

CHAPTER 7   The First Draft   125

CHAPTER 8   References and Plagiarism   146

CHAPTER 9   Figures and Tables   176

CHAPTER 10   Basics of Statistical Analysis   214

Part III   MANUSCRIPTS: RESEARCH PAPERS AND REVIEW ARTICLES   231

A. Research Papers   231

CHAPTER 11   The Introduction   233

CHAPTER 12   Materials and Methods   260

CHAPTER 13   Results   280

CHAPTER 14   Discussion   304

CHAPTER 15   Abstract   329

CHAPTER 16 Titles, Key Words, Footnotes, and Acknowledgments 346

CHAPTER 17 Revising and Reviewing a Manuscript 357

CHAPTER 18 Final Version, Submission, and Peer Review 366

B. Review Articles 383

CHAPTER 19 Review Articles 385

Part IV GRANT PROPOSALS 407

CHAPTER 20 Proposal Writing 409

CHAPTER 21 Letters of Inquiry and Preproposals 424

CHAPTER 22 Proposal Abstracts and Specific Aims 440

CHAPTER 23 Background and Significance 459

CHAPTER 24 Innovation 476

CHAPTER 25 Preliminary Results 484

CHAPTER 26 Approach/Research Design 499

CHAPTER 27 Budget and Other Special Proposal Sections 516

CHAPTER 28 Revision and Submission of a Proposal 539

Part V POSTERS AND PRESENTATIONS 557

CHAPTER 29 Posters and Conference Abstracts 559

CHAPTER 30 Oral Presentations 583

Part VI JOB APPLICATIONS 617

CHAPTER 31 Job Applications and Interviews 619

Appendix A: Commonly Confused and Misused Words 657

Answer Key 673

Glossary of English Grammar Terms 704

Brief Glossary of Scientific and Technical Terms   708

Bibliography   712

Credits and References   716

Index   724

CONTENTS

Preface   xvii

Part One   SCIENTIFIC WRITING BASICS: STYLE
AND COMPOSITION   1

CHAPTER 1   **Science and Communication**   3
1.1   The Scientific Method   3
1.2   Science Communication and Ethics   5
1.3   About Readers   6
1.4   About Writers   8
1.5   Scientific Writing versus Science Writing   10
1.6   Mastering Scientific Writing   12
        Summary   13

CHAPTER 2   **Individual Words**   14
2.1   The Central Principle   14
2.2   Word Choice   15
2.3   Word Choice: Special Cases   17
2.4   Redundancies and Jargon   21
2.5   Abbreviations   24
2.6   Nomenclature and Terminology   25
2.7   Dictionaries   26
        Summary   30
        Problems   31

CHAPTER 3   **Word Location**   36
3.1   Readers' Expectations   36
3.2   Competition for Emphasis   37
3.3   Placement of Words   39
        Summary   45
        Problems   45

CHAPTER 4   **Technical Sentences**   48
4.1   Grammar and Technical Style   48
4.2   Person   49
4.3   Voice   50
4.4   Tense   51
4.5   Sentence Length   53

4.6 Verbs and Action 55
4.7 Clusters of Nouns 59
4.8 Pronouns 60
4.9 Lists and Comparisons 61
4.10 Faulty Comparisons 63
4.11 Common Errors 65
Summary 75
Problems 76

CHAPTER 5 **Common Grammar Concerns** 82
5.1 Prepositions 82
5.2 Articles 84
5.3 Verbs 85
5.4 Adjectives and Adverbs 88
5.5 Nouns and Pronouns 89
5.6 Grammar References 90
Summary 91
Problems 91

CHAPTER 6 **From Sentences to Paragraphs** 96
6.1 Paragraph Structure 96
6.2 Paragraph Organization 99
6.3 Paragraph Coherence 103
6.4 Condensing 110
Summary 116
Problems 117

Part Two PLANNING AND LAYING THE FOUNDATION 123

CHAPTER 7 **The First Draft** 125
7.1 The Writing Process 125
7.2 Prewriting 126
7.3 Authorship 128
7.4 Drafting a Manuscript 130
7.5 Outlining and Composing a Manuscript 133
7.6 Writer's Block? 139
7.7 For ESL Authors 140
7.8 Outside Help 141
Summary 143
Problems 143

CHAPTER 8   **References and Plagiarism**   146

    8.1   About References   146

    8.2   Source Material   147

    8.3   Managing References   150

    8.4   Text Citations   151

    8.5   Plagiarism   155

    8.6   Paraphrasing   157

    8.7   References within a Scientific Paper   162

    8.8   The Reference List   163

    8.9   Common Reference Styles   165

    8.10   Citing the Internet   167

         Summary   169

         Problems   170

CHAPTER 9   **Figures and Tables**   176

    9.1   General Guidelines   176

    9.2   Importance of Formatting and Placement of Information   177

    9.3   Figure or Table?   178

    9.4   General Information on Figures   180

    9.5   Types of Figures   182

    9.6   Formatting Graphs   189

    9.7   Examples of Graphs   193

    9.8   Figure Legends   199

    9.9   General Information on Tables   200

    9.10   Formatting Tables   200

    9.11   Formulas, Equations, Proofs, and Algorithms   204

         Summary   209

         Problems   210

CHAPTER 10   **Basics of Statistical Analysis**   214

    10.1   General Guidelines   214

    10.2   Basic Statistical Terminology   215

    10.3   Distribution Curves   217

    10.4   Statistical Analysis of Data   219

    10.5   Reporting Statistics   223

    10.6   Graphical Representation   225

    10.7   Useful Resources for Statistical Analysis   226

    10.8   Checklist   228

         Summary   229

         Problems   229

Part Three   MANUSCRIPTS: RESEARCH PAPERS AND REVIEW ARTICLES   231

A.   Research Papers   231

CHAPTER 11   **The Introduction**   233
11.1   Overall   233
11.2   Components and Format   234
11.3   Elements of the Introduction   235
11.4   Special Case: Introductions for Descriptive Papers   240
11.5   Important Writing Guidelines for the Introduction   241
11.6   Signals for the Reader   243
11.7   Common Problems of Introductions   244
11.8   Sample Introductions   250
11.9   Revising the Introduction   254
Summary   255
Problems   256

CHAPTER 12   **Materials and Methods**   260
12.1   Overall   260
12.2   Components   261
12.3   Format   265
12.4   Important Writing Guidelines for Materials and Methods   266
12.5   Ethical Conduct   269
12.6   Common Problems of the Materials and Methods Section   271
12.7   Sample Materials and Methods Sections   271
12.8   Revising the Materials and Methods Section   272
Summary   273
Problems   274

CHAPTER 13   **Results**   280
13.1   Overall   280
13.2   Components   281
13.3   Format   285
13.4   Important Writing Guidelines for the Results   289
13.5   Signals for the Reader   291
13.6   Common Problems of the Results Section   291
13.7   Sample Results Sections   295
13.8   Revising the Results Section   297
Summary   299
Problems   299

CHAPTER 14 **Discussion** 304
          14.1   Overall   304
          14.2   Components   305
          14.3   Format   305
          14.4   First Paragraph   306
          14.5   Middle Paragraphs   309
          14.6   Last Paragraph   313
          14.7   Important Writing Guidelines for the Discussion   315
          14.8   Signals for the Reader   316
          14.9   An Alternative: Results and Discussion   316
          14.10 Common Problems of the Discussion   318
          14.11 Sample Discussions   319
          14.12 Revising the Discussion   322
                 Summary   323
                 Problems   324

CHAPTER 15 **Abstract**   329
          15.1   Overall   329
          15.2   Components   330
          15.3   Format   330
          15.4   Applying Basic Writing Rules   335
          15.5   Signals for the Reader   336
          15.6   Common Problems of the Abstract   337
          15.7   Reasons for Rejection   341
          15.8   Revising the Abstract   341
                 Summary   343
                 Problems   343

CHAPTER 16 **Titles, Key Words, Footnotes, and Acknowledgments**   346
          16.1   Overall   346
          16.2   Strong Titles   347
          16.3   The Title Page   350
          16.4   Running Title   350
          16.5   Key Words   351
          16.6   Footnotes and Endnotes   352
          16.7   Acknowledgments   352
          16.8   Revising the Title   354
                 Summary   354
                 Problems   355

CHAPTER 17 **Revising and Reviewing a Manuscript**   357
          17.1   Revising the First Draft   357
          17.2   Subsequent Drafts   360

17.3   Reviewing a Manuscript Pre-Submission   362
        Summary   365

CHAPTER 18   **Final Version, Submission, and Peer Review**   366
18.1   General Advice on the Final Version   366
18.2   Submitting the Manuscript   367
18.3   Writing a Cover Letter   367
18.4   The Review Process   370
18.5   Informal Peer Review   371
18.6   Formal Peer Review   372
18.7   Checklist for Peer Review   372
18.8   Letter from the Editor   373
18.9   Resubmission   377
18.10   Paper Accepted   380
        Summary   381
        Problems   382

B.   **Review Articles   383**

CHAPTER 19   **Review Articles**   385
19.1   Overall   385
19.2   Types of Reviews and General Content   386
19.3   Format   388
19.4   Title   391
19.5   Abstract of a Review Article   392
19.6   Introduction of a Review Article   394
19.7   Main Analysis Section of a Review Article   396
19.8   Conclusion of a Review Article   399
19.9   References   401
19.10   Signals for the Reader   401
19.11   Coherence   401
19.12   Common Problems of Review Articles   402
19.13   Revising the Review Article   402
        Summary   403
        Problems   405

Part Four   **GRANT PROPOSALS   407**

CHAPTER 20   **Proposal Writing**   409
20.1   General   409
20.2   Types of Proposals   411
20.3   Choosing a Sponsoring Agency   412
20.4   Federal Agencies   412
20.5   Private Foundations   414

20.6 Corporations and Other Funders **415**
20.7 Preliminary Steps to Writing a Proposal **417**
20.8 Online Resources **420**
20.9 Starting to Write a Grant **421**
20.10 Interacting with the Funder **422**
Summary **423**

CHAPTER 21 **Letters of Inquiry and Preproposals** 424
21.1 General **424**
21.2 Components and Format **425**
21.3 Abstract/Overview **427**
21.4 Introduction/Background **428**
21.5 Statement of Need **429**
21.6 Objective and Specific Aims **429**
21.7 Strategy and Goals **430**
21.8 Leadership and Organization **431**
21.9 Budget **432**
21.10 Impact and Significance **433**
21.11 Cover Letter **434**
21.12 Verbal Proposals **434**
21.13 LOI Outlines **435**
21.14 Revising an LOI/Preproposal **436**
Summary **437**
Problems **438**

CHAPTER 22 **Proposal Abstracts and Specific Aims** 440
22.1 Overall **440**
22.2 Proposal Abstracts **441**
22.3 Specific Aims **448**
22.4 Significance and Impact **449**
22.5 Applying Basic Writing Rules **450**
22.6 Signals for the Reader **451**
22.7 Common Problems **451**
22.8 Reasons for Rejection **453**
22.9 Revising the Abstract and Specific Aims **453**
Summary **455**
Problems **455**

CHAPTER 23 **Background and Significance** 459
23.1 Overall **459**
23.2 Emphasis, Format, and Length **459**
23.3 References **460**
23.4 Elements of the Section **461**
23.5 Sample Significance Section for Federal Grants **466**

23.6 Signals for the Reader   469
23.7 Coherence   470
23.8 Common Problems   470
23.9 Revising the Background and Significance Section   471
Summary   472
Problems   473

CHAPTER 24   **Innovation**   476
24.1 General Remarks on Proposal Sections   476
24.2 Components   477
24.3 Format   478
24.4 Signals for the Reader   480
24.5 Common Problems   480
24.6 Revising the Innovation Section   481
Summary   483

CHAPTER 25   **Preliminary Results**   484
25.1 Function   484
25.2 Content   485
25.3 Format   487
25.4 Important Writing Rules   492
25.5 Signals for Preliminary Results   494
25.6 Common Problems of Preliminary Results   494
25.7 Revising the Preliminary Results   495
Summary   496
Problems   497

CHAPTER 26   **Approach/Research Design**   499
26.1 Overall   499
26.2 Components   500
26.3 Format   501
26.4 Closing Paragraph   506
26.5 Signals for the Reader   508
26.6 Common Problems   509
26.7 Revising the Research Design and Methods Section   512
Summary   513
Problems   514

CHAPTER 27   **Budget and Other Special Proposal Sections**   516
27.1 Budget   517
27.2 Other Special Proposal Sections   522
Summary   538

CHAPTER 28   **Revision and Submission of a Proposal**   539
28.1   General   539
28.2   Before Sending Out the Proposal   540
28.3   Revising the Proposal   540
28.4   Submitting the Proposal   543
28.5   Being Reviewed   545
28.6   Site Visits   550
28.7   Reasons for Rejection   551
28.8   If Your Proposal Is Rejected   552
28.9   Resubmission of a Proposal   552
28.10   If Your Proposal Is Funded   554
Summary   555

Part Five   POSTERS AND PRESENTATIONS   557

CHAPTER 29   **Posters and Conference Abstracts**   559
29.1   Function and General Overview   559
29.2   Conference Abstracts   560
29.3   Poster Components   562
29.4   Poster Format   564
29.5   Sections of a Poster   567
29.6   Photos, Figures, and Tables for Posters   573
29.7   Resources for Preparing and Presenting a Poster   576
29.8   Revising a Poster   577
29.9   Presenting a Poster   578
29.10   Sample Posters   578
29.11   Checklist for a Poster   581
Summary   581

CHAPTER 30   **Oral Presentations**   583
30.1   Before the Talk   583
30.2   Components and Format of a Scientific Talk   584
30.3   Visual Aids   587
30.4   Planning and Preparing for a Talk   596
30.5   Giving the Talk   600
30.6   Voice and Delivery   601
30.7   Vocabulary and Style   603
30.8   Body Actions and Motions   604
30.9   At the End of the Presentation   606
30.10   Questions and Answers   607
30.11   Other Speech Forms   608

30.12 Resources   **610**
30.13 Checklist for an Oral Presentation   **610**
       Summary   **611**
       Problems   **612**

Part Six   JOB APPLICATIONS   617

CHAPTER 31   **Job Applications and Interviews**   759
31.1   Overall   **619**
31.2   Curricula Vitae (CVs) and Résumés   **620**
31.3   Cover Letters   **631**
31.4   Accompanying Documents   **633**
31.5   Research Statements   **634**
31.6   Teaching Statements   **638**
31.7   The Hiring Process and Interview Questions   **643**
31.8   Resources   **649**
31.9   Letters of Recommendation   **650**
31.10   Checklist for the Job Application   **654**
        Summary   **655**

Appendix A: Commonly Confused and Misused
               Words   657

Answer Key   673

Glossary of English Grammar Terms   704

Brief Glossary of Scientific and Technical Terms   708

Bibliography   712

Credits and References   716

Index   724

PREFACE

Clear communication is a requirement, not an option, for a good scientist. Without such communication skills, scientists stand little chance of publishing their work, obtaining funds, or attracting a wide audience when giving a talk. Even the most promising discovery is worth little if it cannot be communicated successfully. *Scientific Writing and Communication: Papers, Proposals, and Presentations*, Fourth Edition, serves as a **comprehensive "one-stop" reference guide to scientific writing and communication.**

Despite the fundamental role of communication in the sciences, most researchers are not formally trained in scientific writing and thus have only a skeletal knowledge of basic scientific writing principles. Although most universities and colleges are aware of this problem, few of them offer formal training to their students and staff.

Scientific Writing and Communication: Papers, Proposals, and Presentations, Fourth Edition, shows you how to write clearly as a scientific author or technical writer and how to recognize shortcomings in your own writing and communication. The book targets a broad audience ranging from upper-level undergraduate students to graduate students, from postdoctoral fellows and faculty to fully fledged researchers and professional writers. Although *Papers, Proposals, and Presentations* can be used as a textbook, it is structured such that it is equally self-explanatory, allowing you to understand how to write English publications or proposals and to present scientific talks without having to take a class.

DESIGN OF THIS BOOK

This book consists of six main parts plus appendices:

   I. Scientific Writing Basics: Style and Composition
   II. Planning and Laying the Foundation
   III. Manuscripts: Research Papers and Review Articles
   IV. Grant Proposals
   V. Posters and Presentations
   VI. Job Applications
      Appendices on Commonly Confused Terms, MS Word, Excel, and PowerPoint tips.

Part I provides an overview of scientific writing and presents 30 basic scientific writing rules of technical style and composition that every scientific writer should know, including word choice, sentence structure, sentence location, and paragraph construction. These rules are generally applicable to all scientific documents and emphasize such fundamental principles as word location, details of grammar/technical style, and reader interpretation. In addition to the rules, the reader will also find guidelines, which apply to specific sections of a scientific document or presentation or to scientific writing in general.

Part II describes the key steps that must be taken to prepare a well-written scientific document. It includes a section on how to start the writing process and discusses authorship. It also explains how to collect, successfully manage, and use references as well as how to avoid plagiarism. In addition, Part II outlines important guidelines for preparing figures and tables. In contrast to the basic scientific writing rules, guidelines—designated by ➤—apply to specific sections of a scientific document or presentation and may not be universally applicable, but rather situation-dependent.

In Part III, authors learn to apply the basic rules of Part I to writing and revising individual sections of a scientific research paper or review article. Authors are introduced to structural guidelines important for writing each section of an article and are given many examples of well-written sections as well as examples of sections that would benefit from revision. In addition, Part III provides an overview of how to revise a manuscript and how to submit it to a journal.

Part IV covers diverse sections on grant writing. These chapters provide information about federal and private funding organizations and guidelines for writing letters of inquiry and the different sections of a grant proposal. In addition, Part IV also provides instructions on how to submit a proposal and how to communicate with the funding agency.

Part V instructs scientists on how to prepare and present effective oral presentations and posters. In these chapters, real examples of slides and posters are provided, as is advice on effective speaking, combating stage fright, and fielding questions. To guide presentations, basic guidelines on visual aids and the content of a talk or poster are given.

Part VI of this book completes this series on scientific writing skills by providing information on job applications. This section includes advice on how to prepare a curriculum vitae (CV) and how to ask for and write letters of recommendation. Examples of well-written research and teaching statements are also included in Part VI.

The appendices round up the information on scientific writing and communication for those looking for a comprehensive list of commonly confused terms or technical tips on computer software, including MS Word, Excel, and PowerPoint, in composing scientific documents or presentations.

Throughout the book, readers will be guided by rules, guidelines, and annotations. Visually, these elements are designated as follows:

★ – Rules that are generally applicable to all scientific documents, emphasizing fundamental principles

➤ - Guidelines that apply to specific sections of a scientific document or presentation or to scientific writing in general

👍 - Good examples of writing and composition

👎 - Bad examples of writing and composition

HALLMARK FEATURES

Scientific Writing and Communication: Papers, Proposals, and Presentations has been used successfully for a number of years in courses on scientific writing at various universities and institutes worldwide. Like its predecessors, the fourth edition presents basic writing principles and applies these to composing research articles, review articles, proposals, job applications, posters, and oral presentations. Proven hallmark features include the following:

- **A practical presentation** carefully introduces basic writing mechanics before moving into manuscript planning and organizational strategies.
- **Relevant and multidisciplinary examples** are taken from real research papers and grant proposals by writers ranging from students to Nobel laureates. Good and bad examples, often annotated, are drawn from a broad range of scientific disciplines, including medicine, molecular biology, biochemistry, ecology, geology, chemistry, engineering, and physics.
- **Extensive end-of-chapter exercise sets** provide the opportunity to practice style and composition principles.
- **Writing guidelines and revision checklists** warn scientists against common pitfalls and equip them with the most successful techniques to revise a scientific paper, review article, or grant proposal.
- **Annotated text passages** bring the writing principles and guidelines to life by applying them to real-world, relevant, and multidisciplinary examples.
- **Many tables with sample sentences and phrases** are given that apply to different sections of a scientific paper, review article, or grant proposal for beginning scientific writers, non-native speakers, those struggling with writer's block, or those preparing to deliver a talk or poster at a conference.
- **Special features for ESL students and researchers** are presented in an easy-to-follow style, appealing to both native and non-native English speakers.

NEW TO THE FOURTH EDITION

Since the publication of the first, second, and third editions, I have heard from many professors and students that they found the text's comprehensive, practical, and hands-on approach to be of great value as they produced a wide range of scientific documents. Listening to their comments, I have revised the text with the goal of expanding these hallmark features and providing a few all-new resources. Specific updates and improvements in the fourth edition include the following:

- **A new section on "media literacy"** guides students and others in evaluating and verifying good versus bad sources.
- **A new section on scientific versus science writing** provides context and understanding of the difference of these genres.
- **A discussion of open access journals and electronic publishing** lays out these new and growing trends in scientific publishing.
- **An expanded section on scientific ethics** discusses the importance of these issues and provides guidance on key questions.

- **An updated section on basic statistical analysis** expands the fundamentals of reporting statistical data and analyses in a scientific context.
- **An expanded section on plagiarism** intends to guide students on avoiding this pitfall and makes them aware of important bioethical issues.
- **An expanded discussion of peer review and revision of draft papers** informs readers on what constitutes constructive comments and evaluations.
- **An updated chapter on job application** expands on the format and content of CVs and résumés.
- **New information on Google docs, sheets, and slides** informs about file sharing and options to work on the same document simultaneously.
- **Updated examples and exercises throughout the book** include current hot topics in the scientific field.
- **Expanded online resources** include an online instructional video guide for appropriate chapters, online solutions for homework assignments, online exercises, and updated online appendices on MS Word, Excel, and PowerPoint.
- **Updated PowerPoint slides** accompany the revisions of the fourth edition of *Scientific Writing and Communication: Papers, Proposals, and Presentations.*

Writing a clear research paper or grant proposal, and presenting an articulate talk, can be difficult for any scientist, but this difficulty is by no means insurmountable. You, the writer, must *practice* writing and thinking within this structure and learn by example from the writings of others. Ultimately, with guidance and practice, any scientist should be able to write a paper or proposal that sparkles with clarity and to deliver an engaging presentation. As you write your own papers or prepare your talks, you will recognize that every project has its unique challenges and that you will need practice and good judgment to apply all the writing and communication principles presented herein. In giving due attention to composition, style, and impact, your communication skills will improve significantly, and this book will have accomplished its purpose.

ACKNOWLEDGMENTS

Scientific Writing and Communication: Papers, Proposals, and Presentations, Fourth Edition, attempts to capture critical ideas in effective scientific writing in one location. As is the case for research papers and grant proposals, this book builds on previous works published on communicating in the sciences and beyond. These tremendously valuable resources are listed in the bibliography at the end of the book.

I would also like to acknowledge my students, friends, and colleagues from the Max-Planck Institute, the Fritz-Haber Institute, the Humboldt University, the University of Carabobo, the University of Massachusetts at Worcester, and Yale University who have shared information and ideas across the sciences. I am particularly thankful to all the students and scientists who were courageous enough to allow me to use draft sentences, paragraphs, or sections as examples or problems in this book. Without these samples the book would not be nearly as effective in exemplifying clear writing.

I am also grateful to all my friends and colleagues who have edited and commented on various draft chapters, and to those who have organized and supported courses, seminars, and workshops. Above all, I would like to thank Betty Liu, Francois Franceschi, Bettina Holzheimer, Lisa DeCrosta, Tracy Plumley, Anita and Peter Todd, James Hagen, Roopashree Narasimhaiah, Gail Emilsson, Zandra Ruiz, Jane Hadjimichael, Paola Crucitti, Riccardo Missich, and Francisco Triana for their encouragement as well as their critical comments and the many, many helpful discussions over the years.

Furthermore, I am grateful to all the reviewers who have edited and commented on various draft chapters of the fourth edition, including:

Romi Lynn Burks, *Southwestern University*
Sudarshan Chawathe, *University of Maine*
David Fisher, *Emory University*
David Garrison, *University of Houston Clear Lake*
Gary Heiman, *Rutgers University*
Cleo Hughes Darden, *Morgan State University*
Patrick Kelly, *CSU Stanislaus*
Elizabeth Kitchens, *University of Alabama*
Melanie Lee-Brown, *Guilford College*
Mark Lesser, *SUNY Plattsburgh*
Marta Maron, *University of Colorado Denver*
Stephen Mech, *Albright College*
Jennifer Nelson, *Indiana University Purdue*
Jennifer Pharr, *University of Nevada Las Vegas*
Eric Potma, *University of California Irvine*
Brian Schilling, *University of Nevada Las Vegas*
Sarah Wyatt, *Ohio University*
Ken Yasukawa, *Beloit College*

I would also like to acknowledge those reviewers of the first, second, and third edition, whose advice and comments were instrumental in establishing the foundation of this text: Allison Abbott, Daniel Abel, Stephen Amato, Gavin E. Arteel, Brian Avery, David Baumgardner, James Bednarz, Kathy Bernard, Debra Biasca, Erin J. Burge, J. Harrison Carpenter, Mark Clarke, Daun Daemon, Carleton DeTar, Jeffrey A. Donnell, Carlton Erickson, Joseph W. Francis, Christine Freeman, Cathryn Frere, David Gangitano, Mariëlle Hoefnagels, Alan Hogg, Todd Hurd, Katherine Kantardjieff, Diana Sue Katz Amburn, Thomas Kolb, Marilyn James-Kracke, Michael Laughter, Theo Light, David Major, William Matter, Susan McDowell, Curtis Meadow, Anthony A. Miller, Gene Ness, Steve Nizielski, Daniel O' Connor, Katherine Palacio, David Penetar, Florence Petrofes, John Placyk, Scott Pleasant, Roger A. Powell, Susan Semple-Rowland, Irving Rothman, Tony Schountz, Bernard G. Schreurs, Gloria Sciuto, John Thomlinson, Anne Windham, Dan Williams, Patricia Weis-Taylor, Linda L. Werling, Timothy Wright, Sarah Wyatt, Ken Yasukawa, and Marianna J. Zamlauski-Tucker.

Finally, I would like to express appreciation to everyone at Oxford University Press: Jason Noe, Senior Editor; Katie Tunkavige, Assistant Editor; Petra Recter, Director of Content and Digital Strategy; John Challice, Publisher and Vice President; Chris Bowers, Director of Marketing; Tina Chapman, Marketing Manager; Michelle Chang, Marketing Assistant; Theresa Stockton, Production Manager; Brad Rau, Production Editor; and Michele Laseau, Senior Designer.

PART ONE

Scientific Writing Basics

STYLE AND COMPOSITION

CHAPTER 1

Science and Communication

- The scientific method
- Scientific ethics
- The basic precept of writing
- Science writing versus scientific writing
- Expository writing versus technical writing
- The importance of practice

1.1 THE SCIENTIFIC METHOD

➤ Understand the scientific method

Science studies the natural world. It includes many different fields, from the life sciences to earth sciences to physical sciences; it also encompasses other specialized disciplines, such as anthropology and psychology. Through observational studies and experiments, scientists try to gather information and derive models to explain diverse phenomena. The series of steps involved in such studies have a similar pattern throughout the diverse scientific fields, although they are not rigid. These steps are collectively known as the scientific method and consist of

- Asking a question/making an observation
- Proposing a hypothesis
- Testing the predictions of a hypothesis through experiments or observations
- Analyzing and interpreting the data/drawing conclusions
- Communicating results

A scientific hypothesis is a highly probable proposition or explanation, based on observations and prior knowledge, that can be investigated further. It explains how or why a natural phenomenon occurs and makes at least one unique, testable prediction (e.g., hypothesis: the boiling temperature of water depends on altitude; prediction: water will boil at a lower temperature at an altitude of 5,000 ft than at sea level). If the hypothesis is wrong, scientists often come up with a new hypothesis. They then repeat the scientific steps to test it by experimentation or observation and publish their findings. If the hypothesis is widely accepted and broad enough in scope, it can turn into a theory and eventually into a law, on which scientists can build to advance research and knowledge further.

It is the last point of the scientific method, communicating results, with which this book is mainly concerned. Without clear communication, even the best results in science mean little. As a science student, it is therefore essential to be trained not only in observation, formulation of a hypothesis, and research methods but also in scientific communication. Learning how to communicate in science will prepare you well for the professional world. Without these skills, it will be difficult to succeed as your career will depend on publications, successful proposals, and clear presentations and posters. Communication in the professional scientific world includes

- Original scientific research articles—to communicate findings to other scientists and to the public
- Review articles—to glean and communicate in-depth interpretations of current topics published in research articles
- Grant proposals—to apply for funding of research
- Posters and oral presentations—to present your work visually and orally
- Science news articles, blogs, social media postings, or lectures—to communicate science effectively to the public, students, and so forth
- Evaluation of the work of others—for example, as a reviewer of a manuscript or a grant proposal
- Letters of recommendation
- Progress reports
- Cover letters
- Job applications

This book is meant to prepare you for these professional skills by teaching you how to research and use references, present and report on data, write research articles, compose reviews, draft grant proposals, and prepare presentations, posters, and even job applications. To generate such documents and presentations, you need a good foundation, including basic writing skills from the smallest units (words) to the larger ones (sentences, paragraphs, and full articles), all of which are covered in this book.

1.2   SCIENCE COMMUNICATION AND ETHICS

➤ Strive to communicate your findings clearly

Many readers think that scientific documents are generally hard to read because scientific concepts and topics are complex. However, this complexity does not need to result in difficult communication. It is important that readers accurately perceive what you as the author had in mind and that you avoid misunderstandings. To become an effective and successful author, you can and should strive to communicate clearly without oversimplifying scientific issues.

➤ Be ethical

Good science does not only need to be communicated well and effectively, it also has to pass scientific ethics, which apply to all scientists in their research and professional endeavors. Ethical norms cover a wide range of topics, from use of human subjects, such as fetal tissue, to fraud, sponsorship of research, and plagiarism (see also Chapter 8, Sections 8.5 and 8.6, and Chapter 12, Section 12.5), and are governed by standards in diverse scientific disciplines, law, and business. Such norms are important for the following reasons:

- To ensure accuracy and truth (and avoid fabrication, fraud, falsification, and misrepresentation of data)
- To ensure mutual respect, fairness, and trust (and thus make collaborations possible while protecting intellectual property rights, copyright, patenting rules, and authorship recognition)
- To hold scientists accountable (and avoid research misconduct, conflict of interests, human or animal harm)

Accordingly, codes of ethical conduct for scientists have been established by diverse professional organizations, journals, universities, institutes, and government agencies, such as the National Institutes of Health (NIH), the Food and Drug Administration (FDA), and the Environmental Protection Agency (EPA). In medical research, key agreements include the Declaration of Helsinki, a set of ethical principles for research on human subjects developed by the World Medical Association (available at https://www.wma.net), and the Nuremberg Code, a set of ethical principles for experimentation on humans established after World War II (http://ohsr.od.nih.gov/guidelines/nuremberg.html).

Ethics in science also extends to scientific writing and communication. Scientific misconduct is federally defined as intended actions, such as fabrication and falsification of data and plagiarism. All researchers view these as unethical. Note, though, that human errors, sloppiness, miscalculations, bias, disparities of methods and interpretations, and even negligence are not classified as misconduct, although scientists strive to avoid them as well. Other deviations not defined as misconduct also are viewed as unethical by most scientists: stealing someone else's ideas or data, submitting or publishing the same papers in different journals, including someone who has not contributed to a project as an author on a paper, filing a patent without informing collaborators, asking sexual favors in exchange for authorship or a grade, exploiting students and postdocs, misrepresenting facts on a curriculum vitae (CV), not following protocols of animal care, and more. Although these offenses are considered serious, they do not fall within the official federal definition of scientific misconduct and are the subject of much ongoing discussions on this topic.

Most academic institutions offer, and even require, training in the responsible conduct of research and in research ethics in the hopes of reducing the rate of serious deviations. Such training allows researchers to understand ethical issues and challenges and informs them on how to handle situations and dilemmas they may encounter.

➤ Become media literate

At a time when sharing and accessing information is easier than ever and when it seems difficult to distinguish peer-reviewed from nonpeer-reviewed articles and valid scientific data and facts from unverified ones, clear and high-quality scientific communication has never been more important. It is therefore imperative that you verify information when you compose your documents. Do not rely on references in other articles, subjective opinions, policy-based evidence, unsupported facts, and nonpeer-reviewed reports or Web sources (see also Chapter 8, Section 8.2 and 8.10). Moreover, do not report your opinion as fact. Rather, clearly identify your opinion and conclusions (*Our findings indicate that . . .* ; *A possible model of X could be . . .*).

Valuing and respecting the scientific method and the peer-review process is essential when it comes to drafting, publishing, and reviewing manuscripts. Use respected, peer-reviewed journals (also known as primary sources) or peer-reviewed open access journals, rather than nonpeer-reviewed journals or websites. Learn how to distinguish peer-reviewed from nonpeer-reviewed articles. Reputable journals are typically found by searching online databases such as MEDLINE®, SCOPUS®, BIOSIS, and the Web of Science as well as on the Directory of Open Access Journals (see Chapter 8, Section 8.2). Reputable journals are also usually indexed in an academic database or search engine and have indicators such as impact factors.

If you are unsure about information you receive or find, make use of the scientific method: gather evidence, check sources, deduce, hypothesize, and synthesize conclusions yourself rather than relying on those of others. Knowing and using these practices will aid you and others in distinguishing reliable primary sources from those that are not.

1.3   ABOUT READERS

➤ Understand how readers go about reading

To understand how best to write clearly, it is important to understand how readers go about reading. Expectations and perceptions of readers have been widely studied in the fields of rhetoric, linguistics, and cognitive psychology. In this book, I provide an overview of these expectations and perceptions and apply them to a broad range of scientific fields, thus giving you the tools you need to become a better writer.

Readers immediately interpret what they read. Let me illustrate how readers go about reading on the example of one word:

Water!

Immediately, on reading this single word, you will have a picture in your mind. Some of you will think of water as in a dangerous flood or tsunami; others will think of going swimming on a hot day; yet others will think of getting water to drink when they are thirsty or the excitement they feel on finding water for which they have been drilling. In other words, different readers interpret this single word differently. Similarly, sentences can be interpreted in different ways, none of which may coincide with the interpretation intended by the author. Your goal as a writer should be to communicate the intended meaning of your writing clearly to as many readers as possible.

How do readers interpret what is written? There is no single answer to this question. When reading scientific papers, readers are affected not only by the content and format of a paper but also by its composition and style. Readers interpret documents based not only on words, sentences, and paragraphs but, above all, on the structural location of these elements. Thus, readers are bothered much more when a sentence is misplaced than when a word is imprecise. For example, they will find a Materials and Methods paragraph that appears in the Introduction of a manuscript more disruptive than a misspelled word. In other words, the logical and structural organization of scientific documents is much more important than perfect grammatical form (Figure 1.1).

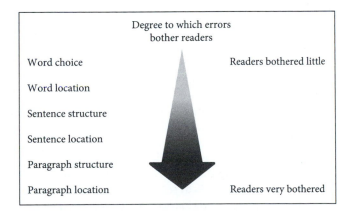

Figure 1.1 Degree to which errors in writing bother readers. Based on the perception of readers, it is more important to logically organize and present one's ideas than to worry about perfect grammatical form or word choice.

As an author, you need to be conscious of these elements when you write. Understanding the correlation of structure and function in a sentence, paragraph, or section is what underlies the science of scientific writing.

Aside from these elements of text composition, communication in science also depends on effective graphics depicting data or communicating complex biological phenomena to complement the writing. These visual elements are often key to providing the required evidence to convince readers of your data interpretations and to provide pictorial models that readers can remember. Understanding the needs of your audience in that respect is thus also important in becoming a successful writer.

Some topics are complicated by nature and can be hard to follow. Reading such science is work. A good writer, however, can make the work much easier.

1.4   ABOUT WRITERS

★ Write with the reader in mind

In the professional world, success in writing is determined by whether your readers understand what you are trying to say. You need to write clearly so that readers can follow your thinking and so that you achieve the greatest possible impact. To "write with the reader in mind" means to consider how the reader interprets what you have written. It requires you to construct your writing clearly, concisely, and at the right level, so that the reader can follow and understand what you want to say immediately. This rule should be viewed as the *Basic Precept* around which all other writing rules revolve.

Most of us can identify unclear writing by others, but we have a harder time recognizing our own mistakes. Scientists who write unclearly rarely think they do, much less intend to. Similarly, your own writing may appear clear to you, and it will come as a surprise when your readers say that it is not. The reason for this insensitivity is simple: Anyone who writes about something and understands its content is more likely to think the passage is clearly written than someone who knows less. However, we all know how difficult it is to understand "legalese" for anybody outside the legal field or "bureaucratese" for anyone other than a bureaucrat. In the same way that legalese and bureaucratese are difficult to understand for most people, "academese" is difficult to understand for most people outside the field of science.

To illustrate how frequent style and composition errors are in scientific writing, I randomly selected and analyzed writing error frequency in 100 prepublication scientific manuscripts by senior graduate students and postdoctoral fellows from several US universities covering a wide range of biological and biomedical topics. Half of the authors were native speakers and the other half were not—a ratio that reflects approximately the ratio of native to nonnative speakers among postdoctoral fellows in the United States today. All manuscripts contained errors

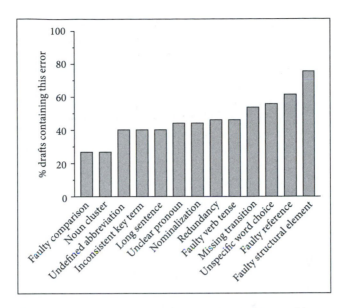

Figure 1.2 The most common grammatical, structural, and stylistic problems.

Explanation of elements:

Faulty comparison	Use of incomplete and ambiguous comparisons (*Example: "Our findings are similar to Frater et al."*)
Noun cluster	Use of two or more nouns in a row (*Example: "filament length variability"*)
Undefined abbreviation	A nonstandard abbreviation is not defined (*Example: "CMMT"*)
Inconsistent key term	Key terms are not repeated or linked (*Example: "nematode," "worm," "C. elegans"*)
Long sentences	Use of too many long sentences
Nominalization	Use of a noun instead of a more interesting verb (*Example: "measurement" instead of "to measure"*)
Redundancy	Ideas are repeated in different ways, or long phrases are used (*Example: "the majority of" instead of "most"*)
Faulty verb tense	Use of present tense instead of past tense
Unspecific word choice	Words are imprecise or unclear (*Example: "The sample was incubated for several hours."*)
Faulty reference	Reference is wrongly cited, inaccurate, missing, or has wrong placement in text (*Example: "It was reported (5) that x is the only element."*)
Faulty structural element	Key component of a paper is missing, misplaced, or obscured (*Example: The purpose of experiment is not stated.*)

in style and composition, but each type of error was scored only once. The percentage of the manuscripts containing the most common grammatical, structural, and stylistic problems is shown in Figure 1.2.

Interestingly, most manuscripts contained faulty structural locations (76%). Faulty or missing references were the second most common error encountered (62%), followed closely by unspecific word choice (56%) and missing transitions (54%). Other errors, such as faulty verb tense, use of redundancies, and nominalization, as well as unclear pronouns, also occurred relatively often (40%–46%). These results demonstrate the variety as well as the amazingly high frequency of writing problems scientists appear to have and show the importance and need for good scientific writing guidelines and instruction.

Why do these problems exist in scientific writing? Many writers simply want to unburden themselves of all the information they have collected and are happy just to get their data onto paper. Others, especially those new to a field or unfamiliar with a topic, learn to write by imitating a particular writing style characteristic of a field without clearly understanding the underlying structure. Some writers plump up their writing to impress readers, convinced that dense writing style sounds sophisticated and reflects deep thinking. The main reason for poor writing style and composition, however, is ignorance and lack of training. Scientific writers are often unaware of how to identify words, sentences, or paragraphs that may give readers a problem. Most are aware of certain "rules" taught to them in high school and undergraduate English composition classes but are not familiar with the basic rules that would benefit them as professionals in science.

To become a successful writer, you should have a firm grounding in basic English composition and an understanding of how readers interpret what has been written. The main goal of this book is to provide you with these basic rules (designated by ★: generally applicable to all scientific documents) and with guidelines (designated by ➤: apply to specific sections of a scientific document or presentation or to scientific writing in general) to ensure that your work has the highest possible impact and is clearly understood by the majority of readers, be they editors, students, or fellow scientists.

1.5   SCIENTIFIC WRITING VERSUS SCIENCE WRITING

➤ Distinguish between science writing and scientific writing

People often get confused about scientific writing versus science writing and mix up these terms. Yet the two are not synonymous. Most science writers are not scientific writers and vice versa. While the rest of this book deals with scientific writing, this brief section outlines the difference between scientific writing and science writing to help make the difference between the two more clear and to aid in guiding those seeking more information on the latter.

Scientific writing is a form of technical writing by scientists for other scientists. Unlike other styles of writing, this style of writing is not meant to be creative but fact-based and objective. It reports on studies, observations, and findings in a specific format. For example, peer-reviewed journal articles, which report on primary research and are published in scientific journals such as *Science* and *Nature*, fall into this category. Grant proposals, which seek funding, and literature review articles, which summarize and interpret published research, also belong to this genre. Example 1-1 is an example of scientific writing.

Example 1-1   Beta-thalassemia, an inherited morbid hemoglobinopathy, is caused by single-point mutations in the gene for beta-globin. Our long-term goal is to diagnose and treat hemoglobinopathies by using high-fidelity editing reagents, such as peptide nucleic acids, carried in targetable nanoparticles. To date, our group has optimized editing reagents using single-stranded DNA to correct the point mutation that causes beta-thalassemia.

Science writing, on the other hand, is nontechnical writing about science for a general audience. It is the form of journalism you find in *Scientific American, National Geographic,* or in articles on science published by *The New York Times.* (See also the website of the National Association of Science Writers at https://www.nasw.org/.)

Science writing is not a how-to manual or a review paper or research article. Science writers explain important and interesting topics in science and technology and lay out the broader social effects of these topics to a wide public audience. For example, a science writer may communicate the effects of climate change or describe an approach to stem the spread of the Zika virus. Science writers often are not trained scientists but humanists interested in science. They are mainly concerned with how to explain and interpret scientific concepts or processes in a way that is clear to a lay audience. They have to use nontechnical language to do so and define terms to ensure that their audience can understand what is being described. Their work may also be critical about scientific findings. Example 1-2 covers the same topic as Example 1-1, but this time in the style of science writing.

Example 1-2   Beta-thalassemia is an often fatal, inherited red blood cell disease, which is caused by a genetic mutation in the beta-globin gene. Scientists plan to repair mutations causing the disease by excising the mutated DNA and replacing it with DNA pieces containing the correct sequence. The correct DNA pieces will be carried into the cells on tiny particles the size of a few nanometers, which are able to cross the cell membrane.

Almost all elements of style found in scientific writing are also valid in science writing, but science writers need to pay particular attention to

- Selecting the key features of a work or process
- Providing background on a process or materials before labeling them

- Employing analogies and metaphors as well as examples of what something is not
- Using active voice and first person rather than passive voice and third person

1.6   MASTERING SCIENTIFIC WRITING

➤ Practice writing

To master the art of scientific writing, use the rules and guidelines of this book to help you grasp fully the necessary format expected for the respective scientific documents. Realize, though, that reading and hearing about these rules and guidelines is not enough. Familiarity with the nuances of these elements will be enhanced as you read scientific literature and pay attention to how professional scientists write about their work. Most important, however, is that you, as the writer, *practice* writing and thinking within this structure. Reading and editing the writing of others, as well as peer review, can also help you toward mastery. You will see improvement in your own writing skills by repeatedly practicing, reading, writing, and critiquing others' work.

This book includes writing exercises that allow you to practice the discussed rules and guidelines. Work through these exercises and give yourself the opportunity to make your own mistakes and achieve your own successes. Know that some of the problems are challenging and may be frustrating. Try not to get stuck in the scientific details of the exercises; what is important is that you recognize mistakes of style and composition and avoid misunderstandings and misinterpretations. You may disagree with some of the answers that have been provided, and many exercises will have more than one correct solution. Some solutions are better than others, and some just a matter of personal style. The goal is to improve the original text so that the material presented is clearer and more easily understood by the reader.

Just like any other skill, mastering the skill of scientific writing and communication takes time—usually years. Ultimately, with guidance and practice, any scientist should be able to write a paper or proposal that sparkles with clarity and to deliver an engaging presentation. As you write your own papers or prepare your talks, you will recognize that every project has its unique challenges and that you will need practice and good judgment to apply all the writing and communication principles presented herein. In giving due attention to composition, style, and impact, your communication skills will receive the necessary foundation and evolve into those needed for professional writing, and this book will have accomplished its purpose.

SUMMARY

BASIC PRECEPT: Write with the reader in mind.

BASIC SCIENTIFIC WRITING GUIDELINES
- Understand the scientific method.
- Strive to communicate your findings clearly.
- Be ethical.
- Become media literate
- Understand how readers go about reading.
- Distinguish between science writing and scientific writing.
- Practice writing.

CHAPTER 2

Individual Words

Word choice in scientific research papers is one of the primary concerns of scientists and editors alike. This chapter provides basic guidelines for the choice of words in scientific papers.

Choosing the best words to write about science is not easy. Over time, the meaning of words may even change, making it even more difficult to select and distinguish between words in English. Authors need to be aware of the exact meaning of words to convey their messages clearly to as many readers as possible.

THIS CHAPTER TEACHES YOU THE FOLLOWING PRINCIPLES AND RULES:

- The central writing principle
- Basic rules of word choice for clarity
- Words and phrases to avoid or omit
- General terminology and nomenclature

2.1   THE CENTRAL PRINCIPLE

★ (1) Write with the reader in mind

Many scientists think that the primary goal in science is to obtain great results, but good science alone will not bring you success. Your collection of data cannot speak for itself—it needs to be communicated and communicated well. Authors have an obligation to their readers to ensure that science is communicated clearly.

In the scientific fields, readers may be reviewers of a paper or proposals, editors, students, Nobel laureates, scientists from a different discipline, or readers whose native language is not English; in fact, probably most of them will be nonnative speakers, as English is being adopted as the global scientific language. Because of this diversity in readership, the burden of clarity rests on you, the author. Your readers need to be able to follow your thinking, so write with your readers in mind. This basic precept is the central principle of this book, and all other writing rules follow from it.

2.2   WORD CHOICE

Precision

★ (2) Use precise words

The problem of many sentences in science is not grammar but word choice. Consider the following three examples.

Example 2-1	a	The current remained increased for <u>several</u> hours.
	b	Nests were observed <u>frequently</u> for signs of predation.
	c	The carbonate layer was prepared <u>with</u> sodium carbonate.

Although the words underlined in these examples can be found frequently in research papers, these word choices are problematic and disliked by editors and reviewers. In each of the three sentences of this example, the underlined words violate the same basic rule: these words are not precise.

You can improve these sample sentences by revising the word choices.

Revised Example 2-1	a	The current remained increased for **6 hours**.
	b	Nests were observed **every 12 hours** for signs of predation.
	c	The carbonate layer was prepared **using** sodium carbonate. or:
		The carbonate layer was prepared **in the presence of** sodium carbonate.

Why are the revised sentences better than those in Examples 2-1a to 2-1c? The revised sentences convey more precisely what the writer is describing. "Several" is imprecise. How many is "several"? Writers should give a quantitative value such as "6 hours." Note also that if you use imprecise words, other scientists may not be able to follow your experiment or thinking.

"Frequently" is also imprecise. How often is frequently? Use a quantitative term such as "every 12 hours," or "at 6 AM and at 6 PM." A quantitative detail such as "every 12 hours" is much clearer than a qualitative term such as "frequently."

Let us look at Example 2-1c more closely. The vague term underlined in this example is "with." This word is one of the vaguest and most ambiguous terms in English. Because "<u>with</u>" can mean so many things, you should use a more precise term whenever possible. Otherwise, the reader has to guess what you mean. Note that "with" does have legitimate uses, such as "in the company of," as in "I went to school with Brian." Another standard meaning is "by the means of," as in "We washed the dishes with soap." "With" can also be used as an attribute, as in "patients with diabetes." Furthermore, some verbs are followed by "with" such as "compared with." However, scientific writers often use "with" instead of a more precise term and thus confuse readers. In the preceding example, it is much more accurate to write "using" or "in the presence of" instead of "with."

Level of Sophistication

★ (3) Use simple words

Words in science should not only be precise, they should also be as simple as possible. Consider the following examples.

 Example 2-2 **a** Fractions of 0.8 ml were collected, <u>reduced to dryness</u>, and dissolved in 3.75% methanol (v/v) prior to being sequenced.

b Our results <u>reflect deviations</u> from thermal equilibrium during desorption.

These sentences are written in a style that appears heavy and dense to the reader and can even be considered portentous. Admittedly, scientific writing has many technical terms. To keep your writing from being too heavy, choose simple words for the rest of the sentence. "Reduced to dryness" can be expressed much simpler by writing "dried" and "reflect deviations" by "deviate."

 Revised Example 2-2 **a** Fractions of 0.8 ml were collected, **dried**, and dissolved in 3.75% methanol (v/v) prior to being sequenced.

b Our results **deviate** from thermal equilibrium during desorption.

The revised sentences are more easily understood by readers because their word choice is much simpler.

Often, young scientists try to mimic more experienced researchers and then write in a style they think experienced researchers would use. However, this can come off as pompous. Here is another such example that needs to be simplified.

 Example 2-3 <u>There is a large body of experimental evidence that clearly shows that members of the genus *Crotalus* congregate simultaneously in cases of prolonged decreased temperature conditions in the later part of the year.</u>

 Revised Example 2-3 **Rattlesnakes come together when it gets cold in the fall.**

Many English as a second language (ESL) authors convert vocabulary of their native language for use in English writing. In other cultures, flowery language is extensively used. In professional English, however, you need to use correct terminology and avoid the grandiose words or phrases. In professional English, statements are rather direct (see Chapter 5 for more on ESL differences). Thus, authors need to pay special attention to avoiding flowery words or unnecessarily complicated phrases.

Regardless of your native language, remember that most of your readers are probably nonnative English speakers. You have to ensure that these readers can understand what you have written. Use simple words. That is, aside from the technical terms, choose a level of words that you would use when talking about your work to a friend; for example, choose "girl" rather than "female child" (see also http://www.userlab.com/Downloads/SE.pdf for more details on using simplified English for an international audience).

2.3   WORD CHOICE: SPECIAL CASES

Misused Words

➤ Watch out for misused words

Words are not always what they seem. Often, words and expressions in science are misused and confused, especially by ESL authors. Some of the words are used incorrectly so often that they sound right even when they are not. Watch out for these misused and confused scientific terms. Use a dictionary if needed or look up definitions on the Internet.

Commonly misused words fall into several categories, including words with suffixes, verbs, adverbs, adjectives, and links.

ESL advice

Suffixes

-ability   Be aware of *-ability* words. Often, the sentence should be rewritten using a stronger verb preceded by the verb *can*.

 Example 2-4   **a**   <u>Changeability</u> of *X* occurs when *Y* is added.

 Revised Example 2-4   **a**   *X* **can change** when *Y* is added.

-ization   Challenge *-ization* nouns. Many writers tend to invent nouns by adding the ending *-ation* or *-ization* onto the verb.

Example 2-4   **b**   <u>Metabolization</u> of phosphates was different than expected.

Revised Example 2-4   **b**   Phosphates were **metabolized** differently than expected.

-ize   Often, nouns or adjectives are wrongly changed to verbs by adding *-ize* to a word.

👎	**Example 2-4**	c	Older patients were <u>prioritized</u>.
👍	**Revised Example 2-4**	c	Older patients were **given priority**.

-ized/-izing   You should also challenge *-ized* or *-izing* adjectives and search for simpler substitutions.

👎	**Example 2-4**	d	<u>Individualized</u> doses were calculated.
			Quantum materials have a <u>transformatizing</u> impact on various technologies.
👍	**Revised Example 2-4**	d	**Individual** doses were calculated.
			Quantum materials have a **transformative** impact on various technologies.
			or, even better:
			Quantum materials transform various technologies.

-ology   This ending means the study of something and is jargon when used in sentences such as the following.

👎	**Example 2-4**	e	No <u>pathology</u> was found. <u>Cytology</u> was normal. <u>Symptomology</u> was severe. <u>Serology</u> was negative.
👍	**Revised Example 2-4**	e	No **pathologic condition** was found. **Cytologic findings** were normal. **Symptoms** were severe. **Serologic findings** were normal.

Verbs

make   Like "to do," "to make" is often overused by ESL writers. Be sure to use the correct terms in context instead of simply substituting "to make" for any unknown term. If you are not sure about the correct terminology, consult an English textbook, journal, or a scientist who is a native speaker.

ESL advice

👎 **Example 2-5**   a   A picture was <u>made</u>.
We <u>made</u> a graph.

👍 **Revised**   a   A picture was **taken**.
Example 2-5   We **graphed** the data. or:
We **constructed** a graph.

affect, effect   "Affect" is usually used as a verb and means to act on or to influence.

👍 **Example 2-5**   b   The addition of KI-3 to MZ1 cells affected their growth rate. [i.e., it could have increased or decreased or induced something else]

More rare, "affect" can also be a noun with a specialized meaning in medicine and psychology: an emotion.

👍 **Example 2-5**   c   People can experience a positive or negative **affect** as a result of their thoughts.
She showed a normal reaction and affect.

"Effect" is usually used as a noun meaning a result or resultant condition.

👍 **Example 2-5**   d   We examined the **effect** of KI-3 on MZ1 cells.

When used as a verb (rarely), "effect" means to cause or bring about.

👍 **Example 2-5**   e   The addition of KI-3 to MZ1 cells **effected** a change in their growth rate. [i.e., it caused or brought about change]

Adverbs and Adjectives

overnext   This word does not exist in English. What you probably mean is "the one *after the next*."

ESL advice

👎 **Example 2-6**   In the <u>overnext</u> slide, we will see . . .

👍 **Revised**   In the **slide after next**, we will see . . .
Example 2-6

significant(ly)   Use only when you are talking about statistical significance and give a *p*-value. Otherwise, use "important," "substantial," "markedly," "meaningful," or "notable."

Linking words

since, because   Use "since" only in its temporal sense, not as a substitute for "because." If you want to indicate causality, use "because."

Example 2-7   **a**   Growth stopped **since** the temperature fell below freezing.
The reaction rate decreased **because** the temperature dropped.

which, that   Sometimes these words can be used interchangeably. More often, they cannot. Use "which" with commas for non-defining (nonessential) sentences.

Example 2-7   **b**   Dogs, which have been domesticated for millennia, are our best friends.

Use "that" without commas for essential sentences. If the section of a sentence introduced by "that" is omitted, the meaning of the sentence is changed or may not be apparent.

Example 2-7   **c**   Dogs that were treated with the antidote recovered.

Be especially careful about words that are easily confused by writers and about words that look similar but mean different things. Examples include those previously described ("affect" and "effect," "since" and "because," and "which" and "that"). More commonly confused words, including *as/like, while/whereas, principle/principal,* and *quantitate/quantify,* are listed in **Appendix A** together with their corresponding meanings.

Handling Language Sensitively

➤ Avoid sexism

Contemporary society prefers gender-neutral (unisex) language to convey inclusion and equality of all sexes. To avoid being accused of chauvinism and insensitivity, carefully consider what you write. If readers get offended, they are likely to stop reading.

Sexism refers to any form of stereotypic attitude, exclusion, or discrimination based on gender. Sexism can be both verbal and visual. It is often unintentionally introduced. Some forms are so subtle that authors might not even notice them unless they are pointed out. Consider the following example.

Example 2-8   a   <u>Man</u> is not the only host for this parasite.

The easiest solution to avoid sexism is to use "unisex" terms.

Revised   a   **Humans** are not the only hosts for this parasite.
Example 2-8

Writing gets more complicated when we have to consider which pronoun to use for singular nouns that do not indicate gender or sexual orientation such as faculty, staff, teacher, scientist, student, and doctor. Although more formal language requires a singular pronoun (its, his, her), it raises the problem of biased language. The change in the English language is toward using a plural "they," and plural verbs for these cases as shown in the next example.

Example 2-8   b   A nurse should double-check her IV settings.

Revised   b   **Nurses** should double-check **their** IV settings.
Example 2-8

2.4   REDUNDANCIES AND JARGON

★ (4) Omit unnecessary words and phrases

Many sentences in science appear complex because they contain redundancies. Redundant words or phrases unnecessarily qualify other words and phrases. Writers should be as brief as possible and should avoid any verbosity by omitting unnecessary words and phrases and jargon (see also Chapter 6, Section 6.4 on condensing). However, if it takes more words to be clear, use more words.

The following are two examples of unnecessarily complex sentences.

Example 2-9   a   High pH values <u>have been observed</u> to occur in areas that <u>have been determined</u> to have few pine trees.

               b   Most galaxies with unusually luminous cores are <u>quite</u> asymmetric <u>in shape</u>.

By cutting out unnecessary words and redundancies, these sentences can be revised to the following.

Revised   a   High pH values occur in areas with few pine trees.
Example 2-9
               b   Most galaxies with unusually luminous cores are asymmetric.

Unnecessary Words

The following individual words can and should be omitted because they add nothing to a text:

actually	basically	essentially	fairly	much	really
practically	quite	rather	several	very	virtually

Other Examples of Redundancies

In the next list, all the words in parentheses are redundant and can be omitted:

(already) existing	at (the) present (time)
(basic) fundamentals	blue (in color)
cold (temperature)	(completely) eliminate
(currently) underway	each (individual)
each and every [choose one]	(end) result
estimated (roughly) at	(final) outcome
first (and foremost)	(future) plans
(main) essentials	never (before)
period (of time)	reason is (because)
(still) persists	(true) facts

Unnecessary Phrases

Many unnecessary words and phrases are used by both native and non-native English speakers. Like commonly misused words, commonly misused phrases can often be avoided to make your writing shorter and clearer.

Certain phrases are often unnecessarily used to introduce previous studies or results. These phrases can almost always be deleted to state the facts more succinctly.

Example 2-10	In a previous study, it was demonstrated that heavy metals can be removed from aqueous solution by sawdust adsorption.
Revised Example 2-10	Heavy metals can be removed from aqueous solution by sawdust adsorption.

Example 2-11	Eddies have been shown to vary depending on the time of year.
Revised Example 2-11	Eddies **vary** depending on the time of year.

Other commonly used unnecessary phrases that can usually be deleted include the following:

there are many papers stating . . .	it is speculated that . . .
it was shown to . . .	it has been found that . . .
it was observed that . . .	it has been demonstrated . . .
it is reasonable to assume that . . .	it has been reported that . . .
evidence has been presented that shows that . . .	it has long been known that . . .

Phrases that can be shortened include:

Avoid	Better	Avoid	Better
A considerable number of	many	in the absence of	without
an adequate amount of	enough	in view of the fact that	because, as
an example of this is the fact that	for example	in the event that	if
as a consequence of	because	it is of interest to note that	note that
at no time	never	it is often the case that	often
based on the fact that	because	majority of	most
by means of	by	no later than	by
considerable amount of	much	number of	many
despite the fact that	although	on the basis of	by
due to the fact that	due to	prior to	before
during the time that	while, when	referred to as	called
first of all	first	regardless of the fact that	even though
for the purpose of	to	so as to	to
has the capability of	can, is able	utilization	use
in light of the fact that	because	with reference/regard to	about (or omit)
in many cases	often	with respect to	about
in order to	to	with the exception of	except
in some cases	sometimes		

A more extensive list of redundancies can be found in Day (1998) and O'Connor (1975).

Jargon

Jargon is the use of terms specific to a technical or professional group. Jargon is often incomprehensible for "outsiders." In science, jargon often includes "laboratory slang" as in the following examples.

Examples of jargon that should be avoided:

Southern blotted   This is laboratory jargon. The correct use is ". . . analyzed by Southern blot . . ."

Western blotted	Similar to "Southern blotted," "Western blotted" is laboratory slang. The correct use is ". . . subjected to Western blot analysis" or ". . . analyzed by Western blot."
electrophorized	the correct usage is "analyzed by or subjected to electrophoresis"
bugs	meaning bacteria, never used in scientific writing
lab	use "laboratory"
prep	use "prepare" or "preparation"
vet	the correct term to use is "veterinarian"
evidenced	use the noun "evidence" instead
vortexed	"vortex" exists only as a noun; use "was mixed by vortex" instead

2.5   ABBREVIATIONS

★ (5) Avoid too many abbreviations

A special type of word choice to consider is the use of abbreviations. Too many abbreviations can be confusing to the reader and should therefore be kept to a minimum. Similarly, nonstandard abbreviations need to be limited or the reader will get lost. Use International System (SI) units when you use standard abbreviations such as kg or m. Standard abbreviations are widely accepted. Check also that you have not used too many abbreviations, even those approved by your target journal. You can legitimately use abbreviations to replace lengthy terms that appear more than about 10 times in a 10-page manuscript or that appear several times in quick succession, but do not use more than four or five such abbreviations in a single paper. Additionally, avoid making sentences indigestible by using too many abbreviations in a short space.

Example 2-12	AI, an important technology for economic activities, is largely focused on <u>ICT</u> and <u>RT</u>, whereas our new <u>BI</u> model is not.

The preceding example may be perfectly intelligible to expert colleagues in the artificial intelligence field but will be unintelligible to most readers.

Define essential abbreviations at their first appearance, in a footnote at the beginning of the paper, or in both places, according to the journal's requirements. Once you have defined an abbreviation, use it whenever you need it—do not switch back to using the full term unless many pages have elapsed since its previous appearance—then you may remind the reader, once, what the abbreviation means. If you use—and define—an abbreviation in the title of a paper (although this is not recommended), redefine it in the text. Do the same for abbreviations used (and defined) in an abstract. If you are using many abbreviations in a long scientific document, consider adding a list of abbreviations with definitions to the document.

Special Abbreviations

Certain Latin-derived abbreviations are used often in science. Note that although the following are Latin derivatives, they are often used without italics:

e.g. = *exempli gratia*—for example
et al. = *et alia*—and others
i.e. = *id est*—that is

2.6   NOMENCLATURE AND TERMINOLOGY

★ (6) Use correct nomenclature and terminology

In science, it is important to use correct vocabulary, nomenclature (taxonomy), and terminology to avoid being misunderstood and to avoid confusing the reader. If you are not sure about a term, do not guess. Rather, take the time to look it up in a dictionary, thesaurus, or other reference book. Dictionaries for the biological, medical, and other scientific fields as well as online dictionaries are listed in Section 2.7.

Common Terminology

In science, all organisms are given a name, consisting of two Latin-derived parts: the genus and the species. This nomenclature is also known as binomial, binominal, or binary nomenclature. The genus name is written first and starts with a capital letter. The species name follows the genus name and starts with a lowercase letter. On first mention, genus names should be written out completely, but in subsequent mentions, the genus name can be abbreviated. For example, *Homo sapiens* can be written *H. sapiens*. Note that scientific names are traditionally written in italics. The names of higher taxa such as families or orders are capitalized but not italicized.

Examples of binominal nomenclature include:

	Genus	*species*
humans	*Homo*	*sapiens*
dog	*Canis*	*lupus*
mouse	*Mus*	*musculus*
apple	*Malus*	*domestica*

Aside from genus and species, genes and proteins of several organisms have also been named on the relevant model organism websites and in scientific journals using formal guidelines. When new information becomes available, scientists often work together to revise the nomenclature as needed. However, alternate names frequently exist and can pose a challenge to effective organization and exchange of biological information.

The most common scientific nomenclature includes the following:

Species and all Latin derivates are in *italics* (*in vivo, Physcomitrella patens*, etc.)
Human genes: all caps and *italics* (*ADH3, HBA1*)
Human proteins: caps, no italics (ADH3, HBA1)
Mouse genes: first letter capitalized, the rest lowercased, *italics* (*Sta, Shh, Glra1*)
Mouse proteins: like genes but no italics (Sta, Shh, Glra1)
Bacterial genes: three lowercase, italicized letters, followed by upper-case letter for different alleles (*rpoB*).
Bacterial proteins: not italicized, first letter capitalized (RpoB).
***Arabidopsis* genes**: three letters, *italics*, lowercase for mutants (*abc*), capital letters for wild type (*ABC*)
***Arabidopsis* protein**: capitalized, no italics (ABC)
***Arabidopsis* phenotypes**: first letter capitalized, rest lowercase, no italics (Abc⁺ for wild type, Abc⁻ for mutant)
Yeast gene: same as for Arabidopsis
Yeast protein: first letter capitalized, rest lowercase, plus number for wild type (Icp1, Icp1p), plus number dash number for mutant (Icp1-1, Icp1-1p)

To distinguish the species of origin for homologous genes with the same gene symbol, an abbreviation of the species name is added as a prefix to the gene symbol. For example, human loci (HSA)*G6PD* where HSA = *Homo sapiens* and homologous mouse loci (MMU)*G6pd* where MMU = *Mus musculus*.

2.7   DICTIONARIES

Dictionaries—Biological and Medical Sciences

Biological Sciences (General)
Hine, R., & Martin, E. (2004). *A dictionary of biology*. New York: Oxford University Press.
Martin, E., & Hine, R. (Eds). (2015). *A dictionary of biology* (7th ed.). New York: Oxford University Press.
McGraw-Hill dictionary of bioscience (2nd ed.). (2003). New York: McGraw-Hill Book Company and Sybil P. Parker.
McGraw-Hill dictionary of scientific and technical terms (6th ed.). (2002). New York: Sybil P. Parker.

Biochemistry
Cammack, R., Atwood, T., Campbell, P., Parish, H., Smith, T., Vella, F., et al. (2006). *The Oxford dictionary of biochemistry and molecular biology*. New York: Oxford University Press.
International Union of Biochemistry. (1992). *Biochemical nomenclature and related documents*. London: Portland Press.

Biotechnology

Bains, W. (2004). *Biotechnology from A to Z.* New York: Oxford University Press.

Sengar, R. S., & Chaudhary, R. (2015). *Dictionary of biotechnology.* New Delhi, India: CBS Publishers.

Cell Biology

Lackie, J. M. (2013). *The dictionary of cell and molecular biology* (5th ed.). New York: Academic Press.

Genetics

King, R. C., Stansfield, W. D., & Mulligan, P. K. (2012). *Dictionary of genetics* (8th ed.). New York: Oxford University Press.

Immunology

Cruse, J. M., & Lewis, R. E. (2009). *Illustrated dictionary of immunology* (3rd ed.). Boca Raton, FL: CRC Press.

Herbert, W. J., Wilkinson, P. C., & Stott, D. I. (1995). *Dictionary of immunology.* New York: Academic.

Playfair, J. H. L., & Chain, B. M. (2009). *Immunology at a glance.* Oxford, UK: Blackwell.

Medical Sciences

Merriam-Webster's medical dictionary. (2016). Springfield, MA: Merriam-Webster

Miller, B. F., Keanne, C. B., & O'Toole, M. T. (Eds.). (2005). *Miller-Keanne encyclopedia and dictionary of medicine, nursing, and allied health.* Philadelphia: Saunders.

Mosby's dictionary of medicine, nursing, and health professions. (2016). St. Louis, MO: C. V. Mosby.

Stedman's word books series. (2001–2004). Philadelphia: Lippincott.

Microbiology

Garrity, G. M. (2005). *Bergey's manual of systematic bacteriology,* Vol. 2 (Parts A, B, and C). New York: Springer.

Gillespie, S. H., & Bamford, K. B. (2012). *Medical microbiology and infection at a glance.* Malden, MA: Blackwell.

Singleton, P., & Sainsbury, D. (2006). *Dictionary of microbiology and molecular biology.* New York: Wiley.

Molecular Biology

Cammack, R., Atwood, T., Campbell, P., Parish, H., Smith, T., Vella, F., et al. (2006). *The Oxford dictionary of biochemistry and molecular biology.* New York: Oxford University Press.

Singleton, P., & Sainsbury, D. (2006). *Dictionary of microbiology and molecular biology.* New York: Wiley.

Plant Biology

Allaby, M. (2012). *A dictionary of plant sciences* (3rd ed.). New York: Oxford University Press.

Beentje, H., & Williamson, J. (2016). *The Kew plant glossary: An illustrated dictionary of plant terms* (2nd ed.). Richmond, UK: Royal Botanic Gardens, Kew.

Mabberley, D. J. (2017). *The plant-book: A portable dictionary of the higher plants.* New York: Cambridge University Press.

Macura, P. (2002). *Elsevier's dictionary of botany.* New York: Elsevier Science.

Virology

Hull, R., Brown, F., & Payne, C. (1989). *Virology: Directory & dictionary of animal, bacterial and plant viruses.* London: Macmillan.

Mahy, B. W. J. (2011). *A dictionary of virology (3rd ed.).* New York: Academic Press.

Dictionaries—Other Scientific Fields

Astronomy

Mitton, J. (2008). *Cambridge illustrated dictionary of astronomy.* Cambridge, UK: Cambridge University Press.

Ridpath, I. (2012). *A dictionary of astronomy* (2nd ed.). New York: Oxford University Press.

Chemistry

Connelly, N. G., Hartshorn, R. M., Damhus, T., & Hutton, A. T. (Eds.). (2005). *Nomenclature of inorganic chemistry: IUPAC recommendations.* London: Royal Society of Chemistry.

Hellwinkel, D. (2010). *Systematic nomenclature of organic chemistry: A directory to comprehension and application of its basic principles.* New York: Springer.

Law, J., & Rennie, R. (Eds.) (2016). *A dictionary of chemistry* (7th ed.). New York: Oxford University Press.

McGraw-Hill dictionary of chemistry (2nd ed.) (2003). New York: McGraw-Hill.

The Merck index: An encyclopedia of chemicals, drugs, and biologicals (15th ed.). (2013). Whitehouse Station, NY: Merck.

Chemical Engineering

Schaschke, C. (2014). *A dictionary of chemical engineering.* New York: Oxford University Press.

Computer Science

Butterfield, A., Ngondi, G. E., & Kerr, A. (Eds.). (2016). *A dictionary of computer science* (7th ed.). New York: Oxford University Press.

Downing, D., Covington, M., & Covington, M. (2017). *Dictionary of computer and Internet terms* (12th ed.). Hauppauge, NY: Barron's Business Dictionaries.

Laplante, P. A. (Ed.). (2000). *Dictionary of computer science, engineering, and technology.* Boca Raton, FL: CRC Press.

Ecology and Evolutionary Biology

Allaby, M. (2010). *A dictionary of ecology* (4th ed.). New York: Oxford University Press.

Daintith, J. (2003). *Dictionary of evolutionary biology.* New York: Facts on File.

Lincoln, R. J., Boxshall, G. A., & Clark, P. F. (1998). *A Dictionary of ecology, evolution, and systematics* (2nd ed.). Cambridge, UK: Cambridge University Press.

Geology

Allaby, M. (2013). *A dictionary of geology and earth sciences* (4th ed.). New York: Oxford University Press.

Neuendorf, K., Mehl, J. P., Jr., & Jackson, J. A. (2005). *Glossary of geology.* Alexandria, VA: American Geological Institute.

Mathematics

Clapham, C., & Nicholson, J. (2014). *The concise Oxford dictionary of mathematics* (5th ed.) New York: Oxford University Press.

Downing, D. (2009). *Dictionary of mathematics terms* (3rd ed.). Hauppauge, NY: Barron's.

Nursing

Martin, E. A., & McFerran, T. A. (2017). *A dictionary of nursing* (7th ed.). New York: Oxford University Press.

Venes, D. (2013). *Taber's cyclopedic medical dictionary*. Philadelphia: F. A. Davis Company.

Physics

Law, J., & Rennie, R. (Eds.). (2015). *A dictionary of physics* (7th ed.). New York: Oxford University Press.

Wertheim, J., Oxley C., & Stockley, C. (2011). *Illustrated dictionary of physics*. London: Usborne.

Statistics

Everitt, B. S. (2002). *The Cambridge dictionary of statistics*. New York: Cambridge University Press.

Online Dictionaries

http://www.medbioworld.com
The largest medical and bioscience resource directory on the Internet
http://www.visualthesaurus.com
Link to a visual thesaurus
http://thesaurus.reference.com/ (or thesaurus.com)
Another visual thesaurus; this site also contains a link to a highly rated iTunes app, Thesaurus Rex by Dictionary.com
http://www.bartleby.com/141/index.html
Stunk and White, the famous short but excellent style guide

http://www.nlm.nih.gov/medlineplus/mplusdictionary.html
National Library of Medicine
http://www.medicinenet.com/script/main/hp.asp
Webster's new world medical dictionary authored by MedicineNet
http://www.biology-online.org/dictionary.asp
Online dictionary of biology terms for the biological and earth sciences
http://www.userlab.com/Downloads/SE.pdf
Contains details on using simplified English for an international audience

Apps

The newest science apps, including dictionaries, may be acquired through, for example,

https://www.bestcollegesonline.com/blog/40-most-awesome-ipad-apps-for-science-students/
http://appsineducation.blogspot.com/p/science-ipad-apps.html
http://download.cnet.com
https://www.geekwrapped.com/posts/the-best-science-apps

ESL Dictionaries and Other Sources

Konstantinidis, G. (2005). *Elsevier's dictionary of medicine and biology: In English, Greek, German, Italian, and Latin.* New York: Elsevier Science.

Long, T. H. (Ed.). (2000). *Longman dictionary of English idioms* (rev. ed.). Harlow, UK: Longman.

Longman dictionary of American English (4th ed.). (2008). White Plains, NY: Longman.

The Oxford dictionary for scientific writers and editors (2nd ed.). (2009). Oxford, UK: Clarendon Press.

SUMMARY

> BASIC RULES—STYLE
> ★ Write with the reader in mind.
> ★ Use precise words.
> ★ Use simple words.
> ★ Omit unnecessary words and phrases.
> ★ Avoid too many abbreviations.
> ★ Use correct nomenclature and terminology.
>
> ALSO: Watch out for misused words. Avoid sexism.

PROBLEMS

PROBLEM 2-1 Precise Words

Find the nonspecific terms in the following sentences. Replace the non-specific choices with more precise terms or phrases. Note that it is generally not necessary to change the sentence structure; just replace the individual words. Guess or invent something if you have to.

1. All OVE mutants showed enhanced iP concentrations.
2. Plants were kept in the cold overnight.
3. Some exoplanets orbit multiple stars.
4. Apart from the discussed main band, weaker emissions were observed.
5. (Last sentence in an Introduction) The present paper reports on continuing experiments that were performed to clarify this surprising effect.
6. The current was greatly affected when temperature was increased.
7. To provide proof of concept for our hypothesis, we studied a virus in its host cells.
8. Only some of the region under study exhibits larger reddening.
9. The first transition state is a little lower in energy than the second transition state.
10. Heating arises after recapture and subsequent equilibration following from the lowest $T- = 0.04$, which is obtained by imaging the gas shortly after release from the trap.
11. The band showing vibrational splitting of 192/cm in Ne with the most intense peak at 444 nm can be identified with the $A \rightarrow X$ transition of the dimer Ag_2.
12. The afterglow of the blast wave was markedly brighter than we expected.

PROBLEM 2-2 Simple Words

Improve the word choice in the following examples by replacing the underlined terms or phrases with simpler word choices. Again, do not change the sentence structure; just change the words.

1. These data <u>substantiate</u> our hypothesis.
2. We <u>utilized</u> UV light to induce *Arabidopsis* for mutations.
3. The differences in our results compared to those of Retter et al. (2015) <u>can be accounted for by the fact that different algorithms were used</u>.
4. <u>For the purpose of</u> examining cell migration, we dissected mouse brains.
5. Our results <u>are in accordance</u> with Seuter et al. (1988) who measured iP in the culture medium of *Physcomitrella* transformed with the agrobacterial isopentenyltransferase gene.

6. We performed a systematic study of the vibrational spectrum of CO_2 using various isotopomers.
7. An example of this is the fact that quantum materials differ substantially.
8. In Swaziland, the number of HIV-infected children increased by an order of magnitude in the past decade.

PROBLEM 2-3 Commonly Confused/Misused Words
Consider the pairs of confused and misused word choices provided for the following sentences. Be sure you understand the difference in word choice. Using the provided word choices, fill in the correct words. It is okay to use Appendix A of this book or a dictionary.

1. **like, as:**
   Plasmids were isolated _____described by Beates (17).
   Carbon dioxide, neon, helium, methane, krypton, and hydrogen are gaseous components of dry air, _____argon.
2. **while, whereas:**
   The first enzyme was added _____ the DNA mixtures were incubating at 37°C.
   In weathered soils, little P is available for biological uptake, _____in young soils little N is available for uptake.
3. **varying, various:**
   _____ water levels in a pond are often the result of climate conditions.
   Each student received _____concentrations of NaCl solution for the experiment.
   Electrodes can be of_____sizes.
4. **effect, affect:**
   Nutrition concentration was the most important factor _____ population size.
   Ozone causes cellular damage inside leaves that adversely _____plant production.
5. **that, which:**
   Fish _____live in caves show many adaptations to living in darkness.
   The value of the standard electrode potential is zero, _____ forms the basis for calculating cell potentials.
   It is still a challenge to produce layered black phosphorus nanosheets, _____ have shown promising applications in electronics.
6. **include, consist of:**
   Her research interests _____ all areas of biochemistry and structural biology.
   Components of Hyperion's crust _____solid H_2O and CO_2.

7. **represents, is:**
   25 mg of ketamine _____ an overdose of anesthetic for mice.
   The Schrödinger equation _____ a fundamental equation in quantum mechanics.

8. **infers, implies:**
   Both curves are of an identical shape, which _____ a constant front profile as well as a constant velocity.
   The Intergovernmental Panel on Climate Change has been criticized for _____ that climate-envelope models are more precise than they actually are.

9. **can, may:**
   It _____ appear that Table 1 contains an essentially complete summary of patterns that occur in electrochemical systems.
   Huge numbers of species _____ be at risk of extinction from climate change.

PROBLEM 2-4 Redundancies and Jargon
Edit the phrases shown; change any redundancies to a shorter and better expression.

absolutely essential	along the lines of
a large number of	as a consequence of
despite the fact that	for the purpose of
in a position to	in close proximity to
in connection with	in order to
in the event that	in view of the fact that
the majority of	it is worth pointing out that

PROBLEM 2-5 Redundancies and Jargon
Improve the word choice of the underlined words in the following examples by removing any redundancies, jargon, and unnecessary words and phrases. Do not change the sentence structure.

1. The doubling rate appeared to be <u>quite short</u>.
2. The data of the analysis on cell cycle parameters <u>are shown</u> in Fig. 1. They have revealed that the cell cycle is controlled by factor X.
3. After 2 hr of <u>incubation</u> of CO_2 on an Ag(110) surface, we ended the incubation procedure.
4. The effect of temperature on conductivity <u>was examined and found not to</u> change dramatically.
5. Often, jewel weed <u>can be found to grow in close proximity to</u> poison ivy. (Two corrections needed.)
6. After infecting a host cell, a herpes viral DNA genome enters the nucleus where it is <u>transcribed from DNA to RNA</u> to both messenger RNAs (mRNAs) and noncoding RNAs (ncRNAs).

7. Transduction efficiencies *in vivo* were much higher than <u>transduction efficiencies</u> *in vitro*.
8. Although transition metals <u>have the capability of</u> forming bonds with six shared electron pairs, only quadruply bonded compounds can be isolated as stable species at room temperature.
9. Upon heat activation, filament size increased, and the number of buds decreased. Both <u>the increase in filament length and the decrease in the number of buds</u> were only seen for cytokinin mutants.
10. Larger mammals are more likely to get extinct than smaller ones <u>due to the fact that</u> larger animals are fewer in number and are disproportionately exploited by humans.

PROBLEM 2-6 Redundancies and Jargon
Identify and remove the jargon and other redundancies in the following sentences.

1. It is also worth pointing out that collagen synthesis returned to normal 3 days post injury.
2. In spite of the fact that our present knowledge on the subject at this point is far from complete, this macromolecular structure can aid in the design of new antibiotics.
3. A substantial proportion of HIV patients also develops tuberculosis.
4. After 3 hr, the old medium was dumped, and the same amount of fresh medium was added.
5. The data in Table 1 are very consistent with Brokl's (1999) model.
6. This appears to indicate that factor A possibly may have a tendency to interact with factor B.
7. In a considerable number of cases, degradation leads to topsoil loss and a reduction in soil fertility.
8. We analyzed helium content in steam escaping from fractures and thermal features of Yellowstone National Park for the purpose of determining the proportions of helium-3 and helium-4 in gas emissions by the super volcano.

PROBLEM 2-7 Abbreviations
Identify the basic writing rule that is violated for the underlined word choices in the following paragraph by describing as best as you can what the general mistake is.

We used the <u>SVWN</u> (Slater, Vosko, Wilk, Nusair) functional as a local-density model and the Becke-Lee-Yang-Parr (<u>B-LYP</u>) exchange-correlation functional for our analysis. All calculations were done in triplicate and adjusted based on the Becke-Perdew (<u>BP</u>) and <u>PW91</u> functionals. We also included the Becke-Lee-Yang-Parr <u>B-LYP</u> functional to overcome the limitations of the <u>BP</u> and <u>PW91</u> functionals. The nonlocal <u>B-LYP</u> functional includes a zero-point energy (ZPE) correction. Configuration-interaction methods (CI) or density-functional theory (DFT) describe the exchange and correlation effects. This is the first report on DFT plane-wave calculations and the SVWN functional for Z. We found that the Z bond is 2.73 Angstroms for PW91 and B-LYP.

PROBLEM 2-8 Mixed Word Choice
Improve the word choice in these examples.

1. A typical scientist spends many long hours, even on the weekend, in his laboratory.
2. We studied the affect of erythromycin on 5 male and 3 female children in three different essays.
3. A graph displaying this data is shown in the overnext slide.
4. We made a picture of the gel we made.
5. We observed a change in cluster size after several minutes.
6. Isolatability of the Nnkla-1 protein was more difficult than expected.
7. Absorbance was measured at varying time points.
8. To make perfect slides of DNA nicks, an electron microscope is absolutely essential.
9. The hiring of new faculty is traditionally overseen by the chairman of a department.
10. To reduce the amount of data points, we tossed out every alternative test point.
11. It has been reported that thiophene was discovered as a contaminant of benzene.

PROBLEM 2-9 Mixed Word Choice
Edit the following passage. Pay attention to word choice. Use precise and simple words. Check for misused and confused terms. Avoid sexism and redundancies.

Sulfonamides were among the first manmade agents used successfully to treat diseases. On account of their broad antibacterial activity, these drugs were in earlier times used almost exclusively in the treatment of a wide assortment of diseases. It is most fortunate that other drugs have supplanted sulfonamides as antimicrobial agents because all pathogenic bacteria are capable of developing resistance to sulfonamides. Sulfonamides prevent the synthesis of folic acid that is a coenzyme important in amino acid metabolism. Although sulfonamides are for the most part readily tolerated, it has been observed that they do have some side affects.

CHAPTER 3

Word Location

A lthough word choice is important for the interpretation of a sentence, readers take the greatest percentage of clues for interpretation not from word choice but from the *location* of words within a sentence. That is, readers expect a certain format in each sentence.

If this format is not met, readers get confused and start paying attention to the organization rather than to the content of a sentence, which increases the possibility of misinterpretation or not understanding. Worse, if readers cannot follow the format, they will lose interest. Thus, you need to pay close attention to word location and to the organizational structure of a text.

THIS CHAPTER EXPLAINS:

- How readers interpret sentences
- How to create good flow
- How to establish importance within sentences

3.1   READERS' EXPECTATIONS

➤ The location of words within a sentence is important for its interpretation

Your task as an author is not only to choose the right words but also the most effective location for your words. You will have to convince most of your readers to interpret your sentences as you intended. There is always a minority, however, who interpret sentences differently from the majority. For this minority of readers, it is even more important to understand the importance of where in a sentence to place what information.

Consider Example 3-1a.

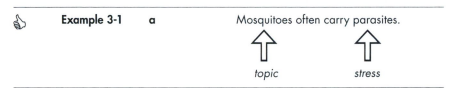

In this sentence, the word "mosquitoes" has been placed at the beginning of the sentence in the topic position, and the word "parasites" has been placed at the end of the sentence in the stress position. This positioning tells the reader that "mosquitoes" are the topic of the sentence and that "parasites" is to be emphasized. To most readers, the format of this sentence implies that the author has talked about mosquitoes before and is about to introduce a new topic, "parasites." Another version of the same sentence presents a different emphasis, as can be seen in Example 3-1b.

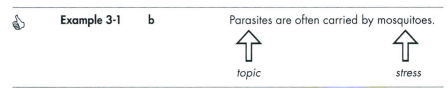

Although the sentence in Example 3-1b uses the same words as the sentence in Example 3-1a, the word locations have been altered. In Example 3-1b, the familiar topic now appears to be "parasites" at the beginning of the sentence, and the emphasized word is "mosquitoes" at the end or stress position of the sentence. Placing "mosquitoes" at the end of the sentence indicates to the reader that you are stressing this term. You may want to stress the term to ensure that the reader immediately understands that the stress is on "mosquitoes" and not on fleas or rats; you may also want to stress the term to ensure that the reader does not miss the introduction of a new topic. Placing "mosquitoes" in the stress position of the sentence guides the reader's attention.

3.2   COMPETITION FOR EMPHASIS

★ (7) Establish importance

To decide on the best placement of words within a sentence, it is crucial that you decide what is important, what is less important, and what is not important before you start writing or revising. When you write, important information can then be stressed, less important information can be subordinated, and unimportant information can be omitted. For example, when reporting on the results of a study conducted by an author in a given year, beginning writers often put the year or author at the start of the sentence as shown in the next example.

Example 3-2	a	In 2018, Smith reported that XXX leads to YYY.
	b	Smith (2018) reported that XXX leads to YYY.

These sentences emphasize the year (Example 3-2a) or author (Example 3-2b) in the topic position, but that is typically not what these writers intended. An alternate version of this sentence would be the following one.

Revised **Example 3-2**	XXX leads to YYY (Smith, 2018).

In the revised example, the findings are now emphasized, and the author and year information, which is less important than the actual findings, is subordinated in parenthesis.

You need to recognize that the format and structure you use to present information will lead the reader to interpret it as more important or less important. In general, readers see the beginning and end position in a sentence as being emphasized, whereby the end position is more emphasized than the beginning position. Furthermore, the main clause is seen as more emphasized than the dependent clause. Thus, a main clause, a clause that is independent and can stand alone as a complete sentence, carries more weight than a dependent clause, which depends on the rest of the sentence for its meaning.

Consider the following four versions of a sentence. In each of these examples, the main clause has been italicized.

Example 3-3	a	Although vitamin B6 seems to reduce the risk of macular degeneration, *it may have some side effects.*
	b	*Vitamin B6 reduces the risk of macular degeneration,* but it may have some side effects.
	c	*Taking vitamin B6 may have some side effects,* but vitamin B6 also reduces macular degeneration.
	d	Although taking vitamin B6 has some side effects, *vitamin B6 reduces macular degeneration.*

For most readers, information in the main clause carries more weight than that in the dependent clause. Furthermore, most people perceive information at the end of the sentence as more important than that at the beginning of the sentence.

In the given example, if readers were to vote on the impact of each sentence, the percentage of readers that would recommend taking vitamin B6 would be highest in version D and lowest in version A. A more detailed analysis shows the following:

Sentence version	News in main clause	News in end position	Perception of vitamin B6 recommendation (%)
A	– negative	– negative	30
B	+ positive	(–) negative (dep. clause)	40
C	– negative	(+) positive (dep. clause)	60
D	+ positive	+ positive	70

Based on these percentages, readers (e.g., physicians) are most likely to recommend taking vitamin B6 after reading sentence D and least likely to recommend it after reading sentence A. The reason for this is that the strongest statement has the positive (+) information in both the main clause and at the end position of the sentence, while the weakest statement contains the negative (–) information both in the main clause and at the end position of the sentence. Thus, even in more complex sentences, word placement, if considered carefully, can help guide and influence readers.

Although word placement is more important than word choice for sentence interpretation by the reader, if a word is strong or extreme enough it can dominate the reader's attention. Let us replace "side effects" with an extreme phrase in the strongest positive sentence above and look at the effect.

Example 3-4   Although taking vitamin B6 may result in <u>serious deformities or even death</u>, vitamin B6 reduces macular degeneration.

In this example, no matter where you put the extreme "serious deformities or even death," it overpowers the structural location of everything else, including that of the stress position.

3.3   PLACEMENT OF WORDS

Complexity

★ (8) Place old, familiar, and short information at the beginning of a sentence in the topic position

★ (9) Place new, complex, or long information at the end of a sentence in the stress position

If information is placed where most readers expect to find it, it is interpreted more easily and more uniformly. Readers expect to see old information that links backward at the beginning of a sentence (or paragraph) and new information at the end of a sentence (or paragraph) where it is emphasized more.

Once sentences are ordered into paragraphs, the importance of word location becomes even more obvious, as writing "flows" much better if the information is linked through word location. Consider the following example.

Example 3-5

Macular degeneration is affected by *diet*. *One of the diet components* that influences the progression of macular

degeneration is *vitamin B6*. Although *vitamin B6* seems to reduce the risk of macular degeneration, it may have some **side effects**.

If the passage were to continue, most people would expect to find information on "side effects" in any subsequent sentence or paragraph. Readers will get confused and misinterpret passages when they do not find the information they are expecting. But why do readers expect to read about side effects next?

Note how the information at the end position of a sentence in the preceding example is placed at the beginning, or topic position, of the next sentence, leading to "jumping word location." In each of these sentences, the new information in the stress position of one sentence becomes old, familiar information in any subsequent sentence and is therefore placed at the topic position in the sentence that follows. The human brain looks for patterns and recognizes that this paragraph is organized by "jumping word location." Because "side effects" has been introduced in the stress position in the last sentence, most people would therefore expect to find information on "side effects" in any subsequent sentence or paragraph.

When you pay attention to word location, as in the preceding example, your writing is perceived to have good flow and continuity. When you do not comprehend the structural needs of readers, readers get confused.

Another way to achieve good flow or continuity is to write a whole paragraph from the point of view of the old information as in Example 3-6.

Example 3-6

Depression in the elderly is thought to affect more than 6.5 million of the 35 million Americans who are 65 years of age and older. *It* is considered to be a disorder that is commonly underdiagnosed, undertreated, and mismanaged by pharmacotherapy both in community dwelling seniors and in those residing in nursing facilities. *Depression* in the elderly has also been closely associated with dependency and disability that presents in both emotional and physical symptoms, thus amplifying the difficulty in diagnosis. *Major depression*, dysthymic disorder, and subsyndromal depression tend to be higher in persons over 65 who live in a long-term care facility.

Note how in this example, the topic "depression" is consistently placed in the topic position of each sentence, providing a link back for the reader. In each of the sentences in the preceding example, new information is always placed at the end of each sentence. Thus, every sentence provides new information, although the writer does not expand on it. If passages are consistently written from the same point of view as in the preceding example, good flow is also achieved.

Not all paragraphs will follow these basic rules of word location as exclusively as shown in Examples 3-5 and 3-6. Many paragraphs display a mixture of the word locations shown in these examples (see also Chapter 6 on paragraph construction). That is okay. What is not okay is to jump back and forth between one point of view and another for no apparent reason.

If we apply these basic rules for old and new information to writing and revising, we quickly realize that although some sentences are easy to write or revise, others are not. It is particularly hard to begin sentences well, especially if they are long and complex.

Which of these two sentences do you prefer?

Example 3-7   a   Outbreaks of limb deformities in natural populations of amphibians across the United States and Canada, especially in wetland associated with agricultural fields, were evaluated in this study.

b   We evaluated outbreaks of limb deformities that occurred in natural populations of amphibians across the United States and Canada, especially in wetland associated with agricultural fields.

Most readers dislike Example 3-7a because it starts with a long and complex subject. Example 3-7b, on the other hand, begins simply and moves toward complexity. Readers prefer to see short information at the beginning of a sentence and long information at the end of a sentence. Thus, you need to consider the length of terms or information when constructing sentences.

Subject

★ (10) Get to the subject of the main sentence quickly, and make it short and specific. If possible, use central characters and topics as subjects

Readers prefer not only to see short information at the beginning of a sentence but also to get to the subject/topic of the main sentence quickly. They understand a sentence more easily if the subject is readily available. When you open sentences with several words before their subject/topic, readers have a hard time understanding what the sentence is about. Thus, avoid long introductory phrases and long subjects.

👎	**Example 3-8**	<u>Due to the nonlinear and hence complex nature of ocean currents</u>, modeling these currents in the tropical Pacific is difficult.
👍	**Revised Example 3-8**	**Modeling ocean currents in the tropical Pacific** is difficult due to their nonlinear and hence complex nature.

The subject in Example 3-8 arrives after the first 11 words, whereas the subject in the revised sentence is immediately available, making the revised sentence more easily understandable.

Readers like to see characters as their subjects. In fact, readers get confused if for no good reason you do not make characters subjects. Consider the following example and its revision.

👎	**Example 3-9**	The reason for rejection on the part of the biochemists was that the focus of the paper was too broad.
👍	**Revised Example 3-9**	The biochemists rejected the paper because it was too broad.

For Example 3-9, most readers consider the revised sentence to be much clearer than the original one because the central characters (biochemists) are the subject of the verb. In the revised version, the subject is also short and specific and much more concise.

Let us look at yet another example.

👍	**Example 3-10**	The cells were incubated at room temperature for two days.

Here, the topic "cells" is also the subject of the sentence. Any possible character such as a biochemist or a laboratory technician is not made the subject because they are not the topic of interest. Instead, "cells" take the place of the live character. This choice is actually preferred in certain sections of a research paper (or grant proposal), such as in the Materials and Methods section. To the reader, sentences appear clearer and more direct if the subject is also the topic of the sentence (and paragraph).

The subject of a sentence does not always state its topic, however, as in the following example.

👍	**Example 3-11**	

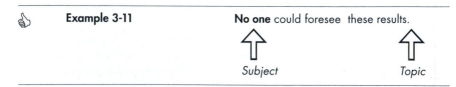

If a subject is deleted entirely, as in the next example, writers create the biggest problem for readers.

👎	**Example 3-12**	A decision was made in favor of the use of dyes, nitrofurans, and amidines as disinfectants.

The author of this sentence may know who is doing what, but the readers do not know and usually need more help than you think. The sentence in Example 3-12 has different interpretations.

👍	**Revised Example 3-12**	We decided to use dyes, nitrofurans, and amidines as disinfectants.
		or:
		They/Researchers decided to use dyes, nitrofurans, and amidines as disinfectants.

For additional discussions about the subject and verb of a sentence, also see Chapter 4, Sections 4.2 and 4.11.

Verb Placement
★ (11) Avoid interruptions between subject and verb and between verb and object Information

General sentence structure in English consists of Subject–Verb–Object/Completer. The verb, or action word, follows the subject (the entity that is doing something) directly, and the object (the entity that is acted upon) immediately follows the verb. Often, sentences are obstructed if the verb does not immediately follow the subject. When readers see the subject of a sentence, they immediately start looking for the verb and pay little attention to any interrupting text between the two. Native and nonnative speakers need to be aware of this rule. English sentences are better understood if their subject and verb are not interrupted, and information is more easily interpreted if it is not obstructed.

ESL advice

Consider the following opening sentence of an introduction.

👎	**Example 3-13**	<u>Rhinovirus</u>, an Enterovirus belonging to the family *Picornaviridae*, which consists of 37 species grouped into 17 genera including pathogens such as Poliovirus, Saffold virus, Coxsackie A virus, and Hepatitis A virus, <u>causes</u> 10–40% of the common cold.

This sentence is obstructed because the grammatical subject ("Rhinovirus") is separated from its verb ("causes") by 30 words. The more words are placed between subject and verb, the more confusion they cause, making the reader sway between finding the verb of the main clause and focusing on the interrupting material. To guide the reader through a sentence, make the sentence flow by placing the verb immediately after the subject and moving the intervening material elsewhere.

Often, an interruption can be moved to the beginning or to the end of a sentence, depending on whether it is connected to old or to new information in the sentence. At other times, you need to consider splitting the information into two sentences or even omitting the interrupting information altogether.

In Example 3-13, it is unclear how important the interrupting material is. If the material is important, the sentence could, for example, be split into two sentences, as shown in the following revision, which allows the important material to be highlighted appropriately.

**Revised
Example 3-13**

Rhinovirus is an Enterovirus belonging to the family *Picornaviridae*, which consists of 37 species grouped into 17 genera including pathogens such as Poliovirus, Saffold virus, Coxsackie A virus, and Hepatitis A virus. **It causes** 10–40% of the common cold.

If the interrupting material is not important and just reflects an interesting side fact, it should be omitted to allow the reader to focus on the main statement, as shown in the following.

**Revised
Example 3-13**

Rhinovirus causes 10–40% of the common cold.

Readers also like to get past the verb to the object of a sentence quickly. Therefore, you should avoid any interruptions between verb and object by placing interrupting passages either at the beginning or at the end of the sentence. In some languages other than English, sentences tend to be complex, and information gets repeatedly interrupted. If English is not your native language, resist the temptation to apply the principles of writing in your native language to writing in English. Avoid interruptions between the verb and its object, as shown in Example 3-14.

Example 3-14

We conclude, based on very simplified models of solar variability, that solar variability is insignificant.

**Revised
Example 3-14**

We conclude that solar variability is insignificant **based on very simplified models of solar variability.**
or:
Based on very simplified models of solar variability, we conclude that solar variability is insignificant.

SUMMARY

BASIC RULES—STYLE
- ★ Establish importance.
- ★ Place old, familiar, and short information at the beginning of a sentence in the topic position.
- ★ Place new, complex, or long information at the end of a sentence in the stress position.
- ★ Get to the subject of the main sentence quickly, and make it short and specific. If possible, use central characters and topics as subjects.
- ★ Avoid interruptions between subject and verb and between verb and object.

PROBLEMS

PROBLEM 3-1 Sentence Interpretation

When scientists submit papers for publication, they often dread the response of reviewers. Here are four sentences that could have been written in different structural arrangements by reviewers to deliver the same news. Which statement is the one most likely resulting in the paper being accepted, and which is most likely the one resulting in rejection? Explain why.

1. Overall, although this manuscript is of interest for structural biologists, a more detailed analysis of ABC should be provided.
2. Although a more detailed analysis of ABC should be provided, this manuscript is of interest for structural biologists.
3. This manuscript is of interest for structural biologists, but a more detailed analysis of ABC should be provided.
4. A more detailed analysis of ABC should be provided, but overall, this manuscript is of interest for structural biologists.

PROBLEM 3-2 Word Placement and Flow

Rewrite one of the following paragraphs. Place words such that the reader can easily follow the logic flow of the message.

(a) Rainwater often picks up carbon dioxide, resulting in a weak solution of carbonic acid. A cave is formed when such rainwater trickles into the ground in areas with a high limestone content. Carbonic acid slowly dissolves the limestone. As more and more limestone dissolves, the cave grows underground. When a cave's ceiling gets eroded and collapses, a sinkhole forms.

(b) A quantum dot is a tiny semiconductor nanostructure. It is made of silicon, cadmium selenide, cadmium sulfide, or indium arsenide. A small number (on the order of 1–100) of conduction band electrons, valence band holes, or excitons are contained in a quantum dot. Colloidal semiconductor nanocrystals are small quantum dots, which can be as small as 2 to 10 nanometers, corresponding to 10 to 50 atoms in diameter.

PROBLEM 3-3 Word Placement and Flow
Write a paragraph using the list of facts provided. To create good flow, place words carefully at the beginning and end positions of sentences.

- fleas transmit plague *bacillus* to humans
- *bacilli* migrate from bite site to lymph nodes
- name "bubonic plague" arises because buboes = enlarged nodes

PROBLEM 3-4 Word Placement and Flow

1. **Construct a paragraph about thermophiles using the list of facts provided. Create good flow of the message through word placement.**
   - microorganisms
   - temperature range for growth between 45°C and 70°C
   - found in hot sulfur springs
   - cannot grow at body temperature
   - not involved in infectious diseases of humans
   - mechanism to resist elevated temperature unclear

2. **What does the reader expect to read next after having read the last sentence of your paragraph?**

PROBLEM 3-5 Subject–Verb–Object Placement
Rewrite the following sentences such that the subject is followed immediately by the verb and interruptions between verb and object are avoided. Place the subject early in the sentence if possible.

1. To date, more than 1,000 exoplanets, some of them with orbits of just a few hours, others with orbits of more than 1,000 years, have been confirmed.
2. Onchocerciasis, with approximately 18 million infected cases worldwide and 80 million more people at risk of infection, is now recognized as one of the major public health and socioeconomic problems in many tropical countries (Murdoch et al., 1996; OEPA, 1998).

3. Aside from protein X, protein Y, with a sequence very similar to a DNA-binding kinase, has been found to be able to bind RNA.

4. Early experiments revealed, as demonstrated by a strong suppression of the transition temperature with impurities (4), the extreme fragility of the superconductivity in the ruthenate superconductor Sr_2RuO_4 (SRO).

5. Earth's primordial atmosphere, consisting of high levels of helium and neon, which are now only present in high quantities in the innermost mantle and core of the Earth, was blown off several times after catastrophic impacts with other space bodies.

6. Recent reports show that digital disease detection systems, which use *big data* sources for information and can lead to early predictions of disease outbreaks, health behavior, and attitudes (4, 7, 8), heavily draw on mobile devices and online sharing platforms.

CHAPTER 4

Technical Sentences

Although this handbook is not an English grammar guide, the book captures the mistakes that are most commonly made by scientific authors, particularly mistakes that tend to reduce the clarity of a scientific manuscript.

IN THIS CHAPTER YOU WILL FIND:

- General advice on style in science
- Issues of grammar and style
- Basics of technical style
- Discussions of first and third person, active and passive voice, past and present tense, sentence length, verbs and action, noun clusters, pronouns, lists, and comparisons
- A list of common errors, including American versus British spelling, numbers versus numerals, capitalization, italics, punctuation, and common grammatical errors

4.1 GRAMMAR AND TECHNICAL STYLE

A paper full of grammatical errors discourages readers as well as reviewers and editors. It may also result in misinterpretation of what has been written. Although logically ordered and clearly expressed ideas are more important than perfect grammatical form, editors, reviewers, and readers will all be grateful if you write not only clearly and concisely but also correctly. Know that editors do not expect perfect English from ESL authors, nor do they expect the ultimate levels of literacy from native English speakers. If you use good technical style and avoid grammatical errors, however, your paper will be clearer and livelier, and you will reach a wider audience.

Many authors (especially native English speakers) are surprised to find certain phrases and sentences of their writing marked by editors because of bad style. A trained writer, however, will be able to recognize common style and grammar problems. Excessive use of third person, passive voice, nominalization, noun clusters, redundancies, and jargon are common causes of wordiness and bad style. Unclear use of tense, pronouns, prepositions, and articles can also confuse readers. All these problems of grammar and technical style are discussed in detail in this chapter.

For additional help with grammar and vocabulary, see, for example, Thurman (2012) or Perelman et al. (1997), which are listed in the references section. A glossary of grammatical terms can be found at the end of this book as well.

4.2   PERSON

➤ (12) Use the first person

Use the first person ("I" or "we") for describing what you did—but do not overuse it; do not use it if the journal (or your supervisor) has banned it or if the focus of the sentence should be on the organism or another topic.

It was once fashionable to avoid using "I" or "we" in scientific research papers because these terms were considered to be subjective, whereas the aim in science is to be objective. However, science is not purely objective. Writing from the point of view of "I" or "we" is appropriate in a scientific research paper wherever judgment comes in as the following examples illustrate.

| **Example 4-1** | a | To determine the mechanism for the direct effect of contrast media on heart muscle mechanics, <u>the study</u> on heart muscles isolated from cats was carried out. |
| | b | <u>The authors</u> show here that two main causes exist for avalanches: snow pack and external stress on the snow pack. |

These sentences taken from two different Introduction sections would be more accurate and more vigorous if the first person "we" were used for the subject instead of the third person: "the study" in Example 4-1a or "the authors" in 4-1b. The advantage of using the first person is that using "we" generally forces the author to use the active voice, which is lively.

| **Revised Example 4-1** | a | To determine the mechanism for the direct effect of contrast media on heart muscle mechanics, **we carried out** the study on heart muscles isolated from cats. |
| | b | **We** show here that two main causes exist for avalanches: snow pack and external stress on the snow pack. |

Although in most of the sections of a scientific document the use of first person is preferred, this use is more controversial in the Materials and Methods section. There, the first person "we" is not usually the topic. Instead, materials, methods, or organisms are usually the topic. In addition, it often may not have been the author(s) who performed a certain experiment but rather a technician or hired helper. Therefore, in the Materials and Methods section, use of third person is usually preferred. In certain fields such as in ecology, however, many journals require the use of first person and active voice, even in the Materials and Methods section.

In some scientific disciplines or as outlined by specific journals, third person continues to be used or even required. In these cases, you may have to go with what is expected, but know that, given a choice, your readers prefer first person.

Rarely is scientific research conducted or a paper authored by only one person. When you have a choice between first person singular ("I") and first person plural ("we"), it generally is better to choose plural for publications ("Here we report . . ."). Not only does it give recognition to the work of others—if indirectly—your reviewers and readers also will not perceive you as working in isolation. In contrast to publications, in grant proposals that clearly involve only you as a researcher, you should feel free to use first person singular.

4.3   VOICE

★ (13) Use the active voice

Use the active voice rather than the passive voice. If the passive voice is used excessively, writing becomes very dull and dense, as in the following examples.

 Example 4-2 **a** Cats <u>are hated</u> by dogs.

b No change in conductivity <u>was observed</u>.

These sentences are much livelier and more interesting when active voice is used.

 Revised **a** Dogs **hate** cats.
Example 4-2

b We **observed** no change in conductivity.

Do not remove the passive voice completely, however; use the passive voice when readers do not need to know who performed the action (e.g., in the Materials and Methods section; see Chapter 12, Section 12.4). You may also have to use the passive voice when the emphasis should be on a specific topic or when word location needs to be considered (see Example 4-4).

Example 4-3   Viral DNA **was isolated** 24 hours after inoculation.

In addition, you should use the passive voice if this allows you to replace a long subject with a short one, gives you a more consistent point of view (i.e., lets you use the same subject in consecutive sentences), or lets you put emphasis on the terms you want to have emphasized.

Example 4-4   Most of the world's diamonds are found in cooled volcanic lava tubes. The stones originally form at great depth under high pressure and temperature. Molten magma that rises up through lava tubes and volcanic pipes often transports them, and thus brings them closer to the surface of the Earth's mantle and crust where they can be mined.

 Revised Example 4-4   Most of the world's diamonds are found in cooled volcanic lava tubes. The stones originally form at great depth under high pressure and temperature. **They are often transported by molten magma that rises through lava tubes and volcanic pipes, and thus brought** closer to the surface of the Earth's mantle and crust where they can be mined.

The passive voice is needed in the preceding revision to keep the focus on the cone rather than shifting to new information. In the revision, word location has been considered. Although the revision leads to use of passive voice, this choice is preferable here to ensure good flow. In general, considering word location (jumping word location or consistent point of view; see Chapter 6, Section 6.3) is more important than use of active voice.

4.4   TENSE

★ (14) Use past tense for observations and specific conclusions

★ (15) Use present tense for general rules and established knowledge

ESL advice

The two main tenses that occur in scientific papers are present tense and past tense. In proposals, future tense is also widely used. Many scientific authors, especially ESL authors, seem to be confused about when to use past tense and present tense. Many are also unsure if past tense and present tense can be mixed in the same sentence or paragraph. Generally, you should use the past tense for observations and specific conclusions. For example, results presented in your paper should be described in past tense because you have done these experiments and your results are not yet accepted "facts." Therefore, the Abstract, Materials and Methods, and Results sections should employ past tense as they refer primarily to your own work.

Example 4-5   a   Higher temperatures **resulted** in less bud formation.

b   The three images **were taken** about 90 minutes apart.

You should use the present tense for general rules, accepted facts, and established knowledge. Thus, results from already published papers should be described in the present tense as published results are generally assumed to be "facts." Similarly, if something is a general rule or fact that is still true in the present, use present tense.

 Example 4-6   a   The newly discovered planet **is** at least as big as Pluto.

b   Most regions where this problem **arises belong** to category X.

c   Fields like public health **explore** the predictive power of Big Data.

If you use past tense for describing results of already published work, you are implying to the reader that you do not consider these results to be "facts" but observations.

Can tense be mixed in the same sentence or paragraph? Certainly, as is apparent in the next example.

 Example 4-7   Brown **reported** that the bacteria *Brucella* **cause** abortion in livestock.

This example describes an experiment that has been completed. "Reported" is therefore written in past tense. "Cause," however, is present tense because this part of the sentence is still true and is considered established knowledge once published.

Like the Introduction section, the Discussion section of a research article can include both past and present tense, depending on whether you are reporting an observation or one-time occurrence or whether you are describing a general rule that continues to hold true. Note that, usually, remarks about the presentation of data should be in present tense, and descriptions of assumptions and theory should also be described in present tense in your paper.

 Example 4-8   The effect of temperature on the microbiome **is shown** in Figure 5.

For findings and events that have been started in the past and are ongoing or have present consequences, use present perfect tense.

Example 4-9   The number of people addicted to painkillers **has been** growing for the past three decades.

Only experiments that you plan to do in the future should be written in the future tense. Future experiments are usually not included in research articles but are described in grant proposals.

 Example 4-10   We **will examine** if parallel universes exit.

4.5   SENTENCE LENGTH

★ (16) Write short sentences. Aim for one main idea in a sentence

Short sentences are easier to understand than long sentences. Generally, the longer a sentence, the more difficult it is to grasp. The average sentence length in many scientific articles is over 30 words per sentence; in most newspapers, it is between 15 and 20 words per sentence (one of the reasons that newspaper articles are easier to understand). Many scientific papers could be strengthened by shorter sentences, although not every sentence should be short. Using *only* short sentences does not result in strong writing but leads to choppy, hard-to-follow passages. Some sentences need to be long to communicate complex ideas. Scientific authors should aim for an average sentence length of about 20 to 22 words. This means that some sentences will be longer and some shorter, but the average number of words per sentence overall will be around 20 to 22.

Short, simple sentences tend to emphasize the idea contained in them. The longer a sentence gets, the more difficult it is for the reader to identify what is of primary importance. Therefore, single-clause sentences have more weight, and thus more importance, than multiclause sentences. Writing a short sentence that highlights the main topic is particularly important at the beginning of a section or paragraph. It ensures that you have the attention of the reader from the outset and lets the reader focus on the main idea.

Similarly, readers assign more importance to sentences that stand on their own (independent sentences) than to a clause that depends on the presence of another clause. Thus, independent sentences have more weight than dependent sentences, which in turn have more weight than phrases. Consider Example 4-11.

Example 4-11   **a**   Rheumatic fever is an autoimmune disease.

   b   *It is generally accepted in the field of medicine that* rheumatic fever is an autoimmune disease.

The words in the sentence of Example 4-11a, "Rheumatic fever is an autoimmune disease," tend to weigh more when they are in their own sentence than when they appear in some longer sentence such as the sentence in Example 4-11b. In addition, in the sentence in Example 4-11b, the same words appear in a dependent clause, which makes the reader perceive them as less important. For both of these reasons, most readers perceive the sentence in Example 4-11a as "weighing more" than the sentence in Example 4-11b.

Many sentences in scientific papers are needlessly complex. As a general guideline, do not present too many ideas in a single sentence. Instead, make sure your sentences do not contain more than one main idea and that they do not wander. The first step to ensure that your sentences do not contain too many ideas is to decide which details in a sentence are important. Only when you have assigned importance will you be able to subordinate less important information and omit unimportant information. Often, you can consider breaking subordinate sentences into separate sentences.

In certain cultures, people write in very complex, indirect ways. If you have this background, be particularly aware that English sentences that are concise and direct are better understood than sentences that are long and contain many different ideas.

It is a good idea to imagine yourself sitting across from an important reader. Write your paper as if you were *telling* this reader about your work. Remember that the purpose of a scientific paper is to inform, not to impress.

Consider Example 4-12.

Example 4-12   **Excessively long sentence**

When central venous IV lines were removed, skin samples from patients with IV line–related bloodstream infections were collected and in 80% of these samples, bacteria with high DNA identity to those found in the bloodstream and IV lines were identified, whereas in 20% of the patients isolated bacteria had no or low DNA identity, suggesting that most bloodstream infections in patients with central venous IV lines arise from contamination of IV lines or needles during insertion.

(76 words/sentence)

In this example, the first idea ends before "and." The second idea ends before "whereas" and the third idea before "suggesting." All of these ideas should be written in separate sentences.

Revised Example 4-12   When central venous IV lines were removed, skin samples from patients with IV line–related bloodstream infections were collected. In 80% of these samples, bacteria with high DNA identity to those found in the bloodstream and IV lines were identified. However, in 20% of the patients isolated bacteria had no or low DNA identity. These observations suggest that most bloodstream infections in patients with central venous IV lines arise from contamination of IV lines or needles during insertion.

(average of 19 words/sentence)

ESL advice

Whereas the original sentence was 76 (!) words long, the revised version has an average sentence length of 19 words. Therefore, the revised version is much easier to understand. The reason is not that the sentences are shorter, but mainly that the ideas are separated into different sentences.

4.6   VERBS AND ACTION

★ (17) Use active verbs

Verbs are perhaps the most important part of an English sentence. With strong and active verbs, your writing enlivens and energizes. Verbs make sentences direct and easy to follow. If you hide verbs by using nominalizations, that is, abstract nouns derived from verbs and adjectives, you will make your writing heavy and much harder to comprehend for readers.

Your readers will perceive your writing as particularly dense when you use many abstract nouns. "Academese" tends to be full of nominalizations. Many nonnative-speaking scientists also excessively use nouns in their native language, which they then translate and apply in English. For better scientific style, avoid nominalizations—use active verbs instead.

ESL advice

👍 Active verb	👎 Buried verb/nominalization
assess	assessment
decide	made the decision
depends on	is dependent on
exist	existence
follows	is following
form	formation
inhibit	inhibition
measure	measurement
remove	removal

In the following example, the action is not in the verb but in the noun.

👎 **Example 4-13**   Their <u>suggestion</u> for us was a different <u>analysis</u> of the data.

In the revision, the actions are all verbs, resulting in a much clearer and less dense style.

👍 **Revised Example 4-13**   They **suggested** that we **analyze** the data differently.

Other examples of verbs and adjectives and their nominalizations include the following.

Verb	Nominalization
analyze	analysis
attempt	attempt
centrifuge	centrifugation
compare	comparison
determine	determination
differ	difference
discover	discovery
discuss	discussion
dissect	dissection
evaluate	evaluation
elute	elution
explain	explanation
fail	failure
hypothesize	hypothesis
increase	increase
isolate	isolation
move	movement
need	need
react	reaction
separate	separation
speculate	speculation

Here, too, use the active verb instead of the nominalization. Note that some nominalizations and verbs are identical, such as graph (verb) and graph (noun).

Science is more interesting particularly if the actions of animals and cells are described in active verbs.

 Example 4-14   Earthworms react to light.
Muscles contract.
Blood flows.

Avoid writing with weak verbs. Weak verbs seem abstract and impersonal and result in boring writing. Examples of weak verbs include the following:

occurred	was seen	was noted	was done	get
was observed	caused	produced	make	

When you find yourself writing using one of these verbs, stop and check if you can use an active verb instead.

Example 4-15   **a**   A 10% <u>increase</u> in temperature <u>occurred</u>.

In this example, the verb ("occurred") is not active but weak. The subject of the sentence ("increase") expresses the action. This noun is also a nominalization of the verb "increase." As a result of the nominalization, the sentence is complicated and indirect. To revise a sentence whose action is buried in a noun, replace the weak verb with the action of the noun.

 Revised   **a**   Temperature **increased** 10%.
Example 4-15

In the revised sentence, when the verb is active and strong, the sentence is simpler, more direct, and more efficient than when the action is nominalized.
Consider Example 4-15b.

 Example 4-15   **b**   This wavelength <u>caused a decrease</u> in the molar absorption coefficient.

This example contains the weak verb "caused." In the sample sentence, the action is buried in the object ("decrease"), and the true object ("molar absorption coefficient") is sidetracked into a prepositional phrase ("in the molar absorption coefficient").

 Revised   **b**   This wavelength **decreased** the molar absorption
Example 4-15   coefficient.

Whereas the original sentence just sits, the revised sentence moves because the verb is strong and active.
    Sometimes writers express action in the object of a preposition instead of in the verb. (Prepositions are words such as "of," "for," "on," "in," "to," and "with.")

 Example 4-15   **c**   <u>Upon early inflammation of organ transplantation, allografts are rejected.</u>

The sentence in this example is heavy and hard to follow because instead of an active verb, the noun "inflammation" is used in conjunction with the preposition "upon."
    To make this sentence easier to read, turn the prepositional phrase into a dependent sentence and use a verb instead of the nominalization. Using an active verb makes this sentence much livelier and easier to understand.

 Revised   **c**   *When organ transplants **become inflamed** early,*
Example 4-15   *allografts are rejected.*

Although most nominalizations in scientific writing can and should be turned into verbs, there are exceptions. Keep a nominalization if it refers to a previous sentence or if it names the object of the verb.

Example 4-16 **a** **This observation** led us to conclude . . .

b An example of this theory is provided by **a delay** in the reaction.

Analysis and Revision

To find and revise sentences that may confuse your readers, analyze your sentences:

1. Underline the first 8 to 10 words in the main sentence, ignoring introductory phrases.
2. For the underlined words, identify the central *character* of the sentence or paragraph.
3. Make the *character* the subject.
4. Look for the action.
5. If the actions are nominalizations, change them into verbs and make the relevant characters their subject.
6. Replace weak verbs with strong, active verbs if necessary.
7. Rewrite the sentence using conjunctions such as "because," "if," "when," "although," "that," or "whether." If necessary, turn a prepositional phrase into a dependent clause.
8. Avoid other nominalizations and abstract nouns in the remainder of the sentence as well—change them to verbs.

Example 4-17 Despite the <u>identification</u> of the AIDS virus by *researchers, there has been a* <u>failure</u> to develop a vaccine for the immunization of those at risk.

Example 4-17 Analysis and Revision

1. The central character of the sentence is *researchers.*
2. Here the nominalization among the first 8 to 10 words is "failure." (Other nominalizations are "identification" and "immunization.")
3. Change the nominalization to a verb: failure > fail (identification > identify, immunization > immunize).
4. Make the character the new subject of the verb, and turn the prepositional phrase into a dependent clause, leading to the following:

Revised Although *researchers* **identified** the AIDS virus, *they* **failed**
Example 4-17 to develop a vaccine to immunize those at risk.

If you revise your sentences using the suggestions and basic rules presented, you will find your sentences not only more concrete, active, and concise but also more coherent and clearer for the reader.

4.7   CLUSTERS OF NOUNS

★ (18) Avoid noun clusters

Noun clusters are nouns that are strung together to form one term. In English, nouns (and adjectives) can be used to modify other nouns. However, when nouns appear one right after the other, it can be difficult to tell how they relate to each other and what the real meaning of the cluster is. When you add more than one modifier in front of a single noun or place additional modifying nouns and/or adjectives in front of an existing noun pair, you may end up with confusing noun clusters. Avoid clusters of nouns, especially if there are more than two or three nouns in the cluster. These noun clusters are awkward and sometimes downright incomprehensible.

Example 4-18	a	porcine tracheal fluid samples
	b	Peter Carri is a <u>condensed matter</u> and <u>quantum many-body theoretical physicist</u>.

Instead, work your way from the back to the front and use prepositions to link the nouns. The prepositions add clarity to the phrase—they show more fully how the nouns relate to one another—and the meaning of your words becomes clearer.

Revised Example 4-18	a	fluid samples *from* the trachea *of* pigs
	b	Peter Carri is a **theoretical physicist studying condensed matter and quantum many-body physics**.

Note that certain noun pairs and clusters—such as "water bath," "cell wall," "egg receptor," and "sucrose density gradient"—are accepted terms. Do not break such terms apart. If you are unsure if a term is accepted as a single word, look it up in a dictionary, a recent journal article, or online.

If a technical name is a noun cluster (example: pyrophosphate dependent phosphofructo-1-kinase apo form structure), there are three ways to help your reader:

1. Use hyphens to show the relationship between the words (pyrophosphate-dependent phosphofructo-1-kinase apo-form structure).
2. Explain the name (the structure of the apo-form of pyrophosphate-dependent phosphofructo-1-kinase).

3. Explain to the reader that you will use a shorter name (the apo-form structure of pyrophosphate-dependent phosphofructo-1-kinase, short PPi-PFK apo-form).

Hyphens in noun clusters are most often used for

- Two-word terms that are used together (high-pressure chamber, ATP-dependent)
- Adjectives that consist of three or more words (four-to-one ratio)
- Terms that contain a capital letter or a number and a noun (C-terminal end, 10-fold increase)

4.8   PRONOUNS

★ (19) Use clear pronouns

Pronouns are words that take the place of nouns.

Examples	it, none, they, these, those, their, them, this, that, which, who, whose

It is essential that you use clear pronouns. Unclear pronouns are one of the most common problems in scientific writing. If the pronoun that refers to a noun is unclear, the reader may have trouble understanding the sentence. An author always knows which term she or he is referring to. A reader is not so lucky. Be sure that the pronouns you use refer clearly to a noun or antecedent in the current or previous sentence. If there are too many possible nouns the pronoun can refer to, repeat the reference noun after the pronoun.

Example 4-19	*Anaerobic organisms* typically live in the *intestines*. Thus, <u>they</u> are of interest to us.
Revised Example 4-19	*Anaerobic organisms* typically live in the <u>*intestines*</u>. Thus, **intestines** are of interest to us. or: **Intestines** are of interest to us because **they** typically contain anaerobic organisms.

Sometimes the noun or antecedent that a pronoun refers to has been implied but not stated. To clarify the reference, explicitly state the implied noun after the pronoun as in the next example.

Example 4-20	If a specimen is frozen in a bath containing dry ice and acetone, the water of the cell can be removed by sublimation to prevent damage to the cell. <u>This</u> is commonly used for preservation of cultures.

Revised	If a specimen is frozen in a bath containing dry ice
Example 4-20	and acetone, the water of the cell can be removed by
	sublimation to prevent damage to the cell. **This tech-**
	nique is commonly used for preservation of cultures.

Consider another example.

Example 4-21	The impact and proliferation of humans are consid-
	ered the main cause for the rate of mass extinction
	today. As this rate increases, life in oceans will be
	affected more dramatically than species on land.
	This can be shown by our model, which calculates
	extinction rate.

Here, "this" refers to an effect on the rate of mass extinction. To clarify the pronoun, you can use an *implied term* such as "trend" that refers to the idea in the sentence before. Using such terms is especially helpful when the implied term is much shorter than expressing the idea to which it refers.

Revised	The impact and proliferation of humans are consid-
Example 4-21	ered the main cause for the rate of mass extinction
	today. As this rate increases, life in oceans will be
	affected more dramatically than species on land.
	This trend can be shown by our model, which cal-
	culates extinction risk.

4.9   LISTS AND COMPARISONS

Parallel Ideas

★ (20) Use correct parallel form

Lists and ideas that are joined by "and," "or," or "but" are of equal importance in a sentence, and so are ideas that are being compared. These ideas should be treated equally by writing them in parallel form. To write ideas in parallel form, the same grammatical structures are used. These grammatical structures can be single words, prepositional phrases, infinitive phrases, or clauses. If parallel ideas are written in parallel form, the reader does not get distracted by the form but can concentrate on the idea.

The next few examples are sentences that contain parallel ideas joined by "and," "or," or "but," which are written in parallel form.

Example 4-22	*Subject*		*Verb*	*Adverb*
For rice,	molybdenum concentrations		were	twice as high,
but	selenium concentrations		were	three times lower
than for wheat.				

In the preceding example, the same parallel form is used for the two ideas that are being compared, namely, the group of words after "but" is in the same grammatical structure as the group of words before "but"—in this case, subject, verb, adverb.

Following are a few more examples.

Example 4-23

Based on our hypothesis, we expected to see	Direct Object	Preposition	Object of Preposition
	a decrease	**in**	**the infection rate**
and	**an increase**	**in**	**survival of patients.**

Example 4-24

	Subject	Verb
	S. franciscanus sperm	**can fertilize**
S. purpuratus eggs, but	**L. pictus sperm**	**cannot.**

Example 4-25

Brown bears were observed	Preposition	Object of Preposition
	at	**the zoo**
and	**in**	**their natural habitats.**

Note that in Example 4-25, "at the zoo" and "in their natural habitats" are in parallel form even though the specific prepositions ("at," "in") are different. For parallel form, it is only important that both terms are prepositions.

Do not confuse the reader by obscuring the logical relationship of parallel ideas.

Example 4-26   Prolonged febrile illness together with subcutaneous nodules in a child could be due to an infection with a Gram+ organism, but it could also be <u>that the child suffers from rheumatic disease.</u>

In this sentence, two equal ideas are connected by "but." However, these parallel ideas are not instantly apparent because the grammatical form of the first part of the sentence is not parallel to that of the second part. Because the second half of the sentence is equal in logic and importance, it should be written in parallel form.

Revised Example 4-26   Prolonged febrile illness together with subcutaneous nodules in a child could be due to an infection with a Gram+ organism, but it could also be **due to rheumatic disease.**

This sentence can be further simplified.

	Revised Example 4-26	Prolonged febrile illness together with subcutaneous nodules in a child could be due to an infection with a Gram+ organism or due to rheumatic disease.

Coordination

➤ Arrange ideas in a list to read from shorter to longer

If a sentence lists two or more ideas, these ideas should not only be in parallel form but they should also be coordinated. Careful writers coordinate ideas that are both grammatically and logically parallel. For good coordination, ideas should be arranged to read from shorter to longer in terms of the number of words contained in the idea. Coordinating ideas in this way makes a sentence more graceful.

Consider the following two sentences.

Example 4-27	**a**	Loss of coral reefs will affect <u>organisms</u>, such as turtles and sea birds that depend on specific habitats for reproduction, <u>and beaches</u>.
	b	Loss of coral reefs will affect **beaches and organisms** such as turtles and sea birds that depend on specific habitats for reproduction.

Example 4-27a seems to end too abruptly with "beaches." Example 4-27b has much better flow because here the two parallel ideas have been arranged from shorter to longer.

4.10   FAULTY COMPARISONS

★ (21) Avoid faulty comparisons

Aside from maintaining parallelism in your comparisons, you should avoid grammatical and logical problems when writing comparisons. These problems result in faulty comparisons, one of the most common problems in scientific writing. Faulty comparisons can arise because of ambiguous comparisons and incomplete comparisons. Faulty comparisons may also be due to the overuse of "compared to." Examples for all of these scenarios are shown in the following sections.

Ambiguous Comparisons

The following example is a typical ambiguous comparison found in scientific papers.

	Example 4-28	Our conclusions are consistent with <u>Tamseela et al. (2013)</u>.

Comparisons such as this are confusing for the reader as they compare unlike things. To avoid such ambiguous comparisons, make sure that you are comparing like items.

| **Revised** | **a** | Our conclusions are consistent with **the conclusions of** Tamseela et al. (2013). |
| **Example 4-28** | | |

This sentence can be written even simpler by using a pronoun to avoid repetition.

| **Revised** | **b** | Our conclusions are consistent with **those of** Tamseela et al. (2013). |
| **Example 4-28** | | |

Incomplete Comparisons

Absolute statements should not be written as comparisons. Information being compared and that with which it is being compared need to be listed completely and in parallel.

| **Example 4-29** | This study tested 24 networks compared to Menon's study. |

Revised	**a**	This study tested 24 networks; Menon's study tested only 8 networks.
Example 4-29		
	b	In this study, the number of networks tested (24 subjects) was three times that of Menon's study (8 subjects).

ESL advice

Such incomplete comparisons may confuse readers because their intended meaning is unclear. In certain foreign languages, incomplete comparisons occur often. Avoid these when writing in English.

Here is another example.

| **Example 4-30** | RNA isolation is <u>more difficult</u>. |

The question the reader naturally asks when reading this sentence is: More difficult than what? To complete the comparison, you need to include the item that RNA isolation is compared with as shown in the revised example.

| **Revised** | RNA isolation is **more difficult than DNA isolation**. |
| **Example 4-30** | |

"Compared to"

Use "than" not "compared to" for comparative terms such as "smaller," "higher," "lower," "fewer," "greater," "more," and so forth.

👎	**Example 4-31**	We found <u>more</u> fertilized eggs in buffer A <u>compared to</u> buffer B.
👍	**Revised Example 4-31**	We found <u>more</u> fertilized eggs in buffer A **than in** buffer B.

Note that "in" is repeated in the revised example to keep parallel form.

Avoid using "compared to" with the words "decreased" or "increased" because the meaning is ambiguous.

👎	**Example 4-32**	K_D increased over time <u>compared to</u> K_A.
👍	**Revised Example 4-32**	K_D increased over time, **but K_A did not.**

4.11   COMMON ERRORS

★ (22) Avoid errors in spelling, punctuation, and grammar

In any type of writing, errors should be avoided. Common errors include (a) spelling, (b) punctuation, (c) words that are omitted, (d) a subject and verb that do not make sense together, (e) a subject and verb that do not agree, and (f) unclear modifiers. When one of these errors appears, the reader is slowed down and may even need to reread the sentence to figure out the intended meaning (see also Chapter 2, Section 2.3). These errors can be avoided by carefully proofreading and double-checking the manuscript. You may choose the grammar and spell checker of MS Word to help do so or consult Grammarly online (https://www.grammarly.com).

Spelling

If you use a computer to prepare your manuscript, make use of a spell checker to avoid some common errors. Such a program will not find all the mistakes, however. For example, the program will not point out words that are wrong but correctly spelled ("from" when you meant "form" or "to" instead of "two").

A spell checker program also will not point out if you spelled the same word in the same way throughout. Compile a word list: Every time you make a decision on spelling, record it, and check your second draft for conformity to the list.

The choice in spelling is sometimes between British and American versions of a word, which is especially confusing for ESL writers. Consult a dictionary to see whether one form is preferred to another or to check which version is British and which is American. Use a dictionary such as the *Dictionary of Contemporary English* that gives both British and American spellings, or search online dictionaries. North American English spelling is more common these days than British English spelling in scientific writing. However, it is usually a good idea to see what style your target journal has adopted. Many journals accept and use both British and American spelling if one of these spellings is used consistently throughout an article. Adjust your choice of punctuation rules accordingly as well. Grammar differences between the two language variations are few, if any. The most common differences in American and British English spelling are listed in Table 4.1 together with scientific words as examples.

Numbers versus Numerals

Various factors determine whether you need to spell out numbers or use numerals. These factors include different style guides, the size of the number, whether it is a scientific number with units, the context of the number, and what it is representing. For example, the American Psychological Association (APA) style manual recommends spelling out single-digit numbers (one through nine) and using numerals for 10 and above. However, *The Chicago Manual of Style* recommends spelling out whole numbers one through one hundred. Exceptions to these rules include:

1. Use numerals to present scientific values (i.e., for values with a unit [6 days, 32 hr, 7.8 g] that do not begin a sentence).
2. Spell out numbers at the beginning of a sentence (e.g., "Sixty-three patients were evaluated"). To avoid beginning a sentence with a number, invert word order if necessary (e.g., "Of these, 34 survived to adulthood").
3. Large numbers, if rounded, are usually spelled out ("Three hundred animals frequented the water hole"), but very large numbers (millions or more) are normally expressed with a mixture of numerals and spelled-out numbers ("The bird population was estimated to be 1.3 million").
4. Although in informal writing you can use numerals and the percent sign (%), as in "3% of the participants," in formal writing you should spell out the percentage, as, for example, in "15 percent of the participants."
5. For a series of mixed numbers greater and less than ten, use numerals as in "5, 10, and 15 hours."

TABLE 4.1 Differences between American and British spelling

AMERICAN SPELLING	BRITISH SPELLING	AMERICAN ENGLISH	BRITISH ENGLISH
-am	-amme	program	programme
-ay	-ey	gray	grey
-e-	-ae-, -oe-	edema, anesthesia, leukemia, pediatric	oedema, anaesthesia, leukaemia, paediatric
-er	-re	center, meter, fiber, liter	centre, metre, fibre, litre
-f-	-ph-	sulfur	sulphur
-ing, -able	-eing, -eable	aging, sizable	ageing, sizeable
im-, in-	em-, en-	imbed, insure	embed, ensure
-ize, -yze	-ise, -yse	analyze, optimize, emphasize, realize	analyse, optimise, emphasise, realise
-l-	-ll-	signaling, labeling, modeling	signalling, labelling, modelling
-ll	-l	enroll, fulfill	enrol, fulfil
-og	-ogue	catalog	catalogue
-or	-our	tumor, flavor	tumour, flavour
-se	-ce	defense	defence
-um	-ium	aluminum	aluminium
-dg or -g	-dge or -gu	judgment, argument	judgement, arguement
-ed	-t	dreamed, learned	dreamt, learnt

Capitalization

Often, authors are confused as to which words to capitalize. Here are some general guidelines.

1. **First word of a sentence**
   a. Capitalize the first word of every sentence.
   b. First words in short lists or bullet points do not need to be capitalized.
   c. First words in longer lists or bullet points should be capitalized. The listed item should also receive a period at the end.

👍 **Example 4-33**   To start writing a research paper, scientists need

1. enough data;
2. valid references; and
3. basic rules.

2. Proper nouns

Capitalize all proper nouns within a sentence. The term "proper noun" refers to a noun that names a specific person, place, time, or thing such as *New York*, *August*, or *Earth*. The following examples are considered proper nouns that should be capitalized:

- Name of a specific person:
  - Fred
  - Celsius
  - Huntington's disease
- Name of a specific place:
  - Canada
  - Beijing, China
- Name of a specific time or period:
  - November
  - Friday
  - the Ice Age
- Name of a specific thing:
  - the Koran
  - the Queen Mary
  - the Kon-Tiki
- Names of cultures or languages:
  - Indian
  - Italian
- Names of institutions or organizations:
  - American Chemical Society
- Professional titles:
  - Director of XXX Institute
  - Chief Financial Officer

In contrast to proper nouns, common nouns, which provide the general names of things (window, planet, or summer), are not capitalized.

3. Scientific names/taxonomy

In taxonomic denotations, capitalize phylum, class, order, family, or genus but not species or subspecies. Do not capitalize common names.

4. Titles

Nouns, adjectives, adverbs, and verbs in titles should be capitalized (*Journal of Biochemistry*, *Handbook of Scientific Communication*). Note that some styles such as the APA style capitalize all words in titles of four letters or more (e.g., "With").

5. Sections of published works

Capitalize the titles of sections of a research article or book and the names of these sections when referred to in a specific sense or in a text reference (Introduction, Materials and Methods, Results, Discussion, Chapter 1,

Appendix A). Do not capitalize these terms if they are discussed in a general sense (methods section, introduction, discussion).

Italics

Certain words and phrases in science, generally Latin derivatives, are placed in *italics*. In addition to genus and species names, subheadings, foreign words, emphasized words, and titles of journals, books, and manuals, the following words are typically written in italics. Note that some styles such as APA style do not follow this rule.

in vivo   *in vitro*   *in organello*   *de novo*   *in vacuo*

Punctuation

To avoid possible misinterpretation of your writing, pay attention to correct punctuation. In contemporary English, the trend is toward less punctuation and more simplicity.

Here are a few important rules for punctuation:

1. Use simple punctuation

ESL advice

The best approach to punctuation is usually the simplest. Avoid sentences with overly complex punctuation. Use simple punctuation instead. Note that unlike some foreign languages, very few exclamation marks exist in English and essentially none in scientific writing.

2. Use a period to end a full sentence

In contemporary scientific writing, avoid semicolons where possible.

	Example 4-34	*Older style:* TS-25 was a heat inducible <u>mutant; thus</u>, it was given the prefix "TS."
👍	**Revised Example 4-34**	*Newer style:* TS-25 was a heat-inducible **<u>mutant. Thus</u>**, it was given the prefix "TS."

3. Use semicolons to connect two independent sentences

In older style writing, semicolons were placed between sentences that contain closely related ideas and are not linked by a coordinating conjunction such as "and," "or," and "but." However, contemporary scientific writing prefers not to use semicolons. One exception is the use of semicolons in complex series, as discussed under point 6.

👍	**Example 4-35**	*Older style:* Some viruses are **<u>deadly; others</u>** are not. or *Newer style:* Some viruses are **<u>deadly. Others</u>** are not.

4. Use commas for clarity and emphasis

Check each sentence for reading errors. Then decide whether a phrase or sentence needs a comma or not. Consider the following example:

☜	**Example 4-36**	Although samples were incubated at 37°C for 24 hr we did not observe any change in the growth pattern.

This sentence has more than one possible interpretation depending on where the comma is placed.

Example 4-36	Interpretation A	Although samples were incubated at 37°C for 24 **hrs, we** did not observe any change in the growth pattern.
	Interpretation B	Although samples were incubated at **37°C, for** 24 hrs we did not observe any change in the growth pattern.

It is important to place a comma at the correct location within the sentence to make the meaning of your writing clear.

Commas are needed under the following circumstances:

(a) For dependent clauses or long phrases before the main sentence. Example 4-36 is an example of this rule.
(b) For transitional words or phrases.

👍	**Example 4-37**	Propanol, ethanol, and butanol are all organic alcohols. **However,** aerosol is not an organic alcohol. (or: Aerosol, **however,** is not an organic alcohol.)

However, commas are not placed after or in front of transition words if the flow of the sentence would be interrupted as in the following case:

👍	**Example 4-38**	This plant grew in acidic soil and therefore has blue flowers.

(c) If the dependent clause is not essential to the meaning of the overall sentence. Set these nonessential clauses off by a comma.

👍	**Example 4-39**	a	Some of the **eggs, which** came from Tennessee, were damaged.
		b	Of the eggs, 10% were damaged upon arrival. The eggs that were damaged were discarded.

Generally, commas are used with the word "which" but not with "that." See also Appendix A on "Commonly Confused and Misused Words" for more information on "which" and "that."

ESL advice

ESL writers are often confused about comma use elsewhere in the sentence. One particular comma placement that gets misused often is between the subject and the verb. Do not separate the subject and verb by a comma.

If you disrupt the subject and verb by a dependent clause, use two commas, one before and one after the clause.

5. Place a comma between the items in a series as well as before the word *and*

Commas should separate items in a series. In American scientific writing, a comma is also placed before the *and* in the series unlike in British English and unlike in American literary writing. Often, this placement is important to the meaning of the sentence.

Example 4-40	**a**	*E. coli* can produce septicemia, **meningitis, and** pneumonia.
	b	Microwave disinfection kills bacteria, yeast, and fungi on dentures (5).

6. Use semicolons and/or numerals to punctuate complex series

Use semicolons and numerals to separate the items within the sentence when items in a series are unusually long and especially when they contain their own punctuation. Use a colon to introduce the list or series.

Example 4-41	Viruses often cause infections and can lead to significant mortality in patients in an ICU. Of concern are especially respiratory viruses, such as respiratory syncytial virus, influenza viruses, parainfluenza viruses, and adenoviruses, which can also lead to gastroenteritis and cystitis.
Revised Example 4-41	Viruses often cause infections and can lead to significant mortality in patients in an ICU. Of concern are especially respiratory viruses, such as (1) respiratory syncytial virus; (2) influenza viruses; (3) parainfluenza viruses; and (4) adenoviruses, which can also lead to gastroenteritis and cystitis.

7. Avoid weak connectors

Scientists are extremely fond of coupling pairs of independent sentences. However, the words "and," "but," "for," "or," and "nor" are weak connectors. If weak connectors cannot be avoided, they should be set off by commas. For stronger writing, consider making each statement a separate sentence.

Example 4-42   The amplitude of the potential pattern in the electrolyte decreases with increasing distance from the electrode and expands parallel to the electrode at the same time (Figure 6a, top), <u>and</u> this expansion is also felt at the electrode/electrolyte interface in the form of changed migration current densities, indicating that the migration coupling encompasses a wide range of the electrodes, sometimes even all of the interface.

Revised
Example 4-42   The amplitude of the potential pattern in the electrolyte decreases with increasing distance from the electrode and expands parallel to the electrode at the same time (Figure 6a, top). This expansion is also felt at the electrode/electrolyte interface in the form of changed migration current densities. Thus, the migration coupling encompasses a wide range of the electrodes, sometimes even all of the interface.

8. Avoid quotation marks

Although it is rare that a direct quote is necessary or even appropriate in scientific writing, whenever you copy another writer's or speaker's exact words, enclose the material in standard (double) quotation marks (although this happens rarely in scientific writing). Use single quotation marks for a quote within a quote. Remember to cite the source of the material, including the page number on which it appears.

When the name of an article or book chapter is cited in the text, enclose it in double quotation marks. Sometimes titles can also be placed in italics.

Example 4-43   An article, "Structure of the 30S Ribosomal Subunit," recently appeared in *Cell*.

Quotation marks, italics, or boldface can also be used to draw attention to words or phrases such as to new terms or terms used in an uncommon way.

9. Avoid hyphenation

Avoid hyphens if possible. Write "cooperation" and "rearranged," but hyphenate two (or more) word clusters such as "English-speaking," "high-pressure," and terms that contain a capital letter or a number and a noun such as "N-terminal" and "100-fold."

10. Abbreviations

It is often not easy to know when or when not to use punctuation for an abbreviation. In scientific English, abbreviations for a single word are spelled with a period as in "Prof." and "fig." Note that in British English, titles do not have a final period if the last letter of the abbreviation corresponds to the last letter of the word itself (Dr, Mr, Mrs, Ms), whereas American English prefers the period after these abbreviations (Dr., Mr., Mrs., Ms.). Units of measure are usually lowercase without a period (kg, min), even when they are acronyms (ppm [parts per million]). Other acronyms not referring to units of measure are usually capitalized without periods (DNA) but can also be a mixture of lowercase and uppercase letters (iRNA, ssDNA). Units of measure named after people are capitalized without a period (C [Celsius], F [Fahrenheit]).

Subject–Verb Correspondence

The subject and verb of a sentence must agree. A singular subject must have a singular verb, and a plural subject must have a plural verb.

 Example 4-44   The *sugar content* and *protein composition* of each sample <u>was</u> determined.

The subject of the sentence is "sugar content and protein composition," not "sample." Because the subject is plural, the verb must be plural as well.

 Revised   The *sugar content* and *protein composition* of each
Example 4-44   sample **were** determined.

Many scientific authors are confused about such words as "data," "spectra," and "media." "Data," "spectra," and "media" are the plural forms of "datum," "spectrum," and "medium" and thus should be treated as plural nouns. (Note that some dictionaries do accept use of both singular and plural verbs for these words.)

 Example 4-45   Our <u>data</u> show that carbonated water increases appetite.

Other singular and plural forms of words commonly used in the biomedical sciences are shown in Table 4.2.

Subjects and verbs should not only agree; they should also make sense together.

 Example 4-46   The *concentration* of the sample was <u>measured</u>.

Unlike temperature or amounts, concentration cannot be measured. It has to be calculated or determined.

 Revised   The *concentration* of the sample was **determined**.
Example 4-46

TABLE 4.2 Singular and plural nouns forms

SINGULAR	PLURAL
alga	algae
analysis	analyses
bacterium	bacteria
criterion	criteria
datum	data
flagellum	flagella
genus	genera
hypothesis	hypotheses
larva	larvae
matrix	matrices
medium	media
nucleus	nuclei
phenomenon	phenomena
serum	sera
spectrum	spectra
stimulus	stimuli

Unclear Modifiers

Unclear modifiers, such as dangling modifiers or misplaced modifiers, are words or phrases that modify an element of a sentence in an ambiguous manner because they could either be modifying the subject or the object of the clause. Avoid dangling or misplaced modifiers.

To avoid confusion, fix unclear modifiers. Place them next to the word they are modifying. Add missing words as necessary as shown in the following examples.

Dangling modifiers most frequently occur at the beginning of sentences (often as introductory clauses or phrases) but can also appear at the end. They often have an "-ing" word (gerund) or an infinitive phrase near the start of the sentence.

Example 4-47   <u>While incubating</u>, we inverted the plates gently.

"While incubating" appears to modify "we." Because "we" were not incubating, this modifier is an unclear modifier. Identifying the appropriate object in the introductory clause avoids this confusion.

 Revised Example 4-47   While incubating **the plates,** we inverted them gently.

Here is another example:

| **Example 4-48a** | <u>Having tested positive</u> for HIV, we disqualified the patients for participation in the study. |

This modifier is also unclear because it modifies "we." It sounds as if "we" tested positive for HIV. Here, identifying the appropriate subject in the main clause clarifies the dangling modifier.

| **Revised Example 4-48a** | Having tested positive for HIV, **the patients** were disqualified for participation in the study.
 or:
 Patients that tested positive for HIV were disqualified for participation in the study. |

Following is another example:

| **Example 4-48b** | <u>While dialyzing</u>, we replaced the dialysate fluid every 20 min. |

"While dialyzing" appears to modify "we." Because "we" were not dialyzing, this modifier is unclear.

| **Revised Example 4-48b** | **While the samples were dialyzing,** we replaced the dialysate fluid every 20 min.
 or:
 We replaced the dialysate fluid every 20 min **while the samples were dialyzing**. |

SUMMARY

> BASIC RULES—STYLE
> * ★ Use the first person.
> * ★ Use the active voice.
> * ★ Use past tense for observations and specific conclusions.
> * ★ Use present tense for general rules and established knowledge.
> * ★ Write short sentences. Aim for one main idea in a sentence.
> * ★ Use active verbs.
> * ★ Avoid noun clusters.
> * ★ Use clear pronouns.
> * ★ Use correct parallel form. If possible, arrange ideas in a list to read from shorter to longer.
> * ★ Avoid faulty comparisons.
> * ★ Avoid errors in spelling, punctuation, and grammar.

PROBLEMS

PROBLEM 4-1 First Person
Rewrite these sentences using first person instead of third person.

1. It is one of the goals of this paper to assess the limitations of the models used.
2. The authors thank Dr. T. J. Guiermo for useful discussions.
3. It is recommended by the authors of the present study that a triple regimen of antibiotics is given to infants with HIV.
4. Based on our results, it is concluded that a computer built from qubits can do many different things in parallel because qubits can be both 0 and 1 at the same time.
5. No delay in feeding was observed.
6. We also studied nucleation of Mg^{2+}. It was determined that binding is dependent on pH and temperature but independent of O_2 concentrations (Figure 2D).

PROBLEM 4-2 Active and Passive Voice
Rewrite the following sentences in the active voice. Condense them if you can.

1. The following results were obtained: . . .
2. Isolated tissues were examined for parasites.
3. The collected specimens and videotape recorded during 25 dives were analyzed by the team.
4. Variations in day length are gauged by plants.
5. Recovery of several cytosine deaminases occurred after 2 days.
6. Insight into complex quantum many-body systems can be obtained through high-temperature superconductors and ultra-cold atoms.
7. No colonies were seen on blood agar.
8. Preliminary findings were unveiled by expedition members.

PROBLEM 4-3 Active versus Passive Voice
Decide whether to use active or passive voice in the sentences in italics. Explain your answer.

1. Deep Water Coral, also known as cold water coral, are most often stony corals belonging to the phylum Cnidaria. One of the most common cold water corals is *Lophelia pertusa*. *Fishing methods such as deep water trawling affect* Lophelia pertusa *negatively.* These methods tend to break corals apart and destroy reefs.
2. The photon is the basic unit of light and all other forms of electromagnetic radiation. It is also the force carrier for the electromagnetic force. The photon has no mass and thus can produce

interactions at long distances. *Both wave and particle properties are exhibited by the photon.* For example, a single photon may undergo refraction by a lens or exhibit wave interference but also act as a particle giving a definite result when its location is measured.

PROBLEM 4-4 Use of Tense
Decide whether to use past or present tense in the following sentences. If there is more than one answer, explain why.

1. In our study, tree size _____ **(increase)** with reduction of pesticides.
2. The larva of the monarch butterfly _____ **(feed)** on milkweed.
3. Females of this species _____**(have)** a unique projection from the dorsal part of the thorax.
4. Recent work by deValt (2001) _____ **(show)** that *nematodes* _____**(reproduce)** readily in the laboratory.
5. Table 3 _____**(show)** that in our study, polychaetes _____ **(be)** most abundant at depths of 10 to 16 m.
6. Baumann (2017) _____ **(find)** that heat production from re-newables _____ **(is)** commercially competitive with conventional energy sources.
7. Sea urchins _____**(spawn)** synchronously in the spring, and seasonal and lunar cycles as well as the presence of phytoplankton _____**(increase)** their spawning behavior.
8. ABSTRACT: The average voltage observed on our K electrodes _____ **(be)** substantially higher than expected under high-pressure conditions.

PROBLEM 4-5 Sentence Length
Rewrite the following overlong or overloaded sentences by breaking them up into shorter sentences.

Example 1 *(52 words per sentence)*
Unlike the bits in classical computers, quantum bits are difficult to maintain in the desired state for even a fraction of a second, and it is a central scientific and engineering challenge to achieve long-lived qubits that still can be manipulated to change their states appropriately in the course of a computation.

Example 2 *(77 words per sentence)*
In the past 540 million years, there have been at least five, and as many as 20, distinct mass extinction events on Earth, whereby in all cases extinction was preceded by major climate changes, quickly changing atmospheric CO_2, and a drop in sea levels, all of which led to up to 97% of all species

being extinct and a recovery period of millions of years, which resulted in an entirely new set of species on the planet.

PROBLEM 4-6 Active Verbs
Put the action in the verbs in the following sentences.

1. Measurements were performed on reaction products by a mass spectrometer.
2. An increase in transplant rejection occurred.
3. Removal of mitochondria was achieved by HPLC.
4. Buffalo, elephant, and black rhino abundance all show a rapid decline after 1977.
5. Two measurements of amyloid plaques were obtained for each brain.
6. Elevations in WBC count did not occur when aspirin and streptokinase were given.
7. Our results showed protection of the dogs by the vaccine.
8. Stanozolol caused prolongation of appetite.
9. To determine whether cell migration is occurring, we dissected mouse brains.
10. Electron liberation occurred through the photoelectric effect.
11. We must understand the central puzzle of how O=O bond formation occurs.
12. Water oxidation leads to the production of biochemical fuel and oxygen.

PROBLEM 4-7 Noun Clusters
Untangle the noun clusters in the following sentences by adding the appropriate preposition(s) and other words as needed. Start at the end of the cluster, and work your way to the beginning.

1. The results are compared using a high-quality plane-wave basis set.
2. The negative penicillin skin test result group showed no allergic rash.
3. The spinach culture callus tissue produced no shoots in 1999.
4. We analyzed a particle flux time series.
5. With this unconventional technique, we could easily define the aromatic hydrocarbon liquid crystal transition temperatures.
6. In 10 of the 15 patients, we observed a chronic depression syndrome.
7. Previous work suggested that vitamin A–rich fish oil diets protect mice against certain mosquitoes.
8. Helium's properties and isotopes make it the ideal, close to absolute zero degree Kelvin freezing point candidate.

PROBLEM 4-8 Pronouns
The following sentences contain unclear pronouns. Improve the clarity of these sentences by repeating one or more words from the previous sentence, rearranging the sentence, or adding a category or implied term instead of the underlined pronoun.

1. When bacterial cells are cured of viruses, <u>they</u> become avirulent.
2. A few microorganisms such as *Mycobacterium tuberculosis* are resistant to phagocytic digestion. <u>This</u> is one reason why tuberculosis is difficult to cure.
3. Patients often suffer from infections of *Corynebacterium* and from toxins released by <u>them</u>.
4. The goal of this study was to identify the reason for the decline in the number of salmon at the Kukso River. To achieve <u>this</u>, we tested mineral concentrations and pollutants A and B at various locations along the migration route of salmon in this river.
5. An action potential triggers calcium channels to open at the synapse. <u>This</u> causes docked synaptic vesicles to fuse to the membrane and release their neurotransmitter content.
6. When the *E. coli* ompA gene was expressed in Sodalis, <u>it</u> displayed a pathogenic phenotype.
7. Another costly and labor-intensive approach is to perform biochemical analysis of the fats in yeast. <u>This</u> has been done for some yeast mutants (47). This involves thin layer chromatography, coupled to gas chromatography.
8. Our findings indicate that binding decreases about 10-fold when temperature increases from 15 to 25°C. (Fig. 1). <u>This</u> suggests that binding among the particles is not due to ionic interactions.
9. Wind or solar power generation combined with pumped hydroelectric storage is being developed. <u>This</u> may contribute to the adoption of renewable energy in isolated networks.

PROBLEM 4-9 Parallelism
Correct the faulty parallelism in the following sentences.

1. The pathogenesis observed in other cells, such as circulating monocytes, may differ from endothelial cells.
2. The stability of these particles appears to be regulated by RNA helicase activity and protein phosphorylation.
3. Dengue hemorrhagic fever can occur in individuals with antibodies from previous dengue virus infections of different serotypes, and severe bleeding, shock, and death results.
4. Diabetes can be affected both by exercise and diet.
5. The pellet was washed with 50 μl of ethanol, recentrifuged, and it was then dissolved in 10 μl of buffer K.

6. Aluminum bulk metallic glasses have a higher specific strength compared to other metal-based metallic glasses but at the same time are malleable like plastics.
7. The internal pressure must not only depend on volume but also the rate of filling.
8. We provide a broad overview of the project, including the target population, the project's primary aim, the problem areas the project addresses, and where the services are being delivered.
9. Biomass constitutes 93% of the total direct heat production from renewables, geothermal 5%, and 2% arises from solar heating.

PROBLEM 4-10 Comparisons
Correct the faulty comparisons in the following sentences. Avoid the overuse of "compared to," ambiguous comparisons, and incomplete comparisons.

1. The kinetics of the protein G accumulation were similar to endogenous proteins (19).
2. Thus, the sequence observed for A proved to be much more diverse compared to that of B.
3. The dendrogram showed that EV109 is closer related to HEV-C.
4. In comparison to eubacteria, archaebacteria are more ancient.
5. We observed a peak for mutant A that was higher than the other mutants.
6. Compared to the other mammals, the male dolphin was larger.
7. The different propagation behavior of electrochemical fronts compared to reaction-diffusion systems is due to a different range of spatial coupling.
8. The observed properties of Au102 nanoparticles were similar to previous works.
9. The Austro-Tai populations were found to be a unity not only in culture but also in genetic structure, compared with a large number of data from other groups in East Asia.

PROBLEM 4-11 Common Errors
Rewrite the following sentences such that they make sense. Ensure that the subject and verb make sense together, that the subject and verb agree, that no helping words are omitted, and that modifiers are clear.

1. Altogether, we measured for 5 days.
2. This structure controls the packaging of the DNA into the phage head and the release after recognizing the receptor on the host.
3. Fig. 7 was obtained during the oxidation of hydrogen at the Pt electrode.
4. Our data indicates that more work needs to be done to prove our hypothesis.

5. The liver, gall bladder, and spleen of each patient was examined.
6. These complexes form *in situ*.
7. Having determined that an anisotropy exists, the full data set was rescanned from 1 January 2004 to 31 August 2007.

PROBLEM 4-12 Punctuation
Correct the punctuation in these sentences.

1. Moreover their results can be explained by the fact that the DNA repair defect of the H2A tail deletion mutant is mainly kinetic.
2. Patients with lnd QT syndrome, can experience seizures arrhythmia and sudden death by ventricular fibrillation.
3. As can be seen in Fig. 3 the amount of amplified product increases in parallel fashion to the amount of inclusions observed.
4. In conclusion our results suggest, that the pathways, that regulate glucose uptake, have even more components than previously thought.
5. The goal was to select the best most efficient deaminases possible.
6. Protostars are difficult to detect in visible light because they are surrounded by a cloud of gas and dust.
7. Thus this system of pathways is termed pioneer metabolism.
8. Two key roadblocks had to be overcome (i) we needed to store the energy of four photons and (ii) we needed to design efficient catalysts.
9. The book, *Handbook of Scientific Writing,* is very helpful.

CHAPTER 5

Common Grammar Concerns

Anyone who wants to be successful in science has to publish internationally and hold talks at conferences. This fact, more often than not, poses a challenge especially for people whose native language is not English. Note, though, that many native speakers also face these issues. These challenges may be few or many and often depend on how familiar you are with the English language or how similar your native language is with English.

ALTHOUGH THIS BOOK POINTS OUT SPECIAL ESL CHALLENGES THROUGHOUT ITS CHAPTERS, CHAPTER 5 IS DEVOTED TO THE MOST COMMON ISSUES ENCOUNTERED ESPECIALLY BY, BUT NOT LIMITED TO, ESL AUTHORS. GRAMMAR ISSUES COVERED HERE INCLUDE:

- Prepositions
- Articles
- Verb forms
- Adjectives and adverbs
- Nouns and pronouns

(Grammar terms are defined under each section in this chapter and also explained in detail—with specific examples of their use—in the Glossary of English Grammar Terms at the end of this book.)

5.1   PREPOSITIONS

➤ Use correct prepositions
Prepositions are "little words" that link nouns, pronouns, and phrases to other words in a sentence, indicating their relationship to the rest of the sentence.

Examples	about	above	after	at	below	by	for
	from	in	into	like	of	off	on
	out	over	than	through	to	up	with

Most verbs can be used with more than one preposition, but you should be sure to choose the preposition that reflects your intended meaning. If you are unsure which preposition to use, consult a dictionary. Be careful: The meanings of corresponding prepositions in English and other languages do not always coincide. Native speakers are also prone to incorrect use of prepositions.

A few words and expressions need special attention.

compared	*Compared* takes the preposition *to* when it refers to unlike things. It takes *with* when two like things are examined.

Example 5-1	**a**	The human brain is sometimes *compared to* a computer.
	b	When we *compared* our results *with* those of Pauling et al. . . .

comparison	The use of the word *comparison* is cause for much confusion, as it can take various prepositions: *comparison* **of** A **and/with** B, or comparison **between** A **and** B.
different	Do not use *different than* when you should use *different from*. *Different than* is acceptable only in sentences such as the following.

Example 5-2	We obtained *different* exfoliation values for our nanosheet *than* Roberts et al. (2012).

But:

Example 5-3	Our exfoliation values were *different from* those of Roberts et al. (2011).

following	Do not use *following* as a preposition. *Following* can be used as an adjective, as in "The following results," but should be avoided when it can be replaced with *after*, which is unambiguous.

Example 5-4	*Following* centrifugation, the supernatant was removed.
Revised Example 5-4	**After** centrifugation, the supernatant was removed.

The most commonly (mis)used prepositions in scientific writing include the following:

in connection **with**	compared **to/with**
in contrast **to**	correlated **with**
similar **to**	analogous **to**

5.2 ARTICLES

➤ Use correct articles

The English language has two kinds of articles, definite (*the*) and indefinite (*a, an*). ESL writers often mix them up, leave them out, or put them in when they are not needed.

Every time you use a noun in English, you need to decide which sort of article to use, if any. Use of the article depends on the noun that follows it. A few general guidelines are shown in Table 5.1.

TABLE 5.1 Use of articles in English

ARTICLE	USE	EXAMPLE
Definite "the"	to show which *particular* item you mean	**The** culture that was contaminated was discarded.
		The Nobel prize winner in 2006 in medicine . . .
	in connection with the phrases	. . . in **the** presence of . . . , . . . in **the** absence of . . .
	before a noun that names something that has already been mentioned	The technician looked at **the** sample we had measured earlier.
		The methods described previously . . .
Indefinite "a" "an"	when the noun is one of a group or new to the reader	**A** scientist needs to write well.
		She used **an** aliquot to determine chemical composition.
	when the noun is generalized	**A** protein can be made up of many different amino acids.
No article	before a plural noun that has not yet been referred to	She collected samples for various analyses.
	before a proper or abstract noun	Berlin was a divided city.
		Degradation occurs after 24 hrs.
		In conclusion, we determined that . . .
"the," "a," "an"	if you use proper or abstract nouns as common nouns	**The** Berlin that was divided is gone.
		The degradation we observed occurred after 24 hrs.
		The conclusion of the discussion should summarize the results.
		A conclusion about X cannot be drawn.

Extensive information about how to determine which article to use, if any, can be found at the following websites:

**https://owl.purdue.edu/owl/general_writing/grammar/using_
    articles.html**
http://esl.fis.edu/grammar/rules/article.htm
http://a4esl.org/q/h/grammar.html
http://www.usingenglish.com/handouts/

Other great resources are Lynch (2008) and Greenbaum (1996).

5.3   VERBS

Plural and Singular Verb Forms

➤ Use correct plural and singular verb forms

For many authors, especially ESL authors, it is not always obvious which subject words are treated as singular and which as plural. To clarify the use of singular and plural verbs, use Table 5.2.

TABLE 5.2 Use of singular and plural verb forms

RULE	SUBJECT	EXAMPLE
Certain indefinite pronouns are singular	one, no one, anyone, each, everything, something, someone, everybody	**Each** fragment was purified.
Certain indefinite pronouns are plural	both, few, many, several	**Both** fragments **were** purified.
Some indefinite pronouns can be singular or plural	most, some, none, part, any, all	**Some** fragments **were** purified. **Some** of the fraction **was** lost.
Collective nouns are singular	audience, class, group, committee, team, politics, news	This **class** of alloys **is** known as . . .
Some collective nouns can be singular or plural	staff, faculty, (fractions such as) one third	**One third** of the patients **were** men. **One third** of the sample **was** used.
Certain abstract nouns are singular despite their plural appearance	news, measles, mumps, physics, kinetics, dynamics	**Measles has** essentially been eradicated in the Western world.
Compound subjects joined by "and" are plural	Subject joined by "and"	**Temperature and time are** both important parameters in polymer formation.
For compound subjects joined by "either, or"; "neither, nor"; "or"; and "not only, but also," the verb must agree with the closest subject	Subject joined by (either, or . . .) (neither, nor . . .) (not only, but also . . .)	Neither the heart nor **the lungs were** inflamed. Not only the boys, but also **the girl was** infectious. X is due to either A or B.

Also pay attention to what is the true subject of a sentence.

Example 5-5   **The role** of cytokinins **has** been studied in detail.

subject   verb

Here, "role" is the subject, not "cytokinins." "Cytokinins" is the object of the preposition "of" and describes or qualifies "role." Objects of prepositions are never the subject of a sentence. Thus, the verb should be singular as it corresponds to "role."

Irregular Verbs

➤ Use correct forms of irregular verbs

Irregular verbs often pose a problem in English writing or speaking, especially for ESL authors. Make sure you use the correct verb forms so mistakes such as the following do not arise.

Example 5-6   Our insert was <u>cutted</u> by EcoRI.

Revised   Our insert was **cut** by EcoRI.
Example 5-6

The most common irregular verbs found in science are listed in Table 5.3. A more complete list of irregular verbs is available from http://www.englishpage.com/irregularverbs/irregularverbs.html and http://www.college-em.qc.ca/prof/epritchard/pastverb.htm. Other great resources are Lynch (2008) and Greenbaum (1996).

Endings of Verbs

➤ Do not omit endings of verbs

ESL speakers who do not pronounce endings of words fully often omit endings of verbs in writing as well. Do not omit "-s," "-es," "-ed," or "-d" endings to use the third person ending of a verb form or to express the past tense or past participle form of a verb.

Example 5-7   a   This journal usually *publish* articles in the field of neuroscience only.
                  b   Before the substrate was added, we *determine* if the enzyme was temperature sensitive.

Revised   a   This journal usually **publishes** articles in the field of neuroscience only.
Example 5-7   b   Before the substrate was added, we **determined** if the enzyme was temperature sensitive.

TABLE 5.3 Irregular verbs and their forms

INFINITIVE	PAST TENSE	PAST PARTICIPLE (USED TO FORM PAST PERFECT TENSE AND PASSIVE VOICE)
arise	arose	arisen
become	became	become
begin	began	begun
bend	bent	bent
bring	brought	brought
choose	chose	chosen
cut	cut	cut
deal	dealt	dealt
draw	drew	drawn
find	found	found
get	got	got/gotten
give	gave	given
grow	grew	grown
hide	hid	hidden
keep	kept	kept
lead	led	led
let	let	let
lie	lay	lain
lay	laid	laid
lose	lost	lost
prove	proved	proven
put	put	put
read	read	read
rise	rose	risen
run	ran	run
see	saw	seen
seek	sought	sought
send	sent	sent
set	set	set
shake	shook	shaken
show	showed	shown
shrink	shrank	shrunk
spend	spent	spent
spin	spun	spun
spread	spread	spread
take	took	taken
write	wrote	written

Gerund and Infinitive

➤ Follow a verb with the correct gerund or infinitive form

Using gerunds and infinitives is a great way to add action to scientific writing and make science come alive. A gerund is a verb form ending in "-ing" (*describing, running*). Gerunds are used as nouns in English. An infinitive is the base form of the verb and is preceded by the word "to," such as in "to describe" or "to run." Some verbs may be followed by a gerund, others by an infinitive, and still others may be followed by either. In addition, for certain verbs, a noun or pronoun must be placed between the verb and the infinitive that follows it. Some more detailed guidelines are listed in Table 5.4.

Note also that verbs used after helping words that are a form of *to do* use the infinitive form and do not show the singular *–s*. This mistake is commonly made by ESL writers.

	Example 5-8	EcoRI does not *cuts* RNA.
		Did the sequence *affects* digestion?
	Revised	EcoRI does not **cut** RNA.
	Example 5-8	Did the sequence **affect** digestion?

TABLE 5.4 Use of gerund and infinitive

FORM	SAMPLE VERBS	EXAMPLE
Verb + gerund	avoid, discuss, imagine, practice, recall, resist, suggest, tolerate	The authors avoided **describing** the problem in detail. Dissecting **requires** steady hands.
Verb + infinitive	agree, decide, want, ask, expect, have, offer, refuse, claim, hope, plan	We decided **to determine** the minimum combustion temperature.
Verb + gerund or infinitive	begin, continue, like, prefer, start	We continued **measuring** . . . We continued **to measure** . . .
Verb + noun/pronoun + infinitive	advise, have, instruct, remind, require, tell	Macromolecules require specific conditions **to form** crystals.
Verb + noun/pronoun + infinitive without "to"	let, make ("force"), notice, see, watch	The bear let us **approach** . . . Notice you **write** better after reading this book.

5.4   ADJECTIVES AND ADVERBS

➤ Distinguish between adjectives and adverbs

Many ESL writers have trouble distinguishing between adjectives and adverbs. Adjectives modify nouns and pronouns and are generally placed in front of the noun or pronoun they modify. Adjectives may also be used to complement a subject and are then placed following a linking verb. Linking verbs are verbs that suggest a state of being or feeling rather than an action.

Example 5-9 a   In our **new** model, we consider not only the number of subunits but also their locations.

  b Our model is **new**.

In Example 5-9a, "new" is the adjective that modifies "model." In Example 5-9b, the adjective complements the subject, describing an attribute of the subject. It follows the linking verb.

    Adverbs modify verbs, adjectives, or other adverbs. They are usually found at the beginning or at the end of a sentence, before or after a verb, or between a helping verb and a main verb.

Example 5-10 a Complete conversion from A to B is **rarely** seen under denaturing conditions.

b Acetylcholine was released in **precisely** controlled amounts using automated injection.

Adverbs should not be placed between a verb and its direct object.

Example 5-11 Neural cells reinternalize *continually* synaptic vesicles.

verb   direct object

Revised Example 5-11 a Neural cells reinternalize synaptic vesicles **continually.**

b Neural cells **continually** reinternalize synaptic vesicles.

5.5   NOUNS AND PRONOUNS

➤ Ensure that every sentence has a subject

In some languages such as Spanish and Italian, the subject of a sentence can be omitted. However, this is not the case in English unless the sentence is imperative. English sentences, and clauses within sentences, need to have a subject. The subject is typically a noun but can also be as simple as the pronoun "it."

	Example 5-12	a	The editor refused their manuscript because have already published a similar paper.
		b	In January, is 25°C in this region.
		c	Is clear that the control experiment was inappropriate.
	Revised Example 5-12	a	The editor refused their manuscript because **they** have already published a similar paper.
		b	In January, **it** is 25°C in this region.
		c	**It** is clear that the control experiment was inappropriate.

5.6   GRAMMAR REFERENCES

Books

Aarts, B. (2011). *Oxford modern English grammar* (1st ed.). New York: Oxford University Press.

Hacker, D. (2012). *Rules for writers* (7th ed.). New York: Bedford/St. Martin's.

Lunsford, A. A. (2016). *The everyday writer* (6th ed.). New York: Bedford/St. Martin's.

Lynch, J. (2007). *The English language: A user's guide.* Newburyport, MA: Focus.

Schrampfer Azar, B., & Hagen S. A. (2014). *Basic English grammar* (4th ed.). New York: Pearson Education.

Online References

Following is a list of Web sources for practicing grammar as well as sites for specific grammar problems:

General grammar, spell-checker and plagiarism detector:
https://www.grammarly.com
This free, online program checks for grammar, spelling, and even plagiarism mistakes. It is currently the most popular tool on these topics.

Other sites for grammar, punctuation, and capitalization:
http://owl.english.purdue.edu/owl
The Owl at Purdue (online writing lab)
http://grammar.ccc.commnet.edu/grammar
Guide to Grammar and Writing
http://www.chompchomp.com
Grammar Bytes (grammar exercises and rules)
http://bcs.bedfordstmartins.com/rules7e/#t_669460
Complete reference for student writers and researchers; includes special section for ESL writers as well as many electronic exercises
http://www.englisch-hilfen.de/en/

For article use:
https://owl.purdue.edu/owl/general_writing/grammar/using_articles.html

http://esl.fis.edu/grammar/rules/article.htm
http://a4esl.org/q/h/grammar.html
http://www.usingenglish.com/handouts/

For verb forms:
http://www.englishpage.com/irregularverbs/irregularverbs.html
http://www.englishpage.com

For grammar practice:
http://englishgrammar101.com/
http://englishteststore.net/index.php?option=com_content&view=
    article&id=11387&Itemid=427

Special English as a Second or Other Language (ESOL) assistance:
http://a4esl.org/
http://www.esl.net
http://www.usingenglish.com/quizzes/

SUMMARY

GRAMMAR GUIDELINES
- Use correct prepositions.
- Use correct articles.
- Use correct plural and singular verb forms.
- Use correct forms of irregular verbs.
- Do not omit the endings of verbs.
- Follow a verb with the correct gerund or infinitive form.
- Distinguish between adjectives and adverbs.
- Ensure that every sentence has a subject.

PROBLEMS

PROBLEM 5-1 Prepositions
Add the correct preposition to the phrases provided. It is okay to use a
dictionary.

1. In connection _____
2. Compared _____
3. In contrast _____
4. Search _____
5. Correlated _____
6. A comparison _____ A _____ B
7. Similar _____
8. To look forward _____
9. Results shown _____ Fig. 3
10. Through the decrease _____
11. Analogous _____

12. Implicit _____
13. Theorize _____
14. Different _____
15. Attempt (n.) _____ attempt (v.) _____
16. _____ respect _____

PROBLEM 5-2 Articles
Add the correct definite or indefinite article where needed.

1. Sea urchin fertilization is _____ model system in developmental biology.
2. Oscillatory behavior was reduced in _____ presence of a ring electrode.
3. Back to _____ nature.
4. Therefore, _____ hypothesis we presented previously is further confirmed.
5. _____ inorganic chemistry proves to be a very interesting subject.
6. Can you name _____ inorganic molecule?
7. *Title:* _____ Role of _____ Physical Activity on _____ Severity of Diabetes.
8. Conversion of _____ CK riboside to the base can be regarded as the last step in the pathway.
9. _____ theoretical studies cited earlier depict a variety of different patterns in oscillating media.
10. Here we discuss _____ physical-chemical mechanisms leading to pattern formation in electrochemical systems.
11. *Figure 4:* _____ effect of _____ temperature on _____ ctx expression.
12. _____ rainfall in _____ Sahara Desert occurs rarely.
13. _____ average rainfall in _____ Sahara Desert is less than 25 mm per year.
14. Under _____ light microscope, _____ organelles are visible in _____ eukaryotes.
15. At the end of its "life," _____ mRNA molecule is degraded.
16. _____ artificial-intelligence technologies can be applied in many fields.

PROBLEM 5-3 VERB FORMS
Add the correct verb form. Ensure that infinitives and gerunds are used correctly.

1. The analysis was limited to _____ (determine) the change in the population number.
2. We sought to _____ (improve) the yield by adding specific catalysts.

3. Impurities in our samples were barriers to _____ (obtain) chemically pure products using standard isolation techniques.
4. Ongoing volcanic eruptions limited the amount of time that we could dedicate to _____ (collect) samples.
5. The reviewers recommend _____ (add) more information about the bear habitat.
6. Reaching temperatures as close to absolute zero Kelvin as possible is central to _____ (design) a quantum computer.

PROBLEM 5-4 Verb Forms
Add the correct verb form. Ensure that singular and plural, irregular verb forms, and endings of verbs are used correctly.

1. The conservation of protein A sequences among the diverse species _____ (suggest) that the protein plays a key role in this metabolism.
2. A crystal structure change in hydrogen or deuterium _____ (have) been _____ (show) to exist via x-ray diffraction, infrared, and nuclear magnetic resonance.
3. *S. multiplicata* females did not _____ (choose) heterospecific mates regardless of water level.
4. Until now, there _____ (be) no experimental confirmation of this hypothesis.
5. Microspheres of like charge were _____ (bind) together with nanoparticles of opposite charge to form clusters of two to nine colloids.
6. Each of the experiments _____ (be) performed at two different photon energies.
7. One third of the mice _____ (be) infected with *P. chabaudi*.
8. Atomic scale resolution through 4D ultrafast electron diffraction, crystallography, and microscopy _____ (have) made it possible to determine complex transient structures.

PROBLEM 5-5 Adjectives and Adverbs
Add the correct adjective or adverb.

1. X behaved _____ (different, differently) than expected.
2. Dark matter, which constitutes most of the mass in the universe, is not _____ (direct) _____ (visible) but can be inferred due to its gravitational properties.
3. Telomeres, DNA sequences at the ends of chromosomes that keep our chromosomes _____ (intact, intactly), shorten with every cell division and are therefore an indication of aging.
4. Species richness in plants is correlated _____ (biological, biologically) and _____ (geohistorical, geohistorically) and increases ecological opportunities.

5. The resolution of telescopes is restricted by the diffraction of light, which cannot _____ (selective, selectively) focus on anything smaller than a plane wave.

6. Mainstream climate science needs to look more _____ (close, closely) at geoengineering.

7. The impact of vaccinating domestic animals on rabies occurrence was evaluated by comparing _____ (ecological, ecologically) field studies with _____ (empirical, empirically) based models.

8. _____ (Recent, recently) studies have shown that _____ (new, newly) imported Australian bees have been linked to the virus responsible for infecting and decimating honeybee colonies.

PROBLEM 5-6 Mixed ESL Errors
Be sure that the following sentences make sense; that correct prepositions, articles, and verb forms are used; that every sentence has a subject; and that adverbs and adjectives are distinguished.

1. Is not clear if the difference in our results was due to the temperature difference in our measurement.

2. We expected observing a large difference in the output.

3. The environment surrounding invasive breast tumors exhibit significant changes.

4. The early universe was much differently from that which we know at present.

5. Although the international community recognizes the importance of species diversity in South American rainforests, deforestation continues.

6. Like an endosperm, the *mycorrhiza* fungus provides needed energy to orchid seeds. Is required to complete the reproduction cycle of these plants as their seeds lack endosperm.

7. Here, we present a model for vaccine development against Ebola virus. However, in support for the model, additional, larger studies of Ebola virus pathogenicity and virulence will be required.

8. Many studies considered *Homo ergaster* to be the common ancestor of *Homo erectus* and modern humans, *Homo sapiens* (Wood, 2000). It is thought that *H. erectus* later evolved into a separate species, which become extinct. However, results from other studies have showed that *Homo erectus* is identical to *Homo ergaster* (O'Connell, 1999).

9. European cave hyena was found throughout Europe and Asia in the Pleistocene, and often competed with Neanderthal man and Cro-Magnon for occupation of caves.

10. In Utah, Gunnison Island is only know breeding site for American white pelicans.

PROBLEM 5-7 Mixed ESL Errors
In the following paragraphs, fill in the correct article, verb form, adverb, or adjective.

a) Polio, a highly contagious disease, has been considered largely eradicated in the developed world. The disease _____ (cause) paralysis in 0.1–1% of patients. Only a few hundred cases are reported _____ (annual/annually) now, compared to 350,000 cases in 1988. Recently, live polio virus was discovered in Israel (Roberts, 2013). Aside from routine childhood vaccinations, _____ (optimal/ optimally) vaccine policies are now needed to prevent transmission within the population and forego an outbreak. Statistical and mathematical models parameterized by surveillance data and survey studies may be able to provide recommendations for the needed shift in policies.

b) Dust storms have been _____ (well, good) studied. Commonly known as "sand storms," these storms in fact contain only few particles big enough to reach the size of sand grains. Instead, they _____ (primary, primarily) contain small particles of dust. This dust typically _____ (do) not arise from _____ (the, no article) deserts as most dust in deserts has been blown away over time. Dust in dust storms usually arises from the fine particles found in dried-up lakes and from areas bordering on deserts, particularly those disturbed through agriculture or farm animals. When wind picks up the dust, the fine particles can end up in _____ (the, no article) upper atmosphere and can be _____ (carry) for thousands of miles, even across oceans and continents, before being deposited. _____ (A, An, The, no article) deposition often occurs in mountain ranges, and sediments on glaciers serve as indicators of past dust storms centuries earlier (Mahowald, 1999). Most depositions peak in the spring and fall. The dust particles in the atmosphere can even influence weather patterns and climate _____ (profound, profoundly) as they act as cloud condensation nuclei (Twohy, 2005).

CHAPTER 6

From Sentences to Paragraphs

A side from paying attention to words and sentences, authors need to construct paragraphs carefully so readers can understand them immediately irrespective of the scientific content.

THE OBJECTIVE OF THIS CHAPTER IS TO UNDERSTAND:

- How to construct a paragraph well, specifically
  - How to organize a paragraph
  - How to make a paragraph coherent
- How to create good flow using word location, key terms, and transitions
- How to emphasize important ideas and signal subtopics
- How to condense text while establishing importance; omitting words, phrases, and sentences; and avoiding writing in the negative and overuse of intensifiers and hedges.

6.1   PARAGRAPH STRUCTURE

If paragraphs are not clearly constructed, a paper that has perfect word choice, word location, and sentence structure can be difficult to understand. Let us look at a paragraph in which the author has not paid much attention to the needs of the reader.

Example 6-1

1 Volcanic ash adsorption poses a great environmental hazard. **2** The deposition of this ash and the subsequent draining of its volatiles is a rapid route by which elements and ions are delivered to the ground (3–5). **3** Due to similar magma types, leachate content from volcanoes in close proximity to each other appears alike. **4** The greatest hazard to the environment is posed by magmas with relatively high halogen content, and many hazardous leachate fluoride concentrations are found in volcanoes with high F/SO_4^{2-} ratios. **5** Particles < 2 mm across seem to have a greater adsorption, and therefore a high leachate hazard exists even with low ashfall (7, 8). **6** Under high humidity conditions, gas accumulation can be further increased as bigger sulphuric acid droplets make contact with ash particles more likely (12). **7** The measuring and reporting of leachate results should be standardized.

(Reference: Witham 2005)

If you catch yourself reading this paragraph more than once, you are not alone. You may think that you have not paid enough attention and start reading it over. Some readers may even consider themselves not smart enough to understand the topic. The individual sentences are intelligently composed and free of grammatical errors. The sentences are also not overly long or complex. The vocabulary is professional but not beyond the scope of the educated general reader. Nonetheless, most of you arrive at the end of the paragraph without fully understanding what the author is saying. The problem lies not with you but with the author because the author has not composed the paragraph with the reader in mind.

When you take a closer look at this paragraph, you can see that the paragraph has not been organized properly. It is neither coherent nor cohesive or consistent. In fact, important information, such as transitions and logical connections, seem to have been left out. Even more, sentences seem to be put together at random and not in any logical order. For readers to follow the logic of a paragraph, its sentences have to be organized.

Let us look at the revised version of Example 6-1.

Revised Example 6-1

1 Volcanic <u>ash adsorption</u> poses a great <u>environmental hazard</u>. **2** The <u>deposition of this ash</u> and the subsequent <u>leaching</u> of its volatiles is a rapid route by which elements and ions are delivered to the ground (3–5). **3'** *Adsorption can be influenced by magma type, particle size, and humidity conditions (7).* **3** *For example,* <u>leachate</u> content from volcanoes in close proximity to each other appears alike due to similar magma types. **4** *In fact,* the greatest <u>hazard to the</u>

environment is posed by magmas with relatively high halogen content, and many <u>hazardous leachate</u> fluoride concentrations are found in volcanoes with high F/SO$_4^{2-}$ ratios. **5** *Aside from magma type,* particles < 2 mm across seem to have a greater <u>adsorption</u>, and therefore a high <u>leachate hazard</u> exists even with low *ash deposition* (7, 8). **6** *In addition,* under high humidity conditions, <u>*adsorption*</u> can be further increased as bigger sulphuric acid droplets make <u>ash adsorption</u> more likely (12). **7** *Ideally,* the measuring and reporting of <u>leachate</u> results should be standardized.

(*Reference:* Witham 2005)

After looking at this revision, you may now recognize the lack of organization and links in the original paragraph. You can see that an important missing link was sentence 3′. It ties sentences 3 through 6 together and links them to the beginning of the paragraph. Adding the transitions "for example," "in fact," "aside from," "in addition," and "ideally" logically connects the ideas in the paragraph. All these connections were left unarticulated in the original paragraph. Replacing "ashfall" with "ash deposition" and "gas accumulation" with "adsorption" helps to keep the reader focused on the topic because terms are used more consistently throughout the paragraph. We can see that most of our difficulty in understanding the original paragraph was due not to any deficiency in our reading skills but rather to the author's lack of knowledge and understanding of our needs as readers.

Although the revision greatly improved the paragraph, we could revise it even more.

Revised Example 6-1

1′ Volcanic <u>ash adsorption</u> poses a great <u>environmental hazard</u> *because adsorbed volatiles can be rapidly <u>deposited</u> and subsequently <u>leached</u> into the ground.* **2′** *Adsorption can be influenced by magma type, particle size, and humidity conditions (7).* **3** *For example,* <u>leachate</u> content from volcanoes in close proximity to each other appears alike due to similar magma types. **4** *In fact,* the greatest <u>hazard to the environment</u> is posed by magmas with a relatively high halogen content and by magmas with high F/SO$_4^{2-}$ ratios in which many hazardous <u>leachate</u> fluoride concentrations are found. **5** *Aside from magma type,* particles < 2 mm across seem to have a greater adsorption, and therefore a high <u>leachate</u> hazard exists even with low <u>*ash deposition*</u> (7, 8). **6** *In addition,* under high humidity conditions, <u>*adsorption*</u> can be further increased as bigger sulphuric acid droplets make <u>ash adsorption</u> more likely (12). **7′** *Unfortunately, currently no uniform <u>leachate</u> analysis methods* *exist and thus data is difficult to compare.* **8** *Ideally, the measuring and reporting of* <u>leachate</u> *results should be standardized.*

(*Reference*: Witham 2005)

The flow of the paragraph has been further improved in second Revised Example 6-1. Sentences 1 and 2 have been combined into sentence 1′, making their relationship clearer through the use of "because." Another link, sentence 7′, has been placed before the last sentence to more logically connect it to the rest of the paragraph.

6.2   PARAGRAPH ORGANIZATION

★ (23) Organize your paragraphs

A paragraph is a group of sentences on a single topic. The sentences within a paragraph are not put together randomly, however. In a well-written paragraph, the sentences need to be logically organized and positioned.

Sentence Positions

Every paragraph contains two important power positions: the first sentence and the last sentence. Usually, the first sentence introduces the topic of the paragraph, whereas the last sentence may be used to summarize, draw a conclusion, or emphasize something of importance.

These power positions are not equal. The first position in a paragraph is considered more powerful than the last position because it gives the reader a direction of where the paragraph is going. Within the first and the last sentence of the paragraph, the psychological geography of the sentence structure is particularly important. The beginning of the sentences should describe familiar information, whereas the stress position within these sentences should highlight significant words to be emphasized (see also Chapter 3).

Topic Sentence

★ (24) Use a topic sentence to provide an overview of the paragraph

Generally, a well-written paragraph gives an overview first and then goes into detail. Note that, most of the time, it is clearest to have only one message per paragraph. The overview is usually provided by the first sentence, the so-called topic sentence. The topic sentence states the central topic or message of the paragraph and guides the reader into the paragraph. The end or stress position of the topic sentence highlights the topics that the author wants readers to follow in the rest of the paragraph. The paragraph then develops that message by using examples, definitions, justifications, contradictions, or by analyzing and solving a problem.

Although a topic sentence may appear anywhere in the paragraph, it is usually the first sentence, that is, the first power position. The first sentence of a paragraph may also contain a transition from the previous paragraph or section. Some paragraphs may even contain more than one topic sentence. If a topic sentence is placed at the end of a paragraph, it receives extra emphasis. This sentence may introduce the topic of the next paragraph(s). It can also serve as a summary or conclusion. In a well-written paper in which all the topic sentences are in the first power position, a reader can simply scan the topic sentences alone and know what the paper is about without having read it entirely.

The Middle of the Paragraph

➤ Arrange the details in the remaining sentences

Details within the paragraph are organized depending on the purpose of the information contained in the paragraph. The pattern of the organization may be listing details from most to least important, least to most important, in an announced order, or chronologically. Other paragraphs may be written in a compare-and-contrast pattern or in a problem-and-solution pattern.

The following examples both begin with a topic sentence. The details found in the remaining sentences are organized logically and consistently to explain the message provided by the topic sentence.

Example 6-2   Volatile organic compounds (VOCs) are released from many manmade and natural sources. Manmade sources range from cars and industrial sources to construction materials, heaters, and other consumer products. Natural sources responsible for biogenic VOC emissions include mainly trees as well as fungi and microorganisms. The exact distribution of manmade versus natural sources of VOCs has not been determined, but clearly depends on locale and environment.

The topic of the preceding paragraph is "VOCs." The pattern of organization, that is, the order of the remaining sentences, is not random but proceeds in the order the items are listed: *manmade* first, then *natural*.

Consider Example 6-3.

Example 6-3   For the preparation of postmitochondrial fractions, placentas were obtained after delivery. Their membranes were removed, and the organs were washed extensively at 4°C with buffer A (10 mM Tris-HCl, pH 7.5, 20 mm Mg(acetate)$_2$, 100 mM K(acetate), 0.4 mM EDTA). The organs were then shock frozen in liquid nitrogen as 20-g aliquots and subsequently stored at –80°C. For postmitochondrial preparation,

20 g of frozen tissue was suspended in 20 ml of buffer A containing 200 mM sucrose through homogenization in a blender. To separate cell debris and nuclei, the homogenate was centrifuged at 1,500 × g for 2 min at 4°C. To obtain the postmitochondrial fraction, the supernatant was then recentrifuged at 20,000 × g for 20 min.

Example 6-3, which is from a Materials and Methods section of a prepublication journal article, also has a topic sentence ("For the preparation of postmitochondrial fractions . . ."). The remaining sentences of the paragraph are organized in a chronological pattern in which the reader can follow step by step how the fractions were prepared from the original tissue. The subjects of the sentences in the preceding example are all tissue or fraction related. The paragraph begins with a goal ("For the preparation of . . .") and ends with a statement that indicates how something was accomplished.

Aside from the general organization of the paragraph, the author also needs to pay attention to word location and to keeping a consistent order of topics and a consistent point of view or person. These points are addressed in more detail in the next sections.

Topic Order

★ (25) Use consistent order

Although parallel form is most often used at the sentence level (see Chapter 4), in scientific papers it can also be used in a larger context. Parallelism helps locate information in paragraphs, sections, chapters, and so forth. Good parallel form puts related ideas together in the same grammatical form and style and thus provides consistency throughout the paper.

If you list items in a topic sentence and then describe them in the remaining sentences of a paragraph, you should not only use parallel form but also keep the same order. For example, if the items in the topic sentence are "dogs," "cats," and "birds," the remaining sentences of the paragraph should explain first "dogs," then "cats," and last "birds." This way the reader's expectation is fulfilled. Make sure you include all the items mentioned in the topic sentence. Avoid interrupting the order of your items by filling in with other information. Also, do not add any items *not* mentioned in the topic sentence.

An example of a paragraph in which consistent order is used is Example 6-4.

Example 6-4   **1** In response to a foreign macromolecule, five different immunoglobulins can be synthesized: IgG, IgM, IgA, IgE, or IgD. **2** IgG is the main immunoglobulin in serum. **3** IgM is the first class to appear following

> exposure to an antigen. **4** IgA, the major immuno-
> globulin in external secretions such as saliva, tears,
> and mucus, serves as a first line of defense against
> bacterial and viral antigens. **5** It is transported
> across epithelial cells from the blood side to the
> extracellular side by a specific receptor. **6** IgE pro-
> tects against parasites. **7** The role of IgD is not
> known.

In this example, the topic sentence lists five items. The remaining sentences of the paragraph explain these items in the same order as they are introduced in the topic sentence and use exactly the same key terms (IgG, IgM, IgA, IgE, and IgD). Note that additional information has been filled in between sentences 4 and 6. Interruption such as that can make paragraphs difficult to read because reading about the last items is delayed. If such an interruption is longer, consider placing the additional information into a separate paragraph.

Consistency in Subject or Person

★ (26) Use consistent point of view

Be consistent in your point of view or person. Switching from one style (point of view or person) to another within a document disorients the reader. ESL authors should pay special attention to this rule, as many tend to switch the point of view within paragraphs unnecessarily.

To create a consistent point of view, and thus good flow of a paragraph or section, use the same subject or person throughout your paragraphs and sections. Changing the subject (e.g., from birds to seeds) or changing the person (e.g., from third to second person) confuses readers.

Here is an example.

ESL advice

Example 6-5   **1** These findings suggest that *patients* returning from tropical countries who show very itchy linear or ser-piginous tracked skin eruptions should be tested for larvae of animal hookworms. **2** *Topical ointments* with an antiparasitic drug such as thiabendazole should be used to treat patients with this disease, known as cutaneous larva migrans. **3** *These patients* often also suffer from complications of the disease such as impetigo and allergic reactions (Heukelbach and Feldmeier, 2008).

The subjects of sentence 1 and 3 are "patients." However, in sentence 2, the point of view is not consistent as the subject of the sentence is not "patients" but "topical ointments." Switches like this are very disruptive to the paragraph and disorienting for the reader.

Revised
Example 6-5

1 These findings suggest that *patients* returning from tropical countries who show very itchy linear or serpiginous tracked skin eruptions should be tested for larvae of animal hookworms. **2 *Patients*** with this disease, known as cutaneous larva migrans, should be treated with an antiparasitic drug such as thiabendazole. **3** *These patients* often also suffer from complications of the disease such as impetigo and allergic reactions (Heukelbach and Feldmeier, 2008).

6.3   PARAGRAPH COHERENCE

Cohesion and Word Location

★ (27) Make your sentences cohesive

Within a paragraph, the sentences not only need to be logically organized, they also need to be cohesive. Sentences are cohesive if they fit neatly and logically together. When authors arrange sentences to be cohesive, they consider word location. Good word location creates good "flow" of a paragraph.

Example 6-6

1 Important pathogens can be found in the genus

Yersinia. **2 *Yersinia*** contains several **species**. **3 One species,**

Y. pestis, is the cause of bubonic **plague**. **4** The **plague** bacillus infects lymph nodes near the site of infection to produce buboes.

The reader conceives these sentences as cohesive because the information provided at the beginning of sentences 2, 3, and 4 relates to the one at the end position of the sentences directly preceding them (see also Chapter 3 on word location.)

If more and more new information is added before the relationship between the two sentences is clear, the continuity of a paragraph is broken. Consider the following example.

Example 6-7

a   *Yersinia* contains several **species**. The cause of bubonic plague, also known as the "black death," is **one species**, *Y. pestis*.

b   *Yersinia* contains several **species**. **One species**, *Y. pestis*, is the cause of bubonic plague, also known as the "black death."

The flow of Example 6-7a is not as smooth as that of Example 6-7b because new information ("bubonic plague," "black death") is introduced before the information of the previous sentence ("*Yersinia*," "species") is repeated.

Example 6-7b is much more direct, and the continuity between the sentences much smoother due to the word location of the key term "species." In this example, the information of the previous sentence has been placed at the beginning of the new sentence due to jumping word location.

Placing information provided at the end of a sentence at the beginning of the next sentence is not the only way to provide paragraph cohesion. Cohesion can also be created by providing a consistent point of view, as in the next example.

 Example 6-8

Rhubarb is a frequently used Chinese herbal medicine.

It is used to treat various ailments including constipation, inflammation, and cancer.[1,2] As a drug, **rhubarb** is made up of the roots and rhizomes of three members of the *Polygonaceae* family, *Rheum officinale, R. palmatum,* and *R. tanguticum.* Different **rhubarb** species show substantial differences in purgative effects and chemical compositions. However, **they** are similar in physical appearance and thus difficult to distinguish.

Here, information in the topic position of each sentence refers back to the topic position of the first (or topic) sentence. In other words, a consistent point of view is kept—the subject in each sentence is the same term or category term.

Many, if not most, paragraphs contain a mixture of these two types of word placement. For some sentences, information in the topic position may refer back to the end position of the previous sentence. For other sentences, information may be written from the point of view of the old information and refer back to the topic position of the topic sentence or subtopic sentence. An example of such a "mixed" paragraph is shown next.

Example 6-9

The **prophylactic** administration of drugs can sometimes be detrimental to patients, especially if prophylaxis is **continued** for several days. **Continued** exposure of the host's microbial flora to antibiotics often leads to resistant strains, and this can lead to **superinfection**. **Superinfection** can be avoided, however, by prophylactically using probiotics together with the antibiotics to restore normal flora. **Prophylaxis** is recommended when the host is subjected to a **treatment**, not involving antibiotics, that can lead to serious disease. One such **treatment** is **surgical procedure**. In **surgical procedures, prophylaxis** is usually directed at preventing staphylococcal infection. In **surgical procedures** that are considered "clean," antibiotics are not recommended.

This paragraph flows well. Notice that every sentence in the paragraph has at least one link to a previous or subsequent sentence. Some sentences will have both, a link to the previous sentence and one to a sentence following. In a few instances, there may be even three links, such as when a sentence introduces a subtopic.

Constructing paragraphs in which mixed word locations occur is okay—as long as the links do not get too muddled to risk losing the reader, as in the next example.

Example 6-10   Animals, particularly domestic animals, are important reservoirs and sources of disease to humans. Salmonella species are normally found in the intestinal tract of animals such as poultry and cattle. When humans ingest contaminated food, the salmonellae can cause disease called salmonellosis. In terms of animal disease transmission, humans often represent a dead end because the disease cannot be transferred from human to human. Salmonellosis may be acquired from animals but the infected human can also serve as a source of disease to other humans.

In this example, old and new information has been misplaced and is therefore not linked clearly. As a result, there are too many links between sentences. Readers become confused because they no longer can distinguish between what the topic is and what the stress is in each sentence or in the paragraph as a whole. As a result of such misplaced information, the paragraph is not very cohesive and does not "flow" well. By checking your own writing for what has been placed in the topic and stress positions in your sentences, you can perceive where potential problems exist and then improve your writing to better meet the reader's expectations.

Misplacement of old and new information, as seen in the preceding example, is one of the main problems in professional writing. Information is usually misplaced because most writers want to capture any important new thought before it escapes. As most writers write linearly, new information is wrongly placed at the beginning of the sentence and old information ends up at the end of a sentence. Only during the revision stage are logical links between sentences considered but may not be caught if the author does not revise or not revise sufficiently enough. Authors who misplace information are attending more to their own need for unburdening themselves of their information than to the reader's need for interpreting what is written.

Another extreme in paragraph construction occurs when an author creates no links between sentences. Links are most often ignored when an author assumes that the reader is familiar with the topic or that logical jumps are clear. Missing links may not be obvious to the author but usually confuse and frustrate readers, as in Example 6-11. Either no links or no clear logical links have been established between sentences 2, 3, 4, and 5. Here, word location has not been considered, and sentences cannot be linked because critical terms, and thus critical information, are missing. Is "megafauna" the same as "ice age species"? Does "megafauna" refer to "Pleistocene epoch"? Does this epoch correspond to the last ice age? The connection between these terms is not clearly established, perhaps because the author assumes that readers are familiar with this terminology. Instead, readers are left guessing at the missing connections between sentences.

    Example 6-11   **1** During the Pleistocene (about 2.6 million to 12,000 years ago), a variety of large mammals, birds, and reptiles were extinct. **2** The megafauna included dire wolves, mammoths, cave bears, giant condors, giant eagles, flightless birds, giant marsupials, giant snakes and lizards, and many other animals. **3** Palaeontologists are very interested in ice age species. **4** Although the migration and expansion of the human population coincides with the extinction of large animals, it is unclear if early hunters caused the extinction of so many creatures. **5** Environmental changes were also associated with the onset of the last glacial cycle.

Coherence and Continuity

Cohesive flow is the first of two steps toward creating continuity for your readers. The other step is to make your paragraphs and passages coherent. A coherent paragraph consists of a series of sentences that lead logically from one to the next, thus creating continuity. The ideas in a sentence need to be linked such that the story flows smoothly from sentence to sentence (and paragraph to paragraph). Readers consider a paragraph to be coherent if they can quickly find the topic of each sentence and if they see how the topics are a related set of ideas.

Along with topic sentences and word location, *key terms* and *transitions* are the main techniques used to create the logical framework, and thus coherence, of a paragraph and of a paper. Repeating and linking key terms ensures that the topic of the work cannot be missed and that relationships between topics are clear. In addition, transitions create continuity by indicating the logical relationship between sentences and/or paragraphs, particularly for sentences that cannot be linked by word location.

Key Terms

★ (28) Use key terms to create continuity. Repeat them exactly and early, and link them

Key terms are words or short phrases used to identify important ideas in a sentence, a paragraph, and the paper as a whole. Usually, key terms are used to identify your main points in the topic sentence. Key terms should be clearly defined and identical throughout the paragraph (and the document). They can be technical terms, such as "kinase" or "HIV-1," or nontechnical terms, such as "mechanism" or "decrease."

Key terms should not be changed but should be kept *exactly* the same, consistently. If you deliberately repeat key terms, your main points are emphasized and you create continuity. If key terms are omitted or substituted by another term, readers can get confused, especially if these terms are not within their scientific discipline.

Example 6-12   To assess original conditions of crystal nucleation and growth in metamorphic rocks, it is necessary to analyze <u>crystal distribution</u> quantitatively. <u>Density</u> could potentially provide insight into the time scale of mineral growth following the thermal peak of metamorphism.

How does "density" relate to the previous sentence? Readers unfamiliar with this particular topic may not know that "crystal distribution" and "density" here mean the same thing. Readers may be confused when two different terms are used. To avoid confusing readers, write with the nonspecialist in mind. Do not change key terms.

Revised Example 6-12   To assess original conditions of crystal nucleation and growth in metamorphic rocks, it is necessary to analyze <u>crystal distribution</u> quantitatively. **Crystal distribution** could potentially provide insight into the time scale of mineral growth following the thermal peak of metamorphism.

In the revised example, "crystal distribution" is the main key term that holds the paragraph together. Because it is repeated exactly, the relationship between the two sentences is clear, and even readers outside the field of geology will understand the passage.

Linking Key Terms

When you need to shift from a category term to a more specific term or the other way around, key terms should be linked so continuity is not lost and the paragraph stays coherent. To link key terms, use the category term to define the specific term.

| **Example 6-13** | Infectious diseases that arise due to travel may be caused by <u>gram-positive organisms</u>. **One such organism**, *Staphylococcus aureus*, can cause cellulitis, purulent arthritis, and suppurative lymphadenitis. |

The term "which is" could be included: "One such organism, which is *Staphylococcus aureus*, can cause cellulitis, . . ." However, because the definition is clear without "which is," it can be omitted. Leaving out the "which is" creates an appositive, a very useful way to define a scientific term while conserving words.

If key terms are not linked, as in the next example, readers stumble and get lost.

| **Example 6-14** | Whales, dolphins, and porpoises are mammals that lived on land before returning to the ocean about 50 million years ago. Today, the bodies of cetaceans are streamlined to glide easily through the water. |

A key term or category term has to be included in these sentences to create a link between them. The term "cetaceans" is the key term in the second sentence. Whales, dolphins, and porpoises are actually "cetaceans," or carnivorous marine mammals. Although this relationship may be clear to the author, readers may not know how these terms are linked. Linking these key terms by a "known as" clause will make the relationship clear to the reader.

| **Revised Example 6-14** | Whales, dolphins, and porpoises, **also known as cetaceans**, are mammals that lived on land before returning to the ocean about 50 million years ago. Today, the bodies of cetaceans are streamlined to glide easily through the water. |

Linking these key terms in the same sentence by a "known as" clause will make the relationship clear to the reader. Once a link has been established, subsequent sentences can then use the key terms or the category term. When key terms are linked and repeated consistently and early, good word location almost always falls right into place.

Transition Words, Phrases, and Sentences

★ (29) Use transitions to indicate logical relationships between sentences

To ensure continuity within a paragraph, a writer must use key terms and also make use of other techniques such as transitions. Transitions ensure that the reader understands what each sentence says and indicate how the sentences and paragraphs are logically related to each other and to the story.

Transitions should be placed at the beginning of a sentence for strongest continuity, usually set off by a comma.

If transitions are missing, the logical relationship between sentences can be unclear and may even be nonexistent. The importance of adding transitions is shown in Example 6-15 and its revisions.

Example 6-15   To determine the effects of solid-solution ratio on K12 adsorption at fixed pH 7, K12 adsorption isotherm experiments were conducted. _____, aqueous carbonate concentrations were measured.

The logical relationship between the first and second sentence of this example is not immediately obvious to the reader. Possible transitions that one could fill in between these two sentences include the following.

Revised   To determine the effects of solid-solution ratio on K12
Example 6-15   adsorption at fixed pH 7, K12 adsorption isotherm experiments were conducted. **In addition**, aqueous carbonate concentrations were measured.

To determine the effects of solid-solution ratio on K12 adsorption at fixed pH 7, K12 adsorption isotherm experiments were conducted. **For this purpose**, aqueous carbonate concentrations were measured.

To determine the effects of solid-solution ratio on K12 adsorption at fixed pH 7, K12 adsorption isotherm experiments were conducted. **First**, aqueous carbonate concentrations were measured.

To determine the effects of solid-solution ratio on K12 adsorption at fixed pH 7, K12 adsorption isotherm experiments were conducted. **Subsequently**, aqueous carbonate concentrations were measured.

When the transition is missing between these two sentences, most readers may guess that the intended relationship is "in addition," but readers should not be guessing.

Example 6-16 is another example.

Example 6-16   We determined sensitivity to external stimulation in hibernating ground squirrels. We measured active, torpid, and interbout arousal states daily by measuring core body temperature.

In Example 6-16, the logical relationship between the first and second sentence is also not clear because a transition is missing. Once the transition is added in, the relationship between the two sentences becomes obvious.

Revised
Example 6-16

We determined sensitivity to external stimulation in hibernating ground squirrels. **For this purpose**, we measured active, torpid, and interbout arousal states daily by measuring core body temperature.

Some transitions that are used in more casual conversation should be avoided in scientific writing:

besides (but not besides X . . .)		additionally		as a matter of fact
suddenly	admittedly	basically	ergo	at once

Use transitions and conjugations but only where appropriate. Note that logical relationships can also be clear without transitions. It is not necessary to place a transition or conjugation in every sentence, as logical relationships are often apparent from the word location within sentences.

Use transitions to link ideas, but do not overuse them. In addition, ensure that you are using the correct transitions, especially when you are an ESL author. ESL writers are prone to using transitions with a different meaning from the one they intend. Check a dictionary rather than guessing when using transitions. Do not trust a thesaurus, however. Certain transitions are considered outdated. Rather, consult a scientific editor, current peer-reviewed articles published by native English speakers in well-respected journals, or a book on scientific style to check whether a transition is still in use in contemporary scientific writing. Avoid outdated terms such as the following:

hitherto	aforementioned	henceforth	lastly
notwithstanding	firstly (use "first" instead)	secondly	

Common transitions include words, phrases, or even sentences. Whereas transition words are standard terms that indicate logical relationships between sentences, transition phrases are either infinitive or prepositional phrases. A transition sentence uses a subject and verb. The subject in the transition sentence, and the object in a transition phrase, usually repeat a key term. Note that transition sentences are stronger than transition phrases, which in turn are stronger than transition words. The farther away from the main point of a paragraph or section, the stronger the transition should be to link back to the main point (see Table 6.1).

6.4   CONDENSING

★ (30) Make your writing concise

A well-written paragraph needs not only to be organized, coherent, and consistent, but it also needs to be concise. Wordiness is a common problem in scientific writing. It can arise within sentences (see also Chapter 2,

TABLE 6.1 Transition words, phrases, and sentences

	EXAMPLE		
USE	**TRANSITION WORDS**	**TRANSITION PHRASE**	**TRANSITION SENTENCE**
Addition	again, also, further, furthermore, in addition, moreover	In addition to X, we . . . Besides X, . . .	Further experiments showed that . . .
Concession	clearly, evidently, obviously, undeniably		Granted that X is . . .
Comparison	also, likewise, similarly, etc.	As seen in . . . In the same way,	When A is compared with B . . . As reported by . . . When compared to . . .
Contrast	but, however, nevertheless, nonetheless, still, yet	In contrast to A . . . On one hand; on the other hand . . . Despite X . . . Unlike X, . . . On the contrary, . . .	One difference is that . . . Although X differed . . .
Example	for example, specifically	To illustrate X . . .	An example of X is that . . . , That is, . . .
Explanation	here, therefore, in short	Because of X . . . In this experiment . . .	One reason is that . . . Because X is . . .
Purpose	for this purpose	For the purpose of . . . To this end, . . . To determine XYZ, we . . .	The purpose of X was to . . .
Result	consequently, generally, hence, therefore, thus	As a result of . . .	Evidence for XYZ was that . . . Analysis of ABC showed that . . .
Sequence/time	after, finally, first, later, last, meanwhile, next, now, second, then, while, subsequently	After careful analysis of X . . . During centrifugation, . . .	After X was completed, . . . When we determined X . . .
Summary	in brief, in conclusion, in fact, in short, in summary	To summarize (our results), . . .	As a summary of our results shows, . . .
Strength of Transition			

Section 2.4) and also within larger structures such as paragraphs, sections, and entire manuscripts. Many journals limit the number of words in various parts of an article or in the article as a whole. It is therefore important to know how to write succinctly.

If you need to condense a paragraph (or paper), do not despair. A wide range of methods is available to make a paragraph more concise without having to remove important material from a paper. Be aware, however, that clarity is always more important than brevity.

Condensing often needs to be done in combination with other techniques:

1. Emphasizing important information
2. Deemphasizing or omitting less important information
3. Replacing or omitting words and phrases

Establishing Importance

➤ Establish importance

Scientific documents often contain much information that is not essential. Thus, it is often difficult for the reader to find the real "meat" of the paper. The key to condensing is to distinguish between important and less important information. Often, authors are too close to their own writing to be able to do so. Instead, they consider everything important and have a hard time prioritizing statements. To help you prioritize, it is critical to have someone else read your draft. The best critics in this respect are trained scientists outside your field. These people are able to guide you in putting levels of importance into better perspective.

Once you have decided on these levels of importance, distinguish information by highlighting important points, deemphasizing less essential ones, and omitting nonessential phrases. To emphasize your important information, either place it in a power position or signal it directly by using statements such as "Most important, . . ." or "The key finding of this study was . . ." Less important information can be deemphasized by subordinating it. This can be done, for example, by placing it in a subordinate clause as shown in the next examples.

	Example 6-17	Information about the relation between WBC count and hospital case fatality rates is limited. A link between WBC count and increased long-term mortality after acute myocardial infarction may exist.
	Revised Example 6-17	*Although* information about the relation between WBC count and hospital case fatality rates is limited, a link between WBC count and increased long-term mortality after acute myocardial infarction may exist.

	Example 6-18	The Hainan aborigines are an ethnic group living at the entrance route to Southeast Asia and have been influenced little by relocation and migrations of other ethnic groups.
	Revised Example 6-18	The Hainan aborigines, *an ethnic group living at the entrance route to Southeast Asia,* have been influenced little by relocation and migrations of other ethnic groups.

When less important information has been reduced or omitted and important information has been emphasized, the reader will be able to see the big picture. See also Chapter 3, Sections 3.2 and 3.3 for placement of important information in sentences.

Words and Phrases That Can Be Omitted

➤ Omit "overview" words, phrases, and sentences

Another important technique in condensing is to replace or to omit unnecessary words and phrases. Aside from redundancies and jargon, all of which are discussed in Chapter 2, other words and phrases can be omitted to condense a document. ESL writers whose native language uses many flowery phrases should be particularly aware of what words and phrases can be omitted in a scientific document.

    Scrutinize your paper for pointless words and phrases ruthlessly. Dissect every sentence. As a rule, when equivalent alternatives exist, choose the shortest one. In scientific writing, every word should count.

ESL advice

Verbs to Omit

Omit verbs such as

describes	noted	was done	reported
noticed	observed	occurred	seen

Just state the facts.

	Example 6-19	Jones et al. reported that intracellular calcium is released when adipocytes are stimulated with insulin.
		(15 words)
	Revised Example 6-19	Intracellular calcium is released when adipocytes are stimulated with insulin (Jones et al., 1996.)
		(10 words)

Phrases to Omit

Omit phrases and sentences that tell your reader what a sentence/paragraph is about.

👎	**Example 6-20**	Products were verified by gel electrophoresis. The results are presented in Figure 1.
		(13 words)
👍	**Revised Example 6-20**	Products were verified by gel electrophoresis (Fig. 1).
		(6 words)

👎	**Example 6-21**	To assess the purity of the gene product, many techniques were employed.
		(12 words)
👍	**Revised Example 6-21**	(Omit)
		(0 words)

Other expressions to omit:

This section describes . . .
In regard to . . .
As far as X is concerned . . .
The experiment was done by . . .
Figure 6 shows that . . .

See also Chapter 2, Section 2.4 for redundancies and jargon.

"It . . . That" Phrases

Aside from jargon and other redundancies, you should replace or omit "It . . . that" phrases. Most of these phrases are pointless fillers and can be omitted entirely. If the idea in the phrase is essential, replace the phrase with a shorter version.

Examples of "It . . . that" phrases:

It is interesting to note that . . .	**omit**
In light of the fact that . . .	**replace** (because)
It is possible that . . .	**reword** (. . . may . . . , perhaps, possibly)
It has been reported that . . .	**omit or replace** (Taylor reported that . . . ; or [reference])

Positive versus Negative Expressions

➤ Avoid writing in the negative

Changing negative expressions to positive expressions usually results in shorter sentences. Moreover, readers prefer to read positive things, not negative things. Avoid writing in the negative. Write instead in the positive. Above all, avoid double negatives, which can easily confuse readers.

Examples of changing from negative to positive:

negative	positive
do not overlook	note
not different	similar
not infrequently	frequently

not many	few
not the same	different
not unimportant	important

Excessive Detail

➤ Omit excessive detail

Detail that can be inferred or is unimportant should be omitted.

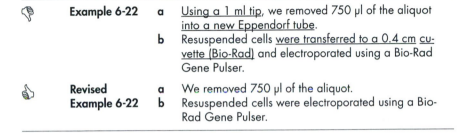

Example 6-22	a	<u>Using a 1 ml tip,</u> we removed 750 μl of the aliquot <u>into a new Eppendorf tube</u>.
	b	Resuspended cells <u>were transferred to a 0.4 cm cu-vette (Bio-Rad)</u> and electroporated using a Bio-Rad Gene Pulser.
Revised	a	We removed 750 μl of the aliquot.
Example 6-22	b	Resuspended cells were electroporated using a Bio-Rad Gene Pulser.

Intensifiers and Hedges

➤ Do not overuse intensifiers or hedges

Intensifiers are adjectives, adverbs, or verbs that are used to strengthen nouns or verbs such as the following:

always	basic	central	certainly	clearly
crucial	prove	quite	show	very

Hedges are cautious adjectives, adverbs, or verbs such as the following:

actually	appear	could	essentially	indicate	many
may	most	often	perhaps	possibly	seem
some	suggest	usually			

Intensifiers and hedges are often overused, especially by ESL writers. Aside from making a sentence wordy, you will sound arrogant and too aggressive if you overuse intensifiers and too cautious or timid if you overuse hedges. Omitting intensifiers and hedges will not only avoid these appearances but also shorten sentences.

Example 6-23	Our results <u>may indicate</u> that chewing gum <u>possibly</u> reduces caries incidence.
Revised Example 6-23	Our results **suggest** that chewing gum reduces caries incidence. or: Chewing gum **may** reduce caries incidence.

👎	**Example 6-24**	Figure 5 <u>clearly</u> shows that the protein was absent in the fraction.
👍	**Revised Example 6-24**	Figure 5 shows that the protein was absent in the fraction.

Figure Legends

Figure legends in particular can often be condensed. Many journals prefer telegram-style figure legends. Usually, articles can be omitted and prepositional phrases can be shortened or omitted not only in figure titles but also in their explanatory notes. See also Chapter 9, Section 9.8 for additional information on figure legends.

👎	**Example 6-25**	Figure 3. <u>Shown here are</u> tangential sections <u>as seen</u> through a barrel cortex of a trimmed mouse. A-C <u>show</u> <u>the</u> ipsilateral barrel cortex; D-F <u>depict the</u> contralateral barrel cortex. <u>The</u> scale bar <u>is</u> 0.2 mm.
👍	**Revised Example 6-25**	Figure 3. Tangential sections through a barrel cortex of a trimmed mouse. A-C, Ipsilateral barrel cortex; D-F, Contralateral barrel cortex. Scale bar, 0.2 mm.

SUMMARY

> BASIC RULES—COMPOSITION
> ★ Organize your paragraphs.
> ★ Use a topic sentence to provide an overview of the paragraph. Arrange the details in the remaining sentences.
> ★ Use consistent order.
> ★ Use consistent point of view (same subject for different sentences within a paragraph).
> ★ Make your sentences cohesive.
> ★ Use key terms to create continuity. Repeat them exactly and early, and link them.
> ★ Use transitions to indicate logical relationships between sentences.
> ★ Make your writing concise:
>   ○ Establish importance
>   ○ Omit "overview" words, phrases, and sentences
>   ○ Avoid writing in the negative
>   ○ Omit excessive detail
>   ○ Do not overuse intensifiers and hedges

PROBLEMS

PROBLEM 6-1 Paragraph Organization

The following paragraph is about the two phases of *P. infestans* infection. Although the first sentence of the paragraph introduces the potato pathogen, the second sentence of the paragraph does not logically link back to the first sentence nor does it clearly introduce the two phases. Rewrite the second sentence to link it to the first sentence and to lead into the description of the two phases. Put parallel ideas into parallel form.

The fungus-like organism *Phytophthora infestans* causes late blight of potatoes and tomatoes. The organism reproduces asexually through sporangia, small sack-like structures containing zoospores, which grow from infected tissues. In the early stage of their life cycle, *P. infestans* infects the plant tissue and develops specialized feeding systems. After this biotrophic phase of the infection, a necrotrophic phase follows during which disease symptoms appear and host tissue is degraded (Whisson, 2007).

PROBLEM 6-2 Paragraph Organization

The following paragraph is about types of glutamate receptors. Although the paragraph has a topic sentence, the remaining sentences of the paragraph do not follow logically from the topic sentence. Add a sentence after the topic sentence to fulfill the expectations of the reader. (What does the reader expect to read about after reading the topic sentence?) Reorganize sentence 2 to make it parallel to sentence 3.

1 Ionotropic glutamate receptors fall into two general categories (7, 8). 2 When NMDA receptors are activated by N-methyl-D-aspartate (NMDA), ion channels are opened, allowing ions to rush into the cell and thus cause an excitatory postsynaptic potential. 3 Non-NMDA receptors like PCP bind to kainate and quisqualate only and prevent ion flow and an excitatory postsynaptic potential, even in the presence of NMDA.

PROBLEM 6-3 Paragraph Organization

The following paragraph is about the migration of salmon. Although sentences 1 and 2 describe the two possible methods salmon may employ to find their way, the paragraph does not have a topic sentence. Write a clear topic sentence for this paragraph. The topic sentence should state the message of the paragraph (the different methods salmon use to find their way during their homeward journey in the fall). In your topic sentence, make the topic the subject of the sentence. Also, make sentence 2 parallel to sentence 1.

1 Salmon use geomagnetic imprinting to return to their freshwater birthplace to spawn. 2 As described by Stabell et al. (15), olfactory cues also play

a role in guiding adult salmon back to the stream of their birth. **3** It is unclear whether salmon also use cues other than geomagnetic and chemical imprinting to orient themselves.

PROBLEM 6-4 Paragraph Consistency
Ensure that the order of the key terms in the topic sentence is consistent with that of the remaining sentences. Ensure also that the same key terms are used consistently and that correct parallel form is kept when signaling the subtopics.

Essential mineral macronutrients for plant growth and survival are nitrogen, potassium, and calcium. These nutrients are absorbed through plant roots. Plants need phosphorous for ATP formation during photosynthesis, respiration, energy storage, cell division, and to grow flowers, seeds, and roots. Phosphorous is often a limiting nutrient in soil but can be supplied by bone meal. Nitrogen is needed for protein formation, metabolic processes, vitamin formation, and, like potassium, for chlorophyll photosynthesis. Legumes supply nitrogen to the soil through nitrogen fixation. Decaying matter also provides nitrogen to the soil and, in turn, this is taken up in the form of NO_3^- or NH_4^+. Potassium is also essential in photosynthesis and protein synthesis, as well as in carbohydrate metabolism. In addition, potassium is needed for water retention and to increase disease resistance. It is highly water soluble and usually readily available in sandy soil.

PROBLEM 6-5 Paragraph Consistency
Keep the organization of the following paragraph in consistent order. In addition, use a transition to indicate the relationship between sentences 2 and 3.

1 Both GAD-positive cell bodies and processes were found in the ventral lateral posterior nucleus and thalamic reticular nucleus. **2** Almost all of the neurons in the thalamic reticular nucleus appeared to contain GAD-immunoreactivity. **3** Only small round cells in the ventral lateral posterior nucleus were GAD positive.

PROBLEM 6-6 Key Terms
The relationship between the two sentences in the following paragraph is not clear because key terms are not linked. Given that heroin and morphine are examples of opioid drugs, link the underlined key terms in the paragraph.

<u>Opioid drugs</u> are avidly self-administered by both humans and laboratory animals (2, 3). <u>Heroin and morphine</u> mimic the endogenous opioid neurotransmitters, known as endorphins, by binding to one or more of the mu or delta opioid receptors in the brain (4).

PROBLEM 6-7 Key Terms
The key terms in the following paragraph are not repeated exactly. Revise the passage such that key terms (life span, fat storage, and the nematode *C. elegans*) are repeated exactly. (Note: A nematode is a type of worm.)

A model system to study the underlying mechanism that connects Type II diabetes, life span, and obesity is the nematode *C. elegans*. Worms are a great system to study the connection of diabetes, age, and obesity because they have a well-conserved insulin signaling pathway that affects life span and fat storage, suggesting that there is an underlying molecular mechanism that connects insulin signaling, aging, and obesity. In addition to the known insulin signaling pathway, the entire genome of the nematode has been sequenced, and many molecular and genetic tools are available that will allow a comprehensive identification of the genes that affect insulin-like signaling, aging, and fat regulation in worms.

PROBLEM 6-8 Key Terms
The following paragraph is about *systolic blood pressure* in teenagers.

1. Ensure that this key term is introduced in sentence 1 and used consistently and exactly throughout the paragraph.
2. The paragraph describes contrasting results (sentences 2 and 3). Signal the results. Rewrite the paragraph so that the contrast is in perfect parallel form.
3. Signal the contrast by using a transition word or phrase at the beginning of sentence 3.
4. Fix the faulty comparison in sentence 4.

1 Using a sphygmomanometer, we determined blood pressures in 1,200 teenagers between the ages of 13 to 18 years. 2 The mean systolic blood pressure for girls was 117 mmHg. 3 Boys had a mean blood pressure of 129 mmHg. 4 Thus, blood pressure in boys was about 10% higher compared to that in girls.

PROBLEM 6-9 Transitions
In the next paragraph, the logical relationship between sentences is not clear. Add a transition word at the beginning of sentences 2, 3, and 6 to make the relationships between sentences clear.

1 Coral rely on two main energy sources. 2 _____ they capture planktonic organisms with their tentacles. 3 _____ they rely on nutrients provided by photosynthetic zooxanthellae, single cell algae that live symbiotically within the coral tissues. 4 Coral bleaching occurs when coral lose their zooxanthellae endosymbionts or zooxanthellae lose their photosynthetic pigments (Glynn, 1996), or both, due to changes in water level, water temperature, UV radiation, sedimentation, pathogens, pollution, and nutrient abundance. 5 Bleaching is reversible if stressors are temporary and not too severe. 6 _____, continuous stress and zooxanthellae depletion leads to coral death.

PROBLEM 6-10 Paragraph Construction

Construct a paragraph on the problems that arise when alien fish are introduced into naturally fishless mountain lakes using the list of facts provided. In your writing, pay attention to writing a good topic sentence and to using good word location. Employ paragraph consistency, key terms, and transitions. Consider also other basic rules such as using parallel form and correct pronouns and prepositions.

Alpine Lake Ecosystem
- Native fauna negatively affected by introduced alien fish species
- Mountain lakes are naturally fishless
- Consequences of fish introduction are particularly bad for the common frog
- Ecological exclusion observed for brook trout in the Gran Paradiso National Park
- Brook trout preys on frog larva
- Solution: eradicate introduced fish to recover amphibian populations in alpine lake

PROBLEM 6-11 Paragraph Construction

Construct a paragraph on cancer cells using the list of facts in the order provided. In your writing, pay attention to writing a good topic sentence and to using good word location. Employ paragraph consistency, key terms, and transitions. Consider also other basic rules such as parallel form and correct pronouns and prepositions.

Cancer Cells
- Are malignant tumor cells
- Differ from normal cells in three ways:
  ◦ They dedifferentiate—for example: ciliated cells in the bronchi lose their cilia
  ◦ Metastasis is possible—travel to other parts of body; new tumor growth
  ◦ Rapid division—do not stick to each other as firmly as do normal cells

PROBLEM 6-12 Paragraph Construction

Construct a paragraph on combination drug therapy using the list of facts provided in any order. In your writing, pay attention to writing a good topic sentence and to using good word location. Employ paragraph consistency, key terms, and transitions. Consider also other basic rules such as using parallel form and correct pronouns and prepositions.

Combination Drug Therapy
- Applied in treatment of tuberculosis
- Can be expensive

- Can increase risk of toxicity and superinfection
- Of important clinical value
- Used against organisms for which resistance to one drug readily develops

PROBLEM 6-13 Paragraph Construction

In the following paragraph, the author has not paid much attention to good word location. Analyze the paragraph with respect to word location by drawing arrows to determine how key words relate the sentences to each other. Then, revise the paragraph by improving word location.

1 Chikungunya, a viral disease spread by mosquitoes, causes severe joint inflammation and pain, high fevers, headache, and rash. 2 The disease is seldom fatal. 3 However, public health officials are concerned about the spread of the mosquito *Aedes albopictus*. 4 Many viral diseases are transmitted by *Aedes albopictus*, which bites not only during dawn and dusk but also during daytime. 5 It is considered an invasive species and was first imported into the United States in tires from Asia in the 1980s (Moore and Mitchell, 1997). 6 Many other nontropical countries have since also been invaded by this vector due to international trade and travel. 7 Italy is the only European country that has had an outbreak (Rezza, 2007). 8 Scientists worry about additional and bigger outbreaks in the developed world.

PROBLEM 6-14 Transitions/Coherence

The following paragraph describes how the liquid crystals change with decreasing temperature. In the paragraph, sentences 2, 3, and 4 are parallel. The subtopics of sentences 2 and 4 are signaled ("At high temperatures," "At even lower temperatures"). However, the subtopic of sentence 3 is not signaled.

1. **Write a signal at the beginning of sentence 3. Your signal should be parallel to the signals in sentences 2 and 4.**
2. **Add a transition word in sentence 6 that connects it to the rest of the paragraph.**

1 Phase changes of liquid crystals can be observed by optical polarizing microscopy. 2 At high temperatures, the orientation of molecules in liquid crystal materials is isotropic or random. 3 _____, these materials pass into the nematic phase where molecules point in the same direction but have no other order. 4 At even lower temperatures, liquid crystals are in the smectic phase, a phase more like a solid than a liquid where molecules point in the same direction while at the same time aligning themselves in layers. 5 Eventually, as temperature decreases, the substances transition into the solid or crystalline form. 6 _____, temperature determines the phase for liquid crystal materials (Styer, 2000; Larsen, 2003).

PROBLEM 6-15 Condensing

Condense the following paragraph. Try to make your revised paragraph less than 35 words.

Our results indicate that between 5° and 25°C, undoped, high-quality diamond as well as diamond covered with chemically bound hydrogen has no conductivity. Undoped, high-quality diamond also shows no conductivity at higher temperatures. However, diamond covered with chemically bound hydrogen clearly displays conductivity at temperatures between 5° and 25°C.

(49 words)

PROBLEM 6-16 Condensing

The following paragraph is an abstract that has far exceeded its permissible length of 100 words. Shorten the abstract to 100 words or less by establishing importance. Omit unimportant information, and subordinate less important information.

Tourette syndrome (TS) is characterized by chronic motor and vocal tics. Habit reversal therapy (HR) is a behavioral treatment for tics which has received recent empirical support. The present study compared the efficacy of HR in reducing tics, improving life-satisfaction and psychosocial functioning in comparison with supportive psychotherapy (SP) in outpatients with TS. [. . .] Thirty adult outpatients with DSM-IV TS were randomized to 14 individual sessions of HR (n=15), or SP (n=15). HR but not SP reduced tic severity over the course of the treatment. Both groups improved in life-satisfaction and psychosocial functioning during active treatment. Reductions in tic severity (HR) and improvements in life-satisfaction and psychosocial functioning (HR and SP) remained stable at the 6-month follow-up. [. . .] Our results suggest that HR has specific tic-reducing effects although SP is effective in improving life-satisfaction and psychosocial functioning. Assessments of response inhibition may be of value for predicting treatment response to HR. *(149 words)*

(With permission from Elsevier)

PART TWO

Planning and Laying the Foundation

CHAPTER 7

The First Draft

The publication of new results in scientific journals is a central component of the scientific process. Communicating your findings makes them available to others and serves as an indication of your expertise and productivity.

THIS CHAPTER DISCUSSES:

- The prewriting and drafting stages of the writing process
- Audience and journal choice
- Instructions to Authors
- Deciding on authorship
- The IMRAD format
- Outlining and composing a manuscript
- Composing the first draft
- Overcoming writer's block
- Getting outside help

7.1 THE WRITING PROCESS

This chapter covers the details of the prewriting and drafting stages for composing a manuscript, while Chapters 17 and 18 cover revising and editing and reviewing and submitting, respectively.

The writing process encompasses several stages:

Prewriting
In the prewriting stage, you decide on the intended audience for your document and choose the journal to which to send your manuscript. You also need to determine authorship.

Drafting

In the drafting stage, you usually review the literature to learn what has been done and what is known in the field of your topic. You have to organize your thoughts and materials. You may consider creating an outline. Then, start drafting the main body of the text as well as the Abstract.

Revising

The key to strong scientific writing is revising your work. Expect to do multiple revisions, which also involve your colleagues.

Editing

After you have drafted and revised your paper, you need to copyedit it for grammar, spelling, and punctuation.

Evaluating

When you have a final draft, submit it to your journal of interest for anonymous peer review. When the editor has received the evaluations from the reviewers, he or she will inform you of the final decision on your manuscript.

Publishing

Once accepted, your manuscript will be published in the corresponding target journal.

7.2   PREWRITING

No experiment will mean much unless it is communicated. But how do scientists trained in research move from doing experiments to communicating what the results mean? Before actually starting to write, you need to consider several things. Specifically, you have to

- Have enough data or ideas to write about
- Know who your readers will be
- Choose a journal to which to submit
- Collect your references
- Decide on authorship
- Know how to start composing the document

Collecting data, researching references, organizing your thoughts, and writing take time and energy. Therefore, make sure that your research warrants publication in a journal. Write an article only if you have something new, important, useful, and comprehensible to tell.

Audience and Journal Choice

➤ Search for the best match of topic, audience, and journal

Once you have a defined message for your paper, you need to decide where to publish your work. This decision will depend on several factors such as the topic, your audience, the journal, the speed of publication, and the journal's acceptance rate. Take the time to look through potential journals and determine who their audience is and what their mission is. Do not hesitate to ask colleagues for advice about which journals might be appropriate. You may also want to check where other papers of a similar topic have been published.

Most scientists choose a journal in a specialized area. Start by looking, for example, at Current Contents, *Index Medicus*, *Journal Citation Reports*, Biological Abstracts, Cumulative Index to Nursing and Allied Health Literature (CINAHL), SCOPUS, or the General Science Index, or at open access journals BioMed Central and PLOS, to get an idea of specific journals that cover your research area (see also Chapter 8, Section 8.2 for more information about open access journals).

Consider consulting the *Journal Citation Reports* in your academic library. These reports will tell you which journals are cited *most*—that is, the impact factor of a journal, which is a measure of the frequency with which the average article has been cited for a given journal in a particular year. Many scientists consider the impact factor a measure of the relative ranking and reputation of a journal. Be aware, however, that numerous journals publish on a broad topic, and many of their papers may not be widely cited (e.g., *Journal of Bacteriology* is a high-tier journal that publishes quality research on any species of bacteria, not just the highly cited ones). There are also variations in how journals count and calculate their citations.

Typically, journals with a high impact factor also have very high rejection rates, some as high as 90%. If you send a manuscript to one of these journals, you may be waiting for weeks just to receive a rejection. If you feel unsure of whether your paper is appropriate for a particular journal, contact the editor and ask if a manuscript such as yours would be considered for publication. This approach is the quickest way to get an answer. Generally, journals that are published more frequently have a faster turnaround time between acceptance and publication. Some may publish articles online first before publishing them in print.

The ever-increasing international readership in science today seeks both an electronic version and a prestigious ranking. Beware, however, of journals that publish only electronically, as many of these journals do not rigorously review their articles and thus are not as prestigious as journals with printed versions. Beware also of new journals; their circulation may be very small.

Aside from choosing a journal, you need to decide on the best format for your writing. Not every study leads to a full article. Other choices include, for example, case reports (articles that analyze a disease, occurrence, prognosis, and so forth) or letters to the editor (a brief report that presents findings of a short study).

Not every work is published in a journal. In the field of computer science, for example, shorter conference publications are valued higher than full-length articles in journals because of the faster reviewing time, higher visibility, and timeliness. This practice has led to much debate as journal publication reviews are considered more detailed and rigorous.

Instructions to Authors

➤ Obtain *Instructions to Authors* and follow them

When you have decided on a target journal, you should obtain the journal's specific *Instructions to Authors*. Read these instructions. Mark important details such as length of abstract, structure and headings of an article, writing style, format of in-text citations, format of references, and electronic format—all of which may differ from journal to journal. Follow the instructions carefully when you are writing, and compare your format to recently published articles in your target journal.

Instructions to Authors are generally available on the website of your target journal as well as in printed volumes of the journal. Web sites that list *Instructions to Authors* for many different journals include the following:

> http://www.icmje.org (Core set of instructions for many biomedical journals)
> http://www.lib.auburn.edu/chemistry/chemjn.php (For chemistry journals)
> http://www-jmg.ch.cam.ac.uk/data/c2k/cj/ (For science journals)

7.3   AUTHORSHIP

➤ Decide on authorship before starting to write

One of the most common concerns of young scientists is who to include as an author on a publication and in which order. Every person who contributed substantially to the research, to the experimental design, and to writing the paper should be included as a coauthor. Someone who did his or her day-to-day job assisting in running assays, collecting data, creating a figure, or revising the manuscript usually is not included as a coauthor. See also the criteria for authors as described by the International Committee of Medical Journal Editors (http://www.icmje.org/news-and-editorials/new_rec_aug2013.html).

Having coauthors can be beneficial but also problematic. Coauthors can help in designing the experiments and in seeing them through as well as in interpreting the data and writing the paper. Disagreement over authorships, however, can easily result in wrecked friendships. A good piece of advice is to decide who is going to be an author as soon as you can—if possible, *before* you start writing the paper (if not before you start the work).

Most people want to be the first author (or as close to the first author position as possible) because typically only the name of the first author appears in citations and reference lists. In most contemporary manuscripts, the first author is the one who did most of the work on the project while the last author is the principal investigator (or head of the laboratory). If there is more than one first author, their equal contribution can be noted in a footnote. Alternatively, if more than one paper arises from the study, the reverse order of authors can be used in the second paper.

The first or last author usually does most of the writing, and one of them is usually identified as the corresponding author, or the author to whom all correspondence should be directed. Being the corresponding author does not give you much recognition. If you decide to become the designated corresponding author, you should ensure that you can be reached through your corresponding address for any follow-up questions.

People who contributed to the work but to a lesser extent than the first author(s) are typically listed based on the level of their involvement. Different institutions or departments may follow different rules regarding who should be included as an author and in what order. For example, in some fields, authors are listed alphabetically; in others, every member of the lab is included regardless of contribution. Large collaborative publications that have many coauthors may exceed a journal's limit on author number. Instead of listing each author, groups often use bylines (e.g., "Schizophrenia Working Group of the Psychiatric Genomics Consortium"). Listing may also include named authors and a byline as shown in the following example.

Example 7-1   Sample authorship

Eur Child Adolesc Psychiatry. 2015 Feb;24(2):141-51. doi: 10.1007/s00787-014-0543-x. Epub 2014 Apr 26.
The Tourette International Collaborative Genetics (TIC Genetics) study, finding the genes causing Tourette syndrome: objectives and methods.
Dietrich A[1], Fernandez TV, King RA, State MW, Tischfield JA, Hoekstra PJ, Heiman GA; TIC Genetics Collaborative Group.
⊕ **Collaborators (48)**
⊕ **Author information**

All coauthors should at least read and approve the final version before submission to a journal. Coauthors also have to agree to any changes made before publication, and the primary author should confer with the coauthors as to the correct spelling of their names.

You may run into a person, such as a senior researcher (investigator or head of department), who asks you to include in the author list a person who has not contributed to the research, possibly even himself or herself. You can try to avoid this situation by diplomatically stating that your target journal has stringent authorship requirements. However, you may not be able to avoid including that person for political reasons. If that happens, at the very least ask that person to help write the article. Conflicts can also arise when more than one person is writing the manuscript. Problems may include not only inconsistencies in language and style but also weak transitions and illogical formats. An example of a paragraph that has been written by more than one author—and in two very distinct styles—is shown in Example 7-2.

 Example 7-2   **Structured abstract**

Numerous antibiotic classes, including macrolides, the streptogramins, and the oxazolidinones, bind to the 50S ribosomal subunit, demonstrating that this is an excellent target for antibiotic drug discovery. Biochemical studies had previously determined that the site of action for all these antibiotic classes was the peptidyl transferase center of the 50S. The crystal structures, however, brought light into the center: they delineated how different antibiotic classes bind to or engage distinct, though often overlapping or adjacent spaces, making the 50S ribosome a "target of targets." <u>Consider the opportunities! Suddenly, one could determine an appropriate position on and a trajectory from an existing antibiotic scaffold to boost affinity or overcome target-based resistance. Likewise, one could study the spaces between two adjacent binding sites and either bridge the two or utilize that as a starting point for a new scaffold. At once, ribosomal drug hunters were freed from the me-too approach of tuning ever-so-slightly the same molecular scaffolds. And so, it was back to the future, targeting the ribosome.</u>

(With permission from Future Medicine Ltd.)

It is important to designate one writer (preferably the first author or best writer) for doing all of the writing or revisions. Alternatively, one person could be designated as the coordinator. This person should oversee the logical framework of the paper as well as its style and consistency.

7.4   DRAFTING A MANUSCRIPT

General Format

➤ Follow the IMRAD format

Most scientific papers have a set format, called the Introduction, Methods, Results, and Discussion (IMRAD) format. This format originates from the health care and natural sciences fields. It reflects the order of the core sections in academic journal articles, which usually include the following:

IMRAD format

Title page
Abstract
Introduction
Materials and Methods
Results
Discussion
Acknowledgments
References
Tables
Figures
Figure Legends

Although the IMRAD format does not corresponds to the actual sequence of research events or thought processes, it presents the most important information for the reader in a well-organized, clear, and logical fashion without unnecessary detail. The format is successful because it allows readers to navigate articles fairly quickly. Papers that do not follow this format or differ significantly from it may disorient their readers. Note though that this format is not used in all scientific fields (e.g., computer science) nor in certain journals. In some papers, for example, the Materials and Methods section appears after the discussion section.

The IMRAD format is also not necessarily the best order in which to draft your manuscript. You can start writing your paper in several different ways. You can write an outline first (simple or detailed), use storyboards or concept maps, or you can start by writing the section you consider the easiest. For some people, this may be the Materials and Methods section; for others, it may be the Results section, including figures and tables, or the Introduction. A few more experienced writers may even choose to write the Abstract first—considering it a short outline. The bottom line is this: Get started somewhere, and keep working toward your end product in any order or by any approach you choose.

Collecting References

➤ Collect, organize, and study your references

To write a scientific document, you will need references. Due to the constant and large flow of information, often compounded by multidisciplinary collaborations, it is almost impossible to keep up with all the publications relevant in your field. Although you may be on a notification list for all the important new material, you still have to devote much time to reading and fully processing published research results.

Even if you are staying abreast with current literature, you may find yourself doing a literature search before writing an article. Search online databases such as MEDLINE or PubMed for relevant references. Ask your colleagues and your advisor for references and reprints, and start reading review articles. They can provide you with great ideas for your introduction and discussion as well as point you to other important references in the field. Importantly, ensure that your sources are relevant and valid (see also Chapter 1, Section 1.2, and Chapter 8, Section 8.1).

High-quality references in your research field provide great examples of how successful papers are written. Study them closely. Also, note the wording and terminology, especially standard phrases that you could use in your own writing. However, be wary of plagiarizing (see Chapter 8, Sections 8.5 and 8.6 on plagiarism and paraphrasing). Do not blindly copy and use entire sentences or paragraphs.

From the start, organize your references and reprints and compile a reference database in EndNote or ReferenceManager. These computer programs can save you much time and frustration later in the writing process, especially when it comes to reformatting references (see also Chapter 8.)

Getting Ready to Write

➤ Decide on the best way to create, share, and edit your documents

When getting ready to write, decide on what program to use to compose your document. Most researchers use either MS Word or Apple Pages. Others, especially in the physical and computer sciences, may prefer LaTeX, a typesetting standard that includes elements for the production of technical and scientific documents, or LyX, an open source document processor based on LaTeX. Alternate, free word-processing programs include downloadable OpenOffice, Abiword, WPS Office Free, and LibreOffice. Online, free alternatives include Office Online, Zoho, and Google Docs. The latter is particularly useful if you are planning to work on a document simultaneously with others as it allows them to view and edit documents in real time. Similar to Google Docs, Google also offers Google Sheets to read and modify spreadsheets online and Google Slides to create and format presentations while working with others. Such file-sharing programs are also very useful for large documents as their use avoids the trouble of sending large documents to others and allows you to store files in the cloud.

➤ Start by writing less

The prospect of writing a manuscript can seem overwhelming, especially for beginning writers. The key is to take the writing process one step at a time. Think of your document as separate sections.

To write a first draft, you need to have a block of time without interruptions. When you start writing, you may not yet know precisely what to write or what information to include. If you do not know what to write at any one point, do not be discouraged. The right words will come to you. Just leave a blank space for the time being and keep going. You or one of your coauthors will eventually fill the gap.

Center your writing on the overall question/purpose of the paper and on its answer (see also Chapter 11, Section 11.2). Then decide what to include in the paper and how to organize it. Be aware, however, that your question and answer might change as you are writing. When you put your thoughts on paper, you will discover the best and most precise way of formulating your question and answer. In the revisions, you can then adjust your writing accordingly.

Do not expect to finish your first draft in one sitting. You may need a week to write the first draft—or more. To make the process less overwhelming, start by writing less. For example, write a very brief Introduction of maybe just one paragraph. You will fill in a lot more during your revisions as ideas come to you or as your colleagues comment. Similarly, in the Discussion, start off by writing 2 paragraphs, not 12.

➤ Worry about basic rules, principles, and guidelines in the revisions

The most important thing for writing first drafts is momentum. Do not worry about any basic rules or any other principles or guidelines at this stage. You can always add or take out any words, sentences, or even paragraphs later. Do not worry about whether your pronouns are clear, whether you use good parallel form, or whether your paragraphs are in the right order. Deal with rules and guidelines during the revisions. For now, just keep moving.

Put your ideas on paper or into the computer so that you have something to work with. Remember, your first draft is not written in concrete. It should contain all you can think of and want to say in one place. Later revisions will allow you to examine, rethink, and rearrange.

7.5   OUTLINING AND COMPOSING A MANUSCRIPT

➤ Pay attention to the overall organization of your document

Before you begin to write, consider the overall organization of your document. An outline may be helpful for some; for others, a more graphic organization of their ideas is useful, such as a storyboard or graphical tools such as concept maps, which start with a main idea and then break down this idea into branches representing specific topics. Even if you do not use an outline, storyboard, or concept map for your writing, consider jotting down the main points for each section before you start.

The same ideas of paying attention to order and organization in your writing process are applicable to more than just journal articles. They apply also to review articles and grant proposals. You need to know the respective content, format, and impact for all of these documents and their sections.

Each section of a research paper, review paper, or proposal has a set internal structure. These structures are described in detail in Chapters 11–18 for research papers, Chapter 19 for review articles, and Chapters 20–28 for grant proposals. Be aware of these internal structures. They differ from each other. For example, the Introduction section of a research paper contains the known or background information of your topic (which can be broken down into general and specific data), the unknown or problem area that your paper aims to explore, the overall purpose of your paper, and a general description of your experimental approach. Example 7-3 provides a general topic outline for an Introduction of a research paper.

 Example 7-3   **Topic outline for Introduction**

    A. Known/background
       1. General
       2. Specific reported data
    B. Unknown/problem
    C. Research purpose/question
    D. Experimental approach

One of the most important ideas to write down relatively early is the purpose or question of your research (see also Chapter 17, Section 17.2). I recommend writing it on a Post-it® note and sticking the note onto the side of your computer screen. In this way, you can always recall your research question immediately to remind you of the central focus of your paper.

 Example 7-4   **Question of study**

> **Question:**
>
> *We wanted to determine how long hummingbirds feed their chicks after they leave the nest.*

If you decide to use a more detailed outline, fill in the topics of your outline with ideas as they come to you and with material you have collected. An outline can be written in full sentences, or it may be written in bullet point form. Be aware, however, that arrangements of data, ideas, and outlines may change.

The following example is an outline for an Introduction.

 Example 7-5   **Full sentence outline**

> **Question:**
>
> To determine how changing CO_2 concentrations affect flowering plants . . .

A. Background:
   1. General: Influence of climate change
      (a) Gas composition of air in past (Liu and Froschauer 2015, p. 80; Dohler et al., 2005)
      (b) Gas composition of air has changed due to increasing temperature of atmosphere (Peters et al., 1990)
         i. O_2 concentration has decreased due to climate change (Joh et al., 1999)
         ii. CO_2 concentration has increased due to climate change (Kim et al., 2000)
   2. Specific reported data: Changing O_2 and CO_2 concentration affects plant populations (Honeck et al., 2003)
      (a) Maple fertility decreases with decreasing O_2 concentration (Weinberger and Shewitz, 2000a)
      (b) Maize growth rate and size increases with increasing CO_2 concentrations (Herbert et al., 2002)
B. Unknown/Problem: Not known how increasing CO_2 concentrations affect the growth of flowering plants
C. Research Purpose/Question: To determine how changing CO_2 concentrations affect flowering plants
D. Experimental Approach: We observed and evaluated the growth rate and size of six different flowering plant species under various CO_2 concentrations.

Note that the author of this outline has kept track of reference sources. This approach will come in handy later in the drafting process.

From Data and Ideas to Composition

➤ Organize your data and ideas before writing

Some people will tell you not to write your manuscript until you have collected all of your data. Others will tell you to write as you go—every day for some time. In all scenarios, the underlying prerequisite to start writing is that you must have accumulated enough data and ideas to start a manuscript.

When you have enough data, lay them out in front of you. Preparing good figures and tables is essential to a research paper. Therefore, include all the necessary information to make your figures and tables informative. It also helps to have a well-organized laboratory notebook where you record your experiments and references daily. In this notebook, you should record the sources of your materials, the protocols used for your experiments, the raw and derived data, and any references. At the same time, write down any ideas that come to you when looking at your notebook or searching the literature. An efficient way to keep track of your ideas and sources is writing them down as bullet points.

When you are ready to write, gather your materials—notebooks, figures, tables, notes, coffee. While some writers are aware of key components and findings from the beginning, others do not notice important facts until they actually begin to write. Regardless of what approach works for you, you will come to realize that you need to organize your material. Laying out the figures and tables will help you start constructing your story.

At the same time, arrange and rearrange your bullet points to determine their best possible presentation. If you have an outline, you may also

---

 Example 7-6   *Bullet Points*

- Fertile soil—supports life
- Plants require macro-, secondary, and micronutrients
- Macronutrients = nitrogen, phosphorus, and potassium—in large amounts
- On very acidic soils, phosphorus forms insoluble complexes
- Secondary nutrients = calcium, magnesium, and sulfur—in lower quantities
- Micronutrients = boron, copper, chlorine, iron, manganese, molybdenum, and zinc—only small amounts
- Soil depletion is usually gradual
- Best soil drains well and retains water
- Ideal soil acidity is between pH 6–7
- Rebuilding fertile soil takes patience
- Crop yields and their nutritional value are reduced if soil is depleted
- Nutrients and organic matter are removed with every harvest
- Fertile soil requires microorganisms to support plant life
- Decomposed plant and animal material are best for replenishing nutrients
- Fertile soils are high in organic matter
- Addition of organic material can improve soil structure
- Organic matter/topsoil improves soil structure and retention of moisture
- Good soil contains many microorganisms that support plant growth
- Depletion = if soil fertility components are removed and not replaced

---

try to fill them into this outline. An example of this approach is the following collection of information on soil fertility.

A collection of bullet points, such as shown in the preceding example, can then be grouped in various ways, for example, by topic and subtopics, by pros and cons, or chronologically. You as the author will need to decide which arrangement will best get the main message of the paper across.

In the preceding example, the information can be roughly divided into two main topics, fertile soil and soil depletion, as shown in the following revised example.

Revised Example 7-6a

Fertile Soil

- Fertile soil—supports life
- Fertile soils are high in organic matter
- Plants require macro-, secondary, and micronutrients
- Macronutrients = nitrogen, phosphorus, and potassium—in large amounts
- On very acidic soils, phosphorus forms insoluble complexes
- Secondary nutrients = calcium, magnesium, and sulfur—lower quantities
- Micronutrients = boron, copper, chlorine, iron, manganese, molybdenum, and zinc—only small amounts
- Fertile soil requires microorganisms to support plant life
- Best soil drains well and retains water
- Ideal soil acidity is between pH 6–7
- Good soil contains many microorganisms that support plant growth

Soil Depletion

- Crop yields and their nutritional value are reduced if soil is depleted
- Soil depletion is usually gradual
- Nutrients and organic matter are removed with every harvest
- Decomposed plant and animal material are best for replenishing nutrients
- Addition of organic material can improve soil structure
- Rebuilding fertile soil takes patience
- Organic matter/topsoil improves soil structure and retention of moisture
- Depletion = if soil fertility components are removed and not replaced

Now that we have somewhat more of an order of the information into topics, you may see other ways to subdivide these topics further, as shown in Revised Example 7-6b

Revised Example 7-6b

Fertile Soil

- General information on fertile soil
  ○ Fertile soil—supports life
- Fertile soil content
  ○ Fertile soils are high in organic matter
  ○ Best soil drains well and retains water
  ○ Ideal soil acidity is between pH 6–7
  ○ Plants require macro-, secondary, and micro-nutrients

- ○ Macronutrients = nitrogen, phosphorus, and potassium—in large amounts
- ○ Secondary nutrients = calcium, magnesium, and sulfur—in lower quantities
- ○ On very acidic soils, phosphorus forms insoluble complexes
- ○ Micronutrients = boron, copper, chlorine, iron, manganese, molybdenum, and zinc—only small amounts
- ○ Fertile soil requires microorganisms to support plant life
- ○ Good soil contains many microorganisms that support plant growth

Soil Depletion

- General information on soil depletion
  - ○ Depletion = if soil fertility components are removed and not replaced
  - ○ Soil depletion is usually gradual
- Causes of soil depletion
  - ○ Crop yields and their nutritional value are reduced if soil is depleted
  - ○ Nutrients and organic matter are removed with every harvest
- Rebuilding fertile soil
  - ○ Decomposed plant and animal material are best for replenishing nutrients
  - ○ Addition of organic material can improve soil structure
  - ○ Rebuilding fertile soil takes patience
  - ○ Organic matter/topsoil improves soil structure and retention of moisture

---

Once your bullets are grouped into their corresponding topics and subsections and have been arranged by importance or chronologically, you can then write these ideas in paragraph form (see also Chapters 3 and 6 on how to construct good paragraphs and how to consider word and sentence location). At any stage, do not be afraid to add or remove information from your lists if needed.

Our example on soil fertility could be composed into the following first draft.

Revised Example 7-6c

Fertile soil supports life and is high in organic matter. Its ideal soil acidity lies between pH 6–7. It drains well but at the same time is able to retain water.

To support plant life and growth, microorganisms need to be present in the soil. Plants also require macro-, secondary, and micronutrients. Macronutrients are needed in large amounts and include nitrogen, phosphorus, and potassium. Secondary nutrients are needed in lower quantities. They include calcium, magnesium, and sulfur. Micronutrients are needed only in small amounts. They include boron, copper, chlorine, iron, manganese, molybdenum, and zinc.

Soil depletion occurs if soil fertility components are removed and not replaced. Depletion is usually a gradual process and happens because nutrients and organic matter are removed with every harvest. If soil is depleted, crop yields and their nutritional value are reduced.

Rebuilding fertile soil takes patience. Whereas addition of organic material can improve soil structure and moisture retention, decomposed plant and animal material are best for replenishing nutrients.

In this example, the point "On very acidic soils, phosphorus forms insoluble complexes" has been taken out, as it did not fit into the paragraph well thematically (i.e., the paragraph is about soil fertility under ideal conditions of pH 6–7; acidic soils would not be ideal). In addition, the first subtopic "General information on fertile soil" contains only a very short first sentence and paragraph, for which another sentence or two may be needed to balance it better. Alternatively, the first paragraph could also be combined with the second paragraph, as shown.

After you have written your paragraphs based on the data and information you collected, arrange the resulting paragraphs according to your tentative outline. Sometimes you may find yourself rearranging your outline. That is okay. Work until you come up with a logical sequence for your paragraphs that ensures a smooth flow of ideas from one paragraph to the next. Subsequent revisions (described in Chapter 17) will serve to fine tune your draft.

7.6   WRITER'S BLOCK?

If you have trouble putting your thoughts into words or finishing your manuscript, you may suffer from writer's block. Most people experience writer's block as a temporary condition in which they may not be able to think of a word or have trouble putting their thoughts into a sentence. For some people, this condition can be a serious impairment. Writer's block is probably responsible for thousands of unwritten or unpublished manuscripts.

Although native English speakers also experience writer's block, if you are an ESL author you most likely are even less confident about writing in English. Lack of confidence can create a major roadblock—but the problem is not without solution. See the following suggestions that can help you get over writer's block.

ESL advice

How to Conquer Writer's Block

1. **Know that you are not alone.**
2. **Organize your materials**—your notebook, your figures and tables, your references.
3. **Buy yourself an ugly, cheap notebook.** Pretty notebooks often make people feel as if they have to write and draw polished things in them. Ugly notebooks let you write down anything that comes to your mind, even things that do not make sense. They also let you try out different figures and tables.
4. **Make your own rules.** Do not follow any preset rules on where or how to start writing. Start wherever you want: the Abstract, the Conclusion, the Reference List. Then, write what you want, not what you think you should.
5. **Use provided sample sentence fragments** as starting points for the different parts of your paper—just add or substitute your data and ideas.
6. **Learn to imitate.** Look at how other authors in the field have worded their Introduction, Materials and Methods, or Results, but do not copy them word by word. Instead, use their ideas and paraphrase.
7. **Write as if you are explaining your work to a friend or parent.** Be informal and use simple words. Do not write as if you are trying to impress someone in your field.
8. **Write one paragraph and one section at a time.** Often, people feel overwhelmed thinking of all the writing in a paper at once.
9. **Set yourself some deadlines** or have someone else set them for you. Many people do not get tasks done unless they feel some pressure, be that self-imposed or from a superior. Setting weekly or daily objectives can help to finish tasks. Be aware that you probably need to double any estimated time for completing a writing task.
10. **Involve your coauthors.** Ask them for comments on drafts. Also, ask them to help in writing different parts of the manuscript, although only one author should put the final version together to ensure consistency in style and format.
11. **Percolate.** Think about your ideas. Daydreaming is often your brain hard at work. Ideas and theories many times need time to be formed.

7.7   FOR ESL AUTHORS

➤ Recognize different cultural practices in publishing

ESL advice

In some departments or foreign countries, it may be common practice to always include the head of the department on a paper. In other cultures, the principal investigator may even expect to be placed in the first author position, possibly for financial or political reasons. The first author place, however, should preferably go to a student or postdoctoral fellow if he or she is the main person drafting the manuscript. These early-career people usually not only did most of the work but also still have to establish themselves and will have a much better chance of doing so if they receive the recognition due to them. Furthermore, people in your field will know who the principal investigator is, and those not in your field will expect to find that principal investigator's name at the end of the author list.

➤ Be careful in applying the same principles of style that are used in your own language

If you are an ESL author (if English is not your first language), choose one of the following methods to write your manuscript:

1. Write the paper in English to the best of your ability (the best method).
2. Write the first draft in your own language and translate it into English yourself.
3. Employ a translator who is familiar with the terminology of your field.

Whichever method you choose, do not apply the same principles of style when you write in English as are used in your language. If you need to, borrow technical phrases such as the ones provided in this book—but never copy whole sentences or paragraphs from articles by English or American scientists in well-edited journals. Instead, learn to imitate the style and wordings of others. (See also Chapter 8 on plagiarism and paraphrasing.)

Ask for help from a scientific colleague, friend, or correspondent whose native language is English. Have a native speaker—preferably a scientist—check any translations. Editors and reviewers will gladly correct minor mistakes, but the English must first be good enough for them to understand what you are trying to say.

7.8   OUTSIDE HELP

Although editors of journals may correct minor mistakes, they usually will not write or edit an article for you. You have to take care of this problem yourself—for grant proposals as well as articles. There are three solutions to writing and editing your manuscript or proposal—all of which may be done, ideally in combination, before you submit your document.

➤ Hire a scientific editor

You may consider asking a scientific editor to write your article outright (this may be quite expensive, as this person will have to do much research before being able to write). A much better approach is to ask a scientific editor to revise your manuscript professionally for style, format, composition, and impact, possibly even content, *after* you have completed your manuscript as much as possible. Ideally, this person should be an editor *and* a PhD scientist within your field or a related one. Native-speaking

ESL advice

English colleagues make good editors only to a point. They may be invaluable for correcting grammar, spelling, and punctuation, but unless they have done extensive editing and are experts in scientific writing, they often fail when it comes to style, format, or impact. Here, a trained *scientific editor* can provide expert help. (Distinguish between *science editors/writers* and *scientific editors/writers*. Science editors/writers are trained mostly in writing or journalism with an emphasis on science, but they generally do not hold a PhD degree in a scientific field while scientific editors do). Scientific editors may be hard to come by but are usually well worth your money.

Ask your peers if they could recommend an editor, or inquire with the American Medical Writers association or the Council of Biology Editors for a list of scientific editors. Some journals also have a list of scientific editors you can approach, or you may find one on the Web by searching under the appropriate key words, possibly together with names in the acknowledgement sections of published articles.

When you work with an editor, view the revision process as a joint effort. Involve the editor well ahead of any deadline, particularly for grant proposal applications. Agree on the level of editing and on charges before you begin to work with the editor.

➤ Collect sample phrases for references

The best source of such sample phrases are articles that have been written in highly known and respected scientific journals by native speakers within the past five years. Again, I am not talking about copying entire passages to be placed into your manuscript but rather individual sample phrases and expressions that can be applied to writing your research article. For example, performing a polymerase chain reaction (PCR) experiment can be described in only so many ways (see also Chapter 8). When you are unsure how to word such an experiment in English, use your collected sample phrases to see how native speakers in known journals describe such an experiment.

➤ Consult books on scientific/technical writing

Consult books on scientific/technical writing for problems of grammar, technical style, good format and composition, and possible sample sentences. This book, for example, presents ample practical sample phrases and wordings for scientific authors and addresses such authors within the international scientific community. Other useful books, especially for native speakers of German or Italian, respectively, are *Langenscheidt Scientific English für Mediziner und Naturwissenschaftler* or *Scientific English—L'inglese scientifico per relazioni e conferenze in medicina, biologia e scienze naturali*. Additional books are described in the Bibliography.

SUMMARY

GUIDELINES FOR PREWRITING AND DRAFTING
- Search journals for the best match of topic, audience, and journal.
- Obtain *Instructions to Authors* and follow them.
- Decide on authorship before starting to write.
- Follow the IMRAD format.
- Collect, organize, and study your references.
- Decide on the best way to create, share, and edit your documents.
- Start by writing less.
- Worry about basic rules, principles, and guidelines in the revisions.
- Pay attention to the overall organization of your document.
- Organize your data or ideas before writing.

TO OVERCOME WRITER'S BLOCK
- Know that you are not alone.
- Organize your materials.
- Use a cheap, ugly notebook.
- Make your own rules.
- Use provided sample sentences as starting points.
- Learn to imitate.
- Write as if you are explaining your work to a friend.
- Write one paragraph at a time.
- Set yourself some deadlines.
- Involve coauthors.
- Percolate.

FOR ESL AUTHORS
- Recognize different cultural practices in publishing
- Be careful in applying the same principles of style as are used in your own language.

OUTSIDE HELP
- Hire a scientific editor.
- Collect sample phrases for references.
- Consult books on scientific/technical writing.

PROBLEMS

PROBLEM 7-1 Outline
Use the following bullet points to create a meaningful outline for a few paragraphs on polar ice reduction.

- Global warming has been a major concern over the past decades

- Ocean levels have risen 15–20 cm in the past century
- By 2100, ocean levels are expected to rise another 13–94 cm (Haug, 1998)
- As our climate warms, ocean temperatures rise
- Water expands at higher temperatures, leading to sea level rise
- In previous interglacial periods, polar ice was at least partially melted and climate conditions were similar to current conditions on Earth
- It has been suggested that in some interglacial periods, ocean levels were more than 20 m higher than now (Haug, 1998)
- Past climate conditions and sea levels are disputed (3–8)
- Microfossil and isotopic data from marine sediments of the Cariaco Basin support the interpretation that global sea level was 10 to 20 m higher than today during marine isotope stage 11 (Poore and Dowsett, 2001)
- Polar ice melt increases temperature rise on Earth; whereas ice reflects sunlight, oceans absorb light and heat
- Glacial-interglacial cycles are affected by variations in the Earth's orbit
- As temperature rises, glaciers and ice sheets melt
- Temperature rise has been accelerated by humans

PROBLEM 7-2 Outline
Use the following bullet points to create a meaningful outline for a few paragraphs on green chemistry: Fischer–Tropsch refining.

- Crude oil is not an infinite resource
- At present, transportation fuels are primarily produced from crude oil
- Liquid fuels can be produced from non-crude-oil carbon sources by direct means, such as direct coal liquefaction, or indirect means, such as biomass gasification followed by hydrocarbon synthesis
- Current transportation sector of the economy is carbon based
- Fischer–Tropsch (FT) synthesis is a key technology for gas-to-liquid (GTL), coal-to-liquid (CTL), and biomass-to-liquid (BTL) conversion
- Assumption: Fischer–Tropsch syncrude can be refined similarly as crude oil
- Syncrude has to be further refined to produce transportation fuels
- There are no specific refining technologies for Fischer–Tropsch syncrude (de Klerk, 2009)
- Little attention has been paid to the refining of Fischer–Tropsch syncrude
- When syncrude is treated as if it is crude oil, refining becomes inefficient, and it violates green chemistry principles

- Responsible syncrude refinery design is important from a commercial and an environmental point of view
- Alternate carbon sources will be important sources for transportation fuel
- Syncrude = a synthetic oil produced by Fischer–Tropsch synthesis
- There are significant differences between Fischer–Tropsch syncrude and crude oil
- Refining crude oil can be less economical than refining syncrude (de Klerk, 2009)

PROBLEM 7-3 Constructing a Paragraph

Compose a short passage using the following bullet points about DEET (N,N-diethyl-meta-toluamide):

- Most common active ingredient in insect repellents
- Developed by the U.S. military in the 1940s
- A few reports of adverse effects in humans (Osimitz and Grothaus, 1995; Sudakin and Trevathan, 2003)
- Easily absorbed into the skin
- Considered safe if used correctly
- Heavy and frequent dermal exposure can lead to skin irritation and in rare cases, death, especially in young children (Osimitz and Grothaus, 1995; Sudakin and Trevathan, 2003)
- Neurological damage and death can result from ingestion (Osimitz and Grothaus, 1995; Osimitz and Murphy, 1997)
- Alternatives: eucalyptus oil containing cineol, 1:5 parts diluted garlic juice, soybean oil, neem oil, marigolds, Avon Skin-so-soft bath oil mixed 1:1 with rubbing alcohol, juice and extract of Thai lemongrass
- True citronella has not been proven to be a very effective mosquito repellent (Revay, 2013)

See Chapters 3 and 6 for additional problems on paragraph construction.

CHAPTER 8

References and Plagiarism

Most of the time, your scientific findings will build on previous studies. Although direct quotations are rarely used in scientific writing, paraphrased versions of source material are very common.

THIS CHAPTER PROVIDES DETAILS ON:

- Selecting, managing, and using references
- Formatting citations and reference lists
- Placement of citations
- Plagiarism
- Paraphrasing
- Reference styles and citations from the Internet

8.1 ABOUT REFERENCES

➤ Reference the ideas and findings of others

Whenever you use the ideas and findings of others, you must cite the source in the text as an in-text citation as well as list it in a Reference List at the end of a research article. References not only give appropriate credit to the contributions of others but also direct readers who want further information to other literature of interest. In addition, references provide editors with a list of potential reviewers and show how familiar you are with your area of specialty.

TABLE 8.1 Definitions and examples of primary, secondary, and tertiary sources

SOURCE	DEFINITION	EXAMPLE
Primary	Original, peer-reviewed publication of a scientist's new data, results, and theories; report results for the first time	Scientific journal articles; theses and dissertations; conference proceedings; speeches
Secondary	Analyze and discuss the information provided by primary sources	Review articles; literary criticisms; some textbooks; commentaries
Tertiary	Compile and reorganize information provided in mainly secondary sources	Textbooks (some may also be secondary); dictionaries; manuals; Wikipedia

Primary, Secondary, and Tertiary Sources

➤ Differentiate between primary, secondary, and tertiary sources

To ensure that specific findings have been accurately conveyed, use a primary source, which is the original, peer-reviewed publication of a scientist's new research and theories. For general overview of a topic, you may also use secondary or tertiary sources (see also Table 8.1). A secondary source, such as a review article, cites, builds on, discusses, or generalizes primary sources. A tertiary source, such as a textbook or dictionary, generalizes and analyzes primary sources while attempting to provide a broad overview of a topic (see also Chapter 19, Section 19.2). Keep in mind you may have to obtain copyright permission to use certain material for republication. Such permissions can often be obtained either from the publisher or author directly or, in case of most journal articles, from the Copyright Clearance Center (http://www.copyright.com). Permissions may be free but may also have to be purchased, depending on who holds the copyright, what portion you plan to use, and for what purpose you intend to use it.

8.2   SOURCE MATERIAL

➤ Select the most relevant references

Most authors identify more than enough references for their manuscripts. The more difficult part is to separate immediately relevant references from all the rest (see also Chapter 1, Section 1.2, and Section 8.1). Potentially relevant references (traditional and open access) can easily be obtained by searching online databases such as MEDLINE®, SCOPUS®, BIOSIS, and the Web of Science using, for example, the author's name, the source, the title, or year.

There are many academic search engines, but some are used more than others. The top 11 science databases on a range of topics from engineering to the natural sciences include:

PubMed	PubMed is a free database developed by the National Center for Biotechnology Information. It contains more than 22 million citations for biomedical literature from MEDLINE, life science journals, and online books. PubMed exists as an iTunes app, **PubMed on Tap** by ReferencesOnTap, which allows you to search for, store, email, and organize references. You may even be able to read some full-text articles.
MEDLINE®	MEDLINE, produced by the National Library of Medicine, covers more than 4,000 journal titles and is international in scope. Broad coverage includes basic biomedical research and clinical sciences. It can also be accessed through a popular app: **PubMed—Unbound MEDLINE.**
SCOPUS®	SCOPUS provides broad international coverage of the sciences and social sciences, indexing 14,000 journals. An alert app exists under **SciVerse Scopus Alert**.
CINAHL	CINAHL covers English-language journals in nursing and allied health fields, indexing 1,200 journals. It is accessible by mobile devices through **CINAHL Plus**.
PsycINFO®	The PsycINFO database covers psychology and related disciplines. Coverage is worldwide, indexing over 1,300 journals and dissertations. See **PsycINFO** for the mobile app.
BIOSIS	BIOSIS is a research database that contains literature references from all of the life sciences with multidisciplinary coverage. The database contains journal articles, books and book chapters, selected US patents, and conference literature. See **BIOSIS Previews** for the app for a mobile menu.
Web of Science	The ISI Citation Databases collectively index more than 8,000 peer-reviewed journals. It provides Web access to Science Citation Index Expanded, which covers 6,300 international science and engineering journals. **Thomson Reuters RefScan**ˢᴹ, powered by Web of Knowledge and EndNote, allows you to scan and capture references directly from Web of Knowledge, and save them to an EndNote Web account.
Current Contents®	Current Contents Science Edition covers all the Science editions of the Current Contents Search® database in one package. Current Contents is a database that is updated weekly and provides access to tables of contents, bibliographic information, and abstracts from the most recently published leading scholarly journals.
Scitation	This database gives online access to publications from top physical sciences publications. It includes journals, conference proceedings, and some blogs. It also contains metrics measuring article impact over time.

| **JSTOR** | JSTOR is a full-text database that gives access to select, leading scholarly journals in a variety of fields, including the life sciences, although the most recent five years are typically not included. |
| **Environment Complete** | Environment complete is a database that provides comprehensive coverage on environment-related subjects, from ecology to public policy. It contains full text articles for 950 journals. |

A more comprehensive list of science databases, ordered by field, can be found at **https://guides.library.georgetown.edu/MajorScienceDatabases**. Other useful sources that have gained popularity in the last few years as a way to search for scientific literature include online databases such as **Google Scholar, Research Gate, Cite Seer, GetCITED, SciFinder, BioOne, ScienceDirect, and the *New Journal of Physics***. Although there may be concern about peer review, databases like Google Scholar have good search capabilities and meta-database coverage. Google Scholar's "instant cite" feature provides a bibliographic entry in many of the most common reference styles (MLA, APA, and Chicago style).

Dissemination of research articles in the form of open access scholarly journals is increasingly common as it allows for faster publication and a potentially larger readership than traditional journals. As for traditional journals, the quality of open access journals varies. Many are not yet well-established or well-known, and thus their impact factor may be unclear or they may not be peer-reviewed. For a directory of reliable, peer-reviewed open access journals, see **https://www.omicsonline.org/top-best-open-access-journals.php** or the Directory of Open Access Journals at **https://doaj.org**. You can also check if the journal belongs to the Open Access Scholarly Publishers Association, for which members undergo a thorough screening process. In addition, like traditional journals, open access journals listed in the Web of Science or similarly respected science databases are typically reputable. Many traditional journals now also make their publications available through open access for an additional charge to the authors, thus providing the best of both: larger accessibility and the reputation of the traditional journal.

Unlike review articles, which contain many references because they have to cover extensive information, research papers should contain only immediately relevant references so as not to overwhelm the reader (20–40 on average). To keep the number of references low, cite original articles and select the most important, the most elegant, or the most recent paper on a subject rather than listing any and all papers published on the topic. Consider citing review articles where possible. Be aware, however, that review articles sometimes contain faulty references and misrepresent or misinterpret data. Always verify the original article if you are citing a specific piece of information from a review article or another tertiary source.

The most relevant references consist of the most significant and the most available references. Such references are generally journal articles followed by books. Although PhD dissertations may also be quoted, they are typically not widely available. Similarly, abstracts for meetings, conference proceedings, personal communications, and unpublished data may be cited in the text in parentheses, but these references are usually not listed at the end of a paper because they are not considered relevant or available. Thus, they should not be used to draw any strong conclusions, only to support findings.

Verifying References

➤ Verify your references against the original document

The citations within the text, the Reference List, and the information you cite must be accurate. Citations and reference lists tend to have a surprisingly high rate of error. Therefore, you need to verify them against the original document. Make sure you have read all references you cite to prevent false representation and that you present the authors' intentions and meaning correctly when you paraphrase (see also Section 8.6). In addition, ensure that every reference in the text is included in the Reference List and every reference in the Reference List is cited in the text. Often, citations and reference lists get out of sync when documents are rewritten or reformatted for a different journal. Use a good reference managing program to help you avoid this problem. Ensure also that citations and references follow the required format as described in the *Instructions to Authors*.

8.3   MANAGING REFERENCES

➤ Manage your references well

Computer programs that help you manage your references are very useful. These include, but are not limited to, EndNote, BibTeX, Zotera, Mendley, RefWorks, or ReferenceManager. Open source reference managers such as the Mozilla Firefox plugin are also available. I strongly recommend using one of these programs to help organize, keep track of, and format your references. You can save yourself much time and frustration if you manage your references from the start. Few aspects of preparing a manuscript are more irritating than painstakingly typing, changing, or correcting the Reference List. Start using a reference managing program right when you download your references from the library. If you are unfamiliar with such a program, inquire at your library. Most libraries offer short classes on reference programs.

In brief, these software systems allow you to download the references from online search engines into your personal database. For example, you can download directly from PubMed, which minimizes errors in the Reference List/Bibliography. There is typically a plugin for your word processing software (e.g., MS Word). When you are working on your manuscript, you just need to click to add a specific reference or group of references and keep writing. This feature will minimize the interruption of the writing process. The software will automatically add a field for the in-text citation (see Section 8.4) and begin building the Reference List/Bibliography (see Section 8.8). Journals typically have different formats for in-text and the Reference List/Bibliography, and these reference programs can quickly and easily adjust and change styles. The online version of EndNote, "EndNote Basic" (http://endnote.com/product-details/basic) is free and also includes video instructions (https://www.youtube.com/user/EndNoteTraining). When you are ready to submit, the reference programs can also remove the fields (as per instructions by some journals).

8.4   TEXT CITATIONS

Form and Order

➤ **Follow the journal's style for details in the reference citation**

➤ **Cite references in the correct form and order**

Text citations list references within the text in short version, such as by number or by name and year. Some common formats for text citations are "(author, year)," "(number)," and "[number]." Two examples of text citations are shown in Examples 8-1a and 8-1b.

Example 8-1   **a**   Too little sleep (less than 6 hours) or too much sleep (more than 9 hours) leads to lower life expectancy **(Miller, 2014).**

b   Too little sleep (less than 6 hours) or too much sleep (more than 9 hours) leads to lower life expectancy **(8).**

Generally, for a paper by one author, cite that author's name when you want to highlight it.

Example 8-2   **a**   . . . described by Popi (20).

For a paper by two authors, cite both authors' names.

Example 8-2   **b**   Daniles and Ebert (9) reported XYZ.

For a paper by three or more authors, cite the first author's name followed by "et al."

Example 8-2   **c**   . . . has previously been reported (Brown et al., 2009a, 2009b; Liu et al., 2013).

(Note that citation rules may differ depending on the style of your target journal. For example, if your target journal uses APA style, "et al." is only used for citing articles of six or more authors in text citations. Thus, be sure to check the *Instructions to Authors* as well as other articles published in your target journal carefully.)

If you cite multiple references for a point in your text and your text citations are by name and year, cite the references chronologically in the text and by alphabetical order in the Reference List.

Example 8-2   **d**   . . . as described in earlier reports (Popi et al., 2010; Lopez et al., 2011).

Reference List:
   Lopez, J., Holz, B., and Homami, H. 2011 . . .
   Popi, J. M., Kalt, A. and Heis, B. 2010 . . .

If your text citations are by number, list multiple references in the text and in the Reference List in numerical order.

Example 8-2   **e**   . . . as described in earlier reports (21, 22, 23).

Reference List
   . . .
   21.   Popi, JM, Kalt, A., and Heis, B. 2010. . . .
   22.   Lopez, J., Holz, B., and Homami, H. 2011. . . .
   23.   . . .

If you cite unpublished data, abstracts for conference proceedings, or personal communications, you should give a publication status in parentheses in the text. Following are a few sample wordings for indicating publication status:

(manuscript submitted)   (manuscript in preparation)
(unpublished data)   (data not shown)
(personal communication)   (manuscript in press)

Purpose of Using References

As a scientific author, you will use sources for varying purposes: to support, refute, compare with, or highlight your own findings or ideas.

Table 8.2 provides you with sample wording when using references for various purposes.

Tone and Style

➤ Remain objective and neutral when citing the work of others

At all times, be professional and courteous when citing other's work. Above all, do not insult the author(s) of previous studies.

Example 8-3   **a**   Unbelievably, Brown failed to consider . . .

b   The study by Aday is without merit.

c   Clearly, Chumsky's theory is wrong.

TABLE 8.2 Use of citation

	TO SUPPORT YOUR FINDINGS (OR THOSE OF OTHERS)	TO REFUTE FINDINGS OF OTHERS	TO COMPARE YOUR FINDINGS WITH OTHERS	TO HIGHLIGHT YOUR FINDINGS
Sample wording	Our results are consistent with those reported previously (5, 8, 10–14).	Different findings have been reported previously (33).	Our findings are comparable to those of Vignanery's (9).	Unlike other previous findings (23–25), our work presents . . .
	Similar results have also been observed by Alton et al. (11).	Our findings show that the previously proposed theory A (17) is not supported by . . .	When compared to recent findings described by (8), our study shows . . .	Our study expands on work of Hui's (33), which reported Y, and highlights the importance of . . .
	Mayer and Bims also determined . . . (Here we confirm Lohn's theory of . . .)	Not all studies agree on this finding. For example, Peters et al. determined . . . (5).	Although Dauh et al. reported recently XXX (6), our study also takes into consideration . . .	Our study sets itself apart from other studies (33–35) in that . . .

Insults will get you nowhere, and you run the risk of having your work rejected if you write nonprofessionally and annoy your editor or reviewers. Remain objective and neutral as in the next examples.

Revised Example 8-3

a   Different theories on this topic exist (27–29).

b   Our study differs from that of Aday (35) in that . . .

c   X has been the topic of much controversy (4, 5, 8).

Placement of Citations

➤ Know where to place references in a sentence

Incorporation of references into the text should be meaningful and done with care. There are two general ways to cite in the text: (a) to emphasize the science, place the citation directly following a concept, idea, or finding; and (b) to emphasize the scientist, place the citation directly following the names of the authors. Citations are typically not accepted anywhere else in the sentence.

Example 8-4

a   Starfish fertilization is species specific (17).

b   Peterson **(17)** reported that starfish fertilization is species specific.

Do not place references in the middle of an idea or after general information of a study, such as after "in a recent study" or "has been reported."

👎	**Example 8-5**	In a previous study (16), bald eagles have been found to increase in size the farther away from the equator and the tropics.
👍	**Revised Example 8-5**	In a previous study, bald eagles have been found to increase in size the farther away from the equator and the tropics. **(16)**.

Often, citations can be found at the end of a sentence when the concept, idea, or finding refers to the entire sentence. Differentiate, however, between your findings and those of others by placing text citations after each corresponding point, as in the following examples.

👎	**Example 8-6**	a	The A virus used in this study was obtained from the supernatant of the 25-3 cell line confluently infected with it [3].
👍	**Revised Example 8-6**	a	The A virus **(3)** used in this study was obtained from the supernatant of the 25-3 cell line confluently infected with it.
👎	**Example 8-6**	b	Compound A can be separated from the mixture by two methods: distillation and HPLC (Koehler et al., 2004; Ramos et al., 2011; Smith et al., 2013).
👍	**Revised Example 8-6**	b	Compound A can be separated from the mixture by two methods: distillation **(Ramos et al., 2011; Smith et al., 2013)** and HPLC **(Koehler et al., 2004)**.

Also, note that references for different points in one sentence have to be cited after the appropriate point rather than grouping all the references together at the end of the sentence.

👎	**Example 8-7**	Three different types of postage stamp adhesives exist: one is gum arabic, the calcium salt of Arabic acid, a water soluble exudate of the acadia tree; another is dextrins, modified starches that are more water soluble than the starches from which they were derived; and the third is polyvinyl alcohols, hydrolyzed PVA resins that are water-soluble (3, 4, 8).
👍	**Revised Example 8-7**	Three different types of postage stamp adhesives exist: one is gum arabic, the calcium salt of Arabic acid, a water soluble exudate of the acadia tree **(3)**; another is dextrins, modified starches that are more water soluble than the starches from which they were derived **(4)**; and the third is polyvinyl alcohols, hydrolyzed PVA resins that are water-soluble **(8)**.

8.5   PLAGIARISM

➤ Ensure that you are not plagiarizing

In scientific writing, direct quotations are rarely used. Instead, information is commonly summarized and paraphrased. In all cases, the source has to be cited. Failing to indicate the source of information in scholarly scientific work is called plagiarism and is a form of academic misconduct. To give credit to the work and ideas of others, you need to acknowledge your sources, even if the writing is not absolutely identical.

To avoid plagiarism, you need to know what it is. Plagiarism includes:

- Using material without acknowledging the source. (This lack of ac-knowledgment is the most obvious kind of plagiarism.)
- Borrowing someone else's ideas, concepts, results, and conclusions and passing them off as your own without acknowledging them—even if these ideas have been substantially reworded.
- Summarizing and paraphrasing another's work without acknowl-edging the source.

The rules apply to both textual and visual information, such as figures, schematics, slides, pictures, and the like. If you are using the Web as a source of information, you must also cite that source (see Section 8.10). If information is copyright protected, you need to obtain permission from the website's owner before using graphics or text for dissemination or republication.

You do not have to document facts that are considered common knowl-edge. Common knowledge is information that can be found in numerous places and is likely to be known by a lot of people, such as the information found in Example 8-8.

    Example 8-8   A vast number of endemic species exists on the Galapagos Islands.

However, information that is not generally known (i.e., information readers outside your discipline would need to look up) and ideas that interpret facts have to be referenced as in Example 8-9.

    Example 8-9   Based on the results of a recent study, the blue iguanas of the Grand Cayman Islands are an endangered species (9).

The finding that "blue iguanas of the Grand Cayman Islands are an endangered species" is not a fact but an *interpretation* as it is "based on the results of a recent study." Consequently, you need to cite your source to show that an actual study has been done and that this study has been accepted as fact in science. Similarly, if you are uncertain whether something falls into the common knowledge cate-gory or if you have to look it up, it is best to document it. Note that information obtained from open access sources follow these same rules: They need to be referenced if it is information that is not common knowledge.

Following are some other examples of common knowledge that do not require a citation.

Example 8-10   a   As phosphorus is a key element for plant growth and is essential for many cell functions, the cycling of phosphorus in the soil has been studied widely.

b   Volcanic eruptions are almost always preceded and accompanied by "volcanic unrest," providing early warning of a possible impending outbreak.

c   Like reptiles, amphibians such as frogs and toads molt regularly.

Statements that contain information and interpretations that result through the work of others need to be cited, as shown in the next examples.

Example 8-11   a   Although low concentrations of phosphorus are often a limiting factor in plant growth, excess phosphorus in the soil is correlated with decreased plant health (19).

b   The eruption of Mount Pinatubo in 1991 was preceded by a relatively short progression of precursory activity before its full-blown outbreak (22).

c   Unlike reptiles, amphibians such as frogs and toads consume their old skin after molting (9).

Writing about the ideas and conclusions of others is a given in science. It is not considered plagiarism to do so, as long as you acknowledge the source in your document. If you cannot verify an original source, the information should not be stated or should be clearly identified as unverified, unpublished, or an opinion (see also Chapter 1, Section 1.2).

It is easy for authors to lose track of cited and verbatim text in a larger work or document, particularly one composed over a longer period. In some cultures, the concept of plagiarism may also not exist or may be much looser than in the Western world. To verify that text is plagiarism-free, software such as **Turnitin** and **PlagScan** and apps such as **Plagiarism Checker** by Plagiarisma.net are available. These tools allow you to screen your papers for plagiarism. Be aware, though, that aside from plagiarism, other forms of ethics violations may also arise. Such ethics violations may include fabrication of data and results, fudging findings, stealing data, and being asked to include an author on a publication although the researcher did not contribute to the project. For more information on research ethics, see Chapter 1, Section 1.2 or one of the following links:

http://www.au.af.mil/au/awc/awcgate/doe/benchmark/ch16.pdf
http://www.niehs.nih.gov/research/resources/bioethics/whatis/
http://www.niehs.nih.gov/research/resources/bioethics/glossary/index.cfm

For details on how to report on ethical conduct in papers, see Chapter 12, Section 12.5.

8.6   PARAPHRASING

➤ Know how to paraphrase

To paraphrase is to express someone else's words, thoughts, or ideas in your own words. Learning how to paraphrase is probably one of the most important skills in scientific writing. In science, you usually have to build on the work and ideas of others, but you need to paraphrase them and reference their work.

It is important that you distinguish between paraphrasing and plagiarizing. Changing a word or two in someone else's sentence or changing the sentence structure while using the original words is not paraphrasing but plagiarizing. The difference between paraphrasing and plagiarizing is outlined in the following example.

☞   **Example 8-12**   **Plagiarized Sentence**

Original:
Grizzly bears (*Ursus arctos* ssp.) encompass all living North American subspecies of the brown bear: the mainland grizzly (*Ursus arctos horribilis*), the Kodiak (*Ursus arctos middendorffi*), and the peninsular grizzly (*Ursus arctos gyas*), but none of the giant brown bear subspecies found in Russia, Northern China, and Korea.

Plagiarized sentence:
Grizzly bears (*Ursus arctos* ssp.) consist of the North American subspecies of the brown bear, including the mainland grizzly (*Ursus arctos horribilis*), the Kodiak (*Ursus arctos middendorffi*), and the peninsular grizzly (*Ursus arctos gyas*), but not the subspecies found in Russia, Northern China, and Korea.

Paraphrased sentence:
Only the three North American brown bear subspecies *Ursus arctos horribilis, middendorffi,* and *gyas* are considered to belong to the grizzly bears. (Brown bears inhabiting Siberia and Northeast Asia are another subspecies.)

In the plagiarized sentence of Example 8-12, only a few words have been changed, omitted, or included, such as the change from "encompass" to "consist of" or from "none of the giant brown bear subspecies" to "not the subspecies." In the paraphrased sentence of the example, on the other hand, the same general idea is presented in an entirely different sentence from the original one.

Following is another example of a paragraph that instead of being paraphrased has been plagiarized.

Example 8-13   **Plagiarized Paragraph**

Original:
Healthy older adults often experience mild decline in some areas of cognition. The most prominent cognitive deficits of normal aging include forgetfulness, vulnerability to distraction and other types of interference, as well as impairments in multi-tasking and mental flexibility (Albert, 1997; Bimonte, 2003). These cognitive functions are the domain of the prefrontal cortex, the most highly evolved part of the human brain. Prefrontal cortical cognitive abilities begin to weaken even in middle age, and are especially impaired when we are stressed. Understanding how the prefrontal cortex changes with age is a top priority for rescuing the memory and attention functions we need to survive in our fast-paced, complex culture.

Plagiarized sample:
In healthy older adults often some areas of cognition decline. The most noticeable cognitive declines of normal aging include forgetfulness, vulnerability to distraction, and problems in multi-tasking. These cognitive tasks are localized in the prefrontal cortex, which is the most highly evolved portion of the brain. Already in middle age prefrontal cortical cognitive functions start to decrease. Such functions are also particularly affected during any type of stress. Studying memory and attention is important to understand how the prefrontal cortex changes with age. It is particularly important to understand these changes in our current fast-paced lifestyle.

In the plagiarized Example 8-13, no sources are cited. Furthermore, only a few words have been changed in any given sentence. In addition, the sequence of sentences has been reordered, but the sentences have essentially remained the same. For each of these reasons, the derived paragraph is considered to be plagiarized.

An acceptable way of paraphrasing the preceding sample paragraph would be the following revised version of the example.

   Revised   *Paraphrased sample:*
Example 8-13   Studies show that the process of aging is accompanied by a decline in cognitive abilities, deficits in working memory, and compromised integrity of neural circuitry in the brain (Albert, 1997; Bimonte, 2003). If these functions of the prefrontal cortex decline, they affect our thinking and eventually our quality of life. To find ways and potential therapies to counteract this process, it is important to understand the underlying mechanisms of aging on neural circuitry.

Revised Example 8-13 is acceptable paraphrasing because the writer accurately relays the information in the original using his or her own words. The writer also lets the reader know the source of the information.

Example 8-14 is another example.

Example 8-14

Plagiarized Paragraph

Original:
It has currently been recognized that both the type and characteristics of the rust layers formed on the steel surfaces are very important because they can determine their protective properties. According to a recent theoretical model developed by Hoerlé et al. (2), the long-term corrosion behavior of iron exposed to wet–dry cycles is largely controlled by the characteristics of the rust layers. Additionally, the differences between the corrosion behavior observed for different types of steels have been related to the rust layer characteristics. Okada et al. (8) have carefully investigated, by using detailed variable temperature Mössbauer spectrometry, the protective rust formed on both weathering and mild steels after 35 years of exposure to a Japanese semirural type atmosphere. They reported that the rust on both steels is composed of goethite (major component), lepidocrocite (minor component) and traces of magnetite.
(With permission from Elsevier)

Plagiarized sample:
Both the type and characteristics of the rust layers formed on the steel surfaces are very important because they determine their protective properties. Recently, Hoerlé et al. developed a theoretical model (2) that the long-term corrosion behavior of iron exposed to wet–dry cycles is largely controlled by the characteristics of the rust layers. The differences between the corrosion behavior observed for different types of steels have also been related to the rust layer characteristics. Using detailed variable temperature Mössbauer spectrometry, the protective rust formed on both weathering and mild steels after 35 years of exposure to a Japanese semirural type atmosphere have been determined by Okada et al. (8) who reported that the rust on both steels is composed of goethite (major component), lepidocrocite (minor component) and traces of magnetite.

In this example, although the author cited the sources, the passage is plagiarized because the writer has only changed around a few words and phrases. An acceptable, paraphrased version of the same passage would be the following revised example.

Revised Example 8-14

Paraphrased sample:
Rust constituents can determine the performance of steels and influence their life expectancy. The characteristics of the rust layers control the long term corrosion behavior of iron exposed to wet–dry cycles (2). For example, after 35 years of exposure in Japan, rust formed on steels under various conditions is composed of mainly goethite, some lepidocrocite, and traces of magnetite (8).

Following is another example of an acceptably paraphrased paragraph.

Example 8-15

Original:
Zika virus was first discovered in 1947 in the Zika Forest of Uganda.[4] It is spread largely by mosquitoes. Initially, it occurred along the equatorial belt from Africa to Asia. Starting in 2007, the virus spread to the Americas, eventually causing the 2015–16 Zika virus epidemic in South and Central America. In most adults, infection by the virus causes no or only mild symptoms. However, the virus can also be transferred from the mother to her unborn child. In these fetuses, infection by the virus can result in severe brain malformations, known as microcephaly, and other birth defects.

Revised Example 8-15

Paraphrased sample:
Zika virus is transmitted to humans mainly through mosquitoes. Its name derives from the Zika Forest of Uganda, where it was first identified in the 1940s. The virus' spread from Africa and Asia to South America led to the 2015–2016 Zika virus epidemic. Although most infected adults experience comparatively mild symptoms, when the virus is transferred from mother to child *in utero*, infection can lead to microcephaly.

Unlike elsewhere in a scientific research paper, many portions of the Materials and Methods section will sound extremely similar to each other, mainly because there are only so many ways one can describe procedures whose technique and setup is essentially identical. Using very similar phrases in such passages, and substituting your variables, would not be considered plagiarism. Therefore, do not desperately try to invent new wordings to describe the same procedure.

Following are some examples of passages that would not be considered plagiarized.

Example 8-16   **a**   *Method description in paper A:*
Real-time fluorescence quantitative PCR was performed in an Applied Biosystems Prism 7000 instrument in the reactions containing an Applied Biosystems SYBR green master mix reagent and oligonucleotide pairs to the endogenous control gene 'A' and cDNA of 'B'. The reagents were denatured at 95°C for 10 min, followed by 40 cycles of 15 s at 95°C and 60 s at 60°C. The primer sequences (5′–3′) were 'A' forward 5′-GACACCTATGCCGAACCGT GAA-3′; 'A'' reverse, 5′-CTGAGTATCAGTCGGCCTT GAA-3′; 'B' forward 5′-GTTCGACGACATCAA CATCA-3′; 'B'' reverse 5′-TGATGACGTCCTTCTC CATG-3′.

b   *Method description in paper B:*
PCR amplification of 'X' sequences was done using the GC RICH PCR System (Roche, Mannheim, Germany). All non-'X' sequences were amplified using Taq DNA polymerase (Promega, Madison, WI). Primers were designed using published sequences for 'X-1' (GenBank: Xxxxxx) and 'x-13' (GenBank: Xxxxxx) (Table 1). PCR thermal cycling conditions were: 2 min at 50°C, 10 min at 95°C, followed by 40 cycles of 15 s at 95°C and 1 min at 60°C. PCR reactions were run with molecular weight standards on 0.8% agarose gels containing ethidium bromide and visualized by UV light. The primers used were: 5′-GGCT CACCAGCATCATATACG-3′ and 5′-GGCTACAATGACGACGTCA-3′.

c   *Method description in paper C:*
Real-time PCR were performed using the TaKaRa SYBR PCR kit and ABI Prism 7000 sequence detection system according to the manufacturer's specifications. The primers for amplification were *abc* (5′-CGCTCCTCTGCATCTAATCAG-3′ and 5′-GA CACTTAGCACGCACTCA-3′) and *def* (5′-GCATCTTCAAGTAAGGACTATC-3′ and 5′-GACTTTCACAGTACCAGATT-3′). Total reaction volume was 50 µl including 25 µl SYBR Premix Ex Taq with SYBR Green I, 300 nM forward and reverse primers, and 2 µl cDNA. The thermal cycler program was 1 cycle at 95°C for 10 s, followed by 40 cycles at 95°C for 5 s and 60°C for 30 s. The PCR products were detected by electrophoresis through a 2% agarose gel stained with ethidium bromide.

For these passages and for similar ones that occur mainly in the Materials and Methods section of a research paper, it may not be a bad idea to collect sample phrases from other articles for your reference. I am not advising you to copy entire passages or paragraphs to be placed into your manuscript, but individual sample phrases and expressions of a few words in length that can be applied to writing your research article.

➤ Keep track of ideas and references

When you compose a document, you can save yourself much time and confusion if you keep track of the sources of information from the start. There is nothing more frustrating than having to identify the origin of ideas and information when you are done writing. Thus, always keep a list of sources.

The best way to avoid plagiarism is to do the following when you collect and use information in scientific writing:

- Keep track of references and write down the information you intend to use whenever you come across a passage that you think may be useful for your document.
- Keep a detailed list of sources in a reference managing program such as EndNote or ReferenceManager.
- If you copy something word by word, put it in quotation marks, but know that writing in the sciences uses direct quotations only rarely. Avoid such quotations. If you have to include direct quotations, ensure that you do not overuse them and that they are not long. Otherwise, when you want to use just a few details or point out the highlights from an original, paraphrase.
- Write down the most important ideas in your own words using bullet points.
- Take notes with the book closed. This way you are forced to put the ideas into your own words.
- Double-check that the reference and information are correct by going back to the original when you compose your document.

8.7   REFERENCES WITHIN A SCIENTIFIC PAPER

➤ Know where to place references in a scientific paper

Abstract	Avoid placing any citations in the Abstract. Start citing sources in your Introduction.
Introduction	Cite the most relevant references only. Although the amount of background information needed depends on the audience, do not review the literature. Limit yourself to the most recent, the most important or original, and the most elegant references. Consider citing review articles when possible.

Materials and Methods	Cite original references for Materials or Methods used in your study. Include references of methods published in widely accessible journals instead of repeating details of those methods. (e.g., "Growth was measured and analyzed according to Billings (1988).")
Results	Usually, statements that need to be referenced, such as comparisons with previous reports, are not written in the Results section. These statements are made in the Discussion section. If a short comparison does not fit smoothly into the Discussion, however, it may be included in the Results section, and then it needs to be referenced.
Discussion	In the Discussion, although your findings are the main topic, your results have to be discussed in a broad context. That means that you have to include references to compare and contrast your findings, studies that provide explanations, or those that give your findings some importance.

8.8   THE REFERENCE LIST

➤ Follow the journal's style for details in the Reference List

The Reference List at the end of a paper displays the full citation of your reference. It should contain a list of only the literature cited in the text. Usually, abstracts for conference proceedings, personal communications, and unpublished data are not available to the public and are therefore not included in the Reference List, even if they were cited in the text. If the work has been submitted and accepted for publication but has not yet been printed, include it in the Reference List with author name and title followed by the journal name (and volume if known) and the term "(in press)." If a paper has been submitted to a journal but has not been accepted, it may or may not be included in the Reference List. If you list such work, use the author name, year of the version, and title followed by the term "Manuscript submitted for publication."

As citation and reference styles vary among journals, be sure to check and follow the reference style guide of your target journal very carefully. These instructions are usually listed in the *Instructions to Authors*. Pay particular attention to where to place and what to include in terms of first name initials, dates of publication, journal volume, and page numbers. Check which elements of the references to italicize, and check on spacing and punctuation between and following names, initials, titles, and numbers. Follow *all* the rules to the letter. Reference programs such as EndNote or ReferenceManager are very useful to get your references into the correct style format. Note that in the Reference List names are inverted: last name, then first name or initial, then middle initial. This custom differs from that used in some countries.

Typically, if you used the "(author, year)" system in the text, references are listed in alphabetical order and are not numbered in the Reference List.

ESL advice

Example 8-17   a   Albandar, J. M., and Tinoco, E. M. (2002). Global epidemiology of periodontal diseases in children and young persons. *Periodontol.* **2000**(29):153–76.

Lackovic, K., Angove, M. J., Wells, J. D., and Johnson, B. B. (2004). Modelling the adsorption of Cd(II) onto goethite in the presence of citric acid. *J. Colloid Interface Sci.* **269**:37–45.

McKee, J. K., Sciulli, P. W., Fofoce, C. D., and Waite, T. A. (2003). Forecasting global biodiversity threats associated with human population growth. *Biol. Conservation.* **115**:161–164.

Tibbetts, J. (2006). Louisiana's wetlands: A lesson in nature appreciation. *Environ Health Perspect.* **114**(1):A40–A43.

If you used the (number) system in the text, references in the list are numbered in the order in which each reference is first cited in the text.

Example 8-17   b   1. McKee, J. K., Sciulli, P.W., Fooce, C. D., and Waite, T. A. (2003). Forecasting global biodiversity threats associated with human population growth. *Biol. Conservation.* **115**:161–164.

2. Albandar, J. M., and Tinoco, E. M. (2002). Global epidemiology of periodontal diseases in children and young persons. *Periodontol.* **2000**(29): 153–76.

3. Tibbetts, J. (2006). Louisiana's wetlands: A lesson in nature appreciation. *Environ Health Perspect.* **114**(1):A40–A43.

4. Lackovic, K., Angove, M. J., Wells, J. D., and Johnson, B. B. (2004). Modelling the adsorption of Cd(II) onto goethite in the presence of citric acid. *J. Colloid Interface Sci.* **269**:37–45.

References to books or book chapters also require great attention to detail. Here, too, there are various formats depending on your target journal.

Example 8-18   Stein, B. A., L. S. Kutner, and J. S. Adam (eds.). 2000. *Precious Heritage: The Status of Biodiversity in the United States.* Oxford University Press, New York. pp. 23–29.

General Conventions

Generally, when you list references in the Reference List at the end of a paper, journal names are abbreviated in the listing. (Note that abbreviations depend on the style preference of your target journal; if APA style is used, for example, then journals listed in the Reference List are not abbreviated.) Journal abbreviations have been standardized for most journals according to the American National Standards Institute. The most common journal title word abbreviations are listed in Table 8.3. Other abbreviations can usually be obtained from listings in databases or from other reference lists.

8.9   COMMON REFERENCE STYLES

If no citation instructions are given, consider using one of the more commonly accepted formats, such as the Modern Language Association (MLA) style, Council of Scientific Editors (CSE; formerly CBE) style, American Psychological Association (APA) style, American Medical Association (AMA) style, or the National Library of Medicine (NLM) style. A few of these are shown in more detail following.

For additional abbreviations, see also

http://images.webofknowledge.com/WOK46/help/WOS/A_abrvjt .html
https://woodward.library.ubc.ca/research-help/journal-abbreviations/
http://library.caltech.edu/reference/abbreviations/

TABLE 8.3 Journal abbreviations

JOURNAL NAME	ABBREVIATION	JOURNAL NAME	ABBREVIATION
Abstracts	Abstr.	Journal	J.
Academy	Acad.	Laboratory	Lab.
Advances	Adv.	Marine	Mar.
American	Am.	Materials	Mater.
Analytical	Anal.	Medical, Medicine	Med.
Annals	Ann.	Molecular	Mol.
Annual	Annu.	National	Natl.
Applied	Appl.	Natural, Nature	Nat.
Archives	Arch.	Nuclear	Nucl.
Biology, Biological	Biol.	Optics	Opt.
Botany, Botanical	Bot.	Pharmaceutical	Pharm.
Cellular	Cell.	Physical, Physics	Phys.
Chemical, Chemistry	Chem.	Proceedings	Proc.
Clinical	Clin.	Research	Res.
Computational	Comp.	Science, Scientific	Sci.
Current	Curr.	Society	Soc.
Developmental	Dev.	Technical	Tech.
Environmental	Environ.	United States	U.S.

European	Eur.	"-ology" words	
Experimental	Exp.	("Psychology")	"-ol." ("Psychol.")
Internal	Intern.	One word titles ("Cell")	Not abbreviated
International	Int.		

👍 **Example 8-19**   **MLA Style**

In Text   . . . as reported previously (1).

Bibliography   **Books**
McCormac, James S., and Gregory Kennedy. *Birds of Ohio*. Lone Pine, 2004.

Journal Article
Jefferson, Thomas A., Pam J. Stacey, and Robin W. Baird. "A Review of Killer Whale Interactions with Other Marine Mammals: Predation to Co-Existence." *Mammal Review,* vol. 21, no. 4, 2008, pp. 151–80.

👍 **Example 8-20**   **CSE Style**

In-Text   . . . observed by Hinter (2008)
(McCormac and Kennedy 2004)
(Meise et al. 2003)
(Hinter 2008)

Bibliography   **Books**
McCormac JS, Kennedy G. 2004. *Birds of Ohio*. Auburn (WA): Lone Pine. pp. 77–78.

Journal Article
Meise CJ, Johnson DL, Stehlik LL, Manderson J, Shaheen P. 2003. Growth rates of juvenile Winter Flounder under varying environmental conditions. Trans Am Fish Soc. 132(2):225–345.

👍 **Example 8-21**   **APA Style**

In Text   . . . as reported by Juls (2009).
Bibliography   Research by Wegener and Petty (2010)
. . .
(Juls, 2009)
(Wegener & Petty, 2004)
(Kernis et al., 2013)

Books
McCormac, J. S., & Kennedy, G. (2004).
Birds of Ohio. Auburn, WA: Lone Pine.

Journal Article
Jefferson, Thomas A., Stacey, Pam J., &
Baird, Robin W. (2008). A review of Killer
Whale interactions with other marine mam-
mals: predation to co-existence. *Mammal
Review 21*(4),151–180.

Online Sources for Common Reference Styles
Certain online sources are a great resource for various scientific style guides:

http://owl.english.purdue.edu/owl/resource/560/02/
http://bailiwick.lib.uiowa.edu/journalism/cite.html

8.10   CITING THE INTERNET

➤ Know how to cite and list references from the Internet
Before you cite sources from the Internet, you should consult the *Instructions to Authors* and recent issues of your target journal. Use of Web citations is not always accepted, but this is a developing area, so check your
target journal's policy early. Most important, ensure that you use respected,
peer-reviewed journals (also known as primary sources) or peer-reviewed
open access journals rather than nonpeer-reviewed journals or websites
(see also Chapter 1, Section 1.2 on becoming media literate, as well as Section 8.1 and 8.2 in this chapter.)

When you cite from the Internet, you can generally decide which citation and reference style to use. However, make sure that the style does not
conflict with that required by your target journal. To cite and list a reference
from the Internet or the Web, use the guidelines in Examples 8-22 to 8-25.

 Example 8-22   Author's name (last name first) Title. Available from: URL:
http://Internet address or Web address. Date of Access.

There are also other good reference styles for citing Internet sources such as
the MLA, CSE (formerly CBE), NLM, and Chicago styles.

 Example 8-23   **MLA Style**

Online document
Author's name (last name first). Document title. Date of Internet
publication. Date of access <URL>.

Book

Bryant, Peter J. "The Age of Mammals." *Biodiversity and Conservation.* 28 Aug. 1999. 20 Oct. 2009 <http://darwin.bio.uci.edu/~sustain/bio65/lec02/b65lec02.htm>.

Article in an electronic journal (ejournal)

Joyce, Michael. "On the Birthday of the Stranger (in Memory of John Hawkes)." *Evergreen Review* 5 Mar. 1999. 20 Oct. 2009 <http://www.evergreenreview.com/102/evexcite/joyce/nojoyce.html>.

Example 8-24   **CSE Style**

Online document

Author's or organization's name. Date of publication or last revision [year month]. Document title. Title of complete work (if relevant). <URL>. Accessed [year month].

Book

Bryant P. 1999 Aug 28. *Biodiversity and conservation.* <http://darwin.bio.uci.edu/~sustain/bio65/ Titlpage.htm>. Accessed March 20, 2016.

Article in an electronic journal (e-journal)

Browning T. 1997. Embedded visuals: student design in Web spaces. *Kairos: A Journal for Teachers of Writing in Webbed Environments* 3(1). <http://english.ttu.edu/kairos/2.1/features/browning/bridge.html>. Accessed 2016 March 20.

Example 8-25   **Chicago Style**

Online document

Author's name. "Title of document." Title of complete work (if relevant). Date of publication or last revision. Accessed [month year] (if required), DOI or <URL>.

Book

Peter J. Bryant. "The Age of Mammals," in *Biodiversity and Conservation.* August 1999. <http://darwin.bio.uci.edu/~sustain/bio65/lec03/b65lec03.htm>, accessed March 20, 2013.

Article in an electronic journal (e-journal)

Tonya Browning. "Embedded Visuals: Student Design in Web Spaces." *Kairos: A Journal for Teachers of Writing in Webbed Environments* 3, no. 1 (1997). http://english.ttu.edu/kairos/2.1/features/ browning/index.html, accessed March 20, 2013.

DOIs

For articles that are published and made available electronically, publishers assign a DOI (digital object identifier) code to link to content and a persistent location on the Internet. This system is especially helpful for electronic documents such as journal articles. DOIs for documents stay fixed over their lifetime, although their location and content might change. Thus, DOIs provide a more stable link than URLs.

For electronic journal articles, DOIs are usually found on the first page. Although DOIs are not typically included when referring to printed scientific references, when available consider including them. When you reference scientific articles that are Web only (e.g., BioMed Central journals), many journals (e.g., *Nature*) require you to include the URL or DOI.

SUMMARY

REFERENCE GUIDELINES
- Reference the ideas and findings of others.
- Differentiate between primary, secondary, and tertiary sources.
- Select the most relevant references.
- Verify your references against the original document.
- Manage your references well.
- Follow the journal's style for details in the reference citation.
- Cite references in the correct form and order.
- Remain objective and neutral when citing the work of others.
- Know where to place references in a sentence.
- Ensure that you are not plagiarizing.
- Know how to paraphrase.
- Keep track of ideas and references.
- Know where to place references in a scientific paper.
- Follow the journal's style for details in the Reference List.
- Know how to cite and list references from the Internet.

PROBLEMS

PROBLEM 8-1 References

When scientists submit papers for publication, references have to be listed as required by the *Instructions to Authors*. The following reference section belongs to a paper submitted to *Molecular and General Genetics*. Its *Instructions to Authors* states the following:

The list of references should only include work cited in the text and that has been published or accepted for publication ("in press"). In the text, references should be cited by author and year (e.g., Hammer 1990a; Hammer 1990b; Hammer 1994; Hammer and Sjöqvist 1995; Hammer et al. 1993). Note that no comma separates author and year, and that only the last name of the first author is indicated. All other authors are abbreviated by et al. If there are two authors, both authors are listed by last name. References should be listed in alphabetical order in the Reference List. Journals should be abbreviated in accordance with Bibliographic Guide for Editors and Authors (BGEA). Note: a correct reference contains one single period (following the article title) and commas only to separate authors.

Examples of the correct styles are shown below:

Article:

Gallie DR, Young TE (2004) The ethylene biosynthetic and perception machinery is differentially expressed during endosperm and embryo development in maize. Mol Gen Genomics 271:267–281

Book:

Pechan P, de Vries G (2004) *Genes on the Menu*. Springer, Berlin Heidelberg New York

Internet:

Doe J (1999) Title of subordinate document. In: *The dictionary of substances and their effects*. Royal Society of Chemistry. Available via DIALOG. http://www .rsc.org/dose/title of subordinate document. Cited 15 Jan 1999

Check the following Reference List to ensure that the references are listed as required. Correct the ones that need to be adjusted.

Bollag RJ, Waldmann AS, Liskay RM (1989) Homologous recombination in mammalian cells. Annu Rev Genet 23:199–225

Capecchi MR (1989a) Altering the genome by homologous recombination. *Science* 244:1288–1292

Capecchi MR (1989b). The new mouse genetics: altering the genome by gene targeting. Trends Genet 5: 70–76

Offringa R, de Groot JA, Haagsman HJ, Does MP, van den Elzen PJM, Hooykaas PJJ (1990) Extrachromosomal recombination and gene targeting in plant cells after *Agrobacterium*-mediated transformation. EMBO J 9:3077–3084

Offringa R, Franke-van Dijk MEI, de Groot MJA, van den Elzen PJM, Hooykaas PJJ (1993) Non-reciprocal homologous recombination between *Agrobacterium*-transferred DNA and a plant chromosomal locus. Proc Natl Acad Sci USA 90:7346–7350

Paszkowski J., Baur M., Bogucki A., Potrykus I. (1988). Gene targeting in plants. EMBO J 7:4021–4026

Puchta H, Hohn B (1996) From centimorgans to base pairs: homologous recombination in plants. Trends Plant Sci 1:340–348

Puchta H, Dujon B, Hohn B (1996) Two different but related mechanisms are used in plants for the repair of genomic double strand breaks by homologous recombination. Proc Natl Acad Sci USA 93:5055–5060.

Roth D., Wilson J. (1988). Illegitimate recombination in mammalian cells. In: Kucherlapati R, Smith GR (eds) Genetic recombination. American Society for Microbiology, Washington DC, pp 621–653

Schaefer DG, Bisztray G, Zrÿd JP (1994) Genetic transformation of the moss *Physcomitrella patens*. Biotech Agric For 29:349–364

Struhl K (1983). The new yeast genetics. *Nature* 305:391–397

PROBLEM 8-2 References

In the following paragraphs, which are part of an Introduction, check that references are cited correctly in the text for the same paper as in Problem 8-1 given the same instructions for publication in MGG.

The integration of foreign DNA into eukaryotic genomes occurs mainly at random loci by illegitimate recombination, even when the introduced DNA has homology to endogenous sequences (Roth et al. 1988; Bollag et al. 1989; Capecchi 1989a; Puchta et al 1994; Puchta and Hohn 1996). Exceptions to this generalization include *Saccharomyces cerevisiae* (Rothstein R., 1991), *Schizosaccharomyces pombe* (Grimm and Kohli, 1988) and some filamentous fungi (Timberlake and Marshall 1989). Direct integration by homologous recombination (gene targeting) is rare compared with non-homologous integration, but would provide a powerful technique for the molecular genetic study of higher organisms and the determination of gene function. Although gene targeting is now a well established tool for the specific inactivation or modification of genes in yeast and mouse embryonic stem cells (Struhl 1983; Capecchi 1989b; Rossant and Joyner 1989; te Riele et al. 1992), the development of similar techniques for plant systems (Puchta and Hohn 1996) is still at a very early stage. Some groups have reported genomic integration by homologous recombination in plant cells using DNA introduced by direct gene transfer or *Agrobacterium*-mediated DNA transfer (Paszkowski et al. 1988; Offringa 1990, 1993; Halfter et al. 1992). The highest rates of gene disruption so far published for flowering plants are between 0.01 and 0.1% (Lee et al. 1990; Miao and Lam 1995; Kempin et al. 1997), and these are too low to allow routine gene disruption. However, it has recently been shown (Schaefer and Zrÿd 1997) that integrative transformants resulting from homologous recombination can be obtained following polyethylene glycol (PEG)-mediated direct gene transfer into protoplasts of the moss *Physcomitrella patens*, at an efficiency comparable to that found for S. cerevisiae.

(With permission from Springer Science and Business Media)

PROBLEM 8-3 Citations

When scientists write papers for publication, they often forget to reference information. In the following passage, indicate where a citation should be made.

The carbon ($\delta^{13}C$), nitrogen ($\delta^{15}N$), and sulfur ($\delta^{34}S$) isotope ratios of humans, other animals, and microbes are strongly correlated with the isotope ratios of their dietary inputs (1–5). There are limited differences ($\leq 1^{0}/00$) between heterotrophic organisms and their diet in either the $\delta^{13}C$ or $\delta^{34}S$ values. Hydrogen ($\delta^{2}H$) and oxygen ($\delta^{18}O$) isotope ratios of organic matter, however, are more useful, because $\delta^{2}H$ and $\delta^{18}O$ values of precipitation and tap waters vary along geographic gradients. Although differences in the $\delta^{2}H$ and $\delta^{18}O$ values of scalp hair have been noted in humans, less is known about diet–organism patterns of $\delta^{2}H$ and $\delta^{18}O$ values. Four potential sources can be important: dietary organic molecules, dietary waters, drinking waters, and atmospheric diatomic oxygen. Hobson *et al.* provided evidence that $\delta^{2}H$ values of drinking water were incorporated into different proteinaceous tissues of quail. Other research showed that the $\delta^{2}H$ values of bird feathers and butterfly wings (both are largely keratin) and water in the region in which the tissue was produced are highly correlated (14, 15). Kreuzer-Martin *et al.* showed that ~70% of the oxygen and ~30% of the hydrogen atoms in microbial spores (~50% proteinaceous) were derived from the water in the growth medium, whereas the remainder was derived from the organic compounds supplied as substrate.

(With permission from National Academy of Sciences, USA)

PROBLEM 8-4 Citations

In the following paragraph, which is part of an Introduction, assume that citations have been verified against the original. Ensure that the text citations have been placed at appropriate locations within the sentences.

Ostracodes are small bivalved Crustacea that form an important component of deep-sea meiobenthic communities along with nematodes and copepods (10). Crustaceans (10, 11) are dense and diverse in the deep sea and are one of the most representative groups of the whole deep-sea benthic community. Pedersen et al. as well as Jackson et al. reported (12–14) that ostracode species have a variety of habitat and ecology preferences (e.g., infaunal, epifaunal, scavenging, and detrital feeders), representing a wide range of deep-sea soft sediment niches. Furthermore, Ostracoda is the only commonly fossilized metazoan group in deep-sea sediments. Thus, fossil ostracodes are considered to be generally representative of the broader benthic community. The distribution and abundance of deep-sea ostracode taxa in the North Atlantic Ocean are influenced by several factors (14, 15), among them, temperature, oxygen, sediment flux, and food supply. Several paleoecological studies suggest (1, 16) that these factors influence deep-sea ecosystems over orbital and millennial timescales.

(With permission from National Academy of Sciences, USA)

PROBLEM 8-5 Paraphrasing
Given the following information from another article (McCaffrey, R. Global frequency of magnitude 9 earthquakes, *Geology 36*(3), 2008), **incorporate this information in the proposed introduction on earthquake predictions shown following the original.**

Original from another article:
Most great earthquakes occur at submarine subduction zones where one tectonic plate slides at a gentle angle (10°–30°) beneath another. In this paper I look at the probabilities of the largest earthquakes and, in the context of recurrence times and our short history of observation, argue that we cannot rule out an Mw ≥ 9 earthquake at any subduction zone. Simulations using recurrence times of the maximum size earthquakes (called M9) at subduction zones suggest that 1–3 M9 earthquakes should occur within any 100 yr span. The five M9 events that have occurred since 1952 probably represent temporal clustering and not a long-term average.
(With permission from the Geological Society of America)

Proposed introduction on earthquake predictions
Humans have long sought to predict seismic activity in an effort to forecast time, location, and magnitude of future earthquakes. Studies in this field have been based either on past patterns or the identification of specific precursors. Patterns that are studied to predict earthquakes include deformations in the Earth's crust, past earthquake characteristics, seismicity patterns, and gaps in seismic activity. Teams that look at earthquake precursors assess animal behavior, radon emissions from rocks, and electromagnetic variations.

To date, these approaches have been controversial, largely due to contradictory reports, lack of baseline data, a large number of variables, and difficulty in rigorous statistical evaluation. However, probabilities of occurrence have been calculated, particularly for earthquakes of magnitude larger than 5. For example, . . . [insert paraphrased citation here].

PROBLEM 8-6 Paraphrasing
Paraphrase the following passage.

The highly pathogenic avian influenza A (H5N1) virus (short H5N1) was first reported in humans in Hong Kong in 1997. Humans have been infected through contact with sick poultry, but outbreaks among humans have been only sporadic so far. Since 2003, 664 cases of avian influenza A (H5N1) have been found, and 60% of the infected people have died (391 cases) (WHO, 2014). Most infections have occurred in Asia and in the Middle East. In the United States, no infections have been described in birds or humans to date. However, scientists fear that the virus could spark a global human pandemic, which may kill millions of people. Unlike other flu viruses, H5N1 does not yet have the ability to spread between humans, but work has shown that the virus only needs five favorable mutations to become transmissible among ferrets (Herfst, 2012). Although the number of human cases has been declining since 2006, the number of outbreaks among birds and poultry remains high. In addition, the virus is constantly evolving. Therefore, WHO recommends monitoring the disease worldwide (WHO 2014), and the H5N1 vaccine has been stockpiled in the United States (CDC, 2014).

PROBLEM 8-7 Paraphrasing
The following passage comes from a research paper publication. Read through the original to get an understanding of its central points and then determine whether the student paraphrases are plagiarized or paraphrased versions.

Original
Killer Whales are well-known as predators of other marine mammals, including the large Sperm and baleen whales. Members of all marine mammal families, except the river dolphins and manatees, have been recorded as prey of Killer Whales; attacks have been observed on 20 species of cetaceans, 14 species of pinnipeds, the Sea Otter, and the Dugong. Ecological interactions have not been systematically studied, and further work may indicate that the Killer Whale is a more important predator for some populations than previously believed. Not all behavioural interactions between Killer Whales and other marine mammal species result in predation, however. Some involve "harassment" by the Killer Whales, feeding by both species in the same area, porpoises playing around Killer Whales, both species apparently "ignoring" each other, and even apparently unprovoked attacks on Killer Whales by sea lions. These non-predatory interactions are relatively common. We conclude that interactions between Killer Whales and marine mammals are complex, involving many different factors that we are just beginning to understand.

Jefferson, T. A., Stacey, P. J., & Baird, R. W. (2008).
Mammal Review, 21(4), 151–180)

Student Version A
Killer Whales prey on other marine mammals such as the large Sperm and baleen whales, and on members of all marine mammal families, except the river dolphins and manatees. Killer Whales have also attacked cetaceans, pinnipeds, the Sea Otter, and the Dugong. Some interactions between Killer Whales and other marine mammal species do not result in predation, for example, "harassment" by the Killer Whales, feeding by both species in the same area, porpoises playing around Killer Whales, and even attacks on Killer Whales by sea lions. These non-predatory interactions are relatively common. Thus, interactions between Killer Whales and marine mammals are complex, and involve many different factors that we are just beginning to understand, [and] further work is necessary to understand the ecological interactions of Killer Whales.

Student Version B

Ecological interaction of Killer Whales with other marine mammals can be predatory or non-predatory. Predatory interactions involve Killer Whale attacks on all marine mammals. Non-predatory interactions involve Killer Whales harassing marine mammals, feeding in the same area, playing around each other, and even attacks on Killer Whales by the other species (Jefferson et al. 2008). Although these interactions have been observed, further work is needed to fully understand Killer Whale interactions with other marine mammals.

CHAPTER 9

Figures and Tables

Figures and tables are meant to demonstrate evidence visually and therefore should be designed for strong visual impact. Together with their figure legends and table titles, they need to be immediately clear and interesting to readers, as many readers judge papers based on figures and tables.

THIS CHAPTER COVERS:

- Selection and design of figures and tables
- Placement of information in figures and tables
- Formatting figures and tables, including how to highlight data
- Different types of figures
- Figure legends
- Other kinds of supplementary information, including formulas, equations, proofs, and algorithms

9.1   GENERAL GUIDELINES

➤ **Decide whether to present data in a table, a figure, or in the text**

➤ **Use the fewest figures and tables needed to tell a story**

Decide first whether to present your findings in a figure, in a table, or in the text. Editors and reviewers will stringently judge if a figure or table is useful. Each figure or table should therefore be important enough to be included in a document. Be aware that figures and tables take longer to create and may cost more to produce than text. In addition, you will lose readers if you overwhelm them with too many figures and tables. Readers typically only pay attention to a maximum of four to five items.

➤ Design figures and tables to have strong visual impact

The main purpose of figures and tables is to visualize data and to support your results. Not all the data need to be described in the text of your paper, however. Describe only your key findings in the text. These findings should be clearly and immediately identifiable in figures and tables.

➤ Figures and tables should be able to stand on their own

At the same time, figures and their legends as well as tables and their titles must be independent of the text and of each other; that is, readers must be able to understand figures and charts without having to refer to the text. Figures and tables need to be numbered in the order they appear in the document. In addition, the name of the variable, the unit of measurement, and the values should be the same in the text and in the illustration. If possible, provide statistical information in figures and tables.

9.2   IMPORTANCE OF FORMATTING AND PLACEMENT OF INFORMATION

➤ Prepare figures and tables with the reader in mind—place information where the reader expects to find it

In the same way that text should be formatted for the reader, figures and tables should be formatted to meet the reader's expectations. Most readers can understand the intended meaning of what is presented only if the illustration has been formatted for this interpretation.

Scientists can present data in different formats. See Example 9-1 for various possibilities for a set of data.

Example 9-1   **a**   0°C, 0.011% hermaphrodites
6°C, 0.011% herm.
25°C, 51/1000   18°C, 0.028%
water T = 30°C , 0.124%   T = 10°C , 2/100
35°C, 0.152%

b   **Sample Data Display 1**

% HERMAPHRODITES $N = 120$	WATER TEMPERATURE (°C)
0.011	0
0.011	6
0.020	10
0.028	18
0.051	25
0.124	30
0.152	35

c   Sample Data Display 2

Water temperature (°C)	0	6	10	18	25	30	35
% Hermaphrodites	0.011	0.011	0.020	0.028	0.051	0.124	0.152

d   Sample Data Display 3

WATER TEMPERATURE (°C)	% HERMAPHRODITES
0	0.011
6	0.011
10	0.020
18	0.028
25	0.051
30	0.124
35	0.152

Although the exact same information appears in all formats, most readers prefer Example 9-1a because it is the easiest to interpret. The reason for the easier interpretation is twofold: (a) the data is written as a table, and even more important, (b) this table is structured such that the familiar context (temperature) appears on the left whereas the interesting results appear on the right in a less obvious pattern.

Usually, information in a table is more readily available for readers than information in the text (unless there are very few data). However, some tables are easier to interpret than others. For example, readers find tables much harder to follow if they are presented as shown in Example 9-1b, or if the table is horizontally arranged as in Example 9-1c. Readers interpret information more easily if it is placed where they expect to find it. Because we read from left to right, we prefer the familiar context on the left and the new information on the right as in Example 9-1d.

9.3   FIGURE OR TABLE?

➤ Use figures to show trends and relationships and to emphasize data

When you have chosen to use an illustration rather than text, you may then need to decide in what format to present them: figure or table. It helps if you spread your data out on the table and arrange them in all possible combinations. Look for patterns. There are better and worse presentations of data in a paper. Go with a simple pattern if possible.

Generally, choose figures when trends or relationships are more important than exact values or when hidden relationships or trends need to be revealed. Choose tables to report precise numerical information, to compare component groups, or when data are not enough to produce a satisfactory graph.

Consider the next two illustrations. Which of the following presentations would you prefer given the same data?

Example 9-2a   Sample Data Presentation A

TIME (DAYS)	HORMONE A	HORMONE D54
0	200.5	455.8
5	187.1	356.7
10	166.5	321.9
15	201.1	400.6
20	289.8	500.7
25	204.1	489.9
30	189.9	389.4
35	288.9	513.4
40	205.1	499.3
45	182.9	298.5
50	278.8	533.2
55	223.4	498.5
60	199.6	250.6

Example 9-2b

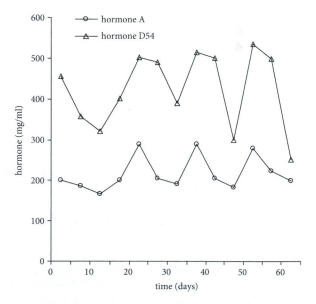

Figure. Sample data presentation B. The same data as in Example 9-2a are presented as a line graph instead of a table.

Most readers would prefer Example 9-2b because the trend and relation-ship of the data are more obvious, and exact numbers seem not as important.

Consider another example in Example 9-3.

Example 9-3

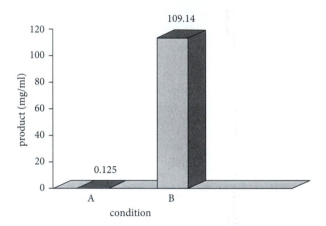

Figure. Bar graph for which data could be presented in the text instead.

Example 9-3 would best be depicted in the text because there are only two data points, and exact values seem important.

9.4   GENERAL INFORMATION ON FIGURES

General Advice

When you prepare a figure, be sure that it carries your point better than the text or a table would. Never use figures simply because they are available. Consider what kind of figure you need. You can choose to present your data in a **photograph**, a **drawing**, a **diagram**, or a **graph**. Select the size and format to fit the journal, poster, or slide.

Overall, figures must be simple and clear enough for readers to get the message immediately. Good figures clarify relationships of complicated data sets and highlight trends or patterns that may not be immediately evi-dent in the text.

The requirements for figures that will appear in print are different from those for oral presentations or posters (see Table 9.1 and also Chapters 29 and 30). Although it is best to design for each medium separately, if you plan carefully you may be able to use the same artwork for a journal article, an oral presentation, and a poster presentation.

TABLE 9.1 Differences between figures in text, slides, and posters

	TITLE	LEGEND
Figure in text	—	✓
Figure on slide	✓	—
Figure on poster	✓	✓
Table in text, on slide or poster	✓	—

Producing Figures

➤ Prepare professional figures

Figures submitted to professional journals must meet high standards. Obtain the *Instructions to Authors* from your target journal, and follow these instructions when creating your figures. Inadequate figures spoil many articles, and few journals redraw or re-letter figures submitted by authors. If you do not have your figures designed professionally, it is a good idea to get help from your colleagues, consult a recommended book, or even take a course on how to prepare professional figures.

You can prepare your own figures by using various computer software: Excel or Lotus are great programs to convert your data and to export it to graphing programs such as Charisma®, CorelDRAW®, Harvard Graphics®, KyPlot®, SigmaPlot®, and DeltaGraphPro® or to slide-making programs such as PowerPoint®, Adobe® Persuasion®, or Astound®. Other great draw, paint, and graphics programs include Adobe® Illustrator®, Adobe® Photoshop®, ChemDraw®, Visio®, and SmartSketch®. You may be able to try out or use some of these programs as free beta versions that can be downloaded from the Web.

If you use a computer graphics program for your figures, the output must be of good enough quality to be reproduced well by conventional printing techniques. Lines must be clearly drawn, with good contrast against a plain white background. Axes, curves, and lettering must be solid and smooth. Symbols must be clear, large, and distinctive. Your figures need to be labeled well, and all components must be identified. Do not clutter your illustrations. Include scale bars on maps, micrographs, and anatomical drawings.

If an artist or photographer who makes figures for you is employed by your organization, note that copyright of the figures belongs to the organization. Thus, you should be able to use the figures without obtaining further copyright permission from the maker. If you employ a freelance artist or photographer, however, the copyright belongs to him or her unless a written agreement is made transferring copyright and other rights to you. An agreement of this kind makes you the owner of a "work made for hire."

It is a good idea to start preparing figures at an early stage, before you even think of drafting the text. First, check whether the journal limits the number or size of figures you may submit. Then, draft your figures and decide which ones to use. Design any others you think you will need, and draft legends for all of them. When you create the final versions of the figures at the revision stage, note that many journals require you to submit figures electronically as tagged image file format (TIFF) files.

Remember that the data you present may be interpreted in more than one way. You can highlight, exaggerate, or even misrepresent a given set of data depending on the way you choose to display it. Thus, constructing a graph is more than just plotting points. Graphing requires that you understand the rationale behind the quantitative methods. It also requires that you are as objective and honest as possible.

Misleading Readers

➤ Do not mislead readers

You should make readers aware of certain trends. However, as a scientist, you have to be responsible enough not to mislead your readers. It may be very tempting to distort information and trends by, for example, deleting data points or by massaging line fits. Resist these temptations. Do not mislead readers.

If you can, provide statistical information in your figures and figure legends. Such information will lend validity to your data. Use error bars if needed (see Example 9-9 or Revised Example 9-17a or 9-18), or show variability of data points through, for example, scatter plots (Example 9-8) or box plots (see Example 9-11). Indicate the statistical test(s) used, the number of times an experiment was repeated, the sample size, p-values, and other statistical information.

Choose these scales carefully, and mark them clearly. Often, readers are misled when different scales are used to compare data in different graphs. For such comparisons, ensure identical scales so readers can compare data directly, and arrange graphs next to each other to help assessment. In addition, do not extrapolate a line graph past the last data point, as such curves can be misleading.

9.5   TYPES OF FIGURES

Photographs

Photographs record results that rely on visual inspection, such as X-rays or electron microscopic presentations. The major advantage of a photograph is realism. If you decide that it is essential to show the actual appearance of a subject or object, choose photographs of the best possible quality. Photographs intended for journal publication should be sharply focused, with good contrast. See also the Council of Biology Editors Scientific Illustration Committee (1988, pp. 219–233) for a more detailed description of photographs.

Never assume that the reader will recognize anything in a photograph. Label everything that is relevant. Some kinds of records do not come out well in photographs. It may be more effective to draw a diagram of a piece of equipment or describe it in the text rather than to include a photograph.

If you are thinking of including color photographs, find out first whether your target journal accepts them (see the *Instructions to Authors* or consult the editor). Be aware that if your target journal accepts color photographs, it is possible that they will charge you for the reproduction.

Always include scale bars on micrographs. This inclusion is preferable to giving magnifications in legends because the scales will remain correct if the figure is reduced during the production process or is later printed at some other size. Alternatively, if you want to state the original magnification, give it in the legend and add the photographic reduction when it is known (see Example 9-4a). Example 9-5 is another example of a photograph together with a line graph.

Example 9-4a

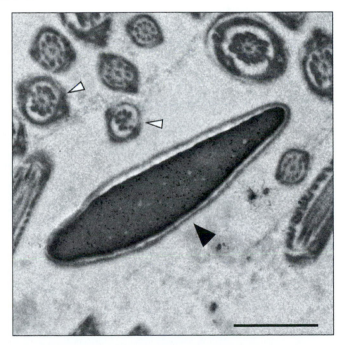

Figure X. Electron micrograph of mouse testis embedded in epoxy resin. Sperm was immuno-labeled with 10 nm gold post embedding. Black arrow, longitudinal section of sperm head; white arrows, cross sections of sperm tails. Bar, 1 μm.

(With permission from Rudolf Lurz)

Example 9-4b

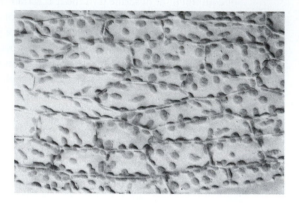

Figure Y. Light micrograph of *Physcomitrella patens* leaf cells with chloroplasts. Scale bar, 10 μm.

(With permission from Rudolf Lurz)

 Example 9-5

Class Description   NDVI-cycle (phenology)

Medium dense
deciduous forest

 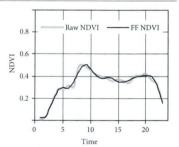

Figure Z. Phenological cycles from 250 m Modis NDVI time-series (version 4). Width of snapshot is about 150 m. Source for snapshot is Google Earth.

(With permission from Roland Geerken)

Drawings and Diagrams

Drawings and diagrams illustrate basic principles or otherwise explain text material. They include flow charts, diagrams, block diagrams, maps, and line art. Because these figures are much more artistic than graphs and usually more difficult to produce, you should seriously consider getting professional help with these types of figures.

The advantage of drawings and diagrams is that you present unusual perspectives while controlling the amount of detail. Ensure that key details and features are clearly and immediately visible. Use arrows, circles, callout boxes, and similar aids to help you draw attention to details if needed. Go for simple rather than complex and overly busy figures to avoid confusing readers. An example of such a diagram is shown in Example 9-6.

Example 9-6

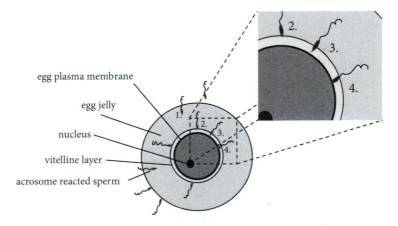

Figure XX. Diagram of *S. purpuratus* egg and sperm fertilization. 1. The egg triggers the sperm acrosome reaction. 2. The sperm attaches to the egg vitellin layer at the acrosomal process. 3. The acrosomal process extends as it penetrates the vitellin layer. 4. The plasma membranes of the egg and sperm fuse.

A specific type of diagram is the block diagram. A block diagram is a diagram of lines and shapes showing the relationship between different components of a system, apparatus, process, or reaction. An example of a typical block diagram is shown in Example 9-7.

Example 9-7   Typical block diagram

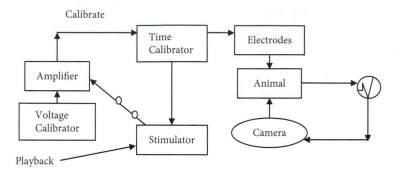

Figure. Block diagram of experimental approach.

Guidelines for Developing Technical Block Diagrams
- Use a maximum of 10 blocks or shapes.
- Label the blocks briefly, using the same key terms as in the text.
- Do not clutter your diagram. Show only major actions and interactions to reduce confusion.
- Use arrows to indicate direction.
- When describing the diagram in the text, follow the direction of flow on the diagram and describe the entire diagram. The same applies when you present such a figure in a talk or poster.

Graphs
Present your data in graphs if the data show pronounced trends, relationships, or patterns. Graphs can be presented as bar graphs, line graphs, pie charts, box plots, or scatter plots. Often, these may be two- and three-dimensional.

Line Graphs

➤ Use line graphs for dynamic comparisons
Line graphs are used for dynamic comparisons, often over time. They are the graphs most commonly encountered in science.

Do not try to cram too much into one figure—but do not waste space either. Three or four curves should be the maximum in a line graph, especially if the lines cross each other two or three times. When curves must cross, show which lines run where by making them of different thickness or different patterns. Draw curves as straight lines between data points or as best fit, smoothed curves. (See Example 9-12.)

Scatter Plot

➤ Use scatter plots to find a correlation for a collection of data

Scatter plots are useful for finding correlations for a collection of data. Scatter plots can be produced in two or in three dimensions. Each mark in the plot represents one data point. If there is an adequate number of marks and these marks are grouped sufficiently close together, a line of best fit can be used to find any mathematical relationship between the collection of data points. The line of best fit, also known as a trendline or regression line, can be curved or linear. It represents a theoretical ideal.

Data points in scatter plots often overlap. That is okay. A scatter plot allows you to find a clear correlation for the points, such as the linear relationship in Example 9-8.

Example 9-8   Sample scatter plot

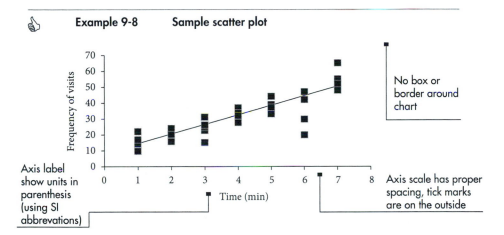

Figure. Well-presented scatter plot. Data stand out clearly, and axes are designed and labeled well.

Bar Graph

➤ Use bar graphs when the findings can be subdivided and compared

Another common graph in science is the bar graph (see Example 9-9). Bar graphs tend to be more effective than line graphs for general audiences. If you use a bar graph, use a vertical rather than a horizontal bar graph because most readers are accustomed to the former. Use bar graphs in preference to line graphs when there is no evidence of a continuum between the experimental points or when the findings can be subdivided and compared in different ways. Keep the widths of the bars, and the space between bars, consistent.

Example 9-9   Sample bar graph

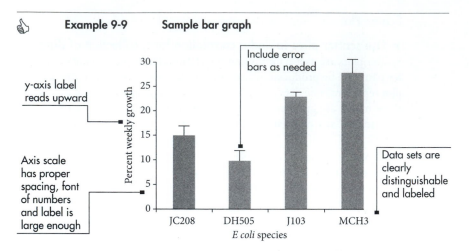

Figure YY. Percentage weekly growth for various *E. coli* cell species in medium A.

Pie Chart

➤ Use pie charts to compare parts of a whole

Yet a different kind of graph is the pie chart. Whereas bar graphs allow you to compare different whole quantities, pie charts allow you to compare parts of a single whole. They are most effective when the segments are arranged from large to small, with the largest segment starting at the top (see Example 9-10).

Example 9-10   Sample pie chart

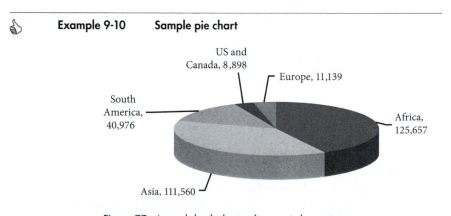

Figure ZZ. Annual death due to disease A, by region.

Box Plot

➤ Use box plots to display differences between data sets, especially in descriptive statistics

Box plots can be drawn either horizontally or vertically. This type of plot divides the data into quartiles, which visually depict the central value, variability, and range of the data set. The variability of the data is indicated by the interquartile range (IQR) of the box. Extreme values (either 1.5*IQR, minimum and maximum data points, or one standard deviation from the mean) are shown as whiskers above and below the box. An example of such a box plot is shown in Example 9-11.

Example 9-11   Sample box plot

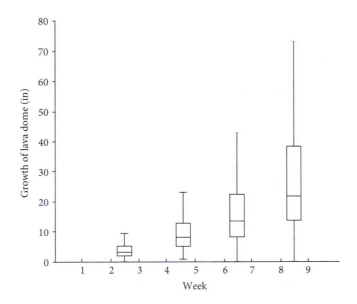

Figure. The plot displays the median (Q2; center black line), inter quartile ranges (box; Q1 and Q3), and extreme values (whiskers) from collected data.

9.6   FORMATTING GRAPHS

➤ Place the independent variable at the *x*-axis and the dependent variable at the *y*-axis

Readers expect to see information in figures at certain places. For line graphs and bar graphs, they expect to find the independent variable at the *x*-axis and the dependent variable at the *y*-axis. If information is placed as expected, readers are much more directed and do not have to spend extra time trying to understand illustrations.

Make the key information immediately obvious. Highlight the most important information in your graphs by making it stand out. Such highlights can be achieved through different colors, line weights, arrows, call-outs,

font size, or labels. At the same time, deemphasize less important information such as axis, axis labels, keys, titles, and figure labels.

Make plotted points stand out well. If they fall on an axis line, break
the axis on each side of the point. Plan graphs so that they need as little lettering as possible. Draft short but informative descriptions for the axes. Use
the same symbols when the same entities occur in several figures, and use
the same coordinates for different figures if values in them are to be compared. If you measured two variables in different ways, do not compare
them on the same axes. Draw the curves for the two variables separately,
using one common axis where appropriate.

An example of a well-constructed line graph is shown in Example 9-12.

Example 9-12   Sample line graph

Figure titles
above the
figure are
typically only
for figures in
posters or
slides

Include a key
on the graph
for 2 or more
data sets

Make data
points easy to
distinguish

Note units in
parentheses
for both axes

Position figure
caption below
the figure;
provide the
figure number, a
title and a brief
description

Figure. Well-organized line graph. Curves and data points are easily distinguishable.
Data stand out well, and axes are designed and labeled clearly.

Note that in a figure title where the word "versus" or its abbreviation "vs." is
used, the convention is to write it as "*y* label vs. *x* label" (or "dependent vs.
independent variable"), as shown in the figure title for Example 9-12.

Readability

➤ Make each figure easy to read
For good first impressions, make each figure easy to read. The lettering and
symbols should be big enough to be legible and discernible *after* the graph is
reduced. For lettering, most journals prefer the type font Arial or Arial
Narrow no smaller than 8 point after reduction. You can check legibility by
reducing the figure to publication size on a photocopier. For symbols, make

them three to five times the width of curves in the graph (see Example 9-12). Do not use X, +, 0, or *, as these symbols are not distinctive enough. Instead, go with the most common symbols, such as circles, triangles, and squares. If possible, keep similarly shaped and colored symbols separate. It may be difficult to distinguish these symbols after reduction. If data points overlap, draw them overlapping. If the points coincide, use just one symbol for these points.

Do not clutter your graphs. If there is no room for curve labels or a key on the face of the graph, define the curves in the figure legend. In addition, avoid using grid lines within the graphs. They usually contribute to clutter. Grid lines are useful when plotting points but only rarely afterward. Furthermore, do not draw boxes around line graphs unless the journal specifically requires this. Example 9-13 is an example of a cluttered graph.

Example 9-13

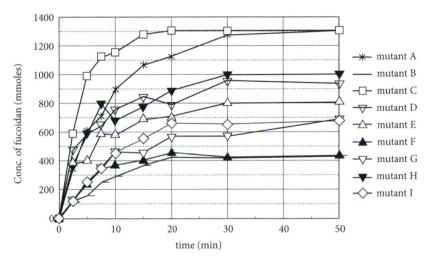

Figure. Cluttered line graph. Curves are superimposed, making it difficult to distinguish individual curves and data points. Grid lines add to the clutter.

Lines and Curves

➤ Differentiate points, lines, and curves well

Differentiate curves by using different symbols for points joined by the same type of line or by using the same symbol joined by different types of lines. When only two curves need to be differentiated, use different types of lines.

Where appropriate, draw a vertical line to show the standard deviation (σ_x, see Chapter 10, Section 10.6), the standard error (the standard deviation of the sample mean $\overline{X}$; standard error = standard deviation$/\sqrt{(sample\_size)}$), or a certain confidence interval (e.g., a 95% interval) for each data point. These lines are usually drawn in pairs in line graphs, one above and one below a data point. For bar graphs, it is really only necessary to draw the top line of each pair.

Make the lines of the error bars thinner than other lines in the main body of the data, and do not let them overlap. In the legend, tell readers what the vertical line represents, and state how many observations each mean is based on.

Rather than using a key, label curves and other components directly where possible (but check your target journal's practice). Keep labels well away from lines and curves, and position them horizontally.

Axis Labels and Scales

➤ Label axes and scales well

Label the axes of graphs as briefly and simply as possible. Write labels for the y-axes horizontally or parallel to the axis, reading upward. Write the label for the x-axis parallel to it, and center the x-axis label below the horizontal line.

For the axis label, name the variable and give the unit of measurement in parentheses (use SI abbreviations). Use lowercase lettering for labeling axes. Reserve capital letters for the first letter of the first word, for any other words usually written with an initial capital, and for abbreviations. Explain all abbreviations in the legends.

Choose scales for axes carefully. For example, for large-scale numbers, use $\times 10^y$. If an axis does not start at zero (or 1 for log scales), mark a break in the scale. Do not extend axis lines beyond the last marked scale point, and do not end them with an arrow pointing away from zero. Mark scale calibrations (tick marks) clearly. Put them outside the axis, and mark and number only as many as are necessary for clarity (see Example 9-14).

Example 9-14

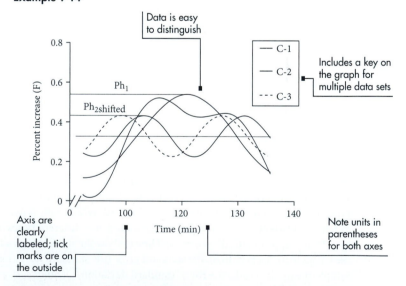

Figure. Clearly designed graph. Curves are easy to distinguish, the graph is uncluttered, and axes are easy to read.

Log-graphs

➤ **Use a logarithmic scaled graph when your data cover a large range over many powers of 10**

Using a logarithmic scale creates a more uniform distribution of data points. In log base 10, each successive whole number is 10 times larger than the preceding whole number.

Semilog graphs use a logarithmic scale on one axis (usually the *y*-axis) and a linear scale on the other axis (usually the *x*-axis). On the logarithmic scale, each cycle division differs by a factor of 10 (or a multiple thereof). So, for data ranging from 1 to 1,000, label the first tick mark as 1 in the first cycle, the one after that as 2, then 3, and so forth until you reach 10, which is the start of the next cycle. The tick mark after 10 would then be labeled as 20, followed by 30, and so forth until you reach 100 in the third cycle. The third cycle would go to 1,000 in multiples of 100 (see Example 9-15).

For data with much lower values, you could also start with 0.001; then the next cycle would be 0.01, then 0.1, and 1.0. For larger data values, you could use 10^3, 10^4, 10^5, and 10^6. What is important is that each cycle division differs by a factor of 10.

Example 9-15   **Semilog grid**

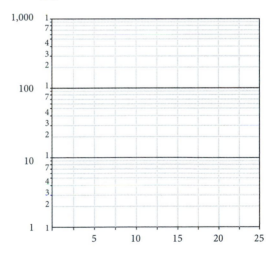

9.7   EXAMPLES OF GRAPHS

The following examples (Examples 9-16, 9-17, and 9-18) display graphs with common faults. Their revised versions show how these figures can be represented better.

Example 9-16

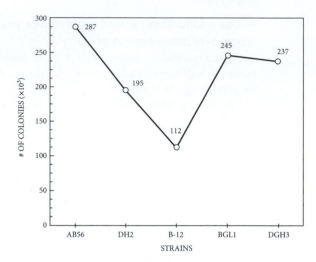

Figure. Graph with some common faults: (1) excessive numbering and tick marks on the y-axis; (2) calibrations (tick marks) face inward where they are either unnecessary or may conflict with the data; (3) heavy black frame is distracting; (4) labels on the x- and y-axes are capitalized, and thus difficult to read; (5) background color is unnecessary; (6) a line graph is plotted, although exact values seem to be compared and of importance.

Revised Example 9-16

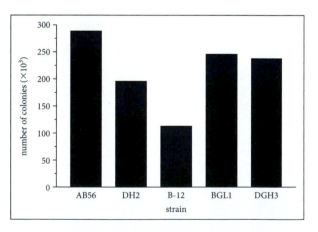

Figure. A more appropriate presentation of the findings in Example 9-16: (1) the number of tick marks has been reduced; (2) tick marks face outward; (3) the box around the graph has been removed; (4) the x- and y-axes labels are easier to read written in lowercase letters; (5) the background color has been removed; last but not least, (6) data are compared in a bar graph rather than a line graph.

Example 9-17a

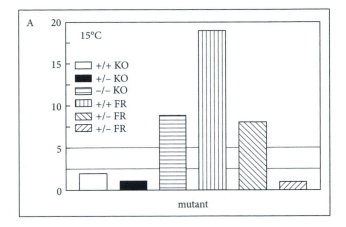

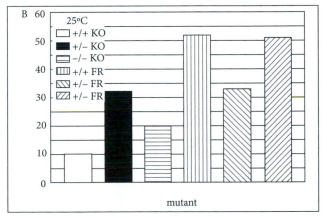

Figure. Graph with several common faults: (1) scales in A and B are different, distorting the information; (2) horizontal grid lines are distracting and make the graph appear cluttered; (3) key within the figure adds to the clutter; (4) box around the graph is unnecessary; and (5) the temperature labels are emphasized too much.

Revised Example 9-17a

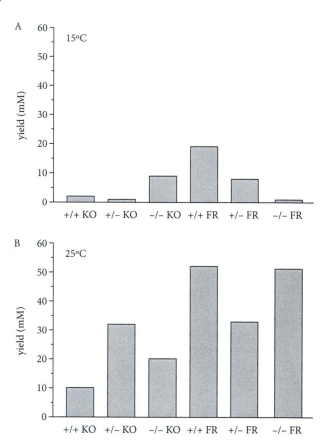

Figure. A better presentation of the data in Example 9-17a. In the revised figure, the two graphs are drawn to the same scale, there is economy of line and lettering, and labels are correctly positions on both the x- and y-axis and arranged to read upward. Grid lines have been removed, and information in the key has been incorporated in the x-axis labeling. In addition, temperature indications are much less pronounced.

In Revised Example 9-17a, the two graphs are drawn to the same scale, there is economy of line and lettering, and labels are correctly positioned on both the *x*- and *y*-axis and arranged to read upward. Grid lines have been removed, and information in the key has been incorporated in the *x*-axis labeling. In addition, temperature indications are much less pronounced.

An alternative to Example 9-17a is to graph the data in two different graphs with the same scale for the *y*-axis but depict the *y*-axis as interrupted for the graph with the data displaying the upper end of the range. Another way to depict widely ranging data on the same graph is by using an insert to enlarge data with much smaller values, such as shown in Example 9-17b.

Example 9-17b

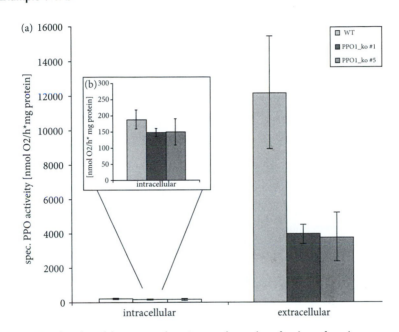

Figure. Graph with widely ranging data. Insert enlarges bars for data of much smaller values.

(With permission from Hanna Richter)

Example 9-18

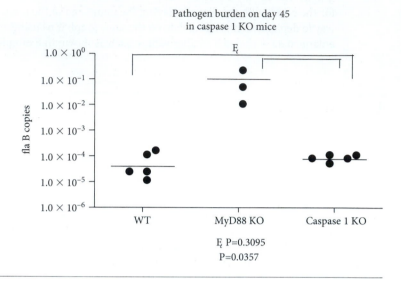

Pathogen burden on day 45
in caspase 1 KO mice

E_t P=0.3095
P=0.0357

A more appropriate presentation of findings in Example 9-18 is shown in the revised example. The data are now represented in a bar graph. Error bars indicate the spread. The title, as well as values for EP and P, has been placed in the figure legend.

Revised Example 9-18

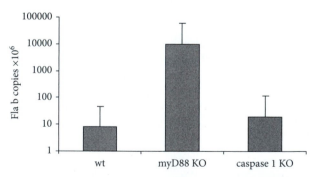

Figure XXX: Pathogen burden on Day 45 in caspase 1 KO mice.
EP = 0.3095, P = 0.0357.

Note that for Revised Example 9-18, if there is considerable spread in the original measurements, a box plot or even the original points, similar to the original Example 9-18, may be more appropriate.

9.8   FIGURE LEGENDS

➤ Describe the figure content in the figure legend

A figure legend, or caption, is typically provided close to each figure in a research or review article, proposal, poster, or research statement. Figure legends do not usually accompany figures in presentations, however. The figure legend should objectively describe the contents of the figure. Include enough information so that the reader can understand the figure easily. A figure and its legend should also be able to stand alone. The legend should contain:

- A title
- Description of figure contents
- Explanations of symbols and abbreviations (written in telegraph style)

In addition to these components, you should provide statistical and experimental details as appropriate and required. Be sure to check the *Instructions for Authors* as well as the content of legends in papers published in your target journal to understand what is required for publication in that particular journal.

The Figure Title

Figures with only figure titles do not normally appear in scientific papers, but they can often be found on slides, posters, or certain state or federal reports. Titles do not normally appear in scientific papers other than in the figure legend. Like the titles appearing in figure legends, the title on a slide or poster figure should be brief and written as an incomplete sentence using a maximum of three to four key terms and no abbreviations. Often, figure titles state important results depicted by the figure.

Description of Figure Content

Draft a set of legends in the style usually used in your target journal. Keep the draft legends and the figures in their preliminary form beside you while you write the first draft of the paper. When you have the final version of the figure, rewrite the legend if necessary.

The descriptive material in a legend may include letters or symbols to explain special abbreviations and symbols shown in the figure. Abbreviations and symbols should be consistent between the figure and its legend, consistent between legends, and consistent with the text as well. Use parallel wording for similar illustrations. Do not write the phrase "see text for explanation."

Example 9-19 shows two different figure legends. The legend in Example 9-19a describes the figure content and that in Example 9-19b describes the experimental details and provides some statistical information in the form of a *p*-value.

Example 9-19   **a**   Figure XXX: Western blot of protein Z deletion analogs. Lane 1 and 10, molecular weight standards. Lane 2, protein Z. Lane 3–6, ammonium sulfate pellets of the processed deletion analogs that exhibited agglutination. Lane 3, N18; lane 4, N74; lane 5, C166; lane 6, C121. Arrows indicate position of protein Z and its deletion analogs.

b   Fig. ZZZ: *In vitro* activity of X. Protein extracts were prepared from cell cultures grown under standard conditions for 14 days in medium Z. X activity was determined by X activity assay using Y as a substrate. Assays were performed in triplicate; $p = 0.05$.

9.9   GENERAL INFORMATION ON TABLES

➤ Prepare tables rather than graphs when it is important to give precise numbers

Like figures, tables exhibit data visually, and the information presented should be independent of the text. Tables present data more exactly than a graph but usually do not show trends within your data. Tables also present facts more concisely than text does while allowing a side-by-side comparison of them. Design separate tables for separate topics. Do not use tables just to show off how much data you have collected. Limit the number of tables, and arrange the data to allow for easy interpretation. Do not repeat data in tables if you plan to put the same data in the text or in a figure, but state the most important data in the text as well.

Check the *Instructions to Authors* to see whether the journal limits the number or size of tables. You will probably make rough versions of tables as your work proceeds. Before drafting the text, decide which tables you need for the paper, and redesign them with publication in mind.

Producing Tables

Tables can easily be prepared electronically. You can use spreadsheets such as Excel to convert your data into a table. Alternatively, you can use MS Word to create a table directly in a word file.

9.10   FORMATTING TABLES

➤ Keep the structure of your table as simple as possible

Keep the structure of your tables as simple as possible. Use tables published in your target journal as models for drafting your tables.

A typical scientific table consists of a title, column headings, row (or side) headings, the body (the rows and columns containing the data), and usually, explanatory notes. Many tables in scientific papers follow the pattern shown in Example 9-20a and 9-20b, with row entries, column headings, and the body of the table.

➤ Place familiar context on the left and new, important information on the right

Because readers read from left to right, they interpret tables more easily if you place familiar information on the left and new, important information on the right.

Example 9-20a   General format of a table

Position table caption above the table and provide a title

Table X. Initial stocking of R. sylvatica, A. maculatum, and A. opacum in pond enclosures

Species	Pond 1 enclosure[a] (number of individuals)	Pond 2 enclosure (number of individuals)[a]
Rana sylvatica	100	50
Ambystoma maculatum	50	100
Ambystoma opacum	10	10

[a]Footnote a.

Headings include units in parentheses where needed

Example 9-20b   Another general format of a table

Table Y. Colony diameter of E. coli, S. aureus, and S. pneumonia under different antibiotic media additions

	Ampicillin (5 mg/l)	Rifampin (10 mg/l)	Streptomycin (10 mg/l)
Diameter of E. coli (mm)	0.21	0.12	0.43
Diameter of S. aureus (mm)	0.45	0.16	0.56
Diameter of S. pneumonia (mm)	0.25	0.96	0

Organize your tables logically to make mental comparisons easy for readers (e.g., pretreatment measurements should precede posttreatment ones). In addition, arrange the column headings and the data in a logical order, and make sure that your listings and numbers are aligned. It is also easier to follow similar components when they are arranged vertically, not horizontally, as in the previous two examples.

Table Titles

➤ Design table titles to identify the specific topic

Draft a concise title for each table. State the point of the table or say which items it compares, and perhaps indicate the experimental design.

Example 9-21   a   Table B. Characteristics of intestinal flora

b   Table C. Soil analysis

The titles in Example 9-21 are not sufficient. Titles should be specific and informative without being wordy.

Revised   a   Table B. Characteristics of intestinal flora found for 21
Example 9-21   patients with HIV-I

b   Table C. Soil analysis of six farm fields near New
Haven, CT

The titles in Revised Example 9-21 are much better titles because they are much more specific. Aim for consistency in style and length of titles.

Column and Row Headings

➤ Label dependent variables in column headings and independent variables in row headings

Columns and rows in a table typically have a heading. Column headings should label dependent variables and row headings, independent variables. Columns and rows may also have subheadings. Headings and subheadings both should include units in parentheses if appropriate.

Set the column and row headings apart from the rest of the table by using a different font, boldface, or capitalization. Capitalize column headings the same as the table title, but use a smaller font. Center one-line column headings, and write all headings horizontally. If the column headings are long, consider placing them at a 45-degree angle reading upward. Alternatively, you may also spread the headings in both columns and rows over more than one line.

To keep column headings brief, and thus save space in the table, use short terms or abbreviations in the column headings and subheadings and explain the abbreviations in footnotes. Note that more abbreviations are used in tables than in the text. If the abbreviation is not defined in a footnote, however, readers who do not know the meaning (and there are always some readers who do not) frustratingly have to search through the text to find the definition.

Keep row headings short as well, and left justify or right justify them. Capitalize only the first letter of the first word in these headings.

Table Size

➤ Avoid overly large tables

A table must contain enough data to justify its existence but not so much as to overwhelm the reader. If your proposed table has only one or two rows of data, it is preferable to present your findings in one or two sentences in the text instead of constructing a table. Similarly, if your table lists descriptions in words rather than numbers, consider whether you really need a table—a few sentences in the text may be better. Table 9.8 in the next example could easily be converted to text.

Example 9-22

Table. Antibiotic targeting of various organisms

ANTIBIOTICS TARGET	ORGANISMS	CELLULAR
I	*Bacillus*	Ribosomes
I	*Saccharomyces*	Mitochondria
II	*Bacillus*	Ribosomes
III	*Streptococcus*	Cell walls
III	*Saccharomyces*	Mitochondria

Revised Example 9-22   *Bacillus* ribosomes were targeted by antibiotics I and II, *Streptococcus* cell walls were targeted by antibiotic III, and *Saccharomyces* mitochondria were targeted by antibiotics I and III.

If you are planning on showing your data in a presentation or in a poster, however, use a table—even if small—rather than text as a table is more memorable and has more visual impact than text. Do not use overly busy tables on slides, however, as the audience can only absorb a limited amount of information in a short time. Thus, very large tables should only be used in written communications and only if really needed.

If possible, avoid oversized tables. Try to design tables that fit easily into one or two journal columns or on a single slide. Do not try to cram a table onto the page by using narrow spacing or tiny print. The copy editor needs room to insert typographical instructions, and too-small print is unsuitable for publication. Make the lettering suitable for reduction to 50% to 75% of the original size. If a table seems too big, ask yourself first whether all the columns and rows are needed. For example, columns that contain only one value can be reported in the text or in a footnote. Columns of less important data can be omitted. If you cannot eliminate any columns, try splitting a very wide table into two smaller tables. Alternatively, if your collection of data is very large or likely to interest only a few readers, send your data to an archive as supplementary material for storage separately from the published paper (see *Instructions to Authors* of your target journal). Make sure, however, that the tables you submit for publication contain enough values for referees to assess what you have done and for other scientists to check their results against yours.

9.11   FORMULAS, EQUATIONS, PROOFS, AND ALGORITHMS

Aside from figures and tables, many scientific fields make use of formulas, equations, proofs, or algorithms. Professional presentation of formulas and equations is an art. Although in this book I do not delve into much detail on these, I describe these additional types of supplementary information in general terms. For a more thorough treatment of formulas and equations, see, for example, Ebel, Bliefert, and Russey (2004).

Formulas and Equations

➤ Treat equations and formulas as part of the text

- Explain all key details of an equation.
- State any assumptions you have made.
- Explain why and how you arrived at a solution.

Often, scientific documents contain formulas or equations. Chemical formulas may have to be created using specialized programs such as Chem-Sketch or ISIS/Draw. Usually, formulas are set apart in a separate line and indented or centered.

Example 9-23

Mathematical formulas or equations are often created as graphic objects that can be pasted electronically into a word-processing file. The best-known programs for mathematical formulas and equations in a word-processing context are Tex and Latex. Other programs that enjoy much popularity with mathematicians are FrameMaker, MathType, and MathML. Equations can also be inserted directly in word-processing programs such as in Microsoft Word (go to the Insert menu, choose Object in the Text box, then select Microsoft Equation).

When writing equations, use mathematical notation correctly and learn how to use symbols properly. Brief equations or definitions may be included in the text if space allows it, although you should not start a sentence with a formula.

Example 9-24   Eligible individuals are vaccinated at a rate of $\kappa_a(\alpha)\phi(\alpha)$, and a vaccine reduces the infection incidence by $1 - \eta = 10\%$ for severe infections and by $1 - \xi = 10\%$ for mild infections.

Lengthy equations or mathematical expressions, and those with larger than normal height, should be set apart in a separate line, indented or centered, and isolated from the surrounding text. If you plan to refer back to them later in your document, number them. Equation numbers are placed at the right margin. An example of such an equation is shown in Example 9-25.

Example 9-25   The average growth spurt per animal over the age interval i to $i + 1$ years can be estimated by (34)

$$\lambda_i = -\ln\left(\frac{1 - \rho_{i+1}}{1 - \rho_i}\right) \qquad \text{(Eq 3.2)}$$

Even if an equation is displayed, it is treated as if it were grammatically part of the text. Punctuation rules apply accordingly. For a sequence of equations, align the = symbol in each line or, if either side is long, align the = symbol with the first operator in the first line.

Note that you need to explain all key details of an equation. State any assumptions you have made, and explain why and how you arrived at a solution. If you are representing quantities and functions with arbitrary letters, define these letters in your formulas.

Equations are often used in proofs. If these proofs are long or if space is limited for the journal article, you may be asked to place these and other elaborate calculations into an appendix. An example of such a proof is shown in the next example.

Example 9-26   **Proof (in an Appendix)**
We use mathematical induction to prove that a skewed distribution as described by Equation 1 in the text is an equilibrium consequence of a variable number of tandem repeats (VNTR) model assuming a minimum size of $\alpha > 0$, a maximum size of Ω, and insertions and deletions that are equally probable and that proceed at a rate proportional to the number of tandem repeats.

Let C_j be the count of genotypes of repeat number j. Because insertion and deletion mutations from each size class are assumed to occur at a rate μ times the number of repeats, the average counts in each size class change each $1/\mu$ generations as $C'_j = C_j + (j-1)C_{j-1} - 2jC_j + (J+1)C_{j+1}$. At equilibrium, $C'_j = C_j$, so that

$$(j-1)\overline{C}_{j-1} - 2j\overline{C}_j + (j+1)\overline{C}_{j+1} = 0 \qquad \text{(A1)}$$

Because $\alpha > 0$, C_1 is a boundary, and transitions from C_1 to C_0 are prohibited. Therefore, from Equation A1 for the case of $j = 1 - \overline{C}_1$ $j = 1, -2 + 2\overline{C}_2 = 0$. Thus,

$$\overline{C}_2 = \frac{1}{2}\overline{C}_1. \qquad \text{(A2)}$$

Substituting Equation A2 into Equation A1 for the case of $j = 2$, $\overline{C}_1 - 4\left(\frac{1}{2}\overline{C}_1\right) + 3\overline{C}_3 = 0$. Thus,

$$\overline{C}_3 = \frac{1}{3}\overline{C}_1. \qquad \text{(A3)}$$

Similarly, substituting Equation A3 into Equation A2 for the case $j = 3$,

$$\overline{C}_4 = \frac{1}{4}\overline{C}_1. \qquad \text{(A4)}$$

A pattern is apparent. To prove that for all positive integers j, $\overline{C}_j = \frac{1}{j}\overline{C}_1$, we assume that for some positive integer k,

$$\overline{C}_{k-1} = \frac{1}{k-1}\overline{C}_1, \qquad \text{(A5)}$$

$$\overline{C}_k = \frac{1}{k}\overline{C}_1, \qquad \text{(A6)}$$

and show that when Equations A5 and A6 are true, it is also true that $\overline{C}_{k-1} = \frac{1}{k-1}\overline{C}_1$.

From Equations A1, A5, and A6,

$$(k-1)\left(\frac{1}{k-1}\overline{C}_1\right) - 2k\left(\frac{1}{k}\overline{C}_1\right) + (k+1)\overline{C}_{k+1} = 0. \text{ Thus,}$$

$$\overline{C}_{k+1} = \frac{1}{k+1}\overline{C}_1, \qquad \text{Q.E.D.} \qquad \text{(A7)}$$

The result, $\overline{C}_j = \frac{1}{j}\overline{C}_1$, is demonstrated for the case $j = 1$ by identity and for integral values of j between 1 and 4 by Equations A2 to A4. It is demonstrated for all integers $j > 4$ by the satisfaction of Equations A5 and A6 for the case $j = 4$ and the proof of Equation A7.

If there is a maximum number of repeats for a VNTR of Ω, beyond which the VNTR may not grow, the equilibrium frequency of size class i, $\bar{P}_i$, is then

$$\bar{P}_i = \frac{\dfrac{1}{i}}{\displaystyle\sum_{j=\alpha}^{\Omega} \frac{1}{i}} = \frac{1}{i \displaystyle\sum_{j=\alpha}^{\Omega} \frac{1}{i}}. \tag{A8}$$

(With permission from Macmillan Publishers Ltd.)

Algorithms

➤ Present algorithms in the text, as a flow chart, or as a list of instructions

Mathematics, computer science, informatics, computational biology, and other related fields often use algorithms for calculation and data processing. An algorithm is a sequence of well-defined, finite instructions for completing a task.

Algorithms can be presented in the text, as a flow chart, or as a list of instructions. When you want to display what the implementation looks like, it is clearest for readers if, like figures and tables, algorithms are labeled and numbered and set off from the main text, for example, by being boxed. However, if you are discussing an algorithm more abstractly, then a written description may be better as not all descriptions lend themselves to a pseudocode representation (structured English for describing algorithms).

Pseudocode describes the entire logic of the algorithm and allows the designer to focus on the logic without being distracted by details of language syntax. Pseudocode is not a rigorous notation. In fact, each individual designer may have their own personal style of pseudocode, but it is helpful to follow a similar style.

The logic within the pseudocode must be decomposed to the level of a single loop or decision, but not all lines within an algorithm are treated equally. All variables and functions are italicized, and all commands are boldfaced. In addition, certain commands that control flow, such as functions, classes, and methods, have associated indentation levels, as do **try/ catch** blocks and the basic constructs that use the commands **for, if, while,** and **do** (counting loops, selections, repetitions). Constructs can be embedded within each other. If such nested constructs are used, they should be clearly indented from their surrounding constructs, and individual loops and branches should be kept at the same indentation distances as is shown in the following example.

Example 9-27   General algorithm

Algorithm 2 mergeSubtrees(StateList *leftList*, StateList *rightList*, node *root*)

Require: *leftList* and *rightList*: the lists of partial states, *root*: a tree node.

Ensure: Set of valid, non-zero probability states combining elements in *leftList* and *rightList*.

```
 1:   mergedList ← emptyList
 2:   for all partial states l in leftList do
 3:     for all partial states r in rightList do
 4:       if compatible(l, r) = true then
 5:         m = merge(l, r)
 6:         if root = initialroot then
 7:           mergedList.add(m)
 8:         else
 9:           for op ∈ {C; D; I; C*; D*; I*} do
10:             if is PossibleUpstream(m,op) then
11:               mergedList.add(addAncestorBranch(m,op))
12:             end if
13:           end for
14:         end if
15:       end if
16:     end for
17:   end for
18:   return mergedList
```

(With permission from Mary Ann Liebert, Inc. Publishers)

References for Formulas, Equations, and Algorithms

References

Alley, M. (1996). *The craft of scientific writing* (3rd ed.). New York: Springer.

Higham, N. J. (1998). *Handbook of writing for the mathematical sciences.* Philadelphia: SIAM.

Kovac, J., & Sherwood, D. W. (2001). *Writing across the chemistry curriculum: An instructor's handbook.* Upper Saddle River, NJ: Prentice Hall.

Krantz, S. G. (1997). *A primer of mathematical writing.* Providence, RI: American Mathematical Society.

Miller, A., & Solomon, P. H. (2000). *Writing reaction mechanisms in organic chemistry.* New York: Academic.

Online Sources

A few online sources provide further insight into formulas, equations, and algorithms.

> http://books.google.com/books?id=MfKmWlzF73UC&pg=
>   PA234&lpg=PA234&dq=Miller,+A+and+Solomon+P.H.+-
>   +Science+%E2%80%93&source=bl&ots=E4c3aayM4M&sig
>   =NM0peu4N2mphVcnjHZZp-SFnwUs&hl=en&sa=X&oi=
>   book_result&resnum=1&ct=result
> http://www.math.niu.edu/~behr/Teaching/writing.html
> http://www.mit.edu/afs/athena.mit.edu/course/other/mathp2
>   /www/piil.html
> http://www.mit.edu/people/dimitrib/Ten_Rules.pdf
> http://web.mit.edu/jrickert/www/mathadvice.html
> http://www.geneseo.edu/~mclean/Dept/JournalArticle.pdf
> http://versita.com/UserFiles/File/Authors/CEJP/CEJP_Paper
>   WritingGuide.pdf

SUMMARY

GENERAL ILLUSTRATION GUIDELINES

- Decide whether to present data in a table, a figure, or in the text.
- Use the fewest figures and tables needed to tell a story.
- Design figures and tables to have strong visual impact.
- Figures and tables should be able to stand on their own.
- Prepare figures and tables with the reader in mind—place information where the reader expects to find it.

GUIDELINES FOR FIGURES

- Use figures to show trends and relationships and to emphasize data.
- Prepare professional figures.
- Do not mislead readers.
- For graphs:
  - Use line graphs for dynamic comparisons.
  - Use scatter plots to find a correlation for a collection of data.
  - Use bar graphs when the findings can be subdivided and compared.
  - Use pie charts to compare parts of a whole.
  - Use box plots to display differences between data sets, especially in descriptive statistics.

> ° Place the independent variable at the x-axis and the dependent variable at the y-axis.
> - Make each figure easy to read.
>   ° Differentiate points, lines, and curves well.
>   ° Label axes and scales well.
>   ° Use a logarithmic scaled graph when your data cover a large range over many powers of 10.
>   ° Describe the figure content in the figure legend.
>
> GUIDELINES FOR TABLES
> - Prepare tables rather than graphs when it is important to give precise numbers.
> - Keep the structure of your table as simple as possible.
> - Place familiar context on the left and new, important information on the right.
> - Design table titles to identify the specific topic.
> - Label dependent variables in column headings and independent variables in row headings.
> - Avoid overly large tables.
>
> GUIDELINES FOR FORMULAS, EQUATIONS, PROOFS, AND ALGORITHMS
> - Treat equations and formulas as part of the text
>   ° Explain all key details of an equation.
>   ° State any assumptions you have made.
>   ° Explain why and how you arrived at a solution.
> - Present algorithms in the text, as a flow chart, or as a list of instructions.

PROBLEMS

PROBLEM 9-1 Figure, Table, or Text

What format (table, graph, text) would you use for the following examples? Justify your choices.

1. You have studied ice location, thickness, and consistency in Greenland and have collected a series of data on these. What is the best way to present these data?
2. Your lab has assessed the rate of mortality for different mammals exposed to the West Nile virus in all the counties of Connecticut. What type of table or figure should you use to present your findings?

3. You and your colleagues have assessed the effectiveness of a new Lyme disease vaccine after inoculating mice. You have evaluated changes in their leukocyte and platelet counts with various doses of the vaccine over six months. During the course of this study, you have also taken photographs of the animals and the vivarium. What information should you present in a paper?

4. You plan to publish a paper on a new species of fungus you discovered during a trip to the rainforest in Costa Rica. You have several nice photographs of yourself in the rainforest as well as photographs of four known species of fungus and one of the new species. Which photograph should you include in your publication?

5. You want to explain a new type of apparatus, a rotating cylindrical annulus apparatus, that was used in your experiments. Should you use a schematic or describe it in the text or both?

6. You have constructed a new drug delivery system involving the use of nanoparticles to deliver anticancer agents to specific brain tumors. Should you explain the system using a schematic or describe it in the text?

PROBLEM 9-2 Figure or Table

Given the raw data provided (in triplicate readings), create a table or figure, whichever represents the data best. Time in hours, units of measurement in ml.

$$\text{Time0, KA} = \left\{ \begin{array}{l} 6.7(1.\text{reading}) \\ 8.1(2.\text{reading}) \\ 7.0(3.\text{reading}) \end{array} \right. \quad \text{KB} = \left\{ \begin{array}{l} 3.9 \\ 3.9 \\ 3.8 \end{array} \right. \quad \text{TFR} = \left\{ \begin{array}{l} 0 \\ 0 \\ 0.1 \end{array} \right.$$

$$\text{Time2, KA} = \left\{ \begin{array}{l} 7.2 \\ 7.4 \\ 6.9 \end{array} \right. \quad \text{KB} = \left\{ \begin{array}{l} 5.0 \\ 5.1 \\ 4.7 \end{array} \right. \quad \text{TFR} = \left\{ \begin{array}{l} 1.2 \\ 2.1 \\ 1.4 \end{array} \right.$$

$$\text{Time3, KA} = \left\{ \begin{array}{l} 5.8 \\ 6.7 \\ 6.1 \end{array} \right. \quad \text{KB} = \left\{ \begin{array}{l} 5.4 \\ 5.5 \\ 5.3 \end{array} \right. \quad \text{TFR} = \left\{ \begin{array}{l} 1.4 \\ 1.0 \\ 1.0 \end{array} \right.$$

$$\text{Time4, KA} = \left\{ \begin{array}{l} 5.8 \\ 6.0 \\ 5.7 \end{array} \right. \quad \text{KB} = \left\{ \begin{array}{l} 5.9 \\ 5.9 \\ 5.8 \end{array} \right. \quad \text{TFR} = \left\{ \begin{array}{l} 2.0 \\ 1.0 \\ 1.1 \end{array} \right.$$

PROBLEM 9-3 Figure

Evaluate the following electron micrograph. Is this a good or bad figure? Why or why not?

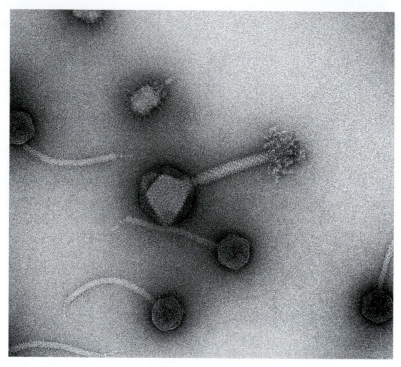

Figure A. Electron micrograph of three types of *Bacillus subtilis* viruses: SPP1 (Siphoviridae), SPO1 (Myoviridae), phi29 (Podoviridae). The phages are negatively stained with uranyl acetate.

(With permission from Rudolf Lurz)

PROBLEM 9-4 Graph or Table

Given the raw data provided, create a table or figure, whichever best represents the data.

Time 0:
Population A: 123 members
Population B: 54 members
Population C: 99 members
Time 1 month later:
Population A: 133 members
Population B: 28 members
Population C: 98 members
Time 2 months later:
Population A: 142 members
Population B: 6 members
Population C: 98 members

PROBLEM 9-5 Figure

Evaluate the following graph. How could this graph be improved? Make a list.

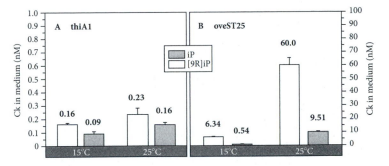

Fig. 1: Thermal induction of cytokinin overproduction in the mutant oveST25. Comparison of cytokinin concentration measured by HPLC-ELISA in the culture medium of thiAl (auxotrophic wt) and its temperature sensitive mutant oveST25. iP (isopentenyladenine) and [9R]iP were measured in the medium of liquid cultures cultivated continuously at 15°C or 25°C. The age of the culture was 3 weeks; 2.1–3.5 mg chloronema tissue were used per ml medium.

(With permission from the American Society of Plant Biologists)

PROBLEM 9-6 Figures

Evaluate the following graphs. Explain why this figure is confusing.

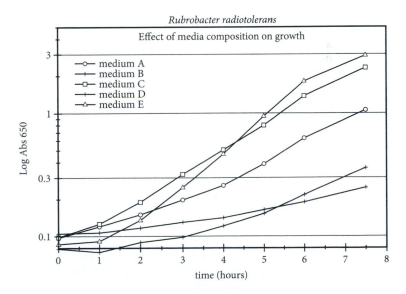

PROBLEM 9-7 Table or Figure from the Literature

Find a badly represented table or figure in the scientific literature, and discuss why it is bad and how it could be improved.

CHAPTER 10

Basics of Statistical Analysis

A basic understanding of statistics is necessary and important in the sciences. Similar to students wanting to know the class average and high and low scores of an exam, scientists seek to gain and compare statistical data to be able to gauge the significance of findings. Although this chapter also reviews basic statistical tests and fundamental analyses, most important, it serves as a guide on how to report these in scientific reports.

THIS CHAPTER DISCUSSES:

- Basic statistical terminology
- The most common distribution curves
- Statistical analysis, including a simple decision tree
- Reporting and presenting statistics
- Resources for statistics

10.1   GENERAL GUIDELINES

➤ Use statistical analysis to determine data trends and chance occurrence

Experimental results often include statistical tests to describe the overall study sample, to analyze data trends for an experiment, and/or to determine whether observed differences, if any, might be due to chance. Results from statistical analyses inform the level of precision around findings to back up your interpretations and conclusions convincingly. Knowing how to report the most common statistical tests is important for reproducibility of methods and adequate interpretation of variation in your data. Statistical analysis enables you to judge objectively how typical or unusual an event is.

It also allows you to determine the probability of an observed difference happening randomly or because of an underlying general cause.

10.2   BASIC STATISTICAL TERMINOLOGY

➤ Know the most common statistical terms and calculations

You should be familiar with the most common terms and methods and their meanings. Note that graphing programs and spreadsheets like Excel often have tools to perform statistical analysis (Analysis Toolpak for Excel, for example, is an add-on program that can be downloaded). These tools make it easier for students to determine statistical values.

mean (x̄ or μ)	the arithmetic average of a set of values, that is, the sum of the values divided by the number of the values. The mean is usually reported together with the number of measurements or sample size (n) and with the *standard deviation* or *range*. The mean is usually not as robust as the median, as it is influenced by extreme data points.

Example 10-1

> Values: 2, 4, 8, 12, 16, 18
>
> $n = 6$
>
> mean = $(2 + 4 + 8 + 12 + 16 + 18)/6 = 10$

median	the middle value in a set of numbers. It separates the higher half and the lower half of the set, and it lets us understand the central tendency of a data set better than the mean as the median is not as sensitive to extreme data points (outliers).

Example 10-2a

> Values: 2, 12, 18
>
> median = 12

Example 10-2b

> Values: 2, 4, 8, 12, 16, 18
>
> As there is no middle number in this set of values, the *median* equals the average of the two middle numbers = 10.

mode	the number(s) that is/are repeated most often in a set of values.

Example 10-3

> Values: 2, 12, 12, 18
>
> mode = 12
>
> If no numbers are repeated, there is no mode.

range

the difference between the maximum and minimum value.

Example 10-4

Values: 2, 4, 8, 12, 16, 18

range = 18 – 2 = 16

deviation

the difference between a data point value and the mean; note: deviation is not the same as *standard deviation* (see the definition of standard deviation)

Example 10-5

Values: 2, 4, 8, 12, 16, 18

$n = 6$

mean = 10

Deviation for the first data point = 2 – 10 = –8

variance (σ^2) and **standard deviation (σ_x)**

the standard deviation provides information about the variability of the data, that is, how far numbers are spread out from their average value. The *variance* is calculated by summing up the squared deviations of each data value and dividing the sum by the number of values in the data population.

Example 10-6

Values: 1, 2, 3, 4, 5

$n = 5$

mean = 3

variance = $((3 – 1)^2 + (3 – 2)^2 + (3 – 3)^2 + (3 – 4)^2 + (3 – 5)^2)/5 = (4 + 1 + 0 + 1 + 4)/5 = 2$

The *standard deviation* is the square root of the variance.

Standard deviation = $\sqrt{2}$ = 1.414

(Note that the variance described here is that for a population, that is, if all data points or members covered by a study. For a subset or sample of the total population [e.g., when you calculate the weight of male black bears in North America, you will use a sample of bears, not all male bears on the continent], the sample variance would need to be calculated by dividing the sum of the squared deviations by $(n – 1)$ to compensate for the lack of information on the population data.)

standard error

the standard deviation of a parameter (i.e., the mean or regression coefficient) in multiple samples. It is a measure of how precise an estimate is from sample to sample. As the number of samples increases, the standard error decreases. As the standard deviation of the samples increases, the standard error increases.

SE = square root of σ^2/n where n = number of samples and σ^2 is the variance of the population

Example 10-7

Values: 1, 2, 3, 4, 5

$n = 5$

Standard deviation = 1.4

Standard error = $1.4/\sqrt{5} = 0.6$

95% confidence interval   estimated range of values, derived from the standard error, which is likely to include the true population parameter (e.g., the mean of multiple samples). The 95% confidence interval indicates the level of probability that the true population parameter is contained in the reported interval of values.

10.3   DISTRIBUTION CURVES

➤ Understand the most important distribution curves

To know what statistical tests to apply to your data, you need to have a basic understanding of different types of data and their distribution patterns also known as distribution curves. Only when you have determined what distribution curves your data fall under can you decide which analysis to apply and what statistical values to report. The most frequently found distribution curves follow.

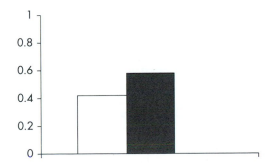

Figure 10.1 Example of a binomial distribution.

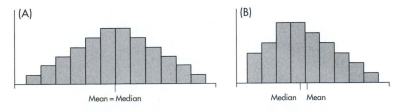

Figure 10.2 Examples of Poisson distributions. A, symmetrical; B, skewed.

Binomial Distribution

The binomial distribution (Figure 10.1) is the most well-known discrete distribution. The distribution is discrete (meaning it has a limited number of potential outcomes, that is, the data have categorical or discrete characteristics), and random variables can be listed as whole numbers. Binomial distributions are used for samples with two possible outcomes (yes/no; dead/alive; sick/healthy; etc.), such as in a coin toss, where the outcome can be either heads or tails. Each trial, known as a Bernoulli trial, has the same probability (p), and trials are independent of each other.

The mean of a binomial distribution is $\mu = np$ (where n is number of samples or trials), and the standard deviation (σ) for binomial distributions is defined as $\sigma = \sqrt{np(1 - p)}$.

Poisson Distribution

The Poisson distribution (Figure 10.2) is also a discrete probability distribution. It describes the probability of a given number of outcomes in a specific space, volume, or time if the average rate of occurrence is known and independent of each other. For example, dogs have on average a litter of seven puppies. Sometimes, however, there are fewer and sometimes more. The Poisson distribution describes the probability that the count is 1, 4, 8, 13, and so on for a given litter. Thus, it calculates the spread of occurrence. When sample size (n) is large and probability (p) is small, the Poisson distribution approaches a binomial distribution. For large n and small p, the variance of a Poisson random variable is the same as its mean.

Other discrete distribution curves include the geometric distribution, the hypergeometric distribution, the negative binomial distribution, and multibinomial distributions. Again, to learn more about these distributions, consider taking a statistics class or refer to a full statistics book such as *Statistics for Engineers and Scientists* by William Navidi or some of the other resources listed in Section 10.7.

Normal Distribution and Standardized Normal Distribution

The normal distribution, also known as the Gaussian distribution, is the most important and most commonly used continuous distribution in statistics. It is continuous rather than discrete, meaning that the potential outcomes can only be described by an interval of real numbers. Continuous random variables are unaccountably infinite, such as, for example, the amount of rain per year in a given location.

The normal distribution is often used in hypothesis tests to determine significance levels and confidence intervals. The distribution appears as a smooth, bell-shaped, symmetrical curve when values are graphed based on their frequency. For a normal distribution, the mean, median, and mode are identical as deviations within the data set are of equal variance. The normal distribution is thus described by only two parameters: the mean μ and the standard deviation σ. The probability for any value can be determined based on the number of standard deviations between the mean and the value.

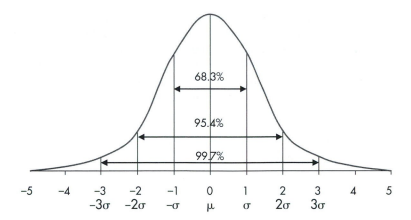

Figure 10.3 Standardized normal distribution curve.

The normal distribution can be standardized by making the mean $\mu = 0$ and the standard deviation $\sigma = 1$. This allows the standard deviation to become the unit of the horizontal axis.

In a standardized normal distribution, 68.3% of the data fall within –1 and 1, or one standard deviation, of the horizontal axis; 95.4% fall within –2 and 2, or two standard deviations; and 99.7% fall within –3 and 3, or three standard deviations (see the standardized normal distribution curve in Figure 10.3). Thus, the standard deviation gives a good criterion for judging the extent of error contained in the data.

Other continuous distribution curves include lognormal distribution, the exponential distribution, the gamma distribution, and uniform distributions. Here, too, it would be meaningful to glean details on these from a full statistics book such as *Statistics for Engineers and Scientists* by William Navidi or any of the other resources listed in Section 10.7.

10.4   STATISTICAL ANALYSIS OF DATA

Statistics is intertwined with the scientific method. To be able to analyze your experiment statistically, you need to

➤ Know the steps for performing statistical analysis
1. Design the experiment and decide on the population to be studied.
2. Identify the key variable to be studied.
3. Collect the data.
4. Determine the correct distribution.
5. Chose appropriate statistical test based on the type of variable and distribution.
6. Draw a conclusion to answer research questions.

➤ Understand the null hypothesis and statistical significance
A null hypothesis assumes that there is no difference among two or more data sets, or between results obtained and results expected. For example, in

an experiment your null hypothesis might be that there is no difference between the growth rate of chives grown at 15°C and those grown at 16°C.
Although you might observe a higher growth rate for 16°C than for 15°C,
there is still the possibility that your results may be purely due to chance,
particularly if the variability in each sample is large. To obtain a measure of
likelihood of wrongly rejecting the null hypothesis, you need to conduct
particular statistical tests.

➤ Acquire sufficient background knowledge

➤ Understand your experimental design

Before any statistical analysis, you need to understand your experimental
design to apply the correct statistical analyses. For example, understand if
your variables are categorical/discrete or continuous, know if you are working with a population or a subset or sample thereof, and know if variables
are random (values that are drawn from a larger population of values) or
fixed (remains constant and identical between studies.)

➤ Know which statistical analysis to choose

Before conducting any parametric analysis (which assumes that data has
come from a probability distribution and has defining properties), you need
to determine which test is the most appropriate to use based on how well
your data fit certain assumptions. To perform any statistical test properly,
however, it will be essential to take a statistics course, to read a book on statistical analysis, or to work with an actual statistician.

An overview of the most common statistical analyses is given in the
following list. Keep in mind that the tests listed are ultimately more complicated than described here. For instance, t-tests can be paired or unpaired and variables can be fixed or random. The analysis must also
consider if the data should have a Gaussian distribution or equal variance,
and if the p-value should be one-tailed or two-tailed. These considerations
can alter the validity of the analysis, which is often hotly contested by scientists in the literature.

Student's t-test	parametric test used to analyze quantitative data for a two-sample case in order to compare means and determine the probability (p) that the means are *statistically* different from each other; the lower p, the higher the probability that values are different from each other; if $p < 0.05$, results are generally considered *statistically* significant. The test can be subdivided into one- and two-sample t-tests, or paired and unpaired t-tests, depending on the number of variables.
Analysis of variance (ANOVA)	parametric test used to analyze quantitative data for two or more groups in order to compare means and calculate the probability (p) that the means are *statistically* different from each other. The test can be subdivided into various ANOVA tests (one-way, two-way, repeated measures, factorial), depending on the number of factors and sets.

Nonparametric tests	used for quantitative data with different variation; data do not have a normal probability distribution; generally less powerful than parametric tests.
Chi-square test	compares data obtained with data anticipated based on a hypothesis. The test determines if there is a significant difference between the expected and the observed data in order to find if this difference is due to sampling variation or a real difference. Chi-square (χ^2) is the sum of the squared difference between observed (o) and the expected (e) data, divided by the expected data. The test is usually used for independent normally distributed data. It can be further subdivided into chi-square test for goodness of fit and chi-square test for independence, depending on the number of variables (one or two, respectively).
Regression analysis	used to study the form of the relationship between two variables, such as in a dose–response relationship; in regression, one of the variables (x) is usually experimentally manipulated.
Correlation analysis	measures the strength of the relationship between two variables that are measured.

Often, scientists use a decision tree to determine which statistical test to use. Figure 10.4 provides a simplified example of such a decision tree. Trees tend to be based on general guidelines that should not be construed as rules. Your data usually can be analyzed in multiple ways, and each path can lead to a legitimate answer. Appropriate statistical tests are selected based on the type of data, whether the data follow the normal distribution, and the overall goal of your study.

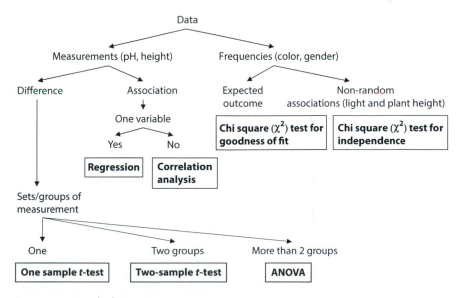

Figure 10.4 Sample decision tree.

For more resources, including websites, to learn about decision trees and how to analyze your data using statistical tests, see Section 10.7.

Consider the following example, which displays the number of bacterial colonies from two different stocks growing on the same medium.

Example 10-8

	Stock 1 # colonies	stock 2 # colonies
plate 1	9	13
plate 2	10	15
plate 3	6	18
plate 4	12	12
plate 5	9	16
plate 6	8	11
mean	*9*	*14.2*

Although the calculated means appear quite different (9 for stock 1 and 14.2 for stock 2), suggesting that stock 2 grows more colonies on average than stock 1, it is important to confirm (or reject) this hypothesis through statistical means as the variability in each sample can be large. Confirmation can be done using a *t*-test or ANOVA. Given that you are comparing only two different groups and no more, the simplest test to use is the *t*-test. (ANOVA would be used for more than two groups.) The *t*-test gives you a statistic that you can compare in a given statistical table (along with the degrees of freedom, which can be determined based on the sample size and the number of calculations of the study) to obtain the probability or *p*-value. *P*-values below 0.05 are considered statistically significant. In our example, $p < 0.01$, and therefore the mean growth of stock 2 is significantly different from that of stock 1.

Note, though, that often *p*-values can get manipulated when data points are selected until results become significant or when various statistical analyses are performed until one is found that leads to a statistically significant result. This type of manipulation is known as *p*-hacking or "selective reporting" and essentially reports a false positive. Do not fall victim to this approach.

➤ Understand correlation and causation

Distinguish between correlation and causation. Many studies assess whether a correlation exists between two variable. If it does, further study may be needed to explore whether one action causes the other, that is, if there is a time order to the variables and any alternative explanation can be ruled out. For example, a scientist may find that there is a strong correlation between green geckos and temperature. This does not mean that temperature causes the geckos to be green. There may be an underlying confounder such as location. Suppose that these geckos are found primarily in the tropical rain forest where there is much green vegetation. In this case, the camouflage pigmentation of the geckos is responsible for its green color and the gecko just happens to live where it is hot. Thus, although color and temperature are correlated, temperature is not the causation for color.

➤ Understand how to perform statistical analysis

To conduct any statistical test properly, it is essential to take a statistics course, to read a book on statistical analysis, or to work with an actual statistician (see also Section 10.7). The previously listed tests are ultimately more complicated than described here. For instance, t-tests can be paired (for comparing subjects or groups of data that are somehow related, such as for the same set of patients before and after treatment) or unpaired (for comparing groups of data that are not related, such as infection in group A versus infection in group B). The analysis must also consider whether the data should have a Gaussian distribution or equal variance and whether the p-value should be one-tailed (tests if the mean is significantly different from x in only one direction; i.e., only one tail of the normal graph) or two-tailed (tests if the mean is significantly greater or less than x; i.e., it tests both tails of a normal graph). Moreover, you may have to conduct additional tests, such as post hoc tests after ANOVA (to determine relationships between population subgroups). These considerations can alter the validity of the analysis, which is often hotly contested by scientists in the literature.

10.5   REPORTING STATISTICS

➤ Know how to report statistical data

Report statistical results in the Results section, in tables or their footnotes, or in figures and figure legends where data for relevant tests are illustrated (see also Chapter 12, Section 12.2 for placing statistical information into the Materials and Methods section; Chapter 13, Section 13.2 for reporting statistical data in the Results section; and Section 10.6 for presenting statistical results graphically). Do not duplicate reporting; in other words, if you report the results in a figure legend, do not repeat it in the text and vice versa unless you are specifically instructed to do so.

When you report statistics, pay close attention to italics, spacing, and decimal places. Typical summary statistics that are reported include the mean and the standard deviation or the standard error of the mean. These statistical values are typically rounded to the same number of decimal places as the original data values. If original data values are integers, however, then the calculated statistical values are reported to one decimal place. Note, though, that some p-values are reported to more than two decimal places. Often, the number of measurements or samples (n) or the frequency of an occurrence is also provided.

Example 10-9	a	The average height of tomato hybrid X-34 ($n = 50$) was **155 cm** (σ = **11.2 cm**).
	b	The average height of tomato hybrid X-34 ($n = 50$) was **155 cm +/– 11.2 cm**.

Example 10-10	a	**About 88%** of tomato hybrid X-34 grew to over 150 cm, but **12%** reached less than 140 cm (Table 2).
	b	Nearly twice as many boys (**66%**) than girls (**34%**) displayed symptoms.

Example 10-11	**The average growth** of tomato hybrid X-34 was **26.9 cm per month**.

➤ Know how to report statistical significance

To help understand if your results are significant, support your findings (or those of others) with statistics (p-value, χ^2, t-statistic, F-value, correlation coefficient, and/or regression coefficient). Remember though that your biological results are most important, so subordinate any statistical findings by placing them in parentheses.

Significance levels/p-values	report p-values either within the Results section, in a table, or in a figure legend as appropriate (e.g., "$p < 0.01$" or "$p = 0.02$"). Note that the term "significant" should only be used for statistical significance; otherwise, use "markedly" or "substantially."

Example 10-12

We observed a *substantial* increase in HIV infections. Disease was *significantly* delayed (10.8 years +/– 2.3 years) under treatment option A (**$p < 0.12$**).

Chi-square	report degrees of freedom and sample size in parentheses and separated by a comma following χ^2; also report the p-value. Round numbers to two decimals.

Example 10-13

There was no gender difference among infected age groups ($\chi^2(1, N = 125) = .86, p = 0.25$).

t-tests	Report like chi-squares, but with only the degrees of freedom in parentheses. Also, report the t-statistic (rounded to two decimal places) and the significance level.

Example 10-14

Gender played a significant role in our results, $t(24) = 3.29$, $p < 0.001$, with boys displaying more symptoms than girls.

ANOVAs	indicate between-groups and within-groups of freedom in parentheses and separated by comma following F; also report the p-value. Round numbers to two decimals.

Example 10-15

We observed a significant effect under treatment option A, $F(1, 95) = 3.22$, $p = 0.01$, and treatment option A + C, $F(2, 98) = 5.84$, $p = 0.02$.

Correlations

report with the degrees of freedom $(N - 2)$ in parentheses and the significance level. Report the correlation coefficient, r, as a number between -1 and 1, where $r = 1$ indicates a perfect correlation, $r = -1$ indicates a perfect inverse correlation, and $r = 0$ indicates no linear correlation between variables. Indicate also the p-value and possibly the sample size.

Example 10-16

The two variables were strongly correlated, $r(95) = .39$, $p < 0.05$.

Regression

Can be presented in the text, in a table, or in a graph. When you describe the results of your regression analysis, report the regression coefficient (β), the squared correlation coefficient (R^2), the F-value (F), degrees of freedom, the significance level (p), and the corresponding t-test. You may also want to report the regression equation.

Example 10-17

With lack of treatment, recurrence of conditions could be predicted, $\beta = -.24$, $t(105) = 8.58$, $p < 0.01$. Lack of treatment also explained a significant proportion of variance in recurrence, $R^2 = .16$, $F(1, 105) = 22.44$, $p < 0.01$.

10.6   GRAPHICAL REPRESENTATION

➤ Know how to depict statistical values graphically

Statistics can also be depicted graphically, which make values visually immediately accessible for the reader. However, you need to indicate in the Materials and Methods section or figure legend what method and software package you used to obtain your statistical values (see Chapter 12, Section 12.2 for inclusion of statistical methods in the Materials and Methods section). Statistical outcomes can be shown graphically in different ways.

Where appropriate, draw a vertical line to show the standard deviation (σ_x), the standard error (the standard deviation of the sample mean $\bar{x}$; standard standard error = standard deviation/$\sqrt{(sample\_size)}$ error = standard deviation/), or a certain confidence interval (e.g., a 95% interval) for each data point or bar. These lines are usually drawn in pairs in line graphs, one above and one below a data point (Example 10-18). For bar graphs, it is really only necessary to draw the top line of each pair (Example 10-19). Make the lines of the error bars thinner than other lines in the main body of the data, and do not let them overlap. In the legend, tell readers what the vertical line represents, and state how many observations each mean is based on.

Example 10-18

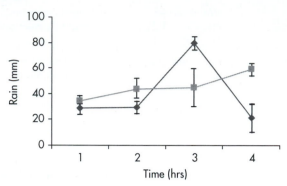

Figure 10.5 Line graph with error bars ($p < 0.5$; $n = 10$ for each data point)

Example 10-19

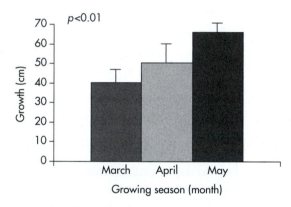

Figure 10.6 Bar graph with error bars ($p < 0.01$). Error bars typically represent one standard deviation from the mean (95% interval).

Other ways to indicate statistical values and relevance is through a scatter plot or through a box plot. Examples of these are shown in Chapter 9, Examples 9-8 and 9-11, respectively.

10.7   USEFUL RESOURCES FOR STATISTICAL ANALYSIS

General statistical analysis
- https://www.khanacademy.org/math/statistics-probability
- https://www.bmj.com/about-bmj/resources-readers/publications/statistics-square-one
- https://www.youtube.com/watch?v=-FtlH4svqx4

- https://www.youtube.com/watch?v=IV_m_uZOUgI
- https://www.youtube.com/watch?v=VK-rnA3-41c

Websites for decision trees

- http://www.ats.ucla.edu/stat/mult_pkg/whatstat//
- http://guides.nyu.edu/quant/choose_test_1DV
- https://upload.wikimedia.org/wikipedia/commons/7/74/Inferential StatisticalDecisionMakingTrees.pdf
- http://www.tnstate.edu/eduadmin/StatisticsDecisionTree%20Use.pdf
- http://www.cbgs.k12.va.us/cbgs-document/research/ Stats%20For% 20Dummies.pdf
- http://www.cios.org/readbook/rmcs/ch19.pdf
- https://onlinecourses.science.psu.edu/stat500/node/68

Textbooks and Reference Books

Berry D. A. *Statistics: A Bayesian Perspective.* New York: Wadsworth, 1996.

Carlberg C. *Statistical Analysis: Microsoft Excel 2016.* Indianapolis: Que, 2017.

Crow E. L., Davis F. A., and Maxfield M. W. *Statistics Manual.* New York: Dover Press, 2011.

Griffiths, Dan D. *Head First Statistics.* Farnham, UK: O'Reilly Media, 2008.

Milton M. *Head First Data Analysis: A Learner's Guide to Big Numbers, Statistics, and Good Decisions.* Farnham, UK: O'Reilly Media, 2009.

Navidi, W. *Statistics for Engineers and Scientists* (3rd ed.). New York: McGraw-Hill, 2011.

Ott, R. L., and Longnecker M. T. *An Introduction to Statistical Methods and Data Analysis* (7th ed.). Boston: Cengage Learning, 2015.

Sokal, R. R., and Rohlf F. J. *Biometry* (4th ed.). New York: Freeman, 2011.

Townsend, J. *Practical Statistics for Environmental and Biological Scientists.* New York: Wiley, 2002.

Vickers A. J. *What Is a p-Value Anyway? 34 Stories to Help You Actually Understand Statistics.* Boston: Addison Wesley, 2009.

Winston W. L. *Microsoft® Office Excel® 2016: Data Analysis and Business Modeling* (5th ed.). Redmond, WA: Microsoft Press, 2016.

Zar J. H. *Biostatistical Analysis* (5th ed.). Boston: Pearson, 2009.

Programs and Software

Several statistical programs exist that can be used to calculate statistical values and to visualize data graphically. Some of these are free open source software, and others are available for purchase. Add-on packages also exist for Excel. The most commonly used programs in the biological sciences include the following:

Analyze-it	This software is largely used by scientists in the life sciences, environmental sciences, and engineering.
GraphPad Prism	Originally designed for biologists in the medical sciences, GraphPad Prism is now used by throughout the biological fields as well as in the social and physical sciences.

IBM SPSS Statistics	Although originally designed for the social sciences, this package has become popular in the health science field as well and is often used for cluster analysis. PSPP is the free open source version of SPSS.
R	R is a free open source statistical software language that is best used for writing custom statistical programs. It has a very steep learning curve but is highly customizable.
SAS (Statistical Analysis Software)	This software is used for statistical analysis of clinical pharmaceutical trials. The software is great for experimental design and ANOVA. The free version of this software is Dap, but note that you should know basic C programming language when using it.
SigmaPlot	SigmaPlot is a scientific graphing program that also incorporates data analysis. It can perform regression and correlation analysis and runs on Microsoft Windows.
SimFiT	This software is good for simulation, curve fitting, statistics, and plotting. It is very user-friendly and even available in Spanish.
Stata	Stata is a general purpose statistical package designed for researchers of all disciplines. It can be used for data management, statistical analysis, graphics, simulations, regression, and custom programming.
STATISTICA	This statistics and analytics package performs data mining, analysis, and visualization and is available in several different languages.

10.8   CHECKLIST

Use the following checklist to ensure that you have addressed all important elements for a figure or a table.

☐   1. When designing your experiment, did you decide on the population to be studied?

☐   2. Did you determine the correct type of distribution curve (binomial, Poisson, normal, lognormal, exponential, etc.)?

☐   3. Did you determine the most appropriate statistical test to use based on your data (*t*-test, ANOVA, nonparametric tests, chi-square, regression analysis)?

☐   4. Did you analyze and summarize the data, such as mean, standard deviation, and statistical significance based on the distribution?

☐   5. Did you subordinate statistical results in the Results section, in tables, or in figure/figure legends?

☐   6. Did you pay close attention to italics, spacing, and decimal places when reporting statistical data?

☐   7. Did you show statistical outcomes graphically?

    ☐   a. Did you draw in correct error bars (lines above and below point for line graphs, lines only above point for bar graph)?

    ☐   b. Did you indicate the number of observations used?

SUMMARY

GUIDELINES FOR STATISTICAL ANALYSIS

- Use statistical analysis to determine data trends and chance occurrence.
- Know the most common statistical terms and calculations.
- Understand the most important distribution curves.
- Know the steps in performing statistical analysis.
  - Design the experiment and decide on the population to be studied.
  - Identify the key variable to be studied.
  - Collect the data.
  - Determine the correct distribution.
  - Chose appropriate statistical test based on the type of variable and distribution.
  - Draw a conclusion to answer research questions.
- Understand the null hypothesis and statistical significance.
- Acquire sufficient background knowledge.
- Understand your experimental design.
- Know which statistical analysis to choose.
- Understand correlation and causation.
- Understand how to perform statistical analysis.
- Know how to report statistical data.
- Know how to report statistical significance.
- Know how to depict statistical values graphically.

PROBLEMS

PROBLEM 10-1 Statistics

Determine the mean, variance, and standard deviation for the following data set:

Values: 12.4, 13.5, 12.7, 14.1, 12.7, 14.2, 13.8, 20.1, 15.3, 13.6, 13.9, 14.8, 12.9, 15.0, 13.8

PROBLEM 10-2 Normal Distribution

For a normal distribution with a mean of 100 and standard deviation of 20, calculate:

a. The percentage of the values between 100 and 120.
b. The percentage of the values between 60 and 80.
c. The percentage of the values between 80 and 140.

PROBLEM 10-3 Normal Distribution

Ethanol is produced by yeast in media whose sugar content has an optimal concentration of 5 mg/ml. At concentrations higher than 8 mg/ml, the yeast dies and the brewing process has to be suspended. If the sugar concentration in the medium is normally distributed with a standard deviation of 1 mg/ml, what percent of the time will the brewing process need to be suspended?

PROBLEM 10-4 Normal Distribution
During your experiment you determine that bacteria X generation times have a normal distribution with a mean of 20 min and a standard deviation of 5 min.

   a. What percentage of bacteria have a generation time of more than 30 min?
   b. Less than 15 min?

PROBLEM 10-5 Reporting Statistics
Are the following statistics reported accurately in a biological lab report or research article? Correct the reporting as needed.

   a. We determined that the average sea surface temperature was 24°C +/- 2°C in July.
   b. For $n = 12$ and $\sigma = 2$, we determined a mean run rate of 11.5 cm/hr on the thin layer chromatography for compound X.

PROBLEM 10-6 Reporting Statistics
You have repeatedly measured the rate of oxygen production for leaf segments from the plant *Arabidopsis thaliana*, and found that oxygen was produced at a rate of 12 bubbles/min when the leaf segment was 5 cm from the light source and at a rate of 10 bubbles/min when the leaf segment was 20 cm from the light source. How would you describe these findings in a sentence if

   a. Your calculations indicate that $p = 0.05$.
   b. Your calculations indicate that $p = 0.75$.

PROBLEM 10-7 Statistical Test
You have collected data on the coloration of grizzly bears in Yellowstone National Park and Denali National Park to determine if coloration is linked to location or average yearly temperature. Use a decision tree, such as the one in Section 10.4, to determine what statistical test you could use for this data.

PROBLEM 10-8 Statistical Test
You have measured the weight of a specific kind of tomato grown at two different elevations in Maine. Based on the decision tree in Section 10.4, what statistical test could you use to examine if there are any significant differences between your data from the two locales?

PART THREE

Manuscripts: Research Papers and Review Articles

A. RESEARCH PAPERS

CHAPTER 11

The Introduction

The Introduction of research articles should draw in the reader and contain all the information needed to provide context. It should also make clear the overall purpose of the work.

THIS CHAPTER COVERS:

- The overall structure and organization of the Introduction for hypothesis-driven and descriptive research articles
- The different elements of the Introduction
- Signaling information in the Introduction
- Common mistakes to avoid
- Revising the Introduction
- Several full and annotated sample Introductions

11.1 OVERALL

➤ Interest your audience, and provide context

The purpose of the Introduction is twofold: to encourage your audience to read the paper and to provide sufficient background information for readers to understand your study independently of other previous publications on the topic. Often, the Introduction also gives an overview of what to expect in the paper, similar to some parts of the Abstract.

11.2   COMPONENTS AND FORMAT

➤ Follow a "funnel" structure
Include:

Background
Unknown/problem
Question/purpose of study
Experimental approach
Optional: Results/conclusion
            Significance

Placing the right elements in the right order in your Introduction ensures that you keep your readers interested and that they can follow your logical argumentation. Readers have relatively fixed expectations about where in a document they will find particular items. Based on the location of these items, they will interpret the text. If writers can become aware of these locations, they can better guide the reader through the document, highlighting and emphasizing various pieces of information depending on the degree of importance.

Most research papers in basic science are investigative. That is, they are based on specific research questions you try to answer or on a particular hypothesis you try to test. Introductions for these papers should contain the following elements:

1. Background	broad and specific background information and previous research in the area
2. Unknown/problem	problems of previous work and unknown factors in the area
3. Question/purpose of study	addition made by your research
4. Experimental approach	approach taken toward this addition

Generally, readers expect the parts of the Introduction to be arranged in a standard structure: a "**funnel**," starting broadly with background information and then narrowing to knowledge on a specific aspect of the topic, to something unknown or problematic, and then to the research question of the paper and its experimental approach (see Figure 11.1; see also Zeiger, 2000).

Although not an absolute necessity, you may consider including your main results and conclusions as well as stating the overall significance of the paper to round up this section. (Note that in certain scientific disciplines, results and conclusion are typically not included in the Introduction. Check how Introductions in your field are written before writing your own.) If you include main results and conclusions, place them at the end of the Introduction. Including results and conclusion in the Introduction will let your readers know what to expect and will let them more easily follow the paper. If your paper deals with a controversial topic in your field, however, you may want to withhold main results and conclusions in your Introduction to encourage as many readers as possible to continue reading your paper and argumentation.

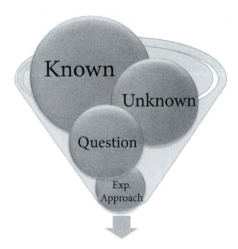

Figure 11.1 Funnel structure of the Introduction.

➤ Keep the Introduction short

The Introduction should be as short as possible but contain all the information needed to lead into the work. Ideally, an Introduction of a typical journal article is one to two double-spaced pages (about 250–600 words). In certain fields, however, Introductions can be substantially longer when a full literature review is expected at the same time.

11.3   ELEMENTS OF THE INTRODUCTION

Background

➤ Provide pertinent background information, but do not review the literature

Start the Introduction by providing some background information. The amount of background information needed depends on how much the intended audience can be expected to know about the topic. You should start very broad to provide some general context of your work. This general background would be information that any article on that topic could include (e.g., AMD disease is characterized by Y; see Revised Example 11-10; many articles on AMD could start with these sentences). Then, write about the specific aspect of the topic that is of interest, mention the existing research in the area, and discuss current beliefs. This specific background (or specific reported data) is information that is pertinent to your particular study (e.g., the pathogenesis of AMD, described in Revised Example 11-10). Reference the work of others as needed.

A good partial Introduction, in which the background starts very broad but then narrows down quickly to the research topic and unknown, is shown next.

Example 11-1a   Partial Introduction showing good funneling of background

The sensory receptors of the auditory system in mammals are the auditory hair cells of the inner ear. Two functionally and structurally different types of mammalian auditory hair cells exist—inner and outer hair cells. While mechanical stimuli are transformed to neural signals in the inner hair cells (Chan and Hudspeth, 2005), outer hair cells do not transmit neural signals to the brain. Instead, when sound enters the inner ear, outer hair cells magnify it mechanically through electromotility, or oscillations at the sound frequency (Brownell, 1985).

Broad background

The molecular basis of this mechanism is thought to be the motor protein prestin, which is embedded in the lateral membrane of the outer hair cells. Mammalian prestin is an 80 kDa, 744 amino acid membrane protein whose function appears to depend on chloride channel signaling (Santos-Sacchi, 2006). Although prestin has been researched intensively, its molecular function has not been fully established.

Specific background

Unknown problem

In this partial Introduction, sensory receptors of the auditory system in mammals are described in the general background. Then, the Introduction narrows to the specific background and from then on focuses only on the mechanism of magnification of mechanical stimuli in the ear. The specific background culminates in a statement of the unknown, the lack of knowledge of the molecular function of prestin.

Note that when you are writing the Introduction of a research paper, you should not review the topic; that is, you should not describe every finding or paper in the field ever obtained, no matter how interesting it may be. Your aim is not to end up with an exhaustive citation of everything on the topic. A summary pertinent to the research you are presenting in the paper should suffice. In the following example, the Introduction of Example 11-1 is shown again, but in this version, the author has reviewed the topic. Too much irrelevant information (underlined) has been included. Consequently, the Introduction does not clearly funnel down to the topic of interest (the molecular function of prestin), and readers get confused because they do not know what aspect of the background information to focus on.

Example 11-1b   Introduction that reads like a review

The sensory receptors of the auditory system and the vestibular in mammals are the auditory hair cells of the inner ear. <u>Auditory hair cells are found in the auditory portion of the inner ear, known as the cochlea.</u> They are

<u>named after a structure known as the hair bundle or stereocilia found on the apical surface of the cell, which extends into the scala media within the cochlea. Damage to the hair cells results in sensorineural hearing loss.</u>

Two functionally and structurally different types of mammalian auditory hair cells exist—inner and outer hair cells. In inner hair cells, the stereocilia are deflected mechanically, thus opening gated ion channels and allowing positively charged ions such as potassium and calcium to pass into the cell (Müller, 2008). The entering of these ions depolarizes the cell, resulting in a receptor potential that subsequently triggers the release of neurotransmitters at the basal end of the cell. The neurotransmitters in turn trigger action potentials in nerve cells, thus transforming the mechanical stimuli into neural signals (Chan and Hudspeth, 2005).

Background starts broad, but then goes into irrelevant details that are not important for the research topic

Outer hair cells, which have evolved only in mammals, do not transmit neural signals to the brain. Instead, when sound enters the inner ear, outer hair cells magnify it mechanically through oscillations at the sound frequency (Brownell, 1985). The amplification is powered by an electrically driven motility of their cell bodies and is known as somatic electromotility. As a result of evolution, outer hair cells have not advanced hearing sensitivity, but have increased hearing frequency to roughly 200 kHz (highest frequency in some marine mammals) (Wartzog and Ketten, 1999). They have also enriched frequency recognition, thus enabling sophisticated human speech.

The specific background also contains some irrelevant information

The molecular basis of the electrically driven motility of outer hair cells is thought to be the motor protein prestin, which is embedded in the lateral membrane of the outer hair cells. Mammalian prestin is an 80 kDa, 744 amino acid membrane protein whose function depends on chloride channel signaling (Santos-Sacchi, 2006). Prestin is compromised by the common marine pesticide tributyltin (TBT) as has been shown by high concentrations of prestin in Orcas and toothed whales (Santos-Sacchi, 2006). Although prestin has been researched intensively, its molecular function has not been fully established.

The Unknown/Problem

➤ State the problem or unknown

After discussing general background and specific aspects of existing research, describe what the problems with the existing research are or what is unknown. The problem or unknown is clearest if you signal it by stating it directly, for example, "X is unknown" or "Y is unclear." You can also use other phrases to state the problem or unknown outright: "has not been established" or "has not been determined." Alternatively, you can imply rather than state the unknown by using a suggestion or a possibility ("Previous findings suggest that . . ."; see also Table 11.1 in Section 11.6 on "Signals for the Reader").

Use an objective tone when criticizing previous work. Avoid antagonistic phrases:

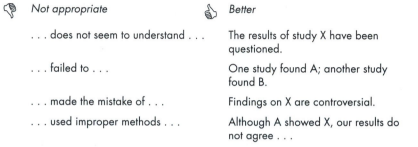

👎 Not appropriate	👍 Better
. . . does not seem to understand . . .	The results of study X have been questioned.
. . . failed to . . .	One study found A; another study found B.
. . . made the mistake of . . .	Findings on X are controversial.
. . . used improper methods . . .	Although A showed X, our results do not agree . . .

Also, do not blame individual authors or teams. You may end up creating your own enemies who one day may be reviewers of one of your papers or grant proposals.

The Question/Purpose

➤ State the central point (question/purpose) precisely

The most important element in a research paper is the research question or purpose of the work. The question/purpose is the "central point" of your Introduction and of the paper as a whole. It therefore needs to be worded very carefully. If the central point is stated precisely, the reader immediately has an idea of what to expect in the paper. Furthermore, the reader can read the paper in a directed way rather than blindly, and the experiments make more sense.

Because the question/purpose provides an overview of the entire paper, and every paragraph and sentence in your paper relates to it, consider writing your research question/purpose on a Post-it® note before you start composing your manuscript. Place this note on the side or top of your computer screen where it cannot be overlooked. It will remind you to keep your writing focused on the question/purpose of the paper.

The research question/purpose of a research paper should name the variables studied as well as the main features of the study. Note that the question/purpose is usually not written in the form of a question but as an infinitive phrase or as a sentence, using a present tense verb, as in the next examples.

Note that the question/purpose and the unknown or problem have to make sense together. The question/purpose should therefore be the statement one would assume following that of the unknown or problem, which in itself derives from what has been presented in the background of the Introduction.

Example 11-2   **Phrasing of question/purpose**

a   To examine the cold and heat tolerance of hibernating chipmunks, . . .

b   Here we asked how rheumatic fever influences heart rate.

c   In this study, we show that a sequential scheme of phosphorylation and dephosphorylation can generate circadian oscillations.

d   To determine if the triggered cellular processes affect the rRNA structure and folding dynamic *in vivo*, . . .

Experimental Approach

➤ State the experimental approach briefly

In the Introduction of your research paper, you should also briefly indicate your experimental approach. In general, the experimental approach is short—usually one sentence and, at most, two or three sentences. The experimental approach should be signaled so readers can identify it immediately. Examples of how to signal the experimental approach are shown in Example 11-3 (see also Table 11.1 in Section 11.6 on "Signals for the Reader").

Example 11-3   **Signals for the experimental approach**

a   **We analyzed X by** agarose gel electrophoresis.

b   To investigate brain size distribution in frogs, we **used** morphological analysis.

c   The structures of the compounds **were characterized by** UV, IR, 1H NMR, 19F NMR spectra, and HRMS.

Results and Conclusion

After the experimental approach, you may briefly state your main results and conclusion (about one to five sentences). Although their inclusion is not a must, know that readers like to read about the main results and conclusion of your work in the Introduction. Most readers dislike having to read the whole paper, waiting and searching for the answer to the research question. In some journals, results and conclusions are delayed until the Discussion section of the paper. If you have the chance to include results and conclusions in your Introduction, however, do so. Readers will be thankful for it. See also Table 11.1 in Section 11.6 on "Signals for the Reader" for ways to signal results in the Introduction.

Significance and Implication

Consider stating why your findings are important. Do so by stating what the significance or implication of the study is, as shown in the following example.

Example 11-4   **Stating the significance or implication**

   a   X is an important addition to . . .

   b   . . . , which aids in the elucidation of . . .

If you state the significance or implication at the very end of your Introduction, it not only rounds up this section nicely but also provides the overall perspective of your work for the reader. See also Table 11.1 in Section 11.6 on "Signals for the Reader" for additional ideas on how to signal the significance or implication of your article.

11.4   SPECIAL CASE: INTRODUCTIONS FOR DESCRIPTIVE PAPERS

➤ For descriptive papers, include

Background
Discovery statement
Experimental approach (if appropriate)
Description
Implication

Some research papers are not written to answer specific questions or to test a hypothesis but rather to describe a new finding such as a new organism, an unknown disease, or a novel apparatus. This category also includes methods papers. Introductions for these papers follow a slightly different "funnel" structure because elements that readers expect to find in the Introduction of a descriptive paper differ from those of an investigative paper. Elements that should be contained in the Introduction of a descriptive paper include:

1.	**Background**	background information and previous research in the area
2.	**Discovery statement**	new discovery made by your research
3.	**Experimental approach (optional)**	approach taken toward analyzing this discovery
4.	**Description**	description of the new element
5.	**Implication**	importance of the findings

In descriptive papers, the Introduction should funnel from the general background to the specific aspect (if it is known), to the discovery statement, and then to the description and implication. For some descriptive papers, the reason for the discovery or previous problems before discovery can also be included in the Introduction. Most descriptive papers do not include an unknown or problem statement as found in other investigative research papers.

The most important element in a descriptive paper is the discovery statement. This statement is usually written as a sentence, which should follow logically from the background information provided. The verb tense of the discovery statement can be either present tense or past tense, depending on how the statement is worded.

☞   **Example 11-5**   **Phrasing the discovery statement**

a   Here we describe a novel bacterial protease, Ecpro-3.

b   We discovered a new toxigenic serogroup, O210, of *Vibrio cholerae*.

c   We examined the structure of the bacterial ribosome in complex with tetracycline.

After the discovery statement, provide an overview of the novelty being reported. You may also include information on how the discovery was made or tested. You may even point out what the advantages and disadvantages of the discovery are. The description of the discovery should be written in present tense ("X is heat resistant to over 2,000°C"; "The new methane clathrate deposit found in Lake X may cap an even larger deposit of gaseous methane."). Note that for descriptive papers, an overview of an experimental approach may not be needed. However, stating the implication is particularly important at the end of the Introduction.

☞   **Example 11-6**   **Phrasing the implication**

a   The deposit is the first such deposit found in the United States and its vast reservoir may provide an enormous economic resource.

b   Solar cells made entirely of carbon can greatly improve resistance to extreme temperature.

c   Virus X appears to be the causative agent for disease A.

11.5   IMPORTANT WRITING GUIDELINES FOR THE INTRODUCTION

➤ Distinguish between past and present tense

In the Introduction, you will find a mix of verb tenses, often in the same sentence. The general rule about verb tense applies (Chapter 4, Section 4.4).

When reporting completed actions, such as when you are referring to previously reported studies, use past tense. For statements of general validity and those whose information is still true, that is, if you consider a finding to be a general rule, use present tense. When stating the question/purpose of your research, use present tense as well, but when describing the experimental approach, use past tense. An example of a mixed tense sentence is shown below.

☞   **Example 11-7   Use of tense**

Herbert et al. (9) **found** that peanut butter **can be contaminated** with Salmonella.

This example shows that the completed action ("found") is described in past tense, whereas the information that still holds true ("can be contaminated") is described in present tense.

Note that previously established knowledge in the Introduction should be described in present tense. If you use past tense for describing results of already published work, you are implying to the reader that you do not consider these results to be true "facts" but observations.

➤ Use strong verbs and short sentences

The initial part of the Introduction often is written in more lay terms than the rest of your manuscript. As this part of the Introduction is also more often read by a larger lay audience, it is particularly important to use short sentences and strong verbs instead of nominalizations. (See also Chapter 4, Sections 4.5 and 4.6.)

➤ Ensure good cohesion and coherence

Throughout your Introduction, you need to achieve good flow so that readers can clearly understand what you are trying to say on first read. Achieving this goal depends largely on good cohesion and coherence of your writing. Aside from creating a cohesive story in each paragraph, you need to weave the paragraphs together logically and ensure that all elements (background, unknown, research question and/or discovery statement, experimental approach) are clearly identifiable in the order given.

For clear cohesion and coherence, employ all the basic rules of good composition discussed in this book (see Chapter 6):

- Sentence location
- Topic sentences
- Word location
- Key terms
- Transitions

11.6   SIGNALS FOR THE READER

➤ Signal all the elements of the Introduction

Generally, all the parts of the Introduction should be signaled so the reader does not have to guess about the information provided. The signals vary depending on how the known, unknown, question, and experimental approach are phrased. Numerous variations on these signals are possible. Some examples are listed in Table 11.1.

ESL advice

These examples provide great starting points when you write your first draft and may be particularly useful for those authors who have writer's block or for those whose native language is not English.

TABLE 11.1 Signals of the Introduction

BACKGROUND	UNKNOWN	QUESTION, PURPOSE, OR DISCOVERY	EXPERIMENTAL APPROACH	RESULTS	IMPLICATION
X is . . .	. . . is unknown	We hypothesized that . . .	To test this hypothesis, we . . .	We found . . .	. . . consistent with . . .
X affects . . .	. . . has not been determined	To determine . . .	We . . .	. . . was found	. . . indicating that . . .
X is a component of Y	The question remains whether. . . .	To study . . . To examine . . . To assess . . . To analyze . . . In this study we examined . . .	We analyzed . . . For this purpose, we . . .	We determined . . .	. . . make it possible to . . .
X is observed when Y happens	. . . is unclear	Here we describe . . .	. . . by . . . . . . using . . .	Our findings were . . .	. . . may be used to . . .
X is considered to be . . .	. . . does not exist	Here we report . . . This report describes . . .	For this study, we . . .	We observed that . . .	. . . is important for . . .
X causes Y	. . . is not known	We examined whether X is . . . We assessed if . . . We determined if . . . We analyzed Y . . .	To evaluate . . . , we . . . To answer this question, we . . .	Based on our observations . . .	Our analysis implies/ suggests . . . Our findings indicate that . . .

11.7   COMMON PROBLEMS OF INTRODUCTIONS

The most common problems of Introductions include

- Missing elements (unknown/problem, question, experimental approach) (Section 11.3)
- Obscured elements (Section 11.5)
- Excessive length (Section 11.2)
- Context/background is too narrow (Section 11.3)
- Overview sentences

Missing or Obscured Elements of the Introduction

➤ Do not omit any element of the Introduction

If one or more of the elements of the Introduction is missing or obscured, readers get confused and frustrated because they come away with the feeling that they have not fully understood what the paper is about. The most common elements that are missing are the unknown and the experimental approach. Sometimes even the question/purpose is not stated or not stated clearly. Obscured elements commonly arise when authors have not clearly signaled the unknown/problem, question/purpose, or experimental approach.

Consider, for example, the following introduction in which the unknown and research question is missing.

☞	**Example 11-8**	**Introduction with missing unknown and question**

Unknown not stated		NcaII is a newly identified *N. carnea* type IIE restriction endonuclease recognizing the DNA recognition site 5′GAGTTCTC3′ (1). NcaII requires two copies of the rec-
Question/ purpose not stated	Background	ognition sequence, 45–80 nucleotides apart from each other, and cooperatively binds both by forming an inter-
	Experimental approach	mediate DNA loop on a linear DNA substrate. Protein-protein interactions suggest a two-domain structure of NcaII (2). When we exposed NcaII to limited proteolysis using chymotrypsin with or without DNA, we found that a stable fragment of 52 kD was obtained when NcaII was digested with chymotrypsin in the absence of DNA, while a fragment of about 55 kD was obtained in the presence of DNA. The N-terminal sequence of the pro-
	Results	teolytic fragments obtained with DNA was subjected to Edman degradation and found to correspond to the N-terminus of NcaII. The fragment obtained without DNA corresponded to the C-terminus of NcaII as determined by its N-terminal amino acid sequence. Based on these findings, NcaII appears to have two functional domains, one at the N- and the other at the C-terminus of the
	Conclusion	enzyme.

When readers read this Introduction, they are left in the dark as to why the proteolysis was done. The concluding sentence at the very end of the Introduction hints at the reason ("functional domains"), but readers will not know for sure. If the unknown and question would have been stated, readers could have read the Introduction in a directed way.

Revised **Example 11-8**	NcaII is a newly identified *N. carnea* type IIE restriction endonuclease recognizing the DNA recognition site 5'GAGTTCTC3' (1). NcaII requires two copies of the recognition sequence, 45–80 nucleotides apart from each other, and cooperatively binds both by forming an intermediate DNA loop on a linear DNA substrate. Although protein-protein interactions suggest a two-domain structure of NcaII (2), <u>the functional domains of NcaII remain unclear. To study the functional domains of the endonuclease,</u> we exposed NcaII to limited proteolysis using chymotrypsin with or without DNA, and we found that a stable fragment of 52 kD was obtained when NcaII was digested with chymotrypsin in the absence of DNA, while a fragment of about 55 kD was obtained in the presence of DNA. The N-terminal sequence of the proteolytic fragments obtained with DNA was subjected to Edman degradation and found to correspond to the N-terminus of NcaII. The fragment obtained without DNA corresponded to the C-terminus of NcaII as determined by its N-terminal amino acid sequence. Based on these findings, NcaII appears to have two functional domains, one at the N- and the other at the C-terminus of the enzyme.
Background Unknown	
Question/ purpose	
Experimental approach	
Results	
Conclusion	

Example 11-9 is another example of an incomplete Introduction in which the unknown has not been stated.

	Example 11-9	**Introduction with missing unknown**
		Modeling ocean currents is difficult due to their nonlinear and hence complex nature. Contributing to the nonlinearity are eddies, such as Tropical Instability Waves (TIWs) (1). In the eastern Pacific, TIWs appear as a regular sinusoidal pattern of waves with a 30-day period that feed on currents, sea level, and sea surface temperature. They are considered unstable waves that are maintained by the kinetic and potential energy of the mean currents (Vialard et al., 2003). These waves not only contribute to the climate variability over the Atlantic but also are influenced by it. Reproducing these eddies in a model is essential for obtaining the correct flow structure and momentum balances (2, 3), both of which influence other ocean features such as energy balances and ocean temperatures, thus affecting global climate.
Unknown **not stated**	Background	

	Not all proposed TIW models resolve these waves effectively. Possible reasons include coarse resolutions of models or the use of a simple friction scheme. Friction is often necessary to slow the strong zonal equatorial currents produced in models. However, using a simple friction scheme to damp these currents has the side effect of suppressing some of the eddy activity (4).
__Background__	
__Question/ purpose__	To determine the effect of eddies on depth-integrated transports and friction to the system, we employed a high resolution ocean circulation model with a biharmonic Smagorinsky friction scheme (5) that allows TIWs to be well resolved. We found that under such a scheme the depth-integrated currents can be almost completely described by wind stress and nonlinearity alone, without the need for a large frictional damping. We also explored how the inclusion of TIWs to our model provides its own friction to the system via the nonlinear advection term. Our results indicate that care must be taken in resolving eddies near the equator, if the correct dynamics are to be resolved.
__Experimental approach__	
__Results and significance__	

(With permission from Jaclyn Brown, PhD)

The Revised Example 11-9 is much easier to follow as the unknown is pointed out clearly.

Revised Example 11-9	. . . Not all proposed TIW models resolve these waves effectively. Possible reasons include coarse resolutions of models or the use of a simple friction scheme. Friction is often necessary to slow the strong zonal equatorial currents produced in models. However, using a simple friction scheme to damp these currents has the side effect of suppressing some of the eddy activity (4). **It is unclear how eddies act as a friction to the system and how they affect depth-integrated transports**.
__Unknown__	
__Question/ purpose__	To determine the effect of eddies on depth-integrated transports and friction to the system, we employed a high resolution ocean circulation model with a biharmonic Smagorinsky friction scheme (5) that allows TIWs to be well resolved. . . .

Excessive Length

This problem usually arises when authors review the topic rather than funnel down stringently from background information to the research question. Readers expect the author to guide them through the pertinent information on the topic. If the author reviews the topic, the reader does not know what topic to focus on or which topic is pertinent for the paper.

In the next example, the author has reviewed the topic. The amount of detail not pertinent to the question of the paper distracts and confuses readers.

Example 11-10   **Introduction that reviews the topic**

In age-related macular degeneration (AMD), the macula, a small central part of the retina, deteriorates, destroying central vision. AMD is the leading cause of vision loss in the elderly (>50 years of age). The condition is progressive, often resulting in severe visual impairment. Currently there is no cure for AMD. The disease is likely a mechanistically heterogeneous group of disorders. Clinically, AMD can be divided into two forms, atrophic (dry) or exudative (wet) (de Jong, 2006). The atrophic form is the most common form, representing approximately 90% of all AMD cases. However, atrophic AMD accounts for only about 12% of the severe vision loss associated with AMD.

Although wet AMD represents approximately 10% of all cases of AMD, it is responsible for about 88% of the severe vision loss associated with the disease. Both genetic and environmental factors, such as cigarette smoking, nutrition, obesity, and lipid level, are likely to play a role in the development and progression of AMD (Seddon et al., 2006).

Background too broad— **unclear what the specific aspect of interest is**

The pathogenesis of AMD is related to adverse vascular changes. Indeed, AMD and cardiovascular disease may have common precursors (Snow and Seddon, 1999). Specifically, it has been suggested that elevated plasma concentrations of total homocysteine increase not only the risk of vascular disease but also that of exudative neovascular AMD (Axer-Siegel, 2004).

Background too broad— **unclear what the specific aspect of interest is**

Homocysteine is an amino acid homologue of cysteine and is synthesized by the body from methionine. It can be converted back to methionine or into cysteine with the help of vitamins B-6 and B-12. The total plasma homocysteine level appears to depend on genetic factors, age, and gender (Seddon, 2006). Dietary factors may also contribute to regulating homocysteine metabolism, particularly B-vitamins and folate (Axer-Siegel, 2004). High levels of plasma homocysteine lead to a release of free radicals, thus affecting vascular walls and increasing the risk for thrombosis (Thambyrajah and Townend, 2000). High levels of this amino acid are also linked to a range of other diseases, including atherosclerosis, heart attack, stroke, and even Alzheimer's disease.

Unknown: " . . . is
unclear."

Question/purpose:
"To examine . . ."

Experimental approach:
". . .we studied . . ."

It has been reported that high plasma homocysteine
levels are also associated with AMD (Seddon, 2006).
However, the relationship between homocysteine and
AMD is controversial, and any relationship between homocysteine levels and AMD is unclear (Seddon, 2006).
To examine if there is any correlation between homocysteine levels and AMD, we assessed the effect of plasma
homocysteine levels on AMD in an independent study of
520 men and women 50 years or older.

In the revision, excess detail has been removed. Readers can now focus on
the specific aspect of the previous research that the paper covers.

**Revised
Example 11-10**

Broad background—
now shorter
and more focused

Specific
background—
now focused

In age-related macular degeneration (AMD), the
macula, a small central part of the retina, deteriorates,
destroying central vision. AMD is the leading cause of
vision loss in the elderly (>50 years of age). The condition is progressive, often resulting in severe visual impairment. The pathogenesis of AMD is related to
adverse vascular changes. Indeed, AMD and cardiovascular disease may have common precursors (Snow
and Seddon, 1999). Specifically, it has been suggested
that elevated plasma concentrations of total
homocysteine increase not only the risk of vascular disease but also that of exudative neovascular AMD
(Axer-Siegel, 2004). However, vascular disease has
been associated with AMD in some but not in all epide-
miologic studies (Seddon et al., 2006a).

Specific
background

Homocysteine is an amino acid homologue of cysteine
and is synthesized by the body from methionine. The
total plasma homocysteine level appears to depend on
genetic factors, age, and gender (Seddon, 2006).
Dietary factors may also contribute to regulating homocysteine metabolism, particularly B-vitamins and folate
(Axer-Siegel, 2004). High levels of plasma homocysteine lead to a release of free radicals, thus affecting
vascular walls and increasing the risk for thrombosis
(Thambyrajah and Townend, 2000).

Unknown: " . . . is
unclear."

Question/purpose:
"To examine . . ."

Experimental
approach: ". . . we
studied . . ."

It has been reported that high plasma homocysteine
levels may be associated with AMD (Seddon, 2006).
However, the relationship between homocysteine and
AMD is controversial, and any relationship between
homocysteine levels and AMD is unclear (Seddon,
2006). To examine if there is any correlation between
homocysteine levels and AMD, we assessed the effect
of plasma homocysteine levels and AMD in an independent study of 520 men and women 50 years of
age or older.

Remember also that short sentences carry more weight than long sentences. Writing a short opening sentence in your Introduction will not only be easier to understand for most readers, but it will also be more interesting and grab more of their attention.

Background Is Too Narrow

If the background information provided in your Introduction is too narrow, most readers will not be enticed to keep reading, because they will feel lost from the start. You will also lose your readers if general background information is absent, too abstract, or too technical. The result is that readers will be missing context, that is, a general overview of the topic is lacking.

Consider the following two openings of an Introduction. Which one would you find more interesting as a reader?

Example 11-11a   Opening of Introduction

FR2 is a member of the DExD/H-box family of proteins (1). DExD/H-box family proteins possess NTPase and often helicase activity (1). FR2 exhibits NTPase and helicase activity from its C-terminal helicase domain (FR2hel) (2,3). FR2 also binds to HCV NS4A to form the complex FR2–4A. FR2–4A exhibits serine protease activity from its N-terminal protease domain (4,5) and is localized to the surface of the endoplasmic reticulum via NS4A (6).

Example 11-11b   Opening of Introduction

Hepatitis C, which is caused by the Hepatitis C virus (HCV), infects an estimated 170 million people worldwide and 4 million in the United States. The virus replicates mainly in the hepatocytes of the liver and in peripheral blood mononuclear cells. An essential replicative component of HCV is FR2 (1,2). FR2 is a member of the DExD/H-box family of proteins (3). Like other members of this family, FR2 exhibits NTPase (3). In addition, FR2 also displays helicase activity from its C-terminal helicase domain (FR2hel) (4,5), an activity that is often seen in other DExD/H-box family members (3). Furthermore, FR2 binds to HCV NS4A to form the complex FR2–4A, which exhibits serine protease activity from its N-terminal protease domain (6,7) and which is localized to the surface of the endoplasmic reticulum via NS4A (8).

Most readers prefer Example 11-11b because its opening is much more generally understandable than that of Example 11-11a. Composing particularly the first sentences and paragraphs of the Introduction in more lay terms and relating the topic to common problems or topics of public interest will entice more readers than an opening that jumps right into technical specifications and information or one that assumes prior knowledge of a specific scientific field.

Overview Sentences

On occasion, authors blend introductions of research and review papers. In such mixed introductions, most sentences are informative, but the last sentence is usually an overview sentence that states that something will be described or discussed in the text ("An overview of some ways to reduce recurrence of malaria infections can be found in the Discussion."). Such sentences are not useful in an introduction for a research article and should be avoided.

11.8   SAMPLE INTRODUCTIONS

The following short, one-paragraph Introduction consists of only three sentences, yet it contains all the elements necessary for the reader to follow: background, unknown, research question, and experimental approach. Because this Introduction funnels down through all the required elements, the reader can logically follow the Introduction of the paper.

    Example 11-12   Complete, short sample Introduction

Background	CT-3 has shown marked antiallodynic and analgesic ef-
Unknown	fects in animals (1). However, it has not been determined whether CT-3 also possesses the ability to treat neuro-
Question/ purpose	pathic pain in humans. To examine the analgesic efficacy and safety of CT-3 for chronic neuropathic pain in
Experimental approach	humans, we conducted a randomized, double blind, placebo-controlled crossover study on 21 patients for 5 weeks using two daily doses of 10 mg CT-3 or placebo.

Longer Introductions should also contain all these required elements. For longer Introductions, it is especially important to follow the funnel structure because additional smaller stories are incorporated within the paragraphs. If the funnel structure is not followed, the reader may get confused and frustrated.

The following Introduction of an investigative paper contains all required elements (background/known, unknown, question/purpose, experimental approach). In addition, it also contains main results and describes the overall significance of the paper, rounding up the Introduction nicely.

    Example 11-13   Complete Introduction

Septic shock and sepsis syndrome is one of the leading causes of death in hospitalized patients and accounts for 9% of the overall deaths in the United States annually (1–4). While commonly initiated by a bacterial infection, the pathophysiological changes in sepsis are often not due to the infectious organism itself but rather to the uncontrolled production of pro-inflammatory cytokines produced mainly by macrophages. This over-production can result in an overwhelming systemic inflammatory response that leads to multiple organ failure. Lipopolysaccharide (LPS) endotoxin, a component of the bacterial cell wall, stimulates macrophages to produce proinflammatory cytokines such as tumor necrosis factor α (TNF-α) and interleukin-1β (IL-1β), both of which have been shown to be critical mediators of septic shock (7). It is the excessive production of these pro-inflammatory cytokines that causes systemic capillary leakage, tissue destruction, and ultimately lethal organ failure and death (1–4, 7). Thus, the expression of these pro-inflammatory cytokines needs to be tightly regulated during an inflammatory response.

Regulation of mRNA stability is critical for controlling gene expression as the abundance of an mRNA transcript is also regulated by the rate of mRNA degradation (5, 6, 8). The mRNAs encoding most inflammatory cytokines are short-lived, with instability conferred by an AU-rich element (ARE) in their 3′ noncoding regions (5, 6). ARE promotes rapid degradation of mRNAs (6) and in some cases translation arrest (9). Furthermore, ARE-mRNAs can be rapidly stabilized upon exposure to certain signals, including immune stimulation, UV and ionizing irradiation (5), and heat shock (10). The stability of ARE-mRNAs is controlled by trans-acting AU-rich element binding proteins (6, 16). Hu ARE-binding proteins, such as HuR, stabilize ARE-mRNAs and inhibit mRNA translation (17–19).

In contrast to ARE, AUF1 (or hnRNP-D) (20, 21), TTP (22), BRF1 (23), and KSRP (24) are shown to mediate rapid decay of various cytokine mRNAs. AUF1, an ARE-mRNA destabilizing factor, consists of four isoforms (37, 40, 42, and 45 kD) generated by alternative splicing (20, 21). Increased expression of AUF1 has been correlated with rapid ARE-mRNA degradation in various types of cells (25–27), with p37 AUF1 isoform exhibiting the highest destabilizing activity (25, 26). However, implication of AUF1 is based largely on correlations with *in vitro* binding to AREs (28, 29) and ectopic overexpression studies in cell lines (25, 26). The role of AUF1 in the regulation of cytokine expression *in vivo* **has not been examined** in a pathophysiological context.

To examine the role of AUF1 in promoting inflammatory mRNA degradation *in vivo*, **we generated** AUF1 null

Background

Unknown

Question/
purpose

Experimental approach	mutant mice and studied their response to LPS-induced endotoxemia and their expression of pro-inflammatory cytokines. We observed that *AUF1⁻* mice were acutely susceptible to endotoxin, showed manifestations typical of endotoxic shock, and had a significantly lower survival rate when challenged by LPS. These phenotypes were associated with over-expression of the pro-inflammatory
Results	cytokines TNFα and IL-1β as a result of abnormal stabilization of their mRNAs. Our results provide the first *in vivo* evidence implicating AUF1 in regulating inflammatory response. This regulation occurs through targeted degradation of selective cytokine mRNAs, deregulation of which
Significance	would contribute to the development of endotoxic shock.

(With permission from Cold Spring Harbor Laboratory Press and Robert J. Schneider)

Two additional examples of longer Introductions that include all the essential elements are the next two examples.

   Example 11-14   **Complete Introduction**

Broad background	Remote sensing through satellites is often used for investigating land-surface phenologies. These phenologies, measured as Normalized Difference Vegetation Index (NDVI) data, can provide a baseline from which to monitor changes in vegetation associated with events such as
Problem of research topic	fire, drought, climate fluctuations, and climate change. A known source of uncertainty limiting the use of remotely sensed time-series is the degree of noise found in the signals. Typical causes for noise include atmospheric influences, sensor performance, and efficiency of post acquisition cloud- and compositing-algorithms (1). To minimize this noise, various authors have demonstrated the efficiency of Fourier filtering (2–5). There are, however, limits to this method beyond which the representation of the general shape of a periodic cycle is too inaccurate. Such limits exist particularly when cycles show erratic outliers as compared to easy to correct periodic noise features.
Discovery/ new method	Here, we present an algorithm that classifies vegetation from NDVI time-series according to the shape of the temporal cycle and is minimally affected by atmospheric and sensor effects. The algorithm is derived from examples from the Middle East and Central Asia using 250 m Modis NDVI data and based on the Fourier components magnitude and phase that can be applied to periodic cycles. High shape fidelity can be ensured through several user controls. The algorithm is invariant to cycle

General description

modifications that may be caused by climate, soil, or topography, but are unrelated to the vegetation type. The output is a highly consistent clustering of NDVI-cycles, which can be linked to distinct vegetation types or land use practices. The algorithm allows vegetation changes to be monitored with the possibility to distinguish between pure coverage fluctuations and actual phenological changes.

(With permission from the ISPRS Journal of Photogrammetry and Remote Sensing)

Example 11-15   **Complete Introduction**

Broad background

Specific background

Problem of research topic ". . . not much systematic study."

The biological activities of proteins depend not only on their amino acid sequences, but also on their discrete conformations. Even slight perturbations of the conformations of a protein may render it inactive (1). . . . Several reports have been published regarding the isolation and purification of recombinant proteins from inclusion bodies, but the methods employed for renaturation may not be generally applicable (5–14). A more general method for determining renaturation conditions would be of significant value because in most cases very little is known about the renaturation of the protein of interest.

Many naturally occurring proteins have been the subject of renaturation studies (13, 15). The rate of renaturation and yield of active enzyme depend not only on the mode of denaturation and the enzyme concentration during renaturation, but also on the solvent composition. . . . Although much is known about the role of denaturants, reducing conditions, and protein concentration on refolding, the effect of solvent composition has not received much systematic study.

Discovery/ new method

General description

Implication: "Our results suggest . . . may be applicable . . ."

Here we report a new sparse matrix method employing 50 different solvent systems for establishing initial solvent conditions that facilitate protein renaturation. . . . This screening matrix is based on a set of solutions originally selected for their demonstrated utility for protein crystallization (22). Our results suggest that this screening method may be widely applicable in identifying conditions that support renaturation and that the same conditions which promote protein crystallizations may also promote protein renaturation.

(With permission from Elsevier)

11.9   REVISING THE INTRODUCTION

When you have finished writing the Introduction (or if you are asked to edit an Introduction for a colleague), you can use the following checklist to "dissect" the Introduction systematically. In your revisions and in editing papers, work your way backward from paragraph location and structure to word choice and spelling.

- ☐   1. Does the topic present something new and interesting?
- ☐   2. Are all the components there? To ensure that all necessary components are present, on the margins of the Introduction clearly mark:

Investigative Paper	Descriptive Paper
Background	Background
Unknown	(Reason/problem)
Question/purpose	Discovery statement
Experimental approach	(Experimental approach)
(Results/conclusion)	Description
(Significance)	Implication

- ☐   3. Is the research question stated precisely? (Is it in present tense?)
- ☐   4. Do all the components logically follow each other? (Is the unknown what one would expect after reading about what is known? Is the research question really the question one would anticipate based on the unknown? Does the answer really answer the research question?)
- ☐   5. Is all background information directly relevant to your research question? Did you list only the most pertinent literature and not review the topic? Is the Introduction less than two double-spaced pages?
- ☐   6. Have all elements been signaled clearly?
- ☐   7. Did you keep the Introduction short?
- ☐   8. Are references placed correctly and where needed? (Chapter 8)
- ☐   9. Is the introduction cohesive and coherent? (Chapter 6, Section 6.3)
- ☐ 10. Has sentence location been considered? (Chapter 6, Section 6.1)
- ☐ 11. Are topic sentences used well? (Chapter 6, Section 6.1)
- ☐ 12. Revise for style and composition using the basic rules and guidelines of this book:
  - ☐ a. Are paragraphs consistent? (Chapter 6, Section 6.2)
  - ☐ b. Are paragraphs cohesive? (Chapter 6, Section 6.3)
  - ☐ c. Are key terms consistent? (Chapter 6, Section 6.3)
  - ☐ d. Are key terms linked? (Chapter 6, Section 6.3)
  - ☐ e. Are transitions used, and do they make sense? (Chapter 6, Section 6.3)

☐ f. Is the action in the verbs? Are nominalizations avoided? (Chapter 4, Section 4.6)

☐ g. Did you vary sentence length and use one idea per sentence? (Chapter 4, Section 4.5)

☐ h. Are lists parallel? (Chapter 4, Section 4.9)

☐ i. Are comparisons written correctly? (Chapter 4, Sections 4.9 and 4.10)

☐ j. Have noun clusters been resolved? (Chapter 4, Section 4.7)

☐ k. Has word location been considered? (Verb follows subject immediately? Old, short information is at the beginning of the sentence? New, long information is at the end of the sentence?) (Chapter 3, Section 3.1)

☐ l. Have grammar and technical style been considered (person, voice, tense, pronouns, prepositions, articles)? (Chapter 4, Sections 4.1–4.4)

☐ m. Are words and phrases precise? (Chapter 2, Sections 2.2 and 2.3)

☐ n. Are nontechnical words and phrases simple? (Chapter 2, Section 2.2)

☐ o. Have unnecessary terms (redundancies, jargon) been reduced? (Chapter 2, Section 2.4)

☐ p. Have spelling and punctuation been checked? (Chapter 4, Section 4.11)

SUMMARY

INTRODUCTION GUIDELINES

- Interest your audience, and provide context.
- Follow a "funnel" structure:
  - For investigative papers, include:
    Background
    Unknown/problem
    Question/purpose of study
    Experimental approach
    Optional: Results/conclusion
    　　　　　Significance
  - For descriptive papers, include:
    Background
    Discovery statement
    Experimental approach (if appropriate)
    Description
    Implication
- Keep the Introduction short.
- Provide pertinent background information, but do not review the literature.
- State the problem or unknown.

> - State the central point (question/purpose) precisely.
> - State the experimental approach briefly.
> - Distinguish between past and present tense.
> - Use strong verbs and short sentences.
> - Ensure good cohesion and coherence.
> - Signal all the elements of the Introduction.
> - Do not omit any element of the Introduction.

PROBLEMS

PROBLEM 11-1 Elements of the Introduction

Why does this introduction seem incomplete? Identify the known, unknown, question/purpose, and experimental approach. Are these elements clearly identifiable? Why or why not?

The carotenoid astaxanthin is a red pigment that occurs in specific algae, fish, crustaceans, and in some bird plumages (McGraw and Hardy, 2006). Astaxanthin is an antioxidant and commonly is used as a natural food supplement, food color, and anticancer agent among other disease preventative measures.

Like many carotenoids, astaxanthin is a colorful, fat/oil-soluble pigment, providing a reddish and pink coloration (2). Whereas in certain bird species, all adult members display carotenoid-containing feathers rich in color, many gulls and terns, which normally have white feathers, display an abnormal pink tinge (or flush) in various degrees across their populations (McGraw and Hardy, 2006). It has been suggested that this pink tinge arises during feather growth when these birds ingest abnormally high quantities of astaxanthin, which often occurs close to salmon farms (McGraw and Hardy, 2006). However, the exact relationship between astaxanthin and plumage is not fully understood. Here, we examine this relationship in more detail and discuss its implication.

PROBLEM 11-2 Elements of the Introduction

Why does this introduction seem incomplete? Identify the known, unknown, question/purpose, and experimental approach. Are these elements clearly identifiable? Why or why not?

Methane clathrate, also known as methane hydrate, methane fire, or fire ice, is a crystalline, cement-like solid that consists mainly of methane and water. Deposits of methane clathrate can accumulate within and on top of ocean sediments (Kvenvolden, 1995; Hoffmann, 2006). Methane hydrates are thought to form when methane from decaying organic matter on the ocean floor combines with seawater under appropriate conditions of low temperature and high pressure.

High-resolution, deep-tow multichannel seismic data, together with coring and sampling, were used to assess the presence of methane clathrates along the ocean floor of the Blake Bahama Outer Ridge. Based on the evaluation of the seismic data and geochemical evidence, we were able to identify several deposits of methane clathrates at depths of 500 to 1,000 m, above a bottom simulating reflector (BSR).

PROBLEM 11-3 Elements of the Introduction
Revise the following Introduction. Ensure that the necessary elements (known, unknown, question/purpose, and experimental approach) are present and clearly signaled.

Ehrlichiae are obligatory intracellular bacteria that infect leukocytes and platelets of a wide variety of mammals and are transmitted by ticks (1). Dogs can be infected by *E. canis, E. chaffeensis, E. ewingii,* and *Anaplasma phagocytophilum* (*Ehrlichia equi*) (2). Infection with any of these species can cause a severe disease with indistinguishable hematological and clinical anomalies (3).

E. *canis* was the first species described in dogs (3–5). It is distributed worldwide, particularly in tropical and subtropical regions, and is the causative agent of classical canine monocytic ehrlichiosis, which presents three well-characterized clinical phases. Dogs treated during the acute phase of the disease normally recover rapidly. However, since this stage of the infection can evolve with moderate or imperceptible clinical signs, infected dogs can develop the subclinical phase and some of them reach the chronic phase (6).

E. *chaffeensis* is the etiological agent of human monocytic ehrlichiosis. This potentially mortal disease was reported for the first time in the year 1987 in the USA. *E. canis* was first thought to be the causative agent due to a cross-reaction of sera from patients with an antigen preparation from this species (7). In 1991, the bacterium was isolated and characterized at the molecular level. It was then established that it was a distinct species of *Ehrlichia* (8). The recognized natural reservoir for this bacterium is the white-tailed deer (9).

Ehrlichiae with tropism for monocytes, lymphocytes, neutrophils, and platelets from dogs diagnosed by BCS have been reported since 1982 (10). The presence of these rickettsiae in monocytes and platelets has been also confirmed using transmission electron microscopy (11, 12). In this study, we describe primary cultures of monocytes from the blood of a dog with canine monocytic ehrlichiosis. We also identified *E. canis* and *E. chaffeensis* using nested PCR with DNA samples extracted from the primary cultures and from dogs with natural and experimental infections.

(Veterinary Clinical Pathology 37(3), pp. 258–265, 2008)

PROBLEM 11-4 Condensing the Introduction

1. This Introduction is too long. Condense this Introduction.
2. Identify the individual elements of the condensed Introduction (known, unknown, question/purpose, and experimental approach).

The establishment of the germline generation after generation is crucial for the propagation of a species. Typically, the germline is set apart from the somatic tissues early in development. To maintain the unique ability of the germline to give rise to all tissues of the next generation, the germline must be protected from somatic differentiation signals, which would restrict the fate of the cells. In many species, germ cells contain specialized cytoplasm, known as germ plasm, which contains proteins and RNAs required for germline development. In some cases, this germ plasm includes distinct ribonucleoprotein particles, known as polar granules in *Drosophila* and as P granules in *C. elegans*. At least some protein components of germ granules are conserved, particularly the homologs of *vasa*, an RNA helicase that is found in germ cells in many organisms from *C. elegans* to mammals.

In *C. elegans*, the germline is set apart from the somatic tissue after only four rounds of asymmetric cell division, which give rise to blastomeres P1, P2, P3, and P4 (1). The germline blastomeres inherit P granules in each round of division. P granules are initially dispersed in the cytoplasm in oocytes and in the early embryo. Beginning in the P2 blastomere, P granules associate with the nuclear membrane, an association that is maintained in the adult germ cells. A protein with similarity to receptor tyrosine kinases, MES-1, controls the asymmetric partitioning of P granules during the divisions of P2 and P3 (2). The blastomere P4 marks the point of germline restriction, as this cell divide gives rise to a pair of primordial germ cells that proliferate to produce all future germ cells in the animal. The germline blastomeres are transcriptionally silent during the early divisions until after the birth of P4 (3). Therefore, genes that function in the germline establishment in these blastomeres are likely to be maternally deposited proteins or maternally deposited mRNAs under post-transcriptional control.

Of the proteins present in P granules that have been previously identified, most contain predicted RNA binding domains, suggesting that regulation of RNAs may be a function of P granules. Some proteins localize exclusively to P granules in both embryonic and adult stages, specifically PGL-1 and PGL-3 (4, 5) and the four homologs of *vasa* (1–4) (6). In contrast, other proteins found in P granules localize to additional cellular locations. For example, PIE-1 is in both P granules and the nucleus, where it represses transcription (7). While many P granule components are required for normal germline development, P granules alone are not sufficient to confer a germline fate. Despite the identification of greater than twenty proteins found in P granules, a molecular function for this complex has not been elucidated. Additionally, genetic relationships between many P granule components have not yet been reported.

In this study, we report the identification and characterization of a novel protein in *C. elegans*. This gene is necessary for the development of the germline of the progeny. We have named this gene *egcd-1*. We show that loss of *egcd-1* results in a severe decrease in germ cell proliferation and abnormalities in the germline blastomeres. Our genetic analysis indicates that *egcd-1* functions synergistically with another gene important in P granules.

(Reprinted with permission from Stefanie W. Leacock, PhD)

PROBLEM 11-5 Your Own Introduction
Write a short Introduction based on your work. Be sure to follow the funnel shape. Include known, unknown, question, and experimental approach. Also, consider including the results, conclusions, and significance of the paper. Be sure to signal the parts of your Introduction.

PROBLEM 11-6 Revising an Introduction
Revise either your own Introduction or that of a colleague. Be sure to identify the known, unknown, question, and experimental approach. Also, look for the results, conclusions, and significance of the paper. Ensure that all elements are signaled clearly and that the question/purpose follows directly from the unknown or problem statement. Pay attention to verb tense as well as other elements of grammar and style. Use the checklist provided in Section 11.9 as a guide.

CHAPTER 12

Materials and Methods

Although most readers are not interested in the experimental details of your work and will therefore not read this section, some readers will want to repeat part or all of your procedures and will read the Materials and Methods section in great detail. Reviewers will usually also read this section very carefully to ensure that it contains sufficient detail to evaluate or repeat your work.

THIS CHAPTER DISCUSSES:

- The overall structure and organization of the Materials and Methods section
- The content of the section, including how to describe statistical analysis
- Reporting on ethical conduct and its guidelines
- Signaling information in the Materials and Methods section
- Common problems of the Materials and Methods section
- Revising the Materials and Methods section
- Annotated sample Materials and Methods sections

12.1 OVERALL

The purpose of the Materials and Methods section is to describe the experimental approach used to arrive at your conclusions. You should write this section with great care because if your experimental approach appears faulty, incomplete, or unprofessional, your paper may get rejected.

12.2   COMPONENTS

➤ Provide enough details and references to enable a trained scientist to evaluate or repeat your work

The Materials and Methods section should cover

- Materials (drugs, culture media, buffers, gases, or apparatus used)
- Subjects (patients, experimental materials, animals, microorganisms, plants)

For medical studies include   (a) Total number of subjects
(b) Number of subjects receiving treatment
(c) How subjects were selected
(d) Details such as sex and age if relevant

- Design (includes independent and dependent variables, experimental and control groups)
- Procedure (what, how, and why you did something)

Define the materials and methods as precisely as you can. Do not forget to include your control experiments. Check and follow the detailed specifications found in the *Instructions to Authors* of the journal to which you plan to send your manuscript.

➤ Include materials and methods but not results

The Materials and Methods section is unavoidably linked to the Results section. In the Materials and Methods section, you need to describe how you obtained the results you report. Vice versa, in the Results section, you need to provide results for everything you describe in the Materials and Methods section. Do not make the error of mixing in some of the results in this section except for necessary intermediate results that provide the information needed for the next logical experimental step of your study.

➤ Provide literature references where needed

If your methods have not been reported previously, you must provide all of the necessary detail. If, however, methods have been previously described in a standard journal, you can provide only that literature reference or refer back to the methods in the previously published study and then give a **brief** overview of the approach. In the latter case, the reader does not have to go back to the original paper to get the gist of the methods.

Example 12-1   **Referring to previously described methods**

a   Plasmids were isolated according to Braun (19).

b   Plasmids were isolated according to Braun (19). **Briefly,** *E. coli* cells were lysed using X before plasmid DNA was precipitated with Y . . .

If you modified a previously published method, provide the literature reference and give a detailed description of your modifications.

Example 12-2   **Referring to described methods that were modified**

Plasmids were isolated according to Braun (19) with minor modifications. Instead of dissolving DNA pellets in sterile water, pellets were dissolved in buffer A.

Be sure to quote original references, that is, references that actually provide the method you want to describe. Do not just list a reference that refers the reader to another paper.

➤ Include sufficient technical details to let others repeat your work

In the Materials and Methods section, include details such as temperature, pH, total volume, time, and quantities to ensure that scientists can repeat your work. Identify organisms with full taxonomic names.

Example 12-3   **Providing sufficient detail**

a   To lyze the cells, we used 250 µl of SDS-solubilization buffer (10 mM Tris-HCl pH 7.5, 150 mM NaCl, 1 mM EDTA pH 8.0, 1% SDS w/v, 1:100 100 mM Phenylmethylsulfonyl fluoride (PMSF), and 100× Lysophosphatidic acid (LPA)). After lysis, cells were resuspended once the supernatant had been removed.

In Example 12-3a, the author provides good detail of the solubilization buffer but fails to do so for the resuspension buffer.

Revised   a   To lyze the cells, we used 250 µl of SDS-solubilization
Example 12-3   buffer (10 mM Tris-HCl pH 7.5, 150 mM NaCl, 1 mM EDTA pH 8.0, 1% SDS w/v, 1:100 100 mM Phenylmethylsulfonyl fluoride (PMSF), and 100× Lysophosphatidic acid (LPA)). After lysis, cells were resuspended in **100 µl sterile saline** after removal of the supernatant.

Here are a two more examples that do not provide sufficient detail:

Example 12-3   **Providing sufficient detail**

b   To identify planetary systems with a high probability of containing planets or moons in the habitable zone, we also used statistical methods.

c   All samples were centrifuged.

**Revised
Example 12-3**

b To identify planetary systems with a high probability of containing planets or moons in the habitable zone, we also used **parametric (*t*-test) and nonparametric (Cochran's Q test) methods**.

c All samples were centrifuged at **5,000 x g for 30 min at 25°C**.

Whereas some Materials and Methods sections contain too little detail, others include unnecessary extra detail, such as the one shown in the next example.

Example 12-4

Unnecessary information in Materials and Methods

CD4⁺ CD44ʰⁱᵍʰ T cell purification
We were interested in examining the gene expression of the anergic portion of T cells in Jak3 KO mice that are CD4± CD44ʰⁱᵍʰ. To obtain the CD4⁺ CD44ʰⁱᵍʰ T cells used for RNA isolation, spleens were removed from Jak3 KO and Jak3 Het mice at 8 to 10 weeks of age. Splenocytes were isolated by homogenizing the tissue with frosted glass slides (Fisher Scientific, Pittsburgh, PA).

In this example, the first sentence of the paragraph is unnecessary. This sentence should have been given in the introduction of the paper. The purpose for the actual portion of the experiment can be found in the second sentence of the paragraph ("To obtain . . ."). Thus, the first sentence can be omitted.

**Revised
Example 12-4**

CD4⁺ CD44ʰⁱᵍʰ T cell purification
To obtain the CD4⁺ CD44ʰⁱᵍʰ T cells used for RNA isolation, spleens were removed from Jak3 KO and Jak3 Het mice at 8 to 10 weeks of age. Splenocytes were isolated by homogenizing the tissue with frosted glass slides (Fisher Scientific, Pittsburgh, PA).

Consider Example 12-5.

Example 12-5

Unnecessary information in Materials and Methods

Cells were scraped out of the wells and resuspended in a 1.5 ml Eppendorf tube.

In the preceding example, we find a description for the tube size used: "in a 1.5 ml Eppendorf tube." Although for some experiments it may be important to mention the manufacturer for certain equipment if it is essential for the success of the experiment, a description of the tube size is usually considered unnecessary detail and should be avoided.

| Revised
Example 12-5 | Cells were scraped out of the wells and resuspended in 100 μl sterile saline. |

➤ Use parentheses for technical specifications

To provide enough details while maintaining good flow in your writing, parentheses are commonly used in the Materials and Methods section. Typically parentheses contain technical specifications including manufacturer's names, lot numbers, names of machinery, statistical information, and additional explanations and specifications.

| Example 12-6 | Use of parentheses |
| | 20 mg/ml trypsin (TPCK, bovine pancrease) dissolved in Z-buffer (10 mM Tris HCl pH 8.0, 120 mM NaCl, 50% (v/v) glycerol) was thermally denatured at 65°C for 3 min. |

➤ Indicate any statistical analysis performed

Often, raw data cannot be interpreted correctly without appropriate statistical analysis. The extent of such analyses typically depends on the author, the topic, or the scientific field. In many cases, in the text a simple statement such as the following will be sufficient.

Example 12-7	Statistical information
	a One-way analysis of variance was performed using the GLM procedure of SAS.
	b Differences between least square means were determined by orthogonal comparisons.

Other approaches and analyses may need more detail on the statistical analysis performed. See also Chapter 10 for a discussion of basic statistical analysis and Chapter 13, Section 13.2 for information on how to incorporate statistical findings in the Results section of a paper.

➤ Place full descriptions of procedures or other lengthy details in an appendix

Rather than putting detailed descriptions of procedures or other lengthy details in the body of the paper, you should place them in an appendix. If so, this must be part of your plan for the paper, not an afterthought submitted with the proofs, because an appendix has to be reviewed with the rest of the paper. Alternatively, you may be able to send lengthy material as supplementary material to an archive recommended by the journal.

Aside from detailed descriptions of procedures, material included in appendices rather than in the main body of a research paper (or proposal) may include detailed calculations, algorithms, proofs, tables, plots, and images or large data sets for meta-analyses and comparisons. Many journals now maintain electronic archives of supplementary material, including original data. This arrangement allows authors to be both thorough overall and concise in the main body of the article.

12.3 FORMAT

➤ Arrange experimental details as protocols grouped in chronological order or by subsections

The Materials and Methods section is usually a long section and typically covers various topics. These topics need to be organized, and methods need to be described in logical order, including the sequence of the procedures for each method.

You can organize your Materials and Methods section by separating each group of actions into one or more paragraphs. Paragraphs on the same type of information can then be grouped into subsections. The sequence of events within these subsections is usually written in chronological order or from most to least important. Each subsection has its own subheading, which functions as a signal, naming the particular material, variable, or specific procedure. Although use of subsections is optional, it usually simplifies and clarifies the presentation for the reader. Subsections can include one or more paragraphs.

Example 12-8   **Materials and Methods subsection**

Cultures. Samples taken from the superficial and the deep fascia in the forearm were prepared as previously described (12). To determine potential sources of infection, swabs from weight room equipment were obtained. In addition, cultures were taken from swabs of the school locker rooms and bathrooms. Isolated bacterial and fungal colonies were identified by standard microbiologic methods.

Check your target journal to find out what subheadings are commonly used in your field, and construct your subsections accordingly. You may even consider using the same subheadings in Materials and Methods and in the Results section. Examples of common subheadings are:

Analysis of X	Antibodies
Cell Cultures	Chemicals and Reagents
Cloning	Data Analysis
Materials	Outcome Measures
Plasmids	Protein Expression and Purification
Sequence Analysis	Statistical Analysis
Study Design	Study Population
Synthesis of Y	Treatment Protocol

➤ Signal and link the different topics

Other ways to signal different topics within the Materials and Methods section can be by topic sentences or transitions. Topic sentences can be used to signal the topic of a paragraph, especially within a subsection. Transitions are often placed at the beginning of the first sentence of a paragraph to link the paragraph to the previous one before introducing the topic of the remaining sentences in the paragraph. No signals are used if the topic becomes apparent from the subject and verb. (See also Chapter 6.)

➤ Explain the purpose for any procedure whose function is not clear

In all cases, it is important to ensure that the reader will understand *why* each procedure was performed and how each procedure is linked to the central question of the paper. Therefore, you should state the purpose or give a reason for any procedure whose function or relation to the question of the paper is not clear. Also, provide any background information that might be necessary to understand the experiments you performed. Statements of purpose or background are usually placed at the beginning of a paragraph and typically serve as topic sentences and transitions.

  Example 12-9   **Topic sentences/statement of purpose**

To purify prolyl 4-hyderoxylase from human placenta, full-term human placentae were collected 30 min after delivery.

12.4   IMPORTANT WRITING GUIDELINES FOR MATERIALS AND METHODS

➤ In the Materials and Methods section, passive voice is often preferred

The Materials and Methods section is the one section in a research paper where often passive voice is preferred over active voice. The reason is twofold: It lets you emphasize materials or methods as the topic of your sentences, and readers do not need to know who performed the action. (See also Chapter 4, Section 4.3.)

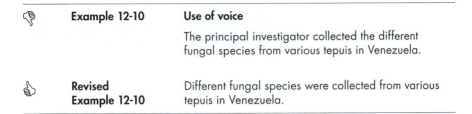

Example 12-10	Use of voice
	The principal investigator collected the different fungal species from various tepuis in Venezuela.

Revised Example 12-10	Different fungal species were collected from various tepuis in Venezuela.

It is easiest to write your entire Materials and Methods section from one point of view. The disadvantage is that if most sentences are written in passive voice, writing becomes dull. If you are more experienced in writing and are taking into consideration word location and cohesion, you may choose to write in both active and passive voice in the Materials and Methods section to make your writing more smooth, interesting, and clear.

Example 12-11	Use of mixed voice
	We stained tissue samples for X using an anti-X antibody (abCam, Cambridge, MA) on tissue slides. For this purpose, the **slides were de-paraffinized** and **rehydrated** as previously reported (34). After rinsing slides in Tris-buffered saline, **we applied** primary antibodies to the tissue sections and **incubated** them for 30 min at room temperature. Secondary antibodies **were then detected** using diaminobenzedine tetrahydrochloride.

➤ Do not switch from one point of view to another for no apparent reason

What you should avoid at all costs is changing back and forth from one point of view to another within one paragraph for no apparent reason. Such switches will unnecessarily confuse and distract readers.

Example 12-12	Use of voice
	We measured the produced ATP levels and activity of X complex using a commercialized kit. Decreased ATP production <u>was expected</u> under decreased oxidative phosphorylation.

ESL advice

In Example 12-12, the author switches from active to passive voice for no apparent reason. These types of switches are seen particularly often for ESL writers. It is easier for the reader to follow a passage if the same voice is used consistently, as in the revised example.

Revised	We measured the produced ATP levels and activity of
Example 12-12	X complex using a commercialized kit. **We expected** a decreased ATP production under decreased oxidative phosphorylation.

➤ Distinguish between past and present tense

In the Materials and Methods section, the general rule about verb tense (Chapter 4) applies. When reporting completed actions, use past tense (see previous Example 12-12 and Revised Example 12-12). However, use present tense for statements of general validity and for those whose information is still true or if you are referring to figures and tables.

Example 12-13	**Use of tense**
	a Because mud volcanoes **emit** incombustible gases such as helium in close proximity to lava volcanoes, we collected gaseous samples from Lusi.
	b Criteria used in selecting subjects **are listed** in Table 2.

Note that in many descriptive papers, especially in computational biology, the Materials and Methods section is written in present tense.

➤ Choose your words carefully

In the Materials and Methods section, exact and specific items are being described. Therefore, avoid jargon and other redundancies. In particular, colloquial phrases or words are inappropriate in scientific texts.

Example 12-14	**Choice of words**
	After 3 hours, the old medium was <u>dumped</u> and the same amount of fresh medium was added.

In Example 12-14, the use of the word "dumped" is jargon and not appropriate.

Revised	After 3 hours, the old medium **was replaced by an**
Example 12-14	**equivalent amount of fresh medium.**

ESL advice

Some jargon terms have been used so often that particularly nonnative speakers view them as "normal" English usage. Examples include "bugs" for bacteria, "overnext" instead of "the one after next," and "Western blotting" instead of "Western blot analysis." ESL writers should be particularly careful and have a native-speaking scientist or scientific editor review their writing.

Precise use of English is also a must in specific word choice. For example, distinguish between "determine," "measure," "calculate," "quantitate," and "quantify."

Determine	to find by investigation, calculation, experimentation, survey, or study
Measure	to find the size, length, amount, degree, etc.
Calculate	to work out or find out something by using numbers, to compute
Quantitate	to measure something precisely
Quantify	to measure the quantity of something

Example 12-15   **Precise use of English**

a   We **measured** the absorbance of XYZ at OD_{560}.

b   The percent error was **calculated**.

c   To **determine** absorbance of XYZ at OD_{560} and percent error, samples were weighed, dissolved in 1 ml buffer A, and incubated at 25°C for 2 hr.

12.5   ETHICAL CONDUCT

➤ Follow guidelines on ethical conduct

Research investigators should be aware of the ethical, legal, and regulatory requirements for research on human subjects and for animal experimentation in their own countries as well as applicable international countries with which they may collaborate (see also Chapter 1, Section 1.2 for more information on research ethics). Check requirements before reporting findings. Most journals require that submitted manuscripts reporting the findings of human and animal research conform to respective policies and mandates. Journal editors may ask authors to produce written approval of their research by an ethics committee (see Table 12.1).

➤ Indicate ethical conduct for human experimentation

Depending on the journal, authors may be requested to indicate within the published article (usually within the Materials and Methods section) whether the procedures followed were in accordance with the ethical standards of the responsible committee on human experimentation (institutional and national) and/or with any other mandate, such as the Declaration of Helsinki, for reporting experiments on human subjects. Furthermore, disclosure of patient data in scientific articles usually requires informed consent of the patients concerned. Patients normally are anonymous such that

TABLE 12.1 Guidelines on ethical conduct

GUIDELINES/ MANDATE	SOURCE	PURPOSE	WEB SITE
Uniform Requirements for Manuscripts Submitted to Biomedical Journals: Writing and Editing for Biomedical Publication (Vancouver rules)	International Committee of Medical Journal Editors (ICMJE)	Statement of ethical principles in the conduct and reporting of research and recommendations relating to specific elements of editing and writing	http://www.icmje .org/
Declaration of Helsinki	The World Medical Association	Statement of ethical principles to provide guidance to physicians and other participants in medical research involving human subjects	htpp://www.wma .net/en/ 30publications/ 10 policies/b3/17c .pdf
Guide for the Care and Use of Laboratory Animals	Institute for Laboratory Animal Research of the National Research Council	Book describing the ethical care and use of laboratory animals	http://grants.nih.gov/ grants/olaw/Guide- for-the-care-and-use- of-Laboratory- animals.pdf

people other than the patients themselves are unlikely to recognize them. If this is not possible, explicit written consent of the patients is required.

Example 12-16	**Phrasing of written consent**
	Written informed consent was obtained on enrollment, and the study was approved by the institutional review boards of the five centers and conducted according to the procedures of the Declaration of Helsinki.

➤ Indicate guidelines followed for care and use of laboratory animals

Authors are also often asked to indicate whether the institutional and national guidelines for the care and use of laboratory animals were followed when reporting experiments on animals. The place within a scientific article that should contain information of this kind is the Materials and Methods section.

Example 12-17	**Phrasing of compliance with guide for animal care**
	All procedures were approved by Z University Animal Care and Use Committee and followed the guidelines of the National Institutes of Health Guide for the Care and Use of Laboratory Animals. All efforts were made to minimize the number of animals used and their suffering.

12.6 COMMON PROBLEMS OF THE MATERIALS AND METHODS SECTION

The most common problems involving the Materials and Methods section of a research paper are

- Insufficient details (see Section 12.2 and Example 12-18)
- Omission of the purpose for an experiment (see Section 12.3)
- Unjustifiably switching between passive to active voice (see Section 12.4)
- Unjustifiably switching between past tense and present tense (see Section 12.4)

12.7 SAMPLE MATERIALS AND METHODS SECTIONS

The following is a good example of a subsection of a Materials and Methods section. It contains sufficient technical detail to allow a scientist to repeat the work. All experimental details have been clearly explained, and technical specifications have been given.

Example 12-18	**Materials and Methods subsection**
	Expression and purification of Salmonella Gol1
Exact cell strain is given	Recombinant plasmids containing the gene for wild type or mutant Gol1 were transformed into *E. coli* K-12 TO3
Necessary background is stated	as described previously (6). *E. coli* K-12 TO3 lacks the *gol1* gene and contains the pRARE plasmid that had been isolated from Rosetta cells (Novagen). Rosetta host cells are derived from BL21 cells and increase eukaryotic protein expression for rare *E. coli* codons.
Technical details (concentration, time, temperature) are stated	*E. coli* cultures were grown in the presence of ampicillin and chloramphenicol to mid-logarithmic phase and were then induced by adding IPTG to a final concentration of 0.4 mM. After induction, cell growth proceeded for 12 hr at 30°C. The protein purification procedure was
Reference given for full details	performed as described previously for the preparation of Salmonella Gol4 a (7).

Unlike the preceding example, the next example does not contain sufficient detail. It is a very carelessly written Materials and Methods section.

☞ | **Example 12-19** | **Subsection without sufficient detail**

In situ **hybridization**

Details for PCR amplification missing

We performed *in situ* hybridization for all samples as described by Chabi et al. (7). cDNA extending from nucleotides 10–1266 after the start codon of gene X was cloned into the pGL2 vector. The resulting plasmid was used to obtain probes for *in situ* hybridization of the developing brain. To amplify the cloned cDNA, PCR was performed in a Biometra thermocycler using 50 ng of the plasmid and 1 µM of primers Y and Z. To label single-stranded sense and anti-sense probes following the PCR amplification, repeated primer extension reactions were achieved using digoxigenin-11-dUTP on 500 ng of the PCR product. For hybridization reactions, the obtained probes were diluted 1:5 in hybridization buffer, added to the brains and hybridized at 25°C for 24 to 30 hr. Subsequently, alkaline-phosphatase-conjugated antibody (11) was added to the tissue, followed by incubation at 0°C for 12 hours. Brain samples were analyzed by fluorescence microscopy after staining with BCIP/NBT according to the protocol provided (12).

Unclear which probes

Unclear how they were achieved

Step missing— What happens to probes and hybridization buffer after hybridization?

12.8   REVISING THE MATERIALS AND METHODS SECTION

When you have finished writing the Materials and Methods section (or if you are asked to edit a Materials and Methods section for a colleague), you can use the following checklist:

- ☐ 1. Do the listed materials and methods describe all procedures done to obtain the results presented?
- ☐ 2. Are sufficient details and/or references provided?
- ☐ 3. Are protocols logically grouped and organized?
- ☐ 4. Are topics signaled and linked?
- ☐ 5. Did you pay attention to voice (mainly passive)?
- ☐ 6. Did you pay attention to correct use of past and present tense?
- ☐ 7. Did you choose your words carefully?
- ☐ 8. Did you ensure that major results are not stated in the Materials and Methods section?
- ☐ 9. Did you pay attention to ethical conduct?
- ☐ 10. Is the purpose stated for any procedure whose function is not clear?
- ☐ 11. Has sentence location been considered?
- ☐ 12. Is the point of view consistent?

☐ 13. Revise for style and composition using the basic rules and guidelines of this book:

☐ a. Are paragraphs consistent? (Chapter 6, Section 6.2)

☐ b. Are paragraphs cohesive? (Chapter 6, Section 6.3)

☐ c. Are key terms consistent? (Chapter 6, Section 6.3)

☐ d. Are key terms linked? (Chapter 6, Section 6.3)

☐ e. Are transitions used and do they make sense? (Chapter 6, Section 6.3)

☐ f. Is the action in the verbs? Are nominalizations avoided? (Chapter 4, Section 4.6)

☐ g. Did you vary sentence length and use one idea per sentence? (Chapter 4, Section 4.5)

☐ h. Are lists parallel? (Chapter 4, Section 4.9)

☐ i. Are comparisons written correctly? (Chapter 4, Sections 4.9 and 4.10)

☐ j. Have noun clusters been resolved? (Chapter 4, Section 4.7)

☐ k. Has word location been considered? (Verb following subject immediately? Old, short information at the beginning of the sentence? New, long information at the end of the sentence?) (Chapter 3, Section 3.1)

☐ l. Have grammar and technical style been considered (person, voice, tense, pronouns, prepositions, articles)? (Chapter 4, Sections 4.1–4.4)

☐ m. Are words and phrases precise? (Chapter 2, Sections 2.2 and 2.3)

☐ n. Are nontechnical words and phrases simple? (Chapter 2, Section 2.2)

☐ o. Have unnecessary terms (redundancies, jargon) been reduced? (Chapter 2, Section 2.4)

☐ p. Have spelling and punctuation been checked? (Chapter 4, Section 4.11)

SUMMARY

MATERIALS AND METHODS GUIDELINES

- Provide enough details and references to enable a trained scientist to evaluate or repeat your work, but do not include unnecessary detail.
- Include materials and methods but not results.
- Provide literature references where needed.
- Include sufficient technical details to let others repeat your work.
- Use parentheses for technical specifications.
- Indicate any statistical analysis performed.
- Place full descriptions of procedures or other lengthy details in an appendix.

- Arrange experimental details as protocols grouped in order or by subsections.
- Signal and link the different topics.
- Explain the purpose for any procedure whose function is not clear.
- In the Materials and Methods section, passive voice is often preferred.
- Do not switch from one point of view to another for no apparent reason.
- Distinguish between past and present tense.
- Choose your words carefully.
- Follow guidelines on ethical conduct.
- Indicate ethical conduct for human experimentation.
- Indicate guidelines followed for care and use of laboratory animals.

PROBLEMS

PROBLEM 12-1 Scientific Style and Format in the Materials and Methods Section
For each pair of sentences provided, select the better version and explain why you chose it.

1. a. As reported by Chasse et al. (17), tissue samples were collected within 30 min of organ removal, flash-frozen in buffer A, and stored at −80°C until further use.
   b. Tissue samples were collected upon organ removal as reported by Chasse et al. (17).
2. a. Seedlings were grown in continuous light before being collected.
   b. Seedlings were grown in continuous light at 15°C for 21 days before being collected.
3. a. Here, we provide a description of how subjects were chosen for our study. We selected only healthy males between 60 and 80 years.
   b. For our study, we selected only healthy males between 60 and 80 years.
4. a. Study subjects were presented with a list of potentially traumatic events, and we asked them to use three response categories (*yes, no, unsure*) to indicate if they had ever experienced them.
   b. Study subjects were presented with a list of potentially traumatic events and were asked to use three response categories (*yes, no, unsure*) to indicate if they had ever experienced them.
5. a. The Stress Index Short Form (SI/SF), which is a 36-item questionnaire, was used to assess stress after natural disasters (26).
   b. The Stress Index Short Form (SI/SF), which was a 36-item questionnaire, was used to assess stress after natural disasters (26).

PROBLEM 12-2 Applying Writing Rules

The following sentences have all been taken from prepublication Materials and Methods sections. Each one violates a basic scientific writing rule. Revise the sentences.

1. The analyses were performed on an Agilent series 1100 HPLC instrument (Agilent, Waldbronn, Germany) equipped with a quaternary pump, a diode-array detector (DAD), an autosampler, and a column compartment.
2. Immunoblotting was performed following standard procedure.
3. Adaptitude to changes in light was determined by . . .
4. Longer DNA fragments were detected by acrylamide gel electrophoresis using kb ladder as a standard.
5. After centrifugation, 10× buffer was added, and the samples were incubated for 2 min on ice.

PROBLEM 12-3 Evaluation of a Materials and Methods Subsection

Evaluate the following Materials and Methods subsection. Is it apparent for what purpose this test was performed? Why or why not? Suggest ways to rewrite the passage.

Wisconsin Card Sorting Test (WCST)
Participants sorted cards according to color, shape, or number stimuli depicted on the card. Sorting the cards by color was initially verbally reinforced. After a participant responded correctly for 10 consecutive cards in that color category, the participant continued sorting by form and numbers without verbal stimuli or reinforcement. Subsequently, the ratio of correct responses to errors was computed, and the numbers of trials, errors, perseverative responses, and perseverative errors were analyzed. In the WCST (Berg, 1948), test subjects have to correctly identify, implement, and remember sorting rules.

PROBLEM 12-4 Materials and Methods Subsection

Evaluate the following Materials and Methods subsection. Identify places where not enough technical specifications have been given to repeat the activity assay.

Determination of Trypsin Activity
Trypsin (TPCK, bovine pancrease) (20 µg/µl) dissolved in 0.01 M HCl, was denatured by adding 8 M urea, 33 mM Tris, pH 8.0, in a 1:4 (v/v) ratio and then boiling the mixture for 20 min. After denaturation, 5 µl denatured trypsin was added to 5 µl of each crystallization buffer, and the mixture was incubated at room temperature. After the first incubation, an equal volume of the crystallization buffer (10 µl) was added and the mixture was incubated again. Addition of equal volume of buffer together with incubation was repeated twice more (20 and 40 µl, respectively). Substrate (0.05 g (1%)

azocasein in 5 ml 10 mM Tris, pH 8) (100 µl) was added, and the mixtures incubated. The mixtures were then precipitated with trichloroacetic acid, and centrifuged. The supernatants were read in the spectrophotometer to determine activity.

(With permission from Elsevier)

PROBLEM 12-5 Evaluation of a Materials and Methods Section

1. On the left margin of this Methods section, write the topic of each paragraph.
2. Which paragraph(s) describe(s) the experiment done to answer the question asked? Write "Experiment" next to this/these paragraph(s).
3. Identify—if possible—the topic sentences (circle them).
4. Identify—if possible—an example of each technique of continuity:
   - Repeated key term (box it)
   - Transition word and transition phrase or clause (underline it)
   - Consistent point of view
   - Parallel form
5. What organization does <u>one</u> of the paragraphs follow (i.e., most to least important, pro–con, etc.)? Are the paragraphs organized well?

Research Question/Purpose: To construct and test a safe, live, attenuated, oral vaccine candidate, IEM108, immune to CTXΦ infection

Construction of the candidate IEM108. The 1.15-kb XbaI fragment containing the upstream regulatory and coding regions of ctxB was recovered from pBR (a pUC19-derived plasmid carrying ctxB and rstR, constructed in our laboratory before) and cloned into the XbaI site of pXXB106, containing *E. coli*-derived thyA (30), resulting in two new constructs, pUTBL1-5 and pUTBL1-6. ctxB and thyA have the same transcriptional direction in pUTBL1-5 and opposite directions in pUTBL1-6 (Fig. 1).

The rstR gene and its upstream sequence were amplified from El Tor strain Bin-43 with primers PrstR1 (CCGAATTCACTCACCTTGTATTCG) and PrstR2 (CGGAATTCTCGACATCAAATGGCATG). The amplified fragment was then cloned into the EcoRI site of pUTBL1-5, yielding new construct pUTBL2. Subsequently, an 0.8-kb PvuI fragment of the bla gene in pUTBL2 was deleted to generate pUTBL3. pUTBL3 was then electroporated into IEM101-T to construct IEM108.

Serum vibriocidal antibody assay. Serum vibriocidal antibody titers were measured in a microassay using 96-well plates. The immunized rabbit sera were inactivated at 56°C for 30 min and diluted 1:5 with PBS before use. The prediluted rabbit sera were added into the first well and then serially diluted threefold in PBS. PBS was added to the last well as a negative control. The plates were incubated for 30 min at 37°C with 25 µl of a solution containing 10^2 CFU of *V. cholerae* Bin-43/ml of culture and 20% guinea pig serum as a complement source in PBS. One hundred fifty microliters of 0.01% 2,3,5-trihenyltetrazolium chloride in LB broth was added to each well, and the plates were further incubated for 4 to 6 h at 37°C until the negative-control wells showed a color change. The reciprocal vibriocidal titer is defined as the highest dilution of serum that completely inhibits growth of Bin-43, i.e., no color change.

Rabbit immunization. Eight adult New Zealand White rabbits (2 to 2.5 kg) were divided into naive, IEM101, and IEM108 groups. The naive group consisted of two rabbits that were not immunized. Each immunization group had three rabbits. After fasting for 24 h, the rabbits in both immunization groups were anesthetized with ether. After the abdominal skin was sterilized with an iodine tincture and alcohol, the abdominal cavity was opened by vertical incision (under sterile conditions). General exploration was performed to find the ileocecal region. This region was ligated to the inner wall of the abdomen. Then, 10^9 CFU of vaccine strain IEM101 or IEM108 were injected into the proximal ileum. Finally, the abdominal cavity was closed. The ligature that tied the ileocecal region to the abdominal wall was removed 2 h later, and the rabbits were given water and feed for 28 days. One rabbit of the IEM101 group died after the operation, probably because of heavy anesthesia. Serum samples were collected from the immunized rabbits prior to the immunization and on days 6, 10, 14, 21, and 28 after the vaccination. The serum titers for the anti-CT antibody and vibriocidal antibody were measured as described above.

Rabbit ileal loop assay and protection model. To evaluate the protection efficacy *in vivo*, the immunized rabbits were challenged with pure CT and four virulent *V. cholerae* strains (395, 119, Wujiang-2, and Bin-43) of different serotypes and biotypes (Table 1) 28 days after the single-dose immunization. Rabbits were anesthetized and their abdomens were opened as described above. Their intestines were tied into 4- to 5-cm-long loops, and then 10^5 to 10^8 CFU of challenge strains or 1, 2, 3, or 4 µg of pure CT were injected into each loop. Normal saline was used as negative control. At 16 to 18 h postchallenge, the rabbits were sacrificed and the accumulated fluid from each loop was collected and measured. The ratio of the volume of accumulated fluid (milliliters) to the length of the loop (centimeters) was calculated for each loop in the challenged rabbits.

(With permission from American Society for Microbiology)

PROBLEM 12-6 Evaluation of a Materials and Methods Subsection

1. **Which paragraph(s) relate directly to the overall purpose of the study?**
2. **Identify—if possible—the topic sentences (circle them).**

3. Identify—if possible—an example of each technique of continuity:
   - **Repeated key term (box it)**
   - **Transition word and transition phrase or clause (underline it)**
   - **Parallel form**
4. **Are the paragraphs of this section organized well?**

Purpose: To evaluate the Child FIRST program, part of the Bridgeport Safe Start Initiative (BSSI), which was established to reduce the incidence and the impact of exposure to violence among children six years old and younger.

Method

Service utilization In collaboration with Child FIRST staff, the evaluation team developed a form that staff used to document the dates, types, and duration (recorded in fifteen-minute increments) of all services provided to children and their families. Examples of services included in-home assessment, classroom assessment and consultation, in-home care coordination, and staff consultation and supervision. Data documented on this form were also used to determine the length of time in the program from entry to discharge.

Family violence and traumatic events The Traumatic Events Screening Inventory–Parent Report Revised, or TESI–PRR (Ghosh-Ippen, Ford, Racusin, Acker, Bosquet, Rogers et al., 2002), is a twenty-four-item semi-structured interview that determines a history of exposure to traumatic events for children six years old and younger. Parents are presented with a list of traumatic events and asked to use three response categories (yes, no, and unsure) to indicate if the child has ever experienced them.

Twelve of the twenty-four TESI-PRR items were used to screen all children entering the program for a history of family violence and thus determine eligibility for inclusion in the evaluation study: separation from a family member; suicide by someone close to the child; physical assault or physical injury or bruising of child by family member; threat of serious physical harm to child; kidnapping by a family member; witnessing by the child of physical fighting or the use of a gun, knife, or other dangerous weapon by a family member; witnessing of verbal threats to seriously harm by a family member; witnessing of arrest of a family member; experiencing inappropriate sexual activity; witnessing inappropriate sexual activity; being yelled at repeatedly in a scary way or told that he or she is no good (verbal abuse); and experiencing neglect. The nonfamily violence items were exposure to serious accident; serious natural disaster; severe illness or injury of someone close; death of someone close; serious medical procedure or life-threatening illness; mugging; animal attack; community violence; direct exposure to war, conflict, or terrorism; exposure to war, conflict, or terrorism via television or radio; or other stressful events. Since the TESI is an inventory-type measurement, it is a reflexive measure in which indicators of internal consistency and other psychometric properties are not suitable.

(With permission from Lyceum Books, Inc.)

PROBLEM 12-7 Your Own Materials and Methods Section
Write a Materials and Methods section based on your work. Above all, arrange your sentences and topics logically. Use topic sentences and transitions. Signal subtopics. Use enough technical detail.

PROBLEM 12-8 Revising a Materials and Methods Section
Revise either your own Materials and Methods section or that of a colleague. Ensure that all elements are signaled clearly and that sufficient details are provided to repeat the experiments. Pay attention to elements of grammar and style. Use the checklist provided in Section 12.8 as a guide.

CHAPTER 13

Results

The Results section is the major scientific contribution of your study. It presents the results of your experiments and points the reader to the data shown in the figures and tables.

THIS CHAPTER COVERS:

- The overall structure and organization of the Results section
- The difference between results and data
- Presenting statistical information
- Signaling information in the Results section
- Common mistakes to avoid in the Results section
- Revising the Results section
- Full, annotated sample Results sections

13.1 OVERALL

In research articles, the Introduction provides background information and states the purpose/question of the paper. The Discussion confers what your data mean when tied in to current knowledge and theories. It is the Results section, however, that represents the core or skeleton of the paper, which may be of interest much longer than any conclusions drawn from your observations.

13.2   COMPONENTS

➤ **Report your main findings as well as other important findings**

➤ **Include control results**

In the Results section, you should report only results that are pertinent to the information provided in the Introduction and to the experiments described in Materials and Methods. Exclude preliminary results and results that are not relevant. Do not forget to include control results, however, and if needed, explain the purpose of an experiment shortly. Also, incorporate results whether or not they support your hypothesis, and explain any contradicting results if necessary.

Other important findings may consist of additional supportive evidence or alternate measurements as well as other results that are meaningful for the paper, even if they may not be part of the main story. You do not need to include every result you obtained in your Results section. For example, results obtained in preliminary experiments may be left out of the Results section or may be included in the Materials and Methods section instead. Concentrate on the most relevant findings, but when deciding what to include, know the difference between leaving out irrelevant results and suppressing contradictory ones. Do not omit the latter.

If you find that you need to collect more data as you write, do so. It is more important to do a thorough job than to submit an incomplete manuscript quickly.

➤ **Point the reader to the data shown in figures and tables**

Although most data should be presented in figures and tables, your main findings should be stated in the text as well, along with your interpretations of all data. When you state your interpretations/results, ensure that you make reference to your data in figures or tables by referencing the figure or table number in parentheses as shown in the next example (see also Revised Example 13-2).

Example 13-1   Table X. Fertilization rate for gametes of individual *S. purpuratus* males and females

	FEMALE 1	FEMALE 2	FEMALE 3	FEMALE 4
Male 1	13%	49%	47%	94%
Male 2	55%	86%	89%	91%
Male 3	82%	86%	96%	88%
Male 4	84%	91%	3%	87%

Text in Results section:

We tested *S. purpuratus* sea urchins for their fertilization efficiency using 4 different males (m1–m4) and 4 different females (f1–f4). Gametes of each individual male were crossed with those of each female in all possible hetero-sexual combinations, and the percent fertilization was determined. Substantial variations in fertilization of less than 60% fertilization success were observed in approximately 30% of the individual urchin crosses **(Table X)**, suggesting that fertilization rate depends on individual female-male combinations.

➤ Interpret your data for the reader

In the Results section, do not just present data, but summarize and interpret their meaning for the reader by presenting them as results. Only data that have been interpreted will be meaningful for your readers.

To present your results to the reader clearly, you need to distinguish between data and results. Data are raw or calculated values derived from scientific experiments (concentrations, absorbance, mean, percent increase). Results interpret data (e.g., "Absorbance increased when samples were incubated at 25°C instead of 15°C").

Example 13-2   **Presenting data without interpretation**

Heart rate was 100 beats per minute after digitalis was added (Fig. 3).

Unless your readers are physicians, they may not be able to put "100 beats per minute" into any relation, especially if no comparative value is given. You need to let them know whether this value is higher or lower than normal.

Revised   a   Heart rate *increased* **to 100 beats per minute** after
Example 13-2   digitalis was added (Fig. 3).

In the revised example, the data have been interpreted and are presented as a result, making the revised example much more meaningful for the reader.

To put the results into relation for nonspecialists in the field, you need to give comparative values as well.

Revised   b   Heart rate *increased from 60 to* **100 beats per minute**
Example 13-2   after digitalis was added (Fig. 3).

When the magnitude of change is given by a comparative value (". . . from 60 to 100 . . . "), the data have been interpreted such that it is understandable for most scientists.

Example 13.3 is another instance of providing data without interpretation or explanation.

Example 13-3   **Presenting data without interpretation**

The sequences for the proteins K 309 and K 415 were compared (Fig. 4).

This example fails to interpret the data provided. The author neither explains nor analyzes the data for the readers but simply refers the reader to a figure. As a consequence, readers do not know if the data are similar or different. Instead, readers are expected to interpret the data themselves. The author should make the point clear so readers do not have to find their own interpretations.

Revised   When the sequences for the proteins K 309 and
Example 13-3   K 415 were compared, **their C-terminal sections were found to be 90% homologous (Fig. 4)**.

The following example is yet another in which an author presents data but fails to interpret them.

Example 13-4   **Presenting data without interpretation**

Among the 785 HIV positive participants in the study group, we found 622 men and 163 women.

To interpret the data for the reader, the author needs to first present the interpretation and then the data that support it.

Revised   We found almost four times as many men (79.2%)
Example 13-4   than women (20.8%) tested positive for HIV in our study group.

Additional examples for interpretation of data within the Results section can be found in Sections 13.6. and 13.7.

➤ Place statistical information with data. Do not use it instead of results

Understanding statistical information in research articles can be problematic. Often, readers blame themselves for not comprehending what has been written. However, the true reason for their comprehension problems lies in the misrepresentation of statistical information.

To avoid confusing readers, make reference to the event to which you are referring. Sentences such as the following in which no reference class is given result in much misunderstanding among your readers.

Example 13-5   **Presenting data without reference**

There is a 20% chance of a big earthquake in California.

Readers interpret this sentence in various ways: 20% *of the area* of California has a big chance for an earthquake, or 20% *of the earthquakes* in California are big, or 20% *of the time* the chances for an earthquake are big. Confusion can be reduced by specifying a reference class, such as time and area, before giving a single event probability.

Even more confusing are sentences that talk about more than one statistical result at a time, such as in the following example.

Example 13-6   **Confusing description of statistical data**

The probability of contracting XDR TB is 80% for HIV patients. For people with XDR TB, the probability that it will be detected through rapid skin tests is 50%. In 10% of the cases, rapid skin tests do not detect XDR TB.

This type of text leads readers to wide misinterpretations, and such misinterpretations, particularly in the medical field, can have severe consequences for patients.

If we were to restate the example using numbers rather than probabilities, the example becomes much easier to understand.

Revised   Out of 100 HIV patients, 80 contract XDR TB. Of
Example 13-6   these, 40 cases will be detected through rapid skin tests. For 8 out of the 80 XDR TB cases (or 1 in 10), rapid skin tests do not detect XDR TB.

Readers usually profit from representing statistical information using numbers or frequencies rather than probabilities or sensitivities.

Many students and novice writers come up with tedious lists of statistical test results rather than a description and interpretation of experimental observations for their Results section. When you report statistical information, include descriptive statistics such as

mean
standard deviation
confidence intervals
p-values
sample size

Also include bivariate analysis such as chi-square or *t*-test or multivariate analysis such as regression analysis.

Ensure that you interpret descriptive statistics for your readers. Do not just list them in your Results section. Statistical analysis should serve as reinforcement for your data and should not replace their interpretation. Therefore, if you can, place statistical information in your figure legends or tables or in parentheses following the description of data.

👍 **Example 13-7**   **Preferred placement of statistical information**

Vaccination rates among the elderly were higher than among younger participants when the risk of flu was high (61.6% vs. 46.8%; OR = 2.67, 95% CI = 1.94–3.67).

For more information on statistical analysis, see Chapter 10. For information on how to indicate statistical tests performed in the Materials and Methods section, see Chapter 12, Section 12.2. Other good resources in providing background material on basic statistics include the following:

Brigitte Baldi. *Study Guide for the Practice of Statistics in the Life Sciences.* New York: W. H. Freeman, 2011.

Myra L. Samuels, Jeffrey A. Witmer, and Andrew Schaffner. *Statistics for the Life Sciences* (4th ed.). London: Pearson, 2011.

Joseph L. Fleiss, Bruce Levin, and Myunghee Cho Paik. *Statistical Methods for Rates and Proportions* (3rd ed.). New York: Wiley-Interscience, 2003.

13.3 FORMAT

➤ Place results that answer the question of the paper at the beginning of the Results section

Start the Results section by presenting your main findings in the first paragraph. Your main findings are the findings used in providing the overall answer/conclusion of the paper. You may also start the first paragraph with a brief overview of your general observations and then move on to the main findings (see Example 13-23). In the latter case, do not devote more than a few sentences to any overview, and ensure that your main findings still appear in the first paragraph, as it is a power position.

➤ Organize the Results section chronologically or from most to least important

In subsequent paragraphs, present your specific observations. The overall structure of this remainder of the Results section is normally either chronological or from most to least important. Use topic sentences to provide an overview of each experiment. Start each subsection or paragraph by explaining the purpose of the experiment, by giving a short background, or by stating the results of an experiment (see "Organization Within Results Segments" for a more detailed explanation). Consider also including a paragraph or two describing specific details of an observation. Readers will understand papers better if specific details are highlighted or given as examples.

➤ **Emphasize your results. Subordinate secondary information**

Throughout the Results section, emphasize the data and their meaning. Subordinate control results and methods.

The following is an example of a paragraph in a Results section. The topic sentence of this paragraph does not present the main findings but rather points the reader to a figure, thus emphasizing the figure rather than the results. Note that, in general, mentioning of a table or figure is best done in parentheses rather than in the text as tables and figures are considered supporting evidence and not results.

Example 13-8	**Results paragraph emphasizing figure rather than results**
Topic sentence does not emphasize results	**Student gender differences.** <u>Figure 3A and B shows the different ratings provided by the faculty toward the applications of the male and a female graduate science student</u>. The faculty viewed the female student as less competent than the male student, despite identical qualifications, similar to ratings reported previously [28]. In addition, the
Details of results	mean starting salary offered to the female student was significantly lower than that offered to the male student. The gender bias appears pervasive among faculty.

To emphasize results, the author should have subordinated the reference to the figure.

Revised Example 13-8	
Topic sentence emphasizes results	**Student gender differences**. Similar to ratings reported previously [28], the faculty viewed the female student as less competent than the male student in our study **(Fig 3A)**, despite identical qualifications. In addition,
Details of results	the mean starting salary offered to the female student was significantly lower than that offered to the male student **(Fig 3B)**. The gender bias appears pervasive among faculty.

Organization Within Results Segments

> ➤ **Organize your Results into different segments. In each segment, state**

- Purpose or background of experiment
- Experimental approach
- Results
- Interpretation of results (optional for descriptive papers)

To organize your Results section, think of it in different segments. Each segment pertains to one set of experiments. Many, if not most, segments will be only one paragraph long; others may be longer. You may even consider dividing your Results section into different subsections, making use of the separate segments.

Information in these Results segments or paragraphs needs to be organized, including in the first paragraph. Each segment that describes results of a specific individual experiment should contain four essential components:

1. Purpose or background of experiment if needed
2. Experimental approach
3. Results
4. Brief interpretation of results

Start your segments or paragraphs by providing a topic sentence. This topic sentence usually indicates the purpose of the experiment performed. It may also provide context in the form of background information. The purpose or background sentence is followed by a short statement of your experimental approach (about half a sentence). The purpose may be written in the form of a transitional phrase or clause, for example.

Follow the experimental approach immediately with your results for the experiment. Signal all of these elements (see Section 13.5). Place important or general results first and less important details later in the segment/paragraph. Last, give a brief interpretation of your results to make them meaningful for the reader. It is important that you do not simply list your data—instead, *interpret* your data and results for the reader. Your interpretations should be limited to one to two sentences in the Results section. Avoid any lengthy interpretations, speculations, or conclusions. Save such detailed discussions for the Discussion section. Note that this brief interpretation sets the stage for a more detailed discussion of your results overall and their interpretation in the Discussion section.

Example 13-9 is a well-formed paragraph/segment of a Results section.

Example 13-9   Well-formed Results segment

Background	**1** Considerable evidence suggests that ATP is needed in the binding of mRNA to the 40S ribosomal subunits (13).
Purpose	**2** To understand the interaction between ATP and mRNP
Experimental approach	(mRNA with bound proteins) particles better, **3** we incubated the mRNP particles with ^{14}C ATP at optimal concentrations for *in vitro* yeast translation. **4** Results indicate that ^{14}C ATP bound to mRNP particles, but the binding
Results	decreased about 4-fold when the temperature increased from 4 to 17°C (Fig. 1). **5** These results suggest that the binding between ATP and mRNP particles may be governed by interactions such as hydrogen bonds or van der Waals that weaken when temperature rises.
Interpretation of results	

In Example 13-9, the sentence 1 gives a short background, sentence 2 states the purpose/question, and sentence 3 explains the experimental approach. Sentence 4 gives the results, while sentence 5 interprets them for the reader. Note that data are presented in a figure (Fig. 1), and the description of the figure is subordinated. Thus, the actual results are emphasized.

Special Case: Format for Descriptions

➤ Describe what you discovered in the Results section of descriptive papers

For descriptive results or papers (i.e., those that describe a new discovery, such as a new organism or technique), list your results and descriptions by describing what you discovered. These descriptions often do not need an interpretation. Instead, provide conclusions and implications in the Discussion section.

Example 13-10   Results segment in a descriptive paper

Purpose	To characterize the protease, various protease inhibitors were tested for their effect on the proteolytic cleavage
Experimental approach	of the fusion protein. None of the inhibitors, antipain, aprotinin, chymostatin, EDTA, leupeptin, p-chloromercurobenzoate, pepstatin, phenylmethyl-sulfonylchloride, or soybean trypsin inhibitor, significantly slowed the rate of cleavage (data not shown). However, the bivalent cations Zn^{2+}, Cu^{2+}, and Co^{2+} were found to inhibit the protease to near completion when added to spheroplasts before lysis (Fig. 2-6). Benzamidine addition resulted in only partial inhibition. Addition of any of the inhibitors or
Results	cations after cell lysis did not inhibit the protease activity.

(With permission from Elsevier)

13.4   IMPORTANT WRITING GUIDELINES FOR THE RESULTS

➤ Pay attention to word choice

Words in the Results section should be chosen carefully. Choose the most precise and descriptive wording that reflects what you want to say, but keep wording simple. Consider the following example.

Example 13-11	Choice of words
	Mg^{2+} binds to the complex and increases complex formation, reaching an optimum at 4 to 10 mM (Fig. 2D). K±, on the other hand, has the *opposite effect* on complex formation, *reaching its optimum at 0 mM K±* (Fig. 2E).

What does the author want to say in the second sentence? What is the "opposite effect," and how can an optimum be *"reached"* at 0 mM? The opposite of *increase* is *decrease* or *inhibition*, and an optimum can never be *reached* at 0 mM, no matter what reagent is added. What the author intends to state here is that K^+ inhibits the formation. Thus, the second part of the second sentence should be omitted entirely because the word *inhibition* already implies the effect. The revision simplifies and clarifies what the author is trying to say.

Revised Example 13-11	Mg^{2+} binds to the complex and increases complex formation, reaching an optimum at 4 to 10 mM (Fig. 2D). **K^+, on the other hand,** *inhibits* **complex formation (Fig. 2E).**

ESL advice   In addition to using simple, precise words and avoiding jargon and repetitive words, pay particular attention to the following specific words and phrases in the Results section. These words are often used carelessly by authors, especially ESL authors, but should be distinguished because of their implied meaning.

Did Not

Choose your words carefully. Use neutral descriptions such as "did not" rather than "could not" or "failed to" when reporting results.

Example 13-12	Choice of words
	We **did not** detect any insulin production. (neutral—no expectation implied)

Clearly/It Is Clear/Obvious

Omit "clearly" and similarly subjective phrases in the Results. "Clearly" makes authors seem arrogant, and they appear as if they are trying to influence the reader.

👎   **Example 13-13**   **Choice of words**

Figure 6 <u>clearly</u> shows that the growth rate of K103 was reduced when Ca^{2+} was added.

👍   **Revised**   Figure 6 shows that the growth rate of K103 was re-
Example 13-13   duced when Ca^{2+} was added.

Significant

"Significant" in science refers to "statistically significant." If you write, for example, "Flow rate decreased significantly," the reader expects statistical details to follow this phrase. If you report results of statistical significance, specify the significance level. Consult a standard statistics textbook for detailed advice.

If you do not plan to provide statistical details, use "markedly" or "substantially" instead of "significantly." It is best, however, to reserve these words for the Discussion section. Also, remember that you should quantify these qualitative words by using precise values or referring to data (e.g., "Flow rate decreased substantially (23%).").

➤ Use past tense for your results but present tense for descriptive papers

Results are usually reported in past tense because they are events and observations that occurred in the past.

👍   **Example 13-14**   **Use of tense**

a   Imidazole **inhibited** the increase in arterial pressure.

b   All hawkmoths **started** feeding from the *A. palmeri* flowers only after nectar from the *D. wrightii* flower **was depleted** (Riffell, 2008).

Exceptions are results of descriptive studies. These results are reported in present tense because the description is still true.

👍   **Example 13-15**   **Use of tense**

a   **The *fgk* gene has several different introns.**

Other exceptions are statements of general validity; that is, if something is still true now or is considered a general rule, it should be written in present tense. (See also Chapter 4, Section 4.4.)

👍   **Example 13-15**   **Use of tense**

b   Their findings **show** that associating nectar with flower type **is** largely due to olfactory learning (Riffell, 2008).

TABLE 13.1 Signals for the results

PURPOSE/QUESTION	EXPERIMENTAL APPROACH	RESULTS	INTERPRETATION OF RESULTS
To determine . . .	. . . we did . . .	We found . . .	. . . , indicating that . . .
To establish if . . .	X was subjected to . . .	We observed . . .	. . . , consistent with . . .
Z was tested . . .	. . . by/using . . .	We detected . . .	. . . , which indicates that . . .
For the purpose of XYZ . . .	ABC was performed . . .	Our results	This observation
	Experiment X showed . . .	indicate that . . .	indicates that . . .
		that . . .	A is specific for . . .

13.5 SIGNALS FOR THE READER

➤ Use signals to highlight diverse elements of the Results

To emphasize different elements of the Results section, consider using signals, such as those shown in Table 13.1.

These examples provide great starting points when you write your first draft and may be particularly useful for those authors who have writer's block or whose native language is not English.

The most important result may be specially signaled to highlight it so the reader cannot miss it.

Example 13-16   **Highlighting important results**

Most interestingly, almost half of the newlywed couples (45.5%) start out sharing everyday household tasks equally or with husbands doing even a greater share than their wives.

13.6 COMMON PROBLEMS OF THE RESULTS SECTION

The most common problems of Results sections include the following:

- Missing components (purpose of experiment, experimental approach, results, or their interpretation) (Section 13.3)
- Inclusion of irrelevant or peripheral information (Section 13.2)
- Excessive experimental details
- Inclusion of comparisons, speculations, and conclusions beyond the interpretation of results

➤ Do not omit any key elements of the Results

Of the four elements that Results segments should contain (purpose of experiment, experimental approach, results, and their interpretation), beginning authors most often forget to include the purpose or the interpretation of the results. Others include only figures with legends/captions or tables and nothing else. If these narrative components are missing in the Results or any of its segments, findings will not be clear to readers.

In the following example, the interpretation of the results has not been included, leaving the reader wondering what these findings mean.

Example 13-17   **Results paragraph without interpretation of results**

Experimental approach— transitional clause	Successful colonization in the small intestines is indicated by prolonged shedding of vibrios in coproculture. Therefore, to test for colonization, we tested for shedding of vibrios in coproculture.
Results—overall	We found that there was a significant difference in shedding between P-5 and the positive control. The coproculture of the rabbits vaccinated with P-5 had a shedding time of 8 days, whereas the shedding time in the rabbits vaccinated with the positive
Interpretation missing Results—details	control was 5.5 days.

When the interpretation is included, as in the following revised version, the reader will gain a much better understanding of what these findings imply.

Revised Example 13-17

Experimental approach— transitional clause	Successful colonization in the small intestines is indicated by prolonged shedding of vibrios in coproculture. Therefore, to test for colonization, we tested for shedding of vibrios in coproculture. We found that
Results—overall	there was a significant difference in shedding between P-5 and the positive control. The coproculture of the rabbits vaccinated with P-5 had a shedding time of 8 days, whereas the shedding time in the rabbits
Results—details	vaccinated with the positive control was 5.5 days,
Interpretation	**indicating that the vaccine candidate strain P-5 efficiently colonizes rabbit intestines.**

➤ Omit peripheral information and irrelevant general statements

Do not confuse the reader, or yourself, by including irrelevant or peripheral information.

Example 13-18   **Irrelevant and peripheral information**

It took 2 hr to process 22,000 molecules and 32 hr to screen the entire ChemBridge database.

Readers are not interested in how long your work took. Statements like these should be omitted. Also, leave out irrelevant general statements of goals or overview sentences such as the one shown next.

Example 13-19   **Irrelevant overview sentences**

To present our results, we first list all components of the macromolecule together with their optima and then describe the outcome of their individual omission.

Overview sentences only add clutter. You do not need to explain how you will proceed in a research paper if your writing is coherent and well organized.

Aside from overview sentences, you should also avoid repeating all the data shown in figures and tables. Instead, describe your results and point the reader to a figure or table by citing this figure or table in parentheses after the description of the results.

 Example 13-20   **Referring to figures and tables**

In 2013, 52 cases of West Nile virus were identified in Connecticut. They are listed in Table 2.

 Revised   In 2013, 52 cases of West Nile virus were identified
Example 13-20   in Connecticut **(Table 2).**

Note that your main results should also be described in the text even if they are shown in a figure or table.

➤ Avoid experimental details

Do not describe experimental approaches in detail again, and do not introduce new experimental setups that were not mentioned in the Materials and Methods section.

 Example 13-21   **Results segment with unnecessary experimental details**

Because *Nocardia* have a unique cell wall and membrane structure, drug permeability should not be overlooked. Growth inhibition can be tested employing the surrogate marker *Nocardia asteroides*, which has high genomic sequence similarity to that of *N. brasiliensis*. To test the inhibitor candidates from the *in vitro* assay, <u>bacteria were grown overnight in LB broth to which 100 mM glucose and 0.01% Tween 20 had been added. (12) The surfactant Tween 20 in the liquid medium allowed *Nocardia* to disperse. Different amounts (1, 5, 10, 15, and 50 µg/ml, respectively) of the five candidate molecules were added to LB broth containing agar, vortexed briefly to mix, and then allowed to solidify. Media without any inhibitor candidate served as a negative control, and media with 50 µg/ml of kanamycin served as a positive control. Cultures were grown on solid agar overnight at 37°C.</u>

Experimental details should not be described in the Results section

> We found that two out of the five candidate mole-
> cules clearly inhibited growth of bacterial cultures
> (data not shown). NB22 demonstrated 100% growth
> inhibition at 15 µg/ml, whereas NB20 inhibited
> growth only weakly at the same concentration.
> Bacterial growth on the negative control plate was
> not affected, and no growth was observed for the
> kanamycin-treated plate (data not shown).

In the preceding example, all experimental details should have been described in the Materials and Methods section. These details should be omitted in the Results section because they make it unnecessarily lengthy and
distract from the actual results.

Revised Example 13-21 Experimental details have been omitted	Because *Nocardia* have a unique cell wall and membrane structure, drug permeability should not be overlooked. Growth inhibition **for the final inhibitor candidates from the *in vitro* assay were tested** employing the surrogate marker *Nocardia asteroides*, which has high genomic sequence similarity to that of *N. brasiliensis*. We found that two out of the five candidate molecules clearly inhibited growth of bac-terial cultures (data not shown). NB22 demonstrated 100% growth inhibition at 15 µg/ml, whereas NB20 inhibited growth only weakly at the same concentra-tion. Bacterial growth on the negative control plate was not affected. No colony was observed for the kanamycin-treated plate (data not shown).

In the revised example, experimental detail has been omitted. In addition,
the second and third sentences have been combined. This paragraph presents the results much clearer and is much more concise than the original
version.

➤ Avoid general conclusions, speculations, or comparisons with other studies

Do not compare your data to that of other studies, speculate on possible
mechanisms, or draw general conclusions. Leave these comparisons, speculations, and conclusions for the Discussion—but remember to interpret
your results briefly at the end of each Results segment, thus setting the stage
for the Discussion section.

Example 13-22	**Results segment with partial discussion**
Background	**Biofilm growth.** Biofilm formation starts with bacterial attachment to the surface, then progresses to auto-aggregation, micro-colonies formation, maturation, and

Purpose	eventually cell detachment. To determine the dynamics of the biofilm growth and the effect on membrane performance in reverse osmosis, biofilm was grown on membranes for 6, 12, 24, or 48 hours post inoculation. Subsequently, specific biovolumes of the viable cells and dead cells were determined (Table 1). Micro-colonies of bacteria were observed 6 hours after inoculation. After 12 hours some dead cells were observed. 24 hours after inoculation, more dead cells were found, and crevices and holes appeared in the biofilm, probably due to detachment of cells and small aggregates from the biofilm.
Interpretation	<u>These rapid changes in biofilm structure do not follow the previously reported biofilm formation stages (12, 43).</u>
Omit in the Results—place into the Discussion	<u>Possible reasons for this rapid change in biofilm structure could be a depletion of nutrients after 24 hours. In addition, previous studies used rich growth media (12, 43), whereas in our study minimal media was applied.</u>

(With permission from Moshe Herzberg)

Revised Example 13-22 **Background**	**Biofilm growth**. Biofilm formation starts with bacterial attachment to the surface, then progresses to auto-aggregation, micro-colonies formation, maturation, and eventually cell detachment. To determine the dynamics of the biofilm growth and the effect on membrane performance in reverse osmosis, biofilm was grown on membranes for 6, 12, 24, or 48 hours post inoculation.
Experimental approach	Subsequently, specific biovolumes of the viable cells and dead cells were determined (Table 1). Micro-colonies of bacteria were observed 6 hours after inoculation. After 12 hours, some dead cells were observed. 24 hours after inoculation, more dead cells were found, and crevices
Results	and holes appeared in the biofilm, **probably due to detachment of cells and small aggregates from the biofilm.**
Interpretation	**This rapid change in biofilm structure differs from that observed previously (12, 43).**

13.7 SAMPLE RESULTS SECTIONS

Example 13-23	**Investigative paper: First paragraph indicating overall results followed by more detailed description of individual experimental results**
Overall background/ purpose and experimental approach	During fertilization experiments performed to determine species specificity of two sea urchin species, *S. purpuratus* and *S. fanciscanus*, we discovered varying degrees of fertilizability when gametes of different individuals of a species were crossed (2, unpublished observation). Furthermore, as many as

Main overall results — one third of the sea urchins examined appeared to be infertile. Substantial variations in fertilization were observed for approximately 30% of the individual urchins (Fig. 2).

Background/purpose — The first set of *S. purpuratus* sea urchins tested for their fertilization efficiency contained 10 different males (m1–m10) and 5 different females (f1–f5).

Experimental approach — Gametes of each individual male were crossed with those of each female in all possible heterosexual combinations, and the percent fertilization was determined. Although each individual cross reached at least 80% fertilization (Fig. 3), the amount of sperm needed to reach maximum fertilization rates varied for the different combination of gametes. For example, about six times as many sperm from individual m10 were required to result in 50% fertilization success of female f4 than females f1 and f2, and about twice as many sperm were required for females f3 and f5 than for f1 and f2 (Fig. 6),

Results

Interpretation — suggesting that eggs from individual females are responsible for the variable success in fertilization.

Although crosses between gametes displayed significantly reduced fertilization efficiency in one cross, they did yield normal levels with gametes of other individuals of the opposite sex. Thus, although gametes of female f4 seemed to require more sperm of m4 and m6 to be fertilized than other females, the males m1, m2, and m9 were able to fertilize eggs of f4 at normal efficiency,

Additional results

Interpretation — suggesting that sperm of different individuals also effect the success of fertilization.

Note how in the preceding example the first paragraph gives an overview of the entire Results section by listing the main results of the study. Subsequent paragraphs describe individual experimental results in more detail and contain the required elements (background/purpose, experimental approach, results, and interpretation). In the last paragraph, the background/ purpose and experimental approach are not repeated as it is a continuation of the setup described in the preceding paragraphs.

Example 13-24   **Results of a descriptive paper**

Results

Overall results provide overview — . . . On the basis of our difference electron density maps and on the available biochemical and functional data, we propose the structural basis for the modes of action of the antibiotics chloramphenicol, clindamycin, erythromycin, clarithromycin, and roxithromycin.

Chloramphenicol

Topic sentence . . . Chloramphenicol has several reactive groups that can form hydrogen bonds with various nucleotides of the peptidyl transferase cavity: two oxygens of the para-nitro (p-NO2) group, the 1OH group, the 3OH group and the 4' carboxyl group.

More detailed results/description One of the oxygens of the p-NO2 group of chloramphenicol appears to form hydrogen bonds with N4 of C2431Dr (C2452Ec) (see Methods for definition), which has been shown to be involved in chloramphenicol resistance14. The other oxygen of the p-NO2 group interacts with O2' of U2483Dr (U2504Ec) (Fig. 1a–c) . . .

Macrolides

Topic sentence In contrast to chloramphenicol and the lincosamides, macrolides of the erythromycin class do not block peptidyl transferase activity (26). Although they bind to the peptidyl transferase ring, the erythromycin group of the macrolides, which includes clarithromycin and roxithromycin, is thought to block the tunnel that channels the nascent peptides away from the peptidyl transferase center (2, 27, 28). Our results confirm this assumption.

More detailed results/description We could unambiguously determine that the macrolides erythromycin, clarithromycin, and roxithromycin all bind to the same site in the 50S subunit of *D. radiodurans*, at the entrance of the tunnel (Fig. 5). Their binding contacts clearly differ from those of chloramphenicol, but overlap those of clindamycin to a large extent (Figs. 3a and 4).

(With permission from Nature Publishing Group)

Note how the topic sentences of the preceding example emphasize the results within each paragraph. Also, note how the topic sentences weave a continuous story throughout the section. Here, the purpose/background and experimental approach are not repeated in every paragraph because they have been provided in the introductory first paragraph. All subsequent paragraphs start immediately with the results, that is, a description of what has been found, as they belong to the same segment.

13.8   REVISING THE RESULTS SECTION

When you have finished writing the Results (or if you are asked to edit a Results section for a colleague), use the following checklist to systematically "dissect" the section:

☐   1. Did you report all main findings as well as other important findings?
☐   2. Are your most important results and their interpretation provided in the beginning of the Results section?
☐   3. Are the data for your most important results also mentioned in the text?

☐ 4. Is the organization from most to least important within the paragraphs?

☐ 5. Does each Results segment or paragraph contain all components (purpose of experiment, experimental approach, results, and their interpretation)?

  ☐ a. Is the purpose of each experiment apparent?
  ☐ b. Is the experimental approach provided?
  ☐ c. Are results interpreted?

☐ 6. Are all components (purpose of experiment, experimental approach, results, and their interpretation) signaled?

☐ 7. Are results emphasized?

☐ 8. Did you place statistical information with data?

☐ 9. Is the reader pointed to figures and tables?

☐ 10. Are control results included?

☐ 11. Are irrelevant statements and peripheral information avoided?

☐ 12. Have general conclusions, speculations, or comparisons with other studies been excluded?

☐ 13. Are the references placed correctly and where needed?

☐ 14. Revise for style and composition based on the basic rules and guidelines of the book:

  ☐ a. Are paragraphs consistent? (Chapter 6, Section 6.2)
  ☐ b. Are paragraphs cohesive? (Chapter 6, Section 6.3)
  ☐ c. Are key terms consistent? (Chapter 6, Section 6.3)
  ☐ d. Are key terms linked? (Chapter 6, Section 6.3)
  ☐ e. Are transitions used and do they make sense? (Chapter 6, Section 6.3)
  ☐ f. Is the action in the verbs? Are nominalizations avoided? (Chapter 4, Section 4.6)
  ☐ g. Did you vary sentence length and use one idea per sentence? (Chapter 4, Section 4.5)
  ☐ h. Are lists parallel? (Chapter 4, Section 4.9)
  ☐ i. Are comparisons written correctly? (Chapter 4, Sections 4.9 and 4.10)
  ☐ j. Have noun clusters been resolved? (Chapter 4, Section 4.7)
  ☐ k. Has word location been considered? (Verb follows subject immediately? Old, short information is at the beginning of the sentence? New, long information is at the end of the sentence?) (Chapter 3, Section 3.1)
  ☐ l. Have grammar and technical style been considered (person, voice, tense, pronouns, prepositions, articles)? (Chapter 4, Sections 4.1–4.4)
  ☐ m. Is past tense used for results and present tense for descriptive papers? (Chapter 11, Sections 11.3 and 11.4)
  ☐ n. Are words and phrases precise? (Chapter 2, Sections 2.2 and 2.3)
  ☐ o. Are nontechnical words and phrases simple? (Chapter 2, Section 2.2)

☐ p. Have unnecessary terms (redundancies, jargon) been reduced? (Chapter 2, Section 2.4)

☐ q. Have spelling and punctuation been checked? (Chapter 4, Section 4.11)

SUMMARY

RESULTS GUIDELINES
- Report your main findings as well as other important findings.
- Include control results.
- Point the reader to the data shown in figures and tables.
- Interpret your data for the reader.
- Place statistical information with data. Do not use it instead of results.
- Place results that answer the question of the paper at the beginning of the Results section.
- Organize the Results section chronologically or from most to least important.
- Emphasize your results. Subordinate secondary information.
- Organize your Results into different segments. In each segment, state
  - Purpose or background of experiment
  - Experimental approach
  - Results
  - Interpretation of results (optional for descriptive papers)
- Describe what you discovered in the Results section of descriptive papers.
- Pay attention to word choice.
- Use past tense for your results but present tense for descriptive papers.
- Use signals to highlight diverse elements of the Results.
- Do not omit any key elements of the Results.
- Omit peripheral information and irrelevant general statements.
- Avoid experimental details.
- Avoid general conclusions, speculations, or comparisons with other studies.

PROBLEMS

PROBLEM 13-1 Scientific Style and Format of a Results Section
Using basic writing rules, improve the following sentences, which were taken from prepublication Results sections.

1. A 10% change in the length of daylight did not cause a significant alteration in the expression level of X studied here.

2. Expression of Y was inhibited in the kidney, lung, and liver of BRF⁻ rats after endotoxin challenge but not in wild-type rats (Figure 2E).

3. Comparison of the cytokinin concentrations after heat induction showed that production of cytokinin was strongly increased in ST25 mutants.

4. In total, 19.7% (542/2,758) of genes located on chromosome 1 and 41.34% (456/1,103) of genes located on chromosome 2 were absent from at least one strain, which seems to indicate higher conservation in chromosome 1.

5. The amount of mutated DNA in A and B strains is much less (0 to A and B mutations for each), exhibiting the highest conservation as described in previous research.

6. The T-test also showed a significant increase in Rh4 expression in the photoreceptor cells.

7. Three of the molecules, CB4, CB6, and CB10, exhibited more than 50% inhibition of the enzyme activity.

PROBLEM 13-2 Elements of the Results Section
Assess the following partial Results section.

1. **Identify**
   - **the purpose or background of the experiment**
   - **the experimental approach**
   - **the results**
   - **the interpretation of the results**
2. **Are all the parts of a paragraph for the Results section provided? Please explain.**

We found that the H384A mutant reduced the k_{cat} value more than 3-fold. The apparent K_m values were increased 7-fold for Fru 6-P and 3.5-fold for PPi. The increase of the K_m values and the reduction of the k_{cat} value of the H384A mutant suggest that the imidazole group of His384 is important for the binding stability as well as for catalytic efficiency of Fru 6-P and PPi substrates.

PROBLEM 13-3 Evaluation and Revision of a Results Section
Assess and revise the following partial Results section. Ensure that all the parts of a paragraph for the Results section are provided and any unnecessary parts are omitted.

To evaluate inhibitory effects of the isolated molecules, 10 mM stock solutions of all isolates were prepared in buffer A. The buffer was previously optimized for the inhibition assay and contained 25 mM Tris-HCl (pH 7.5), 5 mM β-glycerophosphate, 2 mM dithiothreitol (DTT), 0.1 mM Na_3VO_4, 10 mM $MgCl_2$, and 250 μM ATP. For the assay, 0.1 mM of the isolated molecule were used in 200 μl total volume. The reaction mixture was vortexed for 30 sec before incubation at 25°C for one hour. Then, kinase was added to 100 μM. Reaction products were analyzed by 12% PAGE and Western plot analysis.

Six out of 15 isolated molecules inhibited the kinase reaction at 10 μM markedly, and two of the molecules, A3 and A7, exhibited more than 50% inhibition of the enzyme activity. A 10-fold dilution series of these latter samples was prepared to determine the minimal inhibitory concentration. Molecule A3 exhibited 30% and molecule A7 45% inhibition of the kinase reaction at 1 μM. Buffer A was found not to interfere with the enzyme.

PROBLEM 13-4 Evaluating a Partial Results Section
Assess the following partial Results section. Ensure that all essential parts of a Results section paragraph are there. Check that parts are signaled clearly. Improve transitions if needed. Condense where possible.

. . . Previous studies have shown that, in biological networks, hubs tend to be essential [7, 9], and betweenness of a node is correlated with its degree [20]. We found that degree and betweenness are indeed highly correlated quantities in the networks we analyzed (Pearson correlation coefficient of 0.49, $p < 10^{-15}$ for the interaction network; Pearson correlation coefficient of 0.67, $p < 10^{-15}$ for the regulatory network; p-values measure the significance of the Pearson correlation coefficient scores according to t distributions; i.e., many bottlenecks also tend to be hubs). Therefore, we further investigate which one of these two quantities is a better predictor of protein essentiality in both regulatory and interaction networks.

To disentangle the effects of betweenness and degree, we divided all proteins in a certain network into four categories: (1) nonhub–nonbottlenecks; (2) hub–nonbottlenecks; (3) nonhub–bottlenecks; and (4) hub–bottlenecks (see Figure 1). Even though the two quantities are highly correlated, the number of hub–nonbottlenecks and nonhub–bottlenecks is enough for reliable statistics (see Table S1). This is in agreement with the previous observation by Huang and his colleagues, who found that proteins with high betweenness but low degree (i.e., nonhub–bottlenecks) are abundant in the yeast protein interaction network [18].

(PLoS Comput. Biology 3(4), 2007)

PROBLEM 13-5 Evaluation of a Results Section
Assess the following partial Results section.

    1. **Identify**
        - **the purpose or background of the experiment**
        - **the experimental approach**
        - **the results**
        - **the interpretation of the results**
    2. **Are all the parts of a paragraph for the Results section provided? Please explain.**

Sensory deprivation during the first 30 days of development leads to decreased Cat-315 expression in layer IV of the barrel cortex.

Because aggrecan-reactive nets are strongly expressed in the postnatal barrel cortex (Fig. 2A–H), we asked whether altering sensory input from the whiskers would alter the expression of PNs and aggrecan. To evaluate this, whiskers were trimmed from the right whisker pad of mice every other day from birth through P30. Tangential sections through layer IV of the barrel cortex were analyzed using stereological methods. Nissl staining of the barrel cortex illustrated that the gross development of the barrels was not altered by this manipulation and did not differ from controls (Fig. 2A,E,I,M). However, our studies revealed a significant decrease in the number of cells with Cat-315-positive PNs in the sensory-deprived barrel cortex of trimmed animals (Fig. 2N–P) compared with the nondeprived barrel cortex of the same animals (Fig. 2J–L) and compared with both barrel cortices of control animals (Fig. 2B–D, F–H; Table 1). There was a statistically significant reduction in Cat-315 staining in trimmed animals compared with control animals using a two-way repeated-measures ANOVA ($p = 0.0282$). The decrease in Cat-315 in the deprived hemisphere compared with the nondeprived hemisphere in trimmed animals was significant using a paired t test ($p = 0.0034$) (Table 1). There was also a significant decrease in the deprived hemisphere of trimmed animals compared with the left hemisphere of control animals ($p = 0.0015$) (Table 1). However, the nondeprived barrel cortex did not differ significantly from the barrel cortex of control animals ($p = 0.9454$) (Table 1).

(With permission from The Journal of Neuroscience)

PROBLEM 13-6 Evaluating a Results Section
Assess the following partial Results section:

1. **Identify**
   - **the purpose or background of the experiment**
   - **the experimental approach**
   - **the results**
   - **the interpretation of the results**
2. **Are all the parts of a paragraph for the Results section provided? Please explain.**

To explore the forces behind the strong tendency of husbands to decrease their share of housework, and the low tendency to increase their share of housework in the course of marriage, we looked at the findings from the multivariate event-history models. We first explored the role of economic resources in changing couple's division of housework in the course of marriage (Table 1). We investigate the impact of the spouse's relative economic resources on the likelihood of dividing housework either more equally or less equally in the course of marriage. We find that husbands who work a similar number of hours (Husband=Wife), or lower number of hours (Husband<Wife) than their wives, are less likely to decrease their share in household labor, compared to husbands who work longer hours than their wives (Husband>Wife). Equal earning levels between the spouses also seem to reduce the likelihood for husbands to decrease their share of housework in the course of marriage (model 3b). For couples with an 'atypical' female provider earnings ratio (Husband<Wife), the effect is not significant, however. It appears that a winning margin in economic resources does more for the husband than for the wife when housework is redistributed.

We also looked at the effects of family formation on the gender division of household tasks to assess how shifts in economic resources play out in this context (Table 2). We found a pronounced and significant effect for both directions of change. During the first year after childbirth, fathers seem to be about twice as likely to decrease their contribution to housework. In the same period fathers' likelihood to increase their share in housework is reduced by almost 50 percent, compared to childless men. This push towards a more traditional division of housework seems to come to a halt when the youngest child reaches age two. We found no indication, however, that parents readjust back to a more egalitarian division of housework when kids grow older and mothers return to their previous jobs. The time-varying economic indicators (models 2ab and 3ab) do not seem to explain these processes at all. Here, we need to be cautious not to interpret lack of significance in the economic indicators as lack of relevance, especially since the share of parents in the "non-traditional" resource categories is low. The number of mothers out-earning or working longer hours than their husbands is small in this sample and the share of continuously working mothers is low.

(With permission from Daniela Grunow)

PROBLEM 13-7 Your Own Results
Write your own Results section based on the experimental results of your work. Provide the overall question of the paper. Use all basic writing rules studied, and follow the summary on how to write a Results section.

PROBLEM 13-8 Revising a Results Section
Revise either your own Results or that of a colleague. Ensure that the main findings are stated up front. Be sure to identify the elements within each experimental segment and that they are signaled clearly. Pay attention to verb tense as well as other elements of grammar and style. Use the checklist provided in Section 13.8 as a guide.

CHAPTER 14

Discussion

The Discussion is usually the most challenging section to define and to write. This chapter breaks down the organization and content found in Discussion sections into manageable pieces of information.

THIS CHAPTER COVERS:

- How to start and end your Discussion
- What elements to include in the Discussion
- How to deal with discrepancies, unexpected findings, and limitations
- How to signal information
- What common mistakes to avoid in your Discussion
- How to revise your Discussion
- Two full, annotated sample Discussion sections

14.1   OVERALL

Many papers are rejected by editors because of a bad Discussion section. Even though the data may be valid and interesting, their interpretation or presentation in the Discussion may obscure it. Therefore, good style and clear, logical presentation are especially important here.

14.2   COMPONENTS

➤ State and interpret your key findings. Provide the answer to the research question

In the Discussion, provide the answer to the research question. To do so, interpret your key findings overall and draw conclusions based on these findings in the context of the research topic and overall related fields. The Discussion should also explain how you arrived at your conclusion, compare and contrast your findings with existing knowledge on the topic, and state theoretical implications or practical applications. It should give the paper significance by generalizing results while clearly indicating how your study has advanced knowledge. (See also information on common problems of the Discussion in Section 14.10.)

➤ Summarize and generalize

In the Discussion, explain what is new in your work and say why your results are important. You should also include explanations for any results that do not support the answers and discuss other results and hypotheses that are relevant to yours. In addition, you may discuss any possible errors or limitations in your methods, give explanations of unexpected findings, and indicate what the next steps might be. Do not refer to every detail of your work again; repeating the Results section in the Discussion is a common mistake of inexperienced writers. Another common mistake is to start the Discussion with a passage that reads like another introduction. Instead, in your Discussion, summarize and generalize.

➤ Keep in mind who your potential readers will be

Adjust your Discussion according to who your potential readers will be, and make it no longer than necessary. If you are writing for a very specific group of people, stay within their area of interest. If you are writing for a broad audience, you probably need to discuss much broader implications and provide more generalizations and background.

In general, remember that in the related fields of science and medicine, basic scientists and clinicians read each other's papers. So if you write your paper primarily for a scientific audience, do not ignore the clinical implications of your results, and if you are addressing a clinical audience, try to discuss the scientific significance as well. In this way, your work will have much greater impact.

14.3   FORMAT

➤ Organize the Discussion in a pyramid structure

Opposite of the Introduction, which follows a funnel shape, the Discussion follows a pyramid shape. In other words, it moves from specific to general.

Organize the Discussion in a *pyramid structure:*

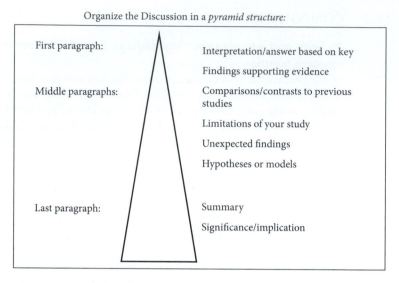

Figure 14.1 Distribution of content in the pyramid structure of the Discussion.

In the first paragraph, briefly tell your readers what your key findings were and what they mean. In subsequent paragraphs, explain how your findings fit into what is known in the field. In the last paragraph, summarize and generalize why the contribution of your study is important overall, in your field, outside your field, and/or for society.

A more detailed explanation of this organizational structure is provided in the next sections of this chapter.

14.4   FIRST PARAGRAPH

➤ State the answer to your research question in the first paragraph

Begin the Discussion with an interpretation of the key finding(s), which present the answer to the question posed in the Introduction. Then, support your answer by stating the relevant results, providing explanations, and/or other data. Readers will not remember everything presented in the Results section. You have to remind the readers about the key results and put the story together for them.

Because the interpretation of your key findings is the most important statement in the Discussion, it should appear in the most prominent position: the first paragraph. The interpretation of your key findings should match the question/purpose for the study stated in the Introduction and answer what the Introduction asked. The interpretation of your key findings should also be repeated in the other power position of this section: the last paragraph. Note that whereas in the Results section you provided brief explanations/interpretations of your data and results, in the Discussion section you need to expand on these interpretations and extrapolate on all findings together in order to arrive at your overall answer to the research question of your paper. Thus, the Discussion contains lengthier interpretations, speculations, and conclusions.

Providing more background information, summarizing your findings, or reporting on limitations and minor results are not good ways to start the Discussion. Instead, begin by directly stating the answer based on your findings in the opening sentence of the Discussion. If you feel this beginning is too abrupt, you can restate the purpose of the study or provide a brief context before stating the answer. Any statements placed before your answer should not exceed more than a few sentences.

If your answer is in the first paragraph of the discussion, the reader is sure not to miss it. Readers typically do not read the whole paper front to back. Rather, they skim over the Abstract, maybe read the Introduction, and then immediately jump to the Discussion. In the Discussion, readers instantly want to find the answer to the research question, that is, the interpretation of the key findings of the study. Readers do not usually take the time to read the whole Discussion. They will typically read the first paragraph and then go on to the last paragraph. Thus, your key findings and their interpretation(s) should be placed in these two power positions.

Example 14-1   **First paragraph of Discussion**

Question/purpose:

Effectiveness of Peridomestic <u>Lyme Disease Protection Measures</u>

Answer/interpretation of key findings:

Answer/ interpretation of key findings	*Our findings emphasize* the need to continue to promote personal <u>protection measures</u> to reduce the risk of <u>Lyme disease</u> infection. *We have identified* three reasonable personal measures that may be <u>protective</u> against <u>Lyme disease</u> when practiced: tick checks, bathing, and insect repellents. Performing tick checks within 36 hr after spending time in the yard may reduce one's risk by as much as 46%. In addition, bathing may reduce one's risk by up to 57%, and the use of insect repellent may be protective against the disease up to 75%.
Supporting evidence	

(With permission from Neeta Connally)

In Example 14-1, the interpretation of the key findings matches the purpose of the study. Note that the same key terms (protection, measures, Lyme disease) appear in the question/purpose as well as in the interpretation or answer to the question. This answer is immediately supported by key findings of the study.

Following is another example.

 Example 14-2   **First paragraph of Discussion**

Question/purpose

Our goal was to determine <u>what part of the bindin polypeptide is responsible for the species-specific egg agglutination activities of the protein.</u>

Answer/interpretation of key findings

Answer to question	*Our results suggest that* <u>the part of bindin responsible for species-specific egg agglutination lies in</u> *the region* of residues 75–121. *We showed that* <u>residues 18–74 and 122–236 can be deleted without loss of egg agglutination activity.</u> All of the biologically active bindin deletion analogs were found to be <u>species-specific</u> by their ability to <u>agglutinate</u> exclusively *S. purpuratus* <u>eggs</u>. Deletion analogs that had any residues of region 75–121 deleted exhibited no significant activity above the bacterial control protein.
Supporting evidence	

(With permission from Elsevier)

In this example, the answer also matches the question/purpose posed. Note the *signals* used to guide the reader through the answer and supporting evidence in both of the previous examples. Note also the different verb tenses. To distinguish between results and the conclusions you draw from them, use the past tense for results and present tense for general statements, interpretations, and conclusions. (See also Section 14.8 for additional signals for the reader.)

Results (past tense)	Interpretation (present tense)
showed	suggest
were found	indicate
exhibited	show

The exception to beginning the Discussion with an interpretation of the key findings is for papers that describe controversial topics and findings, such as embryonic stem cell research. For highly controversial topics, argue your case first by presenting your findings and explaining differences from other findings. This organization will help to logically prepare the reader for your upcoming argument. Present the controversial interpretation of your key findings at the end of the Discussion in these papers.

14.5   MIDDLE PARAGRAPHS

➤ **Organize the topics according to the science or from most to least important**

After stating and supporting your answer, mention other findings that were important. Tell your readers what you think your results mean and how strongly you believe in them. Organize these findings according to the science or from most to least important. To ensure that your Discussion is organized rather than rambling, focus the story on the question/purpose of the paper that was stated in the Introduction.

The presentation of your arguments is a matter of personal style, as is the order of the paragraphs between the first and last paragraph. To develop the middle paragraphs of a Discussion, organize the topics by proceeding from most to least important unless there is a reason for putting one topic before another. Explain any new findings and concepts obtained in your study, but do not present any new data that has not already been mentioned in the Results section. Also, do not repeat any information that has already been presented in other sections of your paper.

➤ **Compare and contrast your findings with those of other published results**

In the Discussion, it is not only important to put your findings in the context of your field but also to make connections to other important implications of the work or to other seemingly related or perhaps unrelated fields. In addition to stating and supporting your answer, you need to explain how your findings fit in with existing knowledge on the topic. You can do so by comparing and contrasting your results with those found by others. One way to get started in comparing and contrasting your findings and interpretations is to prepare lists that contain your findings and those of others. Based on lists like these, you may be better able to see and discuss any similarities or differences between your work and that of previous reports. Remember to signal and reference any and all work done by others. Do not plagiarize, but rather paraphrase, summarize, and generalize when you compare and contrast. The citation style largely depends on your target journal, so pay attention to the *Instructions to Authors* (see also Chapter 8, Sections 8.3–8.7).

When you mention any results that do not support your answer and conclusions, explain these findings as best as you can. If you can explain why a finding conflicts with your results, it is almost always worth the time doing so. If you cannot explain these findings, say so ("We cannot explain why . . . "; "Although the reason for X is not obvious, . . .").

See the following example for an explanation of contrasting findings. In this example, the authors discuss a finding of another study that differs from the answer to their question ("In contrast to our observations . . . "). The authors then discuss why previous findings cannot be directly compared with their results and go on to explain how previous studies had obtained these conflicting results.

 Example 14-3 **Comparing and contrasting findings in the Discussion**

Finding of paper

We observed virtually no size classes of mtDNA molecules. Since the undegraded circular mtDNA molecules were entirely of heterogeneous size, this observed size heterogeneity probably reflects the real situation within plant mitochondria. . . . *In contrast to our observa-*

Signal for contrasting finding

tions, size classes of linear or circular molecules and species-specific differences have been previously reported (24, 25). However, these studies were performed only with a fraction of supercoiled DNA (26), which most likely does not represent the complete set of molecules existing *in organello.* Supercoiled DNA isolated from a *C. album* suspension culture, for example, consisted exclusively of small circular plasmid mp1 DNA. Its oligomers were found in the

Explanation of conflict

open circular form, thus appearing indeed as a few size classes.

(With permission from Springer)

Example 14-4 provides another example of comparing and contrasting findings and ideas in a Discussion.

 Example 14-4 **Comparing and contrasting findings in the Discussion**

Finding of paper

We have shown that annual canine vaccination campaigns achieving 67% coverage in Ngorongoro and 42% coverage in Serengeti should be sufficient to control rabies outbreaks with 95% confidence. These coverage levels are lower than the WHO-recommended annual target of 70% (26). We focused on annual coverage tar-

Comparison to other study

gets, since rabies vaccination in Tanzania is conducted through annual campaigns and since the WHO target is specified as such. However, we also calculated that 39%

Contrasting finding

and 25% coverage consistently maintained in Ngorongoro and Serengeti, respectively, will control rabies outbreaks with 95% confidence. These estimates of required

Comparison to other study

coverage are much lower than previous recommendations of 70% coverage on a consistent basis (11). The difference is possibly due to fact that the parameters of the previous study were drawn from Asia and the Americas, whereas

Explanation of conflict

our model considers rabies dynamics in sub-Saharan Africa (19).

(M. C. Fitzpatrick et al. (2012) PLoS Neglected Tropical Diseases 6(8):e1796)

➤ Explain any discrepancies, unexpected findings, and limitations

Mention any limitations of your study or unexpected findings, and present any new hypothesis or model based on your findings. If useful, include figures to illustrate complex models in the Discussion. Compare and contrast your findings with those of previously published papers, but avoid the temptation to discuss every previous study in your subject area. Stick to the most relevant and most important studies. Explain any disagreements objectively, and credit and confirm the work of others. Give pro and con arguments for your conclusion. Only if you mention both impartially will you sound convincing to the reader. Know that most of the time it is wise to present your opinion carefully rather than too strongly.

Limitations

Limitations of the study as well as assumptions should be explained in the Discussion, although short, minor limitations can also be placed into the Materials and Methods section.

Example 14-5   Explaining limitations in the Discussion

In our modeling of Aβ assembly, we assumed that Aβ monomers are not present in drusen. However, it is possible that Aβ monomers, once polymerized into amyloid fibrils, may accumulate in drusen (40). Such accumulations would result in a lower number of monomers used for calculations in our model than are actually present, and thus
Limitation | a higher risk for the disease than determined based on our assumption.

Unexpected Findings

Aside from comparing and contrasting your work with that of previous studies and describing limitations in your study, unexpected findings may also be mentioned in the Discussion section. Be alert to unexpected findings. Do not automatically assume that your experiment failed or that you made a mistake. Unexpected findings may be important; they may lead to new discoveries and alter the focus of your study.

When you switch from the focus of your study to reporting on something unexpected, provide a signal for the reader ("To our surprise, . . ."; "Unexpectedly, . . ."). Then, report on the finding and give a possible explanation for your observations without going into too much detail.

Example 14-6   Describing unexpected findings

Unexpected finding | *To our surprise* we discovered that the bindin fusion protein was being cleaved during isolation and purification. The proteolysis is remarkably efficient since only small amounts of the unprocessed form remain (Fig. 2, lane 3). We

Description of unexpected finding

purified the cleaved bindin product to homogeneity by reverse phase HPLC and sequenced it to determine the site of cleavage. The predominant product is the mature bindin polypeptide containing an additional 4 amino acids of probindin and a minor product that corresponds to bindin containing a single additional amino acid. Both products contain arginine as the N-terminal amino acid. These results suggest that the fusion protein is cleaved at two sites: the Arg-Arg junction between the factor Xa linker and the probindin coding sequence and within the probindin segment at the Lys-Arg junction.

(With permission from Elsevier)

➤ Provide generalizations where possible

Treat your secondary results as you did your main findings: Summarize and generalize them rather than simply repeating what you found. Do not discuss every single result you obtained. Rather, list and explain any general trends and tendencies and evaluate these.

Sometimes you may be able to generalize your findings and formulate a hypothesis or propose a possible model. Explain how you arrived at your hypothesis or model. Consider illustrating complex models in figures. Describe how the hypothesis or the model works, incorporating a discussion of any figures if needed. If possible, also describe ways to validate your model.

Example 14-7   **Formulating hypotheses**

Hypothesis

We found that the substrate ^{3}H-[9R]iP moves into the cells, where it does not accumulate to concentrations higher than in the medium. However, the mechanism of ^{3}H-[9R]iP uptake is unclear. Because no extracellular activities for the deribolization of ^{3}H-[9R]iP could be detected, we hypothesize that it is metabolized intracellularly to ^{3}H-iP and that the bidirectional transport of iP is based on passive diffusion (Schultz, 2001).

Following is another example that explains a hypothetical model the authors came up with and points to a figure of this model.

Example 14-8   **Formulating a hypothetical model**

Repeated elements located in the amino- and carboxyl-terminus of the protein X vary in sequence for the different species tested. A hypothetical model of how these

| Hypothetical model and figure | repeated elements might interact with the egg receptors is shown in Fig. 7. In this model, protein X is able to interact with complementary sites on its own receptor but not with those of a different species. Protein Y on the other hand contains ligands that interact not only with its own receptor but also with that of another species. This model can explain the unidirectionality in cross-fertilization. |

14.6   LAST PARAGRAPH

➤ **Conclude the Discussion with an analysis of the most important results and the significance of the work**

At the end of the Discussion section, you should provide some closure by writing a one-paragraph conclusion that summarizes your interpretations. Some journals may even request a separate Conclusions section. Readers typically expect to see two things in the summary of a scientific paper: an analysis of the most important results and the significance of the work. The analysis of the most important results is typically provided by the interpretation of your key findings, that is, the answer. Here, too, the answer should match the question/purpose you posed in the Introduction and the answer presented in the first paragraph of the Discussion. Do not bring in new evidence for the summary. Rather, complete the big picture by restating your answer, that is, the interpretation of the key findings.

To highlight the significance of the work, include far-reaching interpretations and conclusions at the end of the Discussion section. Try to generalize your specific findings to other, broader situations. Depending on your level of certainty, significance can range from the practical application to the theoretical proposition. Adding a practical application, giving advice, implying an action, or providing a proposition in the concluding paragraph gives the paper importance. Discuss any theoretical implications, possible applications, recommendations, or speculations based on your findings. If you pose any speculations or implications, base them on solid evidence and make sure that the reader understands that these are your speculations or implications. Do not, however, include wild and random guesses in this section as they are purely of the author's opinion and not based on any actual facts or findings from the results. Instead, it is important to work out the logic behind the ideas discussed and what point is made clearly by the data.

The reader will perceive the level of certainty based on the wording you use. The illustration in Figure 14.2 exemplifies this.

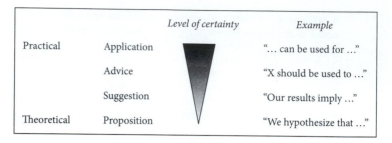

		Level of certainty	*Example*
Practical	Application		"… can be used for …"
	Advice		"X should be used to …"
	Suggestion		"Our results imply …"
Theoretical	Proposition		"We hypothesize that …"

Figure 14.2 Level of certainty and respective sample wording that may be used to indicate the significance of a study.

Because the conclusions are the major message of your paper, you should phrase them with great care. Possible ways to provide some closure of your work at the end of the Discussion section are shown in the next examples.

Example 14-9   **Concluding paragraph with application**

Answer

In summary, our work reveals the functional interactions involved in the binding of antibiotics to the peptidyl transferase cavity of the bacterial ribosome. None of the antibiotics examined show any direct interaction with ribosomal proteins. Chloramphenicol targets mainly the A site, where it interferes directly with substrate binding. Clindamycin interferes with the A site and P site substrate binding and physically hinders the path of the growing peptide chain.

Key findings

Macrolides bind at the entrance to the tunnel where they sterically block the progression of the nascent chain. The structural model of the peptidyl transferase center in com-

Significance indicated by a possible application

plex with the examined antibiotics *can* not only enable a rational approach for antibiotic development and therapy strategies but *can* also be used to identify new target sites on the eubacterial ribosome.

(With permission from Macmillan Publishers Ltd.)

In this example, "In summary, . . ." signals the conclusion. The first sentence of the paragraph is at the same time the topic sentence and the answer to the question. The word "can" in the last sentence indicates the importance of the work by signaling an application.

Other signals for a concluding paragraph are "Taken together, . . ." or "In conclusion, . . ." (see also Section 14.8), or the subheading "Conclusion." Sometimes, even an overall, summarizing question can serve as the signal of the concluding paragraph.

If you have drawn conclusions different from your original hypotheses, you might suggest ways in which these conclusions could be verified in future research. Do not merely say, however, that future research will be needed to clarify the issues without giving the reader any indication of what form this research might take.

The next example shows a conclusion that includes a speculation. This speculation is signaled by the words "may."

Example 14-10   Conclusion with speculation and future direction

Signal of conclusion	*In summary*, we found no statistically significant associations between increased homocysteine (HCY) and age-related macular degeneration (AMD) after analyzing a large and well-characterized population of patients with and without maculopathy from two geographic areas in the United States. An analysis of smoking and HCY tertile subgroups did not show any association between smoking, increased HCY, and increased risk of intermediate or advanced AMD. An association between homocysteine levels and an increased risk of intermediate or advanced AMD *may* exist for patients for whom HCY is above the 90th percentile of HCY, as these patients were more likely to have intermediate or advanced AMD. When subjected to statistical analysis, this observation was found to be not significant however, and only a larger study cohort could determine whether there is any true association.
Answer	
Key findings and conclusions	
Future direction	

14.7   IMPORTANT WRITING GUIDELINES FOR THE DISCUSSION

➤ Make your writing convey confidence and authority

"Beginners" often do very good scientific work but are intimidated by the knowledge and work of "experts." However, when you have collected enough data and are presenting your findings in a paper, you become an expert, too. Show that you are knowledgeable about the subject, and take responsibility for your conclusions. Do not be afraid to take a stand, but do not be boastful either.

Often, ESL authors are also intimidated, but mainly because English is not their native language. Some ESL authors do not realize when they sound too effacing or when they sound too opinionated and their language use is too strong. Ask a native speaker, preferably a scientific editor, to read over your manuscript. Native speakers will be much more aware about the fine nuances of their language.

ESL advice

➤ Use first person and active voice

Do not use third person and passive voice in the Discussion if it can be avoided. Instead, use first person and active voice to make your discussion livelier and interesting. The use of "we" is perfectly okay. If you are a single author, you may also consider using "we." (See also Chapter 4, Section 4.2 and Section 4.3.)

➤ Use past tense for completed actions and results; use present tense for general rules and statements that are still true

In the Discussion, you may find a mix of verb tenses. Remember that when you are referring to your findings and completed actions, use past tense. However, for statements of general validity and those whose information is still true, use present tense. Also, use present tense for the answer to the question and for the statement of significance (see also Example 14-9). (See also Chapter 4, Section 4.4.)

➤ Ensure continuity

It is especially important to provide continuity in the Discussion such that the reader can read through the section without stumbling across unclear passages. To ensure such continuity, use topic sentences, transitions, and key terms (see Chapter 6 for a detailed discussion of these items). Ensure that the topic sentence of each paragraph is logically linked to the paragraph just prior as well as to your research question. Readers often scan topic sentences to identify subtopics quickly.

14.8   SIGNALS FOR THE READER

➤ Signal the elements of the Discussion

Signal the different elements of the Discussion so that readers recognize immediately what they are reading about. Possible signals for unexpected findings, comparisons, conflicting results, limitations, and proposed hypothesis, are listed in Table 14.1 and Table 14.2.

Both of these tables provide great starting and reference points when you put together your first draft, and they may be particularly useful for those authors whose native language is not English.

For unexpected findings, comparisons, conflicting results, and proposed hypothesis, you may use the signals shown in Table 14.2.

ESL advice

14.9   AN ALTERNATIVE: RESULTS AND DISCUSSION

The Results and Discussion sections are sometimes combined into one section. In a combined Results and Discussion section, there is no need for backward page turning. The results are discussed right after they are presented so that the reader can understand why they are important immediately. Note that in a combined Results and Discussion section, *all* results will have to be presented and discussed in a combined section and not just the most important summarized results of your paper.

TABLE 14.1 Signals for the Discussion

ANSWER	KEY FINDINGS	SUMMARY	SIGNIFICANCE	
				Level of certainty
In this study, we have shown that . . .	In our experiments . . .	In summary, . . .	Our findings **can/ will serve** to . . . . . . **can** be **used** . . .	
In this study, we found that . . .	. . . can be attributed to . . . We determined X by . . .	In conclusion, . . . Finally, . . . Taken together, . . .	We **recommend** that X is . . .	
Our study shows that . . .	We found that . . .	To summarize our results, . . .	Y **should** be used for . . . . . . **is probably** . . . Y **indicates** that X **might** . . .	
Our findings demonstrate that . . .	Our data shows that . . . . . . has been demonstrated by . . .	We conclude that . . . [overall question]	These findings **imply** that X **may** . . .	
This paper describes . . .		Overall, . . .	Here we **propose** that . . . . . . we **hypothesize** that . . .	

TABLE 14.2 Signals for the Discussion

COMPARISONS	CONFLICTING RESULTS	LIMITATIONS	UNEXPECTED FINDINGS	PROPOSED HYPOTHESIS
. . . consistent with . . . (ref.)	However, other studies found that . . . (ref.)	. . . was not possible . . .	Surprisingly, . . .	Our results lead to the conclusion that . . .
Similar to . . . (ref.) . . . has also been observed by . . . (ref.) X has been demonstrated . . . (ref.)	. . . is controversial . . . (ref.) . . . does not agree with . . . (ref.) . . . has also been reported . . . (ref.)	. . . could not be measured . . . . . . was limited by . . . Further observations are needed to . . .	To our surprise . . . . . . was not expected.	From these data, we hypothesize that . . . We propose the following new principle . . .

ref. = reference

14.10 COMMON PROBLEMS OF THE DISCUSSION

The most common problems of Discussion sections include:

- The answer/interpretation of your key findings is not provided in the first paragraph (Section 14.4)
- No concluding paragraph is provided; the importance/significance of the study is not clear (Section 14.6)
- Irrelevant or peripheral information is included (Section 14.5)
- Results are repeated or summarized in the Discussion (Section 14.4)

➤ Do not fail to provide the answer in the first paragraph

If you fail to provide the interpretation of your key findings or fail to answer your question in the first paragraph of the discussion, your readers will usually stop reading and jump directly to the concluding paragraph in hopes of finding the answer there.

Providing the answer in the first paragraph can be a problem if your Discussion is divided into subsections with independent headings. For such Discussions, consider providing an overall overview of the study before discussing individual key findings in subsections. Treat this overview paragraph like the most important power position of the Discussion, stating the main, overarching interpretation(s) based on your key findings.

➤ Ensure that the importance of the study is clear

Often, the importance of the results is not discussed or not adequately discussed, which leaves the reader wondering about the significance of the paper. Indicate the significance of your findings at the end of the Discussion section in your conclusion. Take a position—you are the expert in your study now. Remember, the last paragraph(s) is the other power position within this section. Do not waste this position with anything other than the importance of the study (see also Section 14.6).

Although you should discuss how your research fits in with other existing findings in the field, the key is to concentrate on generalizing and summarizing your interpretations, comparisons, and contrasts by providing an overall picture.

➤ Do not include irrelevant information. Do not repeat or summarize your results in the Discussion section

Beginning readers frequently include irrelevant information in the Discussion—often to fill in space and increase the length of the section. Such information may range from summarization of results; to secondary introductions; to discussion of minor experimental setups, limitations, or personal opinions. Do not be afraid to make your Discussion section shorter rather than longer. It is more important to discuss only immediately relevant findings than to fill in space. Irrelevant and repeated information is distracting to readers as it does not allow them to distinguish clearly between what is important and what is less important.

14.11   SAMPLE DISCUSSIONS

Following are two complete discussion sections that show a good organization.

Example 14-11   **Complete Discussion of an informative paper**

Interpretation/ answer ——— Key findings ———	Our findings indicate that fertilization among sea urchins is intraspecies specific and that surface components of both gametes are involved in the fertilization specificity. Results of intraspecies specific fertilization between different gametes of several different individuals of the species *S. purpuratus* displayed varying degrees of fertilization success.
Comparison with previous findings ———	Approximately one third of the examined individuals show significant differences in fertilization. The percent of individual crosses of gametes yielding significantly reduced fertilization specificity corresponds to the previously reported amount of "defective" gametes encountered during fertilization (5) and egg agglutination experiments (9). However, in our study, crosses for gametes of an individual that yield significantly reduced fertilization in one cross reached normal levels with other individuals of the opposite sex, suggesting that a low fertilization success is not due to immature or defective gametes.
Comparison with models of others ——— Suggested hypothesis ———	It seems intuitive that the molecules that mediate sperm–egg interactions may play an important role in speciation since individuals are in separate gene pools if the sperm and egg cannot interact to form a zygote. How these systems evolve so that once common ancestors become reproductively isolated is not so obvious. Geographic isolation is believed to be the principal mechanism of speciation in marine animals (6). However, speciation does not have to be accompanied by major genomic reorganization (5, 7). A few mutations in bindin and its receptor can be sufficient to accomplish reproductive isolation because these proteins are major components of the fertilization mechanism. The hypothesis that bindins and their receptors contain multiple adhesive elements would allow mutations to occur within an individual element of bindin without catastrophic consequences for mutant individuals in the absence of a simultaneous compensating mutation in the receptor.

Summary significance	In summary, fertilization among sea urchins appears intra-species specific due to surface components of both gam-etes. It is conceivable that these surface components contain multiple adhesive elements. Based on our hypoth-esis, mutations in these elements may result in reproductive isolation and speciation.

Example 14-12   Complete Discussion of a descriptive paper

Discovery and implication	Our results suggest that the sparse matrix screen may be of general utility for establishing initial renaturation condi-tions for a wide variety of proteins. Eight of the nine pro-teins tested recovered significant amounts of activity by this method. These proteins included an adhesive protein (bindin), proteins with disulfide bonds (lysozyme, β-gal, trypsin, acetylcholinesterase, HRP, BAP), proteins with multiple subunits (β-gal), and proteins with cofactors (HRP, BAP). The sparse matrix approach has the potential for the elimination of relatively tedious random searching for con-ditions that support renaturation. For example, we ob-
Supporting evidence and explanations	tained bindin renaturation serendipitously with some of the buffers provided by the Crystal Screen Kit in an attempt to crystallize bindin.
Further evidence supporting the discovery	The composition of the crystallization buffer resulting in renaturation differs widely from buffers previously known to be required for biological activity. As observed for bindin, conditions needed for refolding the other proteins tested are different from those required for their maximal enzyme activity. It appears that both crystallization and biological activity require a discrete tertiary conformation of the protein. Thus, solution conditions, which promote crystallization, will also promote the reformation of the native structure.
Description of discovery	The Crystal Screen kit contains three major components: a buffer, salts, and a precipitating agent (see Table 1). In crystal growth the precipitant promotes the filling of an ordered lattice by favoring protein–protein contacts. Since the native protein structure in solution is frequently the same as that determined in the crystal (27, 28), it is per-haps not so surprising that there may be some common-ality between conditions that promote protein renaturation and protein crystallization. This commonality might be the presence of precipitant in the solvent system, as most of the proteins tested do not renature in the absence of this component.

Description of discovery	Once initial renaturation conditions are identified, conditions for renaturation may be optimized by systematically altering other parameters which are known to be important for the refolding of a protein. The most common method for optimizing the yield of active protein is to perform a renaturation/reoxidation in the presence of low concentration of urea or guanidine hydrochloride (13). This situation can be mimicked in the sparse matrix screen by serial two-fold dilution of the urea-denatured enzyme in the crystallization buffers.
Description and application of finding	Extrinsic factors not encoded in the primary amino acid sequence may be of importance for the refolding of a protein (38). In the cell, accessory proteins are involved in regulating the rate of folding and association of a protein. Thus, the addition of molecular chaperones, cofactors, ions, and conjugate components such as carbohydrates, nucleic acids, and lipids to the buffer system may improve the rate and yield of renatured protein.
Limitations	Renaturation of multiple subunit proteins or multiple domain proteins may be especially challenging and may require different conditions for each domain or subunit. Proteins with multiple domains are thought to refold slower than those with single domains (26). Since each subunit may have to refold independently, it may be necessary to screen each subunit for renaturation separately, and then test for the ability of the individually renatured subunits to reassociate in an active form. The screening for these conditions could be simplified by combining subunits from several renaturation systems and testing the mixture for activity. Optimum combination could be determined by testing mixtures containing progressively fewer combinations.
Signal of conclusion	To summarize, here we report a sparse matrix method employing 50 different solvent systems for establishing initial
Key finding	solvent conditions that facilitate protein renaturation. Using this method, we tested nine different proteins for renaturation and observed significant amounts of renaturation for eight of the nine proteins tested. Once initial conditions are obtained, renaturation conditions may be systematically optimized by varying conditions such as pH, temperature, and protein concentration. This screening matrix is based on
Description of new finding	a set of solutions originally selected for protein crystallization. Our results suggest that this screening method may be widely applicable in identifying conditions that support renaturation and that the same conditions, which promote protein crystallization may also promote protein
Speculation	renaturation.

(With permission from Elsevier)

14.12 REVISING THE DISCUSSION

When you have finished writing the Discussion (or if you are asked to edit a Discussion for a colleague), you can use the following checklist to systematically "dissect" the Discussion:

☐ 1. Did you interpret the key findings/provide the answer to your research question?
☐ 2. Is the interpretation/answer to the research question provided in the first paragraph?
☐ 3. Is the answer stated precisely? (And in present tense?)
☐ 4. Is the answer followed by supporting evidence?
☐ 5. Does the discussion follow a pyramid structure?
☐ 6. Is a summary paragraph placed at the end of the Discussion?
☐ 7. Is the significance of the work apparent?
☐ 8. Did you organize the topics according to the science or from most to least important in the middle of the discussion?
☐ 9. Did you compare and contrast your findings with those of other published results?
☐ 10. Did you explain any discrepancies, unexpected findings, and limitations?
☐ 11. Did you provide generalizations where possible?
☐ 12. Did you avoid restating or summarizing the results?
☐ 13. Are all elements signaled?
☐ 14. Revise for style and composition based on the basic rules and guidelines of the book:
   ☐ a. Are paragraphs consistent? (Chapter 6, Section 6.2)
   ☐ b. Are paragraphs cohesive? (Chapter 6, Section 6.3)
   ☐ c. Are key terms consistent? (Chapter 6, Section 6.3)
   ☐ d. Are key terms linked? (Chapter 6, Section 6.3)
   ☐ e. Are transitions used and do they make sense? (Chapter 6, Section 6.3)
   ☐ f. Is the action in the verbs? Are nominalizations avoided? (Chapter 4, Section 4.6)
   ☐ g. Did you vary sentence length and use one idea per sentence? (Chapter 4, Section 4.5)
   ☐ h. Are lists parallel? (Chapter 4, Section 4.9)
   ☐ i. Are comparisons written correctly? (Chapter 4, Sections 4.9 and 4.10)
   ☐ j. Have noun clusters been resolved? (Chapter 4, Section 4.7)
   ☐ k. Has word location been considered? (Verb follows subject immediately? Old, short information is at the beginning of the sentence? New, long information is at the end of the sentence?) (Chapter 3, Section 3.1)
   ☐ l. Have grammar and technical style been considered? (person, voice, tense, pronouns, prepositions, articles; Chapter 4, Sections 4.1–4.4)

☐ m. Is past tense used for results and present tense for descriptive papers? (Chapter 11, Section 11.5)

☐ n. Are words and phrases precise? (Chapter 2, Sections 2.2 and 2.3)

☐ o. Are nontechnical words and phrases simple? (Chapter 2, Section 2.2)

☐ p. Have unnecessary terms (redundancies, jargon) been reduced? (Chapter 2, Section 2.4)

☐ q. Have spelling and punctuation been checked? (Chapter 4, Section 4.11)

SUMMARY

DISCUSSION GUIDELINES

- State and interpret your key findings. Provide the answer to the research question.
- Summarize and generalize.
- Keep in mind who your potential readers will be.
- Organize the Discussion in a pyramid structure:

First paragraph:	Answer based on key findings
	Supporting evidence
Subsequent paragraphs:	Comparisons/contrasts to previous studies
	Limitations of your study
	Unexpected findings
	Hypotheses or models
Last paragraph:	Summary
	Significance/implication

- State the answer to your research question in the first paragraph.
- Organize the topics according to the science or from most to least important in the middle of the discussion.
- Compare and contrast your findings with those of other published results.
- Explain any discrepancies, unexpected findings, and limitations.
- Provide generalizations where possible.
- Conclude the Discussion with an analysis of the most important results and the significance of the work.
- Make your writing convey confidence and authority.
- Use first person and active voice.
- Use past tense for completed actions and results; use present tense for general rules and statements that are still true.
- Ensure continuity.

> - Signal the elements of the Discussion.
> - Do not fail to provide the answer in the first paragraph.
> - Ensure that the importance of the study is clear.
> - Do not include irrelevant information. Do not repeat or summarize your results in the Discussion section.

PROBLEMS

PROBLEM 14-1 Basic Writing Rules in the Discussion
Evaluate the following statements found in different Discussion sections. Use basic writing rules to improve the sentences:

1. Therefore, accuracy of Ebb's data is obviously doubtable and should be reevaluated.
2. A comprehensive analysis for optimal HPV vaccine distribution is necessitated for evaluation of cost-effective and age-appropriate application of this critical disease intervention in teenagers.
3. The influence of counter ions is obvious in Figure 3B.
4. Our future studies may reveal how certain single-point mutations affect the conformation of the protein and thus its interaction with the receptor.
5. The model proposed here however is currently an experimentally underdetermined system.

PROBLEM 14-2 Opening Paragraph
Consider the two different opening paragraphs of a Discussion section about a preventive measure against malaria. Which one is a better first paragraph for a Discussion and why?

Version A
We trapped and counted the number of mosquitoes within the urban environment of the city of Kumasi using conventional carbon dioxide traps. Nearly 70% more adult *A. gambiae* were caught in communities near moist urban agricultural establishments than in rural locations or in areas without irrigated urban settlements. When we evaluated malaria episode reports from people living in various parts of the city, we found that 18% of malaria cases in all seasons were reported by those near urban agricultural sites, whereas only 2% of the control groups reported incidences of malaria per year.

Version B
The results of this study show that open-space irrigated vegetable fields in cities can provide suitable breeding sites for *A. gambiae*. This is reflected in higher numbers of adult *A. gambiae* in settlements in the vicinity of irrigated urban agricultural sites compared to control areas without irrigated urban agriculture. Moreover, people living in the vicinity of urban agricultural

areas reported more malaria episodes than the control group in the rainy as well as dry seasons. Apparently, the informal irrigation sites of the urban agricultural locations create rural spots within the city of Kumasi in terms of potential *Anopheles spp* breeding sites.

(With permission from Elsevier)

PROBLEM 14-3 Conclusion
Consider the two different concluding paragraphs of a discussion about desert frogs. Which one is a better conclusion for a Discussion and why?

Version A
In conclusion, this study shows that desert frogs can avoid death by desiccation by maintaining a high body water content and water storage in their urinary bladder and by rapid hydration when water is available. These measures may be employed in combination with behavioral adaptations such as burrowing and change in pigmentation to minimize stresses tending to dehydrate the animals.

Version B
A limitation of this study was the small number of animals, a single species of frogs, and the location of the study area, which took place in only one oasis in the Mohave Desert. Future studies should be extended to other species, a larger number of animals, and to a greater diversity of locations.

PROBLEM 14-4 Conclusion
Given the following concluding paragraph of a Discussion about a simulation of a Sahel drought in the twentieth and twenty-first centuries, explain why or why not this is a good final paragraph of an article.

We have described a global climate model (CM2) that generates a simulation of the 20th century rainfall record in the Sahel generally consistent with observations. The model suggests that there has been an anthropogenic drying trend in this region, due partly to increased aerosol loading and partly to increased greenhouse gases, and that the observed 20th-century record is a superposition of this drying trend and large internal variability. The same model projects dramatic drying in the Sahel in the 21st century, using the standard IPCC scenarios. No other model in the IPCC/AR4 archive generates as strong a drying trend in the future. Until we better understand which aspects of the models account for the different responses in this region to warming of SSTs and devise more definitive observational tests, we advise against basing assessments of future climate change in the Sahel on the results from any single model in isolation. In the interim, given the quality of CM2's simulation of the spatial structure and time evolution of rainfall variations in the Sahel in the 20th century, we believe that its prediction of a dramatic 21st century drying trend should be considered seriously as a possible future scenario.

(With permission from the National Academy of Sciences, USA)

PROBLEM 14-5 Conclusion
In the following paragraph, identify

—the signal for the conclusion
—overall results and their interpretation
—statement of significance

Conclusions

Habitat heterogeneity, as estimated by an advanced land cover classification, provides a stronger prediction of butterfly species richness in Canada than any previously measured factor. At large spatial scales, virtually all spatial variability (90%) in butterfly richness patterns is explained by habitat heterogeneity with secondary but significant contributions from climate (especially PET) and topography. Patterns of species turnover across the best sampled southern region of Canada are strongly related to differences in habitat composition, supporting species turnover as the mechanism through which land cover diversity may influence butterfly richness. Differences in climate are unrelated to butterfly community similarity at this scale, suggesting that the influences of energy on richness may be indirect or limited to within-habitat diversity. These results have significant conservation implications and indicate that the role of habitat heterogeneity may be considerably more important in determining large-scale species-richness patterns than previously assumed.

(With permission from the National Academy of Sciences, U.S.A.)

PROBLEM 14-6 First and Last Paragraph of Discussion
In this problem, you find the first and last paragraph of a Discussion for a manuscript with the title "Effectiveness of Peridomestic Lyme Disease Prevention Practices."

a. Evaluate the paragraphs (Does the first paragraph state the answer to the question? Does the last paragraph give a good conclusion and indicate the significance of the study?).
b. The opening/first paragraph of this Discussion is rather long. Suggest ways to condense it or to rearrange information.

Discussion

Our findings emphasize the need to continue to promote personal protection measures to reduce the risk of Lyme disease infection. We have identified three reasonable personal measures that may be protective against Lyme disease when practiced. Performing tick checks within 36 hours after spending time in the yard may reduce one's risk by as much as 46 percent. Because studies have suggested it takes more than 24 hours for blacklegged ticks to transmit the etiologic agent[22,23], prompt removal of ticks found attached to the body is a logical method of Lyme disease prevention. The effectiveness of performing tick checks has been suggested previously[8,13], however this is the first time it has been demonstrated in a peridomestic setting. We also found that controls were more likely than cases to shower or bathe within two hours after spending time in the yard, and that bathing may reduce one's risk by up to 57 percent. Frequent bathing is not currently among the commonly recommended Lyme disease prevention measures. Although it is unlikely that bathing will remove ticks that have already attached to the body, taking a shower or bath soon after spending time outside may help to remove ticks that are yet unattached, or may create an opportunity to find ticks attached to the body. In addition, the act of bathing may indirectly prevent tick bites in that it necessitates the removal of clothing that may have blacklegged ticks upon them. Our data also suggest that the use of insect repellent may be protective against disease. This finding refers to all types of repellents, including those that contain N,N-diethyl-3-methyltoluamide (DEET), but not including permethrin insecticide. Responses varied greatly regarding the active ingredients of repellents used, and were often inconsistent with the brands and products named by the respondents. The effectiveness of insect repellents has been shown previously[8,16], but not in the peridomestic environment. We found that wearing clothing treated with permethrin insecticide was rarely practiced. It may be that residents are unaware of permethrin products, or do not want to spend the time treating outdoor clothing (permethrin must be applied to clothing and allowed to dry for several hours before wearing). Clothing that is pre-treated with permethrin and sold in specialty stores may be too expensive for daily use in one's yard.

. . .

In summary, this study sought to evaluate the effectiveness of peridomestic Lyme disease prevention behaviors. Though studies suggest that Lyme disease risk in the northeastern United States is largely peridomestic, members of the study population could have been exposed to ticks outside of their own yards. The analysis attempted to evaluate and control for this non-peridomestic risk by asking participants about their recreational, occupational, and travel exposures to ticks, although it is difficult to determine when and where people are exposed to ticks. Our findings emphasize the need to continue to promote personal protection measures to reduce the risk of Lyme disease infection. In the absence of a vaccine against Lyme disease, it is encouraging that the protective measures identified in our study—tick checks, showering or bathing after spending time in the yard, and use of repellent—are measures that anyone can take at limited expense to reduce their risk of infection. Clinicians and public health practitioners in Lyme disease endemic areas should continue to educate the public about these simple, practical methods for reducing Lyme disease risk after spending time outdoors.

(With permission from Neeta Connally)

PROBLEM 14-7 Your Own Discussion
Write your own Discussion section based on the experimental results of your work. Provide the overall question the paper is meant to ask. Use all basic rules studied, and follow the summary on how to write a Discussion section.

PROBLEM 14-8 Revising a Discussion Section
Revise either your own Discussion or that of a colleague. Identify the elements within the Discussion section. Ensure that the main findings are stated up front together with their overall interpretation, that is, the answer to the question of the paper. Be sure that the answer matches the question of the paper. Ensure also that the Discussion has a conclusion and that the significance is clearly stated. In addition, check that all elements are signaled clearly. Pay attention to verb tense as well as other elements of grammar and style. Use the checklist provided in Section 14.12 as a guide. Make sure references have been placed where needed as well.

CHAPTER 15

Abstract

Knowing how to write an Abstract is one of the most important skills in science, as virtually all of a scientist's work will be judged first (and often last) based on the Abstract.

THIS CHAPTER DISCUSSES:

- Different forms of Abstracts for research papers
- Components of an Abstract
- Format of the Abstract
- Signaling the different elements
- Common mistakes to avoid in composing the Abstract
- Common reasons for rejections of articles based on Abstracts
- Revising your Abstract
- Full, annotated sample Abstracts

15.1 OVERALL

Most people (including editors and reviewers) will read your article only if your Abstract interests them. The Abstract is often the only part of the paper—together with the title (see Chapter 16)—that can be retrieved through a search, such as those done through MEDLINE or Ovid. It is therefore essential that the Abstract interests your reader.

The ability to write a competitive Abstract applies not only to research papers but also to grant proposals, progress reports, project summaries, and conference submissions and proceedings. Therefore, learning this critical skill should not be underestimated.

15.2   COMPONENTS

The Abstract should fully summarize the contents of the paper in one paragraph. The Abstract must also be written such that it can stand on its own without the text. It must be concise, informative, and complete. Do not try to include every finding in your Abstract. Rather, include your key results and all other important details of the paper, but use as few words as possible. Write the Abstract with the nonspecialist in mind. Remember, you want to attract as wide an audience as possible.

The Abstract selects the highlights from each section of the paper. Because the Abstract summarizes all of these highlights, it is easiest to write once your manuscript is complete. Some people prefer to write the Abstract before all other sections, using it as an outline for the manuscript. Regardless of when you write the Abstract, it needs to cover all of the main information in the paper (Introduction, Materials and Methods, Results, and Discussion) in a single paragraph. When you are done writing your manuscript, double-check that the Abstract matches the paper.

In your Abstract, do not include any information or conclusion not covered in the paper. Avoid abbreviations, unfamiliar terms, jargon, and citations. Do not include or refer to tables or figures. Do not include any references, but be sure to include all the important key terms found in the title because the Abstract and the Title have to correspond to each other (see Chapter 6, Section 6.3 for more details on key terms, and distinguish between key terms and key words, also known as indexing terms—for key words see Chapter 16, Section 16.5).

15.3   FORMAT

➤ Use an informative or structured abstract for research articles

Abstracts for research papers (both investigative and descriptive papers) differ from those used for review articles or proposals. Research paper abstracts can be divided into *informative* and *structured* abstracts. The latter form is used mainly for clinical journals. Both informative and structured abstracts are discussed in detail in this section along with a *graphical* abstract, a visual form of what is reported in a paper, and the abstract for a descriptive paper reporting on a new discovery. (Note that in contrast to a research paper, review articles and book chapters use *indicative* abstracts [see Section 15.6 and Chapter 19, Section 19.5], grant proposals use *proposal* abstracts [see Chapter 22], and at conferences you may have to provide a *conference* abstract [Chapter 29, Section 29.2] or a poster abstract [Chapter 29, Section 29.5].)

When you write the Abstract for your research paper, follow the specific instructions from the journal to which you are planning to submit your manuscript. Although there will be some differences among journals, the content of the Abstract remains the same.

➤ The Abstract of an investigative research paper includes

- Background
- **Question or purpose**
- **Experiments**
- **Results**
- **Conclusion (answer to the question)**
- Significance

Aside from the key components (in bold), the Abstract may begin with a sentence or two of background information to help the reader understand the question and end with a sentence indicating the significance of the paper, such as an implication, recommendation, application, or speculation.

Although the Abstract is a mini-version of the paper, following its general organization, the Abstract does not give equal weight to all the parts of a paper. It may include a sentence or two of background information from the Introduction. It has to include the overall question or purpose of the paper found in the Introduction. It typically describes the experimental approach only generally. It includes the main results from the Results section and contains the answer to the research question/purpose, which is found in the Discussion. In addition, the Abstract may end with a sentence stating an implication, a speculation, or a recommendation based on the answer. However, you should avoid general descriptive statements that merely hint at your results or act like a rough table of contents. There should be no surprises or elements of suspense in an Abstract.

Example 15-1   **Abstract of an investigative research paper**

	1 Interleukin 1 (IL-1), a cytokine produced by macrophages and various other cell types, plays a major role in the
Background	immune response and in inflammatory reactions. **2** IL-1 has
Question/ purpose	been shown to be cytotoxic for tumor cells. **3** To determine the effect of macrophage-derived factors on epithelial tumor cells, **4** we cultured human colon carcinoma cells
Experimental approach	(T84) and an intestinal epithelial cell line (IEC 18) with purified human IL-1. **5** Microscopic and photometric analysis
Results	indicated that IL-1 has a cytotoxic effect on colon cancer cells as well as cytotoxic and growth inhibitory effects on
Answer/ conclusion/ implication	intestinal epithelial cells. **6** As IL-1 is known to be released during inflammatory reactions, this factor may not only kill tumor cells but also affect normal intestinal cells and may play a role in inflammatory intestinal diseases.

In Example 15-1, sentences 1 and 2 provide a short background. Sentence 3 signals the question, whereas sentence 4 states the experimental approach. The key results are signaled in sentence 5. Sentence 6 provides the conclusion by stating an implication.

The format and length of the Abstract is generally specified in the *Instructions to Authors*. Usually, Abstracts contain between 100 and 250 words. You should not exceed the maximum allowed, but you may certainly summarize your paper in fewer words.

Structured Abstracts

➤ Use a structured abstract for clinical reports, if requested

➤ Use subheadings for structured abstracts

Structured abstracts are just like informative abstracts except that structured abstracts have subheadings (Background, Methods, Results, and Conclusions) and are often not written in complete sentences. These types of abstracts are most often found in clinical reports and clinical journals. They can sometimes be longer than informative abstracts—many contain a maximum of 400 words. Clinical journals in which these abstracts are required usually provide instructions for authors that state the length and subheadings for the structured abstract.

Example 15-2   **Structured abstract**

Background and question/ purpose	*Background* In infants and children with maternally acquired human immunodeficiency virus type 1 (HIV-1) infection, treatment with a single antiretroviral agent has limited efficacy. We evaluated the safety and efficacy of a three-drug regimen in a small group of maternally infected infants.
Experimental approach	*Methods* Zidovudine, didanosine, and nevirapine were administered in combination orally to eight infants 2 to 16 months of age. The efficacy of antiretroviral treatment was evaluated by serial measurements of plasma HIV-1 RNA, quantitative plasma cultures, and quantitative cultures of peripheral-blood mono-nuclear cells.
Results Answer/ conclusion	*Conclusions* Although further observations are needed, it appears that in infants with maternally acquired HIV-1 infection, combined treatment with zidovudine, didanosine, and nevirapine is well tolerated and has sustained efficacy against HIV-1.

(With permission from the Massachusetts Medical Society)

If you find it difficult to get started writing an abstract, consider starting to write a structured abstract first, following the general setup of subsections. Then, convert the abstract to an informative abstract by connecting the sentences using transitions and word location where needed. Always recheck the abstract after completing a manuscript as your ideas may have evolved in the writing process. You may find that you have to revise the initial abstract to better represent the final paper.

Special Case: Graphical Abstracts

In recent years, some journals have requested the submission of a graphical abstract, a visual representation of the core message of the study. Comparable to the core slide of a presentation (see Chapter 30, Section 30.4), graphical abstracts summarize your take-home message in either a specially designed visual representation or a concluding figure of your article. The message of a graphical abstract should be understandable at a single glance.

Although they take much longer to produce, the concept(s) of these single-figure abstracts can be understood faster and easier by a wider audience than a textual representation. These abstracts also increase the reach of your message to a broader audience as they generally are more favorably viewed and easily shared through media. They allow your readers to understand findings of a study quickly and to judge whether the paper is relevant to them. They also eliminate potential language barriers ESL readers may have. To make your graphical abstract look professional, consider using a graphic designer to assist with its production.

Example 15-3a   **Graphical abstract**

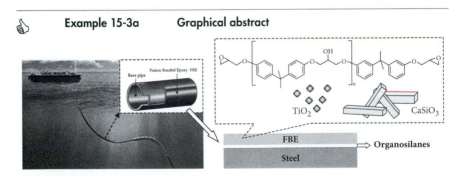

Figure X. Schematic representation of the FBE coatings of steel pipelines for deep-water petroleum exploration.

(With permission from Herman Mansur)

Example 15-3b   **Graphical abstract**

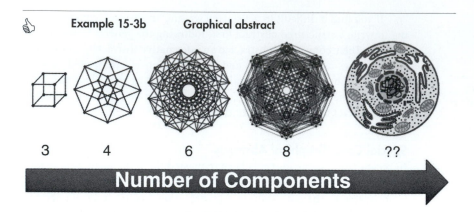

(*Reprinted from Journal of Systems Chemistry 1.6: 2010*)

Special Case: Abstracts for Descriptive Papers

➤ **Abstracts for descriptive research papers should include**

- Descriptive statement
- Description of the new findings
- Conclusion/significance/implication

  Example 15-4a   **Abstract of a descriptive paper**

Descriptive statement	The "sexually deceptive" orchid *Chiloglottis trapeziformis* attracts males of its pollinator species, the thynnine wasp *Neozeleboria cryptoides*, by emitting a unique volatile compound, 2-ethyl-5-propylcyclohexan-1,3-dione, which is also produced by female wasps as a male-attracting sex pheromone.
Description and implied significance	

The preceding example is a very short Abstract of just one sentence, but it contains all the necessary components of an Abstract for a descriptive paper. The Abstract relates the discovery of a "sexually deceptive" orchid in the first half of the sentence. In the second half of the sentence, the compound responsible for the deception is described ("a unique volatile compound, 2-ethyl-5-propylcyclohexan-1,3-dione"), and its significance is implied ("male-attracting sex pheromone").

Longer Abstracts of descriptive papers typically expand on the description of the discovery and on its significance. They may also contain additional background information. An example of a longer Abstract for a descriptive paper is shown next.

Example 15-4b   Abstract of a descriptive paper

Background and problem	Indirect radiative forcing of atmospheric aerosols by modification of cloud processes poses the largest uncertainty in climate prediction. We show here a trend of
Descriptive statement	increasing deep convective clouds over the Pacific Ocean in winter from long-term satellite cloud measurements (1984–2005). Simulations with a cloud-resolving weather research and forecast model reveal that the increased deep convective clouds are reproduced when accounting for the aerosol effect from the Asian pollution outflow, which leads to large-scale enhanced convection and precipitation and hence an intensified storm track over the Pacific. We suggest that the wintertime Pacific is highly vulnerable to the aerosol-cloud interaction because of
Results/ description	favorable cloud dynamical and microphysical conditions from the coupling between the Pacific storm track and Asian pollution outflow. The intensified Pacific storm track is climatically significant and represents possibly the first detected climate signal of the aerosol-cloud interaction associated with anthropogenic pollution. In addition to radiative forcing on climate, intensification of the Pacific storm track likely impacts the global general circulation
Significance/ implication	due to its fundamental role in meridional heat transport and forcing of stationary waves.

(With permission from the National Academy of Sciences)

15.4   APPLYING BASIC WRITING RULES

➤ Apply basic scientific writing rules and guidelines in the Abstract

To write your Abstract, follow the basic rules covered in Chapters 1 through 6. In the Abstract, pay particular attention to using simple words and avoiding jargon (see Chapter 2 for more details). Avoid noun clusters (see also Chapter 4). Also, avoid abbreviations unless a long term occurs repeatedly in the Abstract. Here, it is even more important than in the rest of the paper to keep sentences short, dealing with just one topic each and excluding irrelevant points (see also Chapter 4). It is also important to pay attention to word location, use the same key terms, and provide clear transitions and signals (see also Chapter 6 and Section 15.5).

Pay particular attention to verb tense. The basic guideline is that if a statement is still true, use present tense. For completed actions and observations, use past tense. Thus, when you write about the answer or describe a structure, use present tense because these statements are still true. However, when you describe the results of experiments, use past tense because these events are finished. When possible, use active verbs. See also Chapter 4, Section 4.4 for examples and in-depth discussion of the verb tense.

15.5 SIGNALS FOR THE READER

➤ **Signal the question, the experimental approach, the results, the answer, and the implication**

ESL advice

Because Abstracts are usually written as one paragraph, it helps the reader if you signal the different parts of the Abstract. Examples of signals for the Abstract are shown in Table 15.1. Writers whose native language is not English may find this table particularly useful.

Following is an Abstract with components that have been signaled well.

Example 15-5 **Abstract with signaled components**

	Human papillomavirus (HPV) vaccines provide an opportunity to reduce the incidence of cervical cancer. Optimization of cervical cancer prevention programs requires anticipation of the degree to which the public will
Question	adhere to vaccination recommendations. ***To compare vaccination levels*** driven by public perceptions with
Experimental approach	levels that are optimal for maximizing the community's overall utility, ***we develop*** an epidemiological game-
Results not stated clearly Answer/inter- pretation of results	theoretic model of HPV vaccination. The model is parameterized with survey data on actual perceptions regarding cervical cancer, genital warts, and HPV vaccination collected from parents of vaccine-eligible children in the United States. ***The results suggest*** that perceptions of survey respondents generate vaccination levels far lower than those that maximize overall health-related util-
Implication	ity for the population. Vaccination goals ***may be achieved*** by addressing concerns about vaccine risk, particularly
Another implication	those related to sexual activity among adolescent vaccine recipients. ***In addition***, cost subsidizations and shifts in federal coverage plans ***may compensate*** for per-
Implication	ceived and real costs of HPV vaccination to achieve public health vaccination targets.

(With permission from the National Academy of Sciences)

TABLE 15.1 Signals for the Abstract

QUESTION/ EXPERIMENT	RESULTS	ANSWER/ CONCLUSION	IMPLICATION
To determine whether . . . , we . . .	We found . . . Our results	We conclude that . . .	These results suggest that . . .
We asked whether . . .	show . . .	Thus, . . .	These results may play a role in . . .
To answer this question, we . . .	Here we report . . .	These results indicate that . . .	Y can be used to . . .
X was studied by . . .			

In Example 15-5, the question "To compare vaccination levels . . ." leads the reader directly to the experimental approach, signaled by "we develop . . ." The complete description of the experiment in the sentence following describes the subjects of the study (parents of vaccine-eligible children) and sets the stage for the answer to the question. We know where the answer starts because it is signaled by "The results suggest . . ." The answer is followed by two implications, which are signaled by ". . . may be achieved . . ." and "In addition, . . . may compensate. . . ." Note that the implications are written in similar form and have parallel signals, making it clear that both ". . . may be achieved . . ." and ". . . may compensate . . ." are of equal importance. To strengthen this Abstract, the results could be stated more clearly by, for example, providing the percentages and types of perceptions found in the study and signaling these.

15.6   COMMON PROBLEMS OF THE ABSTRACT

The most common problems of Abstracts include

- Omission of elements (see also Section 15.3)
- Excessive length
- Wrong type of Abstract

➤ Do not omit any parts of the Abstract

If any of the parts (question/purpose, experimental approach, results, or answer/conclusion) are missing or obscured in an Abstract, the reader may have to reread the Abstract several times because it is difficult to understand. The same problem arises when the parts of an Abstract are not signaled.

	Example 15-6	Abstract without clearly stated question
Experimental approach		Using either conventional violet red bile glucose agar or its derivative, medium X, the bacterium *Enterobacter sakazakii* was found in 5 out of 50 commercially available powdered infant milk formulas, 3 out of 30 dried infant foods, and in 4 out of 25 milk powders. Although both media detected *Ent. sakazakii*, the chromogenic Druggan–Forsythe–Iversen medium proved to be about twice as sensitive in the organism's detection than the conventional violet red bile glucose agar. Enrichment of the dry foods in Enterobacteriaceae enrichment broth increased the likelihood of detection almost in order of magnitude. *Salmonella* serovars, the standard organism used for testing food products for the presence of Enterobacteriaceae, were not detected even after enrichment,
Results		

Conclusion/ implication	suggesting that monitoring dry milk and food products solely for *Salmonella* serovars is insufficient for the detection and control of *Ent. sakazakii.*

In Example 15-6, although some experimental approach, the results obtained, the conclusion/implication of the work are stated, the author did not state what the question was. The description of the results lets the reader guess what the question of the paper is, but it should not be the goal to have the reader guess. Another problem with this Abstract is that no context is given for the experiments done.

Revised Example 15-6	**Meningitis, necrotizing enterocolitis, bacteremia, and deaths have been associated with the bacterium *Enterobacter sakazakii* in infants (Iverson and Forsythe, 2004). Although this bacterium has been identified in a variety of food products such as in dairy, produce, meat, cereals, and spices, its presence in powdered milk and dried**
Background	**infant foods has not been investigated. In this study, we**
Experimental approach	**used** conventional violet red bile glucose agar and its derivative, medium X, **to test 50 commercially available**
Question/ purpose	**powdered infant milk formulas, 30 dried infant food products, and 25 milk powders for the presence of *Ent. sakazakii.*** We detected *Ent. sakazakii* in 5 out of 50 powdered infant milk formulas, 3 out of 30 dried infant foods, and in 4 out of 25 milk powders. Although both media detected *Ent. sakazakii,* the chromogenic Druggan–Forsythe–Iversen medium proved to be about twice as sensitive in the organism's detection than the conventional violet red bile glucose agar. Enrichment of the dry foods in Enterobacteriaceae enrichment broth increased the likelihood of detection almost in order of magnitude. *Salmonella* serovars, the standard organism used for
Results	testing food products for the presence of Enterobacteriaceae, were not detected even after enrichment, suggest-
Conclusion/ implication	ing that monitoring dry milk and food products solely for *Salmonella* serovars is insufficient for the detection and control of *Ent. sakazakii.*

In the revision, background information has been added in the first two sentences. The question is stated in the third sentence, following the experimental approach. Results are indicated, and an implication is provided. This Abstract is much easier to follow and to comprehend. Not only does it contain all the elements necessary for a well-written Abstract, it clearly signals the parts as well ("In this study, we used . . . to test . . ."; "We detected . . ."; ". . . suggesting that . . .").

➤ Keep the Abstract short

One of the most common problems for Abstracts is excessive length. When you write the Abstract, consider every word carefully. If you can write your Abstract in fewer words than the maximum allowed, do so. If you find yourself in a situation in which you have to condense your Abstract, follow the suggestions below (see also Chapter 6, Section 6.4).

To condense a long Abstract:

- Omit unnecessary words and combine sentences
- Condense background
- Omit or subordinate less important information (definitions, experimental preparations, details of methods, exact data, confirmatory results, and comparisons with previous results).

An example of an unnecessarily long Abstract is shown next in Example 15-7.

Example 15-7 **Excessively long Abstract**

Altered endothelial cell function appears to be an important part of the pathophysiology during a dengue virus infection. *The effect of such an infection on gene expression in primary human umbilical vein endothelial cells can be studied by Differential Display.* We confirmed altered gene expression detected by differential display by semi-quantitative and real-time fluorogenic RT-PCR. *Using these* Wordy *techniques,* we identified at least nine cDNAs with altered expression in virus-infected cells not previously reported. These cDNAs included the human inhibitor of apoptosis-1, 2'-5' oligoadenylate synthetase, a 2'-5' oligoadenylate synthetase–like cDNA, Galectin-9, Myxovirus protein A and 1, the regulator of G-protein signaling, and the endothelial and smooth muscle cell–derived neuropilin-like protein. *We also performed RT-PCR analyses on the nine* Wordy *cDNA and found* that dengue virus infection also appears to increase the expression of tumor necrosis factor-α (TNF-α), interleukin 1β (IL-1β), and the toll-like receptor 3 in HUVECs. These results point to the possibility of the activation of at least three signaling pathways during dengue Wordy virus infection of HUVECs. *These three* signaling *pathways include* the TNF-α pathway, the IL-1β pathway, and the type-I IFN (IFN-α/β) pathway.

(With permission from Irene Bosch)

This Abstract is 184 words long and describes what the same Abstract, when condensed, says in 133 words.

Revised
Example 15-7   Altered endothelial cell function appears to be an important part of the pathophysiology during a dengue virus infection. We confirmed altered gene expression detected by Differential Display using semi-quantitative and real-time fluorogenic RT-PCR. The identified nine cDNAs included the human inhibitor of apoptosis-1, 2'-5' oligoadenylate synthetase, a 2'-5' oligoadenylate synthetase–like, Galectin-9, Myxovirus protein A and 1, the regulator of G-protein signaling, the endothelial and smooth muscle cell–derived neuropilin-like protein, and the phospholipid scramblase 1. RT-PCR analyses confirmed that dengue virus infection also increased the expression of the tumor necrosis factor-α (TNF-α), the interleukin 1β (IL-1β), and the toll-like receptor 3 in HUVECs. These results point to the activation of at least three signaling pathways during dengue virus infection of HUVECs: the TNF-α pathway, the IL-1β pathway, and the type-I IFN (IFN-α/β) pathway.

In the revised version, long wordy constructions have been replaced with more concise wording. Sentences have been combined, and background information has been condensed. Even though the original, longer Abstract is quite readable, the shorter revision gets the overview across more clearly.

➤ Do not use the wrong type of abstract

Do not confuse or mix your abstract for a research paper with that for a review article or book chapter. Review articles or reports use different types of abstracts called *indicative abstracts* (also known as descriptive abstracts; see Chapter 19, Section 19.5 for examples and a more in-depth discussion of these type of abstracts). Indicative abstracts provide the reader with a general idea of the contents of the paper and include little if any methods or results. Thus, an indicative abstract is essentially a table of contents in paragraph form. Unlike research paper abstracts, indicative abstracts usually are not self-contained and need to be read together with the text of the article. Often, the final sentence of this type of abstract is an overview sentence, listing what will occur in the document, as shown in the second sentence of Example 15-8.

Example 15-8   Indicative abstract

This paper describes how plants adapt to desert environments in differing locales. We present various ways they use to collect and store water as well as a variety of characteristics that allow plants to reduce water loss, and discuss tolerance, evasion, and water storage/succulence in detail.

On occasion, authors blend research paper abstracts and indicative abstracts (also known as descriptive abstracts). In such mixed abstracts, most sentences are informative, but the last sentence is usually descriptive or indicative, stating that something will be described or discussed in the text. Such statements are not useful in an abstract for a research article, however, and should be avoided.

15.7   REASONS FOR REJECTION

The Abstract is usually the first part of a manuscript that the editor and the reviewers read. You are more likely to impress an editor or reviewer (not to mention your readers) if you master the skill of writing simple, clear, and concise Abstracts. You need to spike the interest of the reviewers with your Abstract. Very often, reviewers will be tempted to judge your complete manuscript based on the Abstract alone. The following are common reasons for quick rejection of a paper based on problems in the Abstract:

- **Lack of originality**—you must write about something new or better than what has been presented in previous work.
- **Lack of context**—you need to provide a background of your work as well as its implication. Never assume that the editor or reviewers will be sufficiently familiar.
- **Limited sample size**—few samples in a study may not convince a reviewer of the significance of the paper.
- **Lack of numbers**—the reviewers will not know whether there is anything of value in the paper.
- **Too many numbers/too much data**—you will leave reviewers with the sense that they have to figure out and interpret the results themselves.
- **Lack of conformity**—you should follow the *Instructions for Authors* very carefully so as not to antagonize a reviewer.
- **Wrong style of Abstract**—differentiate between abstracts for research papers and other abstracts such as indicative abstracts for review papers.
- **Too many abbreviations**—this will turn off most reviewers (and readers).

15.8   REVISING THE ABSTRACT

When you have finished writing the abstract (or if you are asked to edit an abstract for a colleague), you can use the following checklist to dissect the abstract systematically:

- ☐ 1. Did you distinguish between a research paper abstract and an indicative abstract?
- ☐ 2. Is the question/purpose stated?
- ☐ 3. Is the experimental approach stated?
- ☐ 4. Are the results indicated?

☐   5. Is the answer/conclusion provided?
☐   6. Are all elements signaled?
☐   7. Is the length within the required limits?
☐   8. Has the Abstract been condensed as much as possible?
☐   9. Is the significance of the work apparent?
☐ 10. Is the context clear?
☐ 11. Is the work original?
☐ 12. Revise for style and composition based on the basic rules and guide-lines of the book:
    ☐ a. Are paragraphs consistent? (Chapter 6, Section 6.2)
    ☐ b. Are paragraphs cohesive? (Chapter 6, Section 6.3)
    ☐ c. Are key terms consistent? (Chapter 6, Section 6.3)
    ☐ d. Are key terms linked? (Chapter 6, Section 6.3)
    ☐ e. Are transitions used and do they make sense? (Chapter 6, Section 6.3)
    ☐ f. Is the action in the verbs? Are nominalizations avoided? (Chapter 4, Section 4.6)
    ☐ g. Did you vary sentence length and use one idea per sentence? (Chapter 4, Section 4.5)
    ☐ h. Are lists and comparisons in sentences kept parallel? (Chapter 4, Section 4.9)
    ☐ i. Are comparisons written correctly? (Chapter 4, Sections 4.9 and 4.10)
    ☐ j. Have noun clusters been resolved? (Chapter 4, Section 4.7)
    ☐ k. Has word location been considered? (Verb following subject immediately? Old, short information at the beginning of the sentence? New, long information at the end of the sentence?) (Chapter 3, Section 3.1)
    ☐ l. Have grammar and technical style been considered (person, voice, tense, pronouns, prepositions, articles)? (Chapter 4, Sections 4.1–4.4)
    ☐ m. Is past tense used for results and present tense for descriptive papers? (Chapter 11, Sections 11.3 and 11.4)
    ☐ n. Are words and phrases precise? (Chapter 2, Sections 2.2 and 2.3)
    ☐ o. Are nontechnical words and phrases simple? (Chapter 2, Section 2.2)
    ☐ p. Have unnecessary terms (redundancies, jargon) been reduced? (Chapter 2, Section 2.4)
    ☐ q. Have spelling and punctuation been checked? (Chapter 4, Section 4.11)

SUMMARY

> ABSTRACT GUIDELINES
> - Use an informative or structured abstract for research articles.
> - Abstracts for research papers include
>   - Question/purpose
>   - Experimental approach
>   - Results
>   - Conclusion
>   - If useful, also include background and significance
> - Use a structured abstract for a clinical report if requested.
> - Use subheadings for a structured abstract.
> - Abstracts for descriptive research papers should include
>   - Descriptive statement
>   - Description of the new findings
>   - Conclusion/significance/implication
> - Apply basic scientific writing rules and guidelines in the Abstract.
> - Signal the question, the experimental approach, the results, the answer, and the implication.
> - Do not omit any parts of the Abstract.
> - Keep the Abstract short.
> - Do not use the wrong type of Abstract.

PROBLEMS

PROBLEM 15-1 Evaluating an Abstract

1. In the following Abstract, check that all required parts are present and signaled correctly. Also, check if basic rules have been considered.
2. Does the Abstract correspond to the Title?

Structural Basis for the Interaction of Chloramphenicol, Clindamycin, and Macrolides with the Peptidyl Transferase Center in Eubacteria

Ribosomes, the site of protein synthesis, are a major target for natural and synthetic antibiotics. Detailed knowledge of antibiotic binding sites is the key to understand the mechanisms of drug action. Conversely, drugs are excellent tools for studying the ribosome function. To elucidate the structural basis of ribosome-antibiotic interactions, we determined the high-resolution X-ray structures of the 50S ribosomal subunit of the eubacterium *Deinococcus radiodurans* complexed with the clinically relevant antibiotics chloramphenicol, clindamycin, and the three macrolides: erythromycin, clarithromycin, and roxithromycin. We found that antibiotic binding sites are composed exclusively of segments of 23S rRNA at the peptidyl transferase cavity and do not involve any interaction of the drugs with ribosomal proteins. Here, we report the details of antibiotic interactions with the components of their binding sites. Our results also show the importance of Mg^{2+} ions for the binding of some drugs. This structural analysis should facilitate rational drug design.

(With permission from Macmillan Publishers Ltd.)

PROBLEM 15-2 Evaluating an Abstract
In the following Abstract, identify all essential components and their signals if provided.

Many insects possess a sexual communication system that is vulnerable to chemical espionage by parasitic wasps. We recently discovered that a hitch-hiking (H) egg parasitoid exploits the antiaphrodisiac pheromone benzyl cyanide (BC) of the Large Cabbage White butterfly *Pieris brassicae*. This pheromone is passed from male butterflies to females during mating to render them less attractive to conspecific males. When the tiny parasitic wasp *Trichogramma brassicae* detects the antiaphrodisiac, it rides on a mated female butterfly to a host plant and then parasitizes her freshly laid eggs. The present study demonstrates that a closely related generalist wasp, *Trichogramma evanescens,* exploits BC in a similar way, but only after learning. Interestingly, the wasp learns to associate an H response to the odors of a mated female *P. brassicae* butterfly with reinforcement by parasitizing freshly laid butterfly eggs. Behavioral assays, before which we specifically inhibited long-term memory (LTM) formation with a translation inhibitor, reveal that the wasp has formed protein synthesis-dependent LTM at 24 h after learning. To our knowledge, the combination of associatively learning to exploit the sexual communication system of a host and the formation of protein synthesis-dependent LTM after a single learning event has not been documented before. We expect it to be widespread in nature, because it is highly adaptive in many species of egg parasitoids. Our finding of the exploitation of an antiaphrodisiac by multiple species of parasitic wasps suggests its use by *Pieris* butterflies to be under strong selective pressure.

(With permission from the National Academy of Sciences, USA)

PROBLEM 15-3 Evaluating an Abstract

1. **In the following Abstract, ensure that all required parts are present.**
2. **Is this an Abstract for an investigative research paper or for a descriptive paper?**

The cyanobacterial circadian pacemaker is an enzymatic oscillator, which orchestrates the metabolism of the bacteria to fit the day and night alternations of this planet. Interactions among KaiA, KaiB, and KaiC, the three components of this oscillator, result in many oscillatory properties *in vitro*, including an overall phosphorylation level of KaiC, an apparent segregation between synchronized phosphorylation and dephosphorylation reactions, and the size and the composition of this oscillator. To explain these properties, we propose here a molecular mechanism for this pacemaker within the framework of a cyclic catalysis scheme.

(With permission from Jimin Wang)

PROBLEM 15-4 Your Own Abstract

1. Write your own Abstract.
2. Revise your own Abstract: Check that all required parts of an Abstract are present and signaled correctly. Also, check if basic writing rules are applied correctly.
3. Have a peer comment on your Abstract.

CHAPTER 16

Titles, Key Words, Footnotes, and Acknowledgments

Many readers will discover your paper by seeing it in the results list of a database query. The title, along with the key words and the Abstract, help readers find your article and determine whether they want to read it.

THIS CHAPTER EXPLAINS:

- The hallmarks of a strong title
- The title page
- The running title
- Selection of key words
- Revising the title
- Footnote, endnotes, and acknowledgments

16.1 OVERALL

➤ Aim to attract readers

The title is the single most important phrase of an article. Most readers will read your Abstract only if the title interests them. They will judge the paper's relevance on the title alone.

➤ State the main topic of your study in the title

To identify the main message of the paper, state the main topic in the title and make the title interesting to attract readers. Because a title in a scientific research journal will be used in indexing systems and bibliographic databases, it should also be informative and accurate.

A title is typically stated as a phrase, but it may also be a complete sentence. However, a full sentence with an active verb is usually not a good title—neither is an overly long phrase.

➤ Use a title that separates your article from others in the field

Ensure that your title separates the article from all other articles in that field. Do not hesitate to try out several versions of a title on colleagues and friends. Look at other good published titles to get ideas on how to phrase yours. When you convey the message of your paper, be assertive, but do not brag. Be exact and clear, or the readers for whom you wrote the paper will never read it.

16.2 STRONG TITLES

➤ Use a strong title. Make it clear and complete but succinct

The first thing readers, editors, and reviewers see is the title. It is therefore very important that your title is strong. A strong title should fulfill three criteria: It needs to be clear, complete, and succinct.

Clarity of a Title

Unclear titles confuse or mislead readers and can give an annoying first impression. Some titles are not clear because word choice is too general, as shown in Example 16-1.

 Example 16-1   Effect of hormones on tumor cells

The title in Example 16-1 is not clear because words in it are unspecific, such as the category terms "hormones" and "tumor cells." What hormones and which tumor cells? The specific hormones and tumor cells should be listed. Otherwise, the title is essentially meaningless. At the same time, this title is unclear because it does not state what specific effect has been observed on the study. The revised example is a much stronger title.

 Revised   Effect of **testosterone and estradiol on the growth and**
Example 16-1   **morphology of rat epithelial** tumor cells

Titles can also be ambiguous because of unspecific words such as *and* and *with*. Avoid these words as they tend to confuse readers.

 Example 16-2   Tracking long-distance migration of gray whales <u>with</u> geolocators

---

In Example 16-2, the relationship between "gray whales" and "geolocators" is not clear. It reads as if gray whales have geolocators. "With" needs to be replaced by a more specific word.

| 👍 | **Revised** **Example 16-2** | Tracking long-distance migration of gray whales **by using** geolocators |

Ambiguity can also arise because of noun clusters, as is shown in the next example.

| 👎 | **Example 16-3** | Involvement of amygdala in the oxotremorine memory enhancing effect |
| 👍 | **Revised** **Example 16-3** | Involvement of amygdala in memory enhancement by oxotremorine |

The original title contains a confusing noun cluster. In the revised title, this noun cluster has been split up, and the word *effect* has been omitted, making the title much clearer.

Aside from unspecific word choices and confusing noun clusters, abbreviations, especially nonstandard abbreviations, can create unclear titles. Avoid abbreviations in titles unless they are well-known abbreviations, acronyms, contractions, or accepted conventions, such as chemical formulas. Examples of such abbreviations include the following: H_2O, CO_2, ATP, DNA, RNA, PCR, spp., and so forth.

Completeness of a Title

A title should not only be clear, it also has to be complete. Therefore, include and highlight the key items of your study in your title (e.g., a specific disease seen in a certain group of people, a novel assay, the species studied). Keep in mind, however, that in the title readers can only absorb three or four details, and that it cannot replace the Abstract. Concentrate on the most distinctive aspect of your work. Details of secondary importance can be presented in the Abstract or Introduction. Announcing the main variables of the paper is stronger than trying to fit all the variables into the title.

Consider the title in the following example.

| 👎 | **Example 16-4** | Dengue virus activates human umbilical vein endothelial cells |

Although this title orients the reader to the area of research, it does not give any specifics as to how the activation takes place. Adding a few more specific words completes the title and sets it apart from others in the field.

| 👍 | **Revised** **Example 16-4** | Dengue virus activates human umbilical vein endothelial cells **via the tumor necrosis factor-α pathway** |

To check that your title is complete, compare your title with the question and answer of your paper (see also Chapter 11). Ensure that you use the same key terms (see Chapter 6) in the title as in the question and answer of the paper.

Succinctness of a Title

Although a title has to be clear and complete, you need to use the fewest words possible to describe your work adequately. Many titles are simply too long and thus difficult for the reader to follow. Short titles are much more memorable and attract much more attention.

Most journals also prefer short titles, typically 100 to 150 characters (including the spaces between the words), which translates to about 10 to 15 words. It is important that you check the *Instructions to Authors* (or journal guidelines) for specific requirements on the length of the title as these can vary substantially. Limit your words by cutting out trivial words and phrases that contribute nothing to the information in the title. The aim is to state the message of your paper clearly and completely in as few words as possible. If needed, rearrange key elements of your title to find the phrasing that is pertinent and provides the right impact for the reader.

To make a title succinct, omit unnecessary words and condense necessary words as much as possible. Unnecessary words and phrases that can be omitted include "Nature of," "Studies on," and the like.

Example 16-5   Examination of the differential response properties of single units within neuronal clusters in the inferior colliculus

---

In this example, "Examination of" is unnecessary. Even "properties" could be omitted.

Revised Example 16-5   Differential response of single units within neuronal clusters in the inferior colliculus

---

Aside from unnecessary words and phrases, articles at the *beginning* of the title can be omitted. However, do not omit articles before singular nouns later in the title.

Example 16-6   The kinetic analysis of the Na-ATPase in the corneal epithelium

Revised Example 16-6   Kinetic analysis of the Na-ATPase in the corneal epithelium

---

Titles can also be shortened by condensing necessary words. You can condense titles through the use of clear noun clusters or through hyphenation of words. Some examples of the latter technique include:

"seizure-induced" instead of "induced by seizures"
"food-deprived mice" instead of "food deprivation in mice"
"ATP-dependent" instead of "dependent on ATP"

An example of a longer title that can be shortened using noun clusters is shown next.

☞	**Example 16-7**	Diabetes promotes cholesterol deposits in **the corneas of the eyes of rabbits**
👍	**Revised Example 16-7**	Diabetes promotes cholesterol deposits in **rabbit eye corneas**

Note that in the revised example, the word *eye* can be dropped if the target journal is highly specific, such as the *Journal of Ophthalmology*, as most readers would know that eye corneas are being referred to.

16.3   THE TITLE PAGE

➤ Record the title of the paper as well as the names and affiliations of the authors on the title page

Titles are usually written on the first page of a manuscript, the title page. As journals have very specific requirements and expectations regarding the content and format of the title page, be sure to check the *Instructions to Authors* of your target journal.

Typically, the title page contains not only the title of the paper but also the names and affiliations of the authors. The names are written in Western-style name order: first name, middle initial, then last name. Superscript numbers after each name identify the footnote that contains the affiliation of the author as well as the corresponding author. The latter may also be identified through another symbol such as an asterisk. Footnotes may also identify authors who contributed equally to the work. See Example 16-9 for a sample title page.

Aside from the title and the authors, some journals may require you to provide a running title, key words, or a total word count on the title page. Check the *Instructions to Authors* for any other rules and requirements.

16.4   RUNNING TITLE

➤ Make the running title identifiable as a short version of the title

A running title is a short version of the complete title. This short version is placed either on the top as a header or on the bottom as a footer of every page (or alternating with the authors' names, on every other page) of a journal to identify the article.

👍	**Example 16-8**	**Title and running title**
		Title: Immunocytochemical localization of seizure-induced increases in ornithine decarboxylase in the rat hippocampus
		Running Title: Ornithine decarboxylase in the rat hippocampus

ESL advice (side margin)

For most journals, the running title cannot exceed 50 to 60 characters (with spaces), but make sure you check the *Instructions to Authors* of your target journal for specific requirements.

16.5   KEY WORDS

> ➤ Select important and specific terms as key words

> ➤ Avoid general single key words that may apply to a very large number of papers

Many journals require an author to provide three to ten key words or phrases. These are placed either on the title page or after the Abstract and will help in cross-indexing the article.

Key words should name important topics in your paper. Select the most important and most specific terms of your paper. Choose words or phrases that you would look up if you were trying to find your own paper and that would attract the readers you hope to reach.

Example 16-9   **Sample title page**

Seasonal variation of ibotenic acid and muscimol in *Amanita muscaria*

Natalya Motka[2], Klaus Malbeck[3], Leng-Fei Wu[1], Alicia Fernández[2], and Samuel Girald[1*]

[1] *Department of Biol. Chemistry, Yale University, New Haven, CT, USA*
[2] *Institute of Experimental Botany, Academy of Sciences of the Czech Republic, Rozvojová 263, CZ-16502 Prague, Czech Republic*
[3] *Max Planck Institute for Plant Biology, Dortmund, Germany*

*To whom correspondence should be addressed
     Telephone: +001 (203) 234 8006
     Fax: +001 (203) 785 3360
     Email: samuel.girald@yale.edu
     Address: Department of Biol. Chemistry,
                   Yale University,
                   New Haven, CT 06520

Running title: Ibotenic acid and muscimol in *Amanita muscaria*
Key words: neurotransmitter, GABA agonist, NMDA glutamate receptor, neurotoxin, psychoactive, mushroom poisoning, muscarine

---

> ➤ Avoid words that appear in the title

Do not use terms that appear in the title already. Terms from the title are used in cross-referencing as well. These terms should not overlap. Using terms different from those in the title will broaden the selection of words and terms that can be looked up. Similarly, given that most searches today are capable of finding words throughout the Abstract as well, you may want to expand your key word selection to words and phrases that do not appear in either the title or the Abstract.

If possible, use terms and phrases instead of individual words. For example, use "cholesterol degradation" rather than "cholesterol" as looking up papers with cholesterol as a search word will produce far too many hits.

If necessary, include a term as an indexing term even if the term does not appear in your paper. For biomedical articles, for example, it is best to select current specific terms from the medical subject headings (MeSH) list of Index Medicus. Science databases can then be searched by such key words (see also Chapter 8, Section 8.2 for a list of top science databases).

16.6   FOOTNOTES AND ENDNOTES

Journals usually do not like to see footnotes or endnotes because they are often expensive to reproduce. Exceptions are footnotes for figures or tables. If explanatory notes in the text still prove necessary to your document, insert a number formatted in superscript following a punctuation mark or the name of the author.

 Example 16-10   Solange Brault[2] and Hal Caswell[2]

List all footnotes on the bottom of the same page of your document on which your superscript appeared.

 Example 16-11   [2]Order of authorship determined by toss of a coin.

16.7   ACKNOWLEDGMENTS

➤ List all the people whose help was important but not enough to warrant authorship

General

Acknowledge any organizations or individuals who provided grants, materials, or financial or free technical assistance. If possible, specify the type of support you received. Also, acknowledge people who contributed ideas, information, critical writing or editing, and advice to your work.

You do not need to acknowledge anyone who received payment, was hired to perform a task, or did their day-to-day regular work, such as a technician preparing media, an editor or consultant you hired, or an assistant who typed in the text of your manuscript. People named in the acknowledgments should give their permission to be named and should approve of the wording of your acknowledgment. This approval process can save friendships, avoid embarrassing situations, and ensure future collaborations.

Typically, start by listing intellectual contributions and then move on to technical support, provision of materials, helpful discussions, and revisions and preparations of the manuscript. Last, list any funds, grants, fellowships, or financial contributions.

Example 16-12   **a**   We thank Dr. J. Holzheimer for his technical advice on crystallization assays. We are also grateful to Drs. Thomas Hugh and Fred Grant for their critical review of the manuscript. This study was supported by grant XXX from the National Institutes of Health.

b   The authors thank Brad Burdick who went out of his way in collecting the HPLC data. The laboratory of Dr. F. Verus provided invaluable data and technical support. Dr. Peter Gress, Belinda Gross, and Paul Sepega contributed immensely to the preparation of the manuscript. This work was made possible in part by a grant from the ABC Foundation. The authors have no conflicting financial interests.

Conflict of Interest Statement

➤ Disclose any possible conflict of interest

Some Acknowledgments sections contain a statement about conflicts of interest (also known as a competing interest statement) as a last sentence in the paragraph. (Conflicts of interest may also be mentioned in a separate section just before or after the Acknowledgments or through the submission of a conflict of interest letter, signed by all the authors.) Conflicts of interest may arise if an author's or coauthor's judgment, action, and interpretation may be affected due to financial and other gains or due to personal relationships and affiliations. Often, people are not aware that their interests pose a conflict, which can range from negligible to conflicts involving a direct stake in the content or competition. To assess a potential conflict of interest, consider if something at work is associated with your private interests and may lead to a personal gain. If so, disclose this possible conflict of interest. Generally, the greater the gain, the more serious the conflict of interest. To avoid any appearance of bias in publications and to ensure their integrity, you have to disclose such conflicts of interest. By disclosing any potential conflict of interest, you can head off negative feedback, future embarrassment, dismissal of your publication, and even potential fines.

Possible ways to phrase a conflict of interest statement are listed in Example 16-13.

Example 16-13   **a**   The authors declare no competing financial interests.

b   A grant (AG0967) has been awarded to DB by Y for the study of ZZZ. No other author has any other financial link to Y.

c   Y University owns a patent, xx/yyy,zzz, that uses the approach outlined in this article and has been licensed to Z.

16.8 REVISING THE TITLE

When you have decided on a title (or if you are asked to edit a paper for a colleague), use the following checklist for the title:

☐ 1. Does the title attract readers?
☐ 2. Does the title state the main topic of your study?
☐ 3. Is the title clearly and unambiguously stated?
☐ 4. Have abbreviations been avoided?
☐ 5. Is the title complete?
☐ 6. Did you use the same key terms as the question and the answer of your paper?
☐ 7. Is the title succinct? Have unnecessary words and phrases been omitted?
☐ 8. Does the title separate your article from others in the field?
☐ 9. Is the running title identifiable as a short version of the title?

SUMMARY

TITLE GUIDELINES
- Aim to attract readers.
- State the main topic of your study in the title.
- Use a title that separates your article from others in the field.
- Use a strong title. Make it clear and complete but succinct.
- Record the title of the paper as well as the names and affiliations of the authors on the title page.
- Make the running title identifiable as a short version of the title.

KEY WORDS GUIDELINES
- Select important and specific terms as key words.
- Avoid general single key words that may apply to a very large number of papers.
- Avoid words that appear in the title.

ACKNOWLEDGMENTS GUIDELINES
- List all the people whose help was important but not enough to warrant authorship

CONFLICT OF INTEREST GUIDELINES
- Disclose any possible conflict of interest

PROBLEMS

PROBLEM 16-1 Clear Title
The following titles are unclear or not specific. Make suggestions on how to improve these titles. Invent additions if needed.

1. Variation of fossil density with Triassic sedimentary deposits
2. Antibacterial, anti-inflammatory, and antimalarial activities of some African medicinal plants
3. Classification of Fowl Adenovirus Serotypes with genome mapping and sequence analysis of the hexon gene
4. Hemolymph-dependent and -independent responses with Drosophila immune tissue
5. Temperature dependence of fir tree carbon dioxide sequestration
6. Stable, immunogenic, and nasal-specific formulation of NoV vaccine using VLP and adjuvant components
7. Microscopic observation drug susceptibility assay for the diagnosis of TB

PROBLEM 16-2 Complete Title
The following titles are not complete. Suggest information to add to make these titles complete.

1. Differences of old-world and new-world species of hazelnuts
2. Preparation of single nanocrystals of platinum
3. Widespread increase of bat mortality rates
4. Transmission of coccidioidomycosis

PROBLEM 16-3 Title Length
The following titles are too long. Condense them to make them more concise.

1. The adaptive value of cued seed dispersal in desert plants: Seed retention and release in *Mammillaria pectinifera* (Cactaceae), a small globose cactus
2. Analysis of historical data of Groundwater flow in the South Wales coalfield used to inform 3D modeling
3. The effect of negative mood on persistence in problem solving
4. The intensity of hurricanes is linked to increased atmospheric dust originating from the Sahara desert
5. A cost-effectiveness analysis is used to determine how best to optimize flu virus vaccination strategies
6. New cellular antioxidant activity assay can be used to quantify antioxidant activity effectively
7. Analysis of temperature and light requirements for seed germination and seedling growth of *Sequoiadendron giganteum*

PROBLEM 16-4 Creating a Title

Ensure that the title for an article on the antioxidant effects of blueberries is specific, complete, concise, and unambiguous. The title should convey the following facts:

- Blueberries have an antioxidant effect
- Eating blueberries increases plasma antioxidant capacity
- If milk/milk protein is digested together with blueberries, no increase in plasma antioxidant capacity is observed

Choose the best title from the list:

1. Antioxidant effect of blueberries reduced with milk
2. Increase of plasma antioxidant capacity of blueberries prevented with milk protein
3. Consumption of blueberries with milk influences plasma antioxidant capacity
4. Milk consumption prevents antioxidant effect of blueberries
5. Milk consumption effects plasma antioxidant capacity
6. Influence of milk protein on antioxidant effect of blueberries

PROBLEM 16-5 Running Title

For the following titles, find a corresponding running title:

1. Transcoronary transplantation of progenitor cells after myocardial infarction
2. Analysis of the surface characteristics and mineralization status of feline teeth using scanning electron microscopy
3. Inflammatory mechanisms in chronic obstructive pulmonary disease
4. The roles of subsurface carbon and hydrogen in palladium-catalyzed alkyne hydrogenation
5. Picosecond coherent optical manipulation of a single electron spin in a quantum dot
6. Structural analysis of *E. coli* hsp90 reveals dramatic nucleotide-dependent conformational rearrangements
7. Ecological correlations to decline and extinction of the endemic Brazilian tree frog *Corythomantis greeningi*
8. Determination of chlorophyll density for corn from spectral reflectance data
9. Rare structural variants disrupt multiple genes in neurodevelopmental pathways in schizophrenia
10. Otoferlin, defective in a human deafness form, is essential for exocytosis at the auditory ribbon synapse
11. NMR imaging of catalytic hydrogenation in microreactors with the use of para-hydrogen
12. Generation and photonic guidance of multi-octave optical-frequency combs

CHAPTER 17

Revising and Reviewing a Manuscript

After you have composed a first draft of your manuscript and added the Abstract as well as the title, you need to revise the document. Revision is the key to successful writing. Once you master revising your own manuscript, you will at the same time have mastered reviewing manuscripts for others.

THIS CHAPTER EXPLAINS:

- Revision of your first draft
- Revision of subsequent drafts
- Pre-submission peer review

17.1   REVISING THE FIRST DRAFT

➤ Check the first draft for content and content location

Once you have written down everything you could think of for a first rough draft, let your draft incubate for at least a day or two. Then, it is time to revise it. When you read your first draft again, you will see passages that you would like to change, recognize portions that need work, think of points to include, and notice those to omit or condense.

When you revise your first draft, check it first for content and organization. Make sure that all the essential points you want to make have been included. Everything you say should contribute in some way to your question and answer (see also Chapters 11 and 14) and no steps should have been left out. You may have included unnecessary material, left out essential evidence, or discussed points in the wrong order. Any irrelevant points need to be removed, and any missing evidence should be included. Place all information in the right order.

In revising, you essentially work your way from the outside in. Therefore, check the individual sections of the paper first, then the paragraphs, and then sentences. Check that your manuscript is logically organized. All parts, paragraphs, and sentences must be in the right order. After verifying the order, you can then revise the style.

The overall structure of your paper should conform to the following outline:

Title:	Three to four important key terms	
Abstract:	Content: question/purpose, experimental approach, results, interpretation/answer, significance	
Introduction:	Format: funnel shape (known, unknown, question, experimental approach) **First paragraphs: Background** **2nd to last paragraph: Unknown** **Last paragraph: Question/purpose and experimental approach. Optional are main results and significance.**	
Materials and Methods:		Organize chronologically, most to least important, or by subsections
Results:	1. Paragraph(s):	**Overview of most important/ interesting result(s)**
	Middle paragraphs:	Describe other results. Organize chronologically or most to least important—every result segment should contain purpose of experiment, experimental approach, results, and their interpretation.
	Last paragraph:	**State interesting result(s) or summarize main findings if Results section is lengthy.**
Discussion:	Format: pyramid shape (interpretation of findings, compare and contrast, models, conclusion, significance)	
	1. Paragraph	**Interpretation of most important results/answer to the question of the paper; support and defend interpretation**
	Middle paragraphs:	Chain of topics, compare and contrast findings, list limitations, etc.
	Last paragraph:	**Conclusion: summary of main findings and significance (future directions)**

Note that in this overall outline, key power positions are written in bold as they indicate the most crucial structural locations of an article. Pay particular attention to the content and location of these power positions throughout your revisions.

Once all of your structural components are in place, make sure that you have not missed any content component. Use the checklists provided for each section of a paper (see end of Chapters 11–16) to double-check that you have included all relevant components in each section. Clearly mark and label important components of each section on a print version of the manuscript to check their completeness using the summaries provided in Chapters 11 through 16. Identifying each essential component will make it clear to you if there is anything missing or not clearly signaled.

Ensure that you have phrased the purpose of the study/question and interpretation of results/answer the *same* in all key locations: Abstract, Introduction, and the first paragraph of the Discussion and Conclusion. Make certain that your question and answer make sense together.

Pass your draft on to coauthors for their comments and suggestions. Let your coauthors know that this is a working first draft.

➤ Check logical organization and flow of sections and subsections

When you are happy with the structural organization and content of the power positions, revise each section for logical organization and flow. Use the checklists at the end of Chapters 11–16. Ensure that headings refer to the text they describe. Look at how the ideas are distributed among the paragraphs, and make sure that your arguments are logical. Is it clear how and why the evidence presented supports the interpretation of your findings? Is it clear why a particular experimental approach or technique is appropriate? Have the main concepts been clearly and logically connected?

To check for logical flow, verify that you have a chain of topic sentences running throughout the paper. When read by themselves, the topic sentences should be sufficient to provide a rough outline of the paper. Add transitions as needed, but note that typically no transitions should follow subheadings.

It may also help to make a reverse outline of your manuscript by going through it paragraph by paragraph. Check that this reverse outline is logically organized. You may even want to compare this reverse outline to your original outline to ensure that you have not missed any content (see also Chapter 7, Section 7.5 for a discussion of outlines).

➤ Revise for style only after you are satisfied with the content and organization

Once you are satisfied with the content and organization of the first draft, revise it stylistically. You will probably see a lot to change. Here, too, start by working your way from the bigger structures toward the smaller ones.

For good flow between sections and between paragraphs, ensure that the transitions between paragraphs and sections are smooth and that they tie the pieces together. Pay particular attention to key terms and transitions within paragraphs as well. Add transition phrases and clauses to create the overview of the story (see also Chapter 6, Sections 6.1 and 6.3). Then, use the basic writing rules discussed in this book to check for paragraph structure, sentence structure, and word choice (Chapters 2–6). Inch through your manuscript sentence by sentence, word by word. Consider word location (see Chapter 3 and Chapter 6, Section 6.3). Check whether you have paid attention to either jumping word location or to a consistent point of view.

➤ Condense where possible

➤ Proofread your manuscript

Look for all possible ways to condense your paper. Omit unnecessary details, unnecessary words, and unnecessary paragraphs (see Chapter 2, Section 2.4 and Chapter 6, Section 6.4). Most readers, editors, and reviewers prefer short, meaty, clear papers. Ensure that you have not repeated any information unnecessarily. Scientists have a tendency to repeat the same information in different sentences or sections using the same or different wording. Your writing will be more concise if you learn to recognize such repetitions. Finally, proofread the text for punctuation, spelling, and typographical errors.

You will not be able to do all this revising on one draft, so revise in stages. Expect that revision of your manuscript will be long. Indeed, you likely will take more time revising than writing the first draft. Do as much as you can on the first revision. When you no longer see anything to change, put the paper in a drawer again for a few days. Then, you are ready to work on the second revision or third draft.

17.2 SUBSEQUENT DRAFTS

➤ Let some time elapse between revisions. Count on ≥6 drafts

➤ Recheck for content, logical organization, and style, and revise if needed

After you have waited a few days, you will be ready to look at your paper with fresh, critical eyes. Start anew by rechecking your draft for content and logical organization, and then recheck for style, especially word location.

➤ Give complete copies of revised manuscript to your coauthors

➤ Show your manuscript also to a colleague in a related field and a friend in a different discipline

➤ Ask for comments and constructive criticism in writing

When you have revised the paper structurally and stylistically as much as possible, give copies to any coauthors and/or your mentor. Show it to colleagues in the same or related fields of work as well as to a friend in a different discipline. Readers unfamiliar with the manuscript are more likely than you to spot inconsistencies, jargon, parts that are not logical, and other flaws. Be aware, though, that too many critics may result in chaos and confusion.

ESL advice

If you are not a native English speaker, ask someone who is fluent in English to review your paper. Reviewers and editors are more likely to reject a manuscript that does not read well or shows poor English.

Give everyone complete copies of the paper. As appropriate, point out specific areas in the manuscript where you may need their expertise or opinion. Provide peer reviewers also with a checklist to guide them through the review process (see Section 17.3 for a sample checklist). Ask for comments in writing. Oral comments are easy to forget, especially if there are a lot of them. If you are only provided with oral comments, immediately sit down and write them out to ensure that you will remember as much as possible about what was said.

Ask for constructive criticism not only of style, spelling, and punctuation but especially for content and logic. Do not be defensive when you receive harsh criticism. You should be concerned if your critics are indifferent, afraid to hurt your feelings, or say very little.

➤ Be prepared to accept the criticism

Be prepared for more revisions. If you consider the criticism valid, incorporate it. If you do not think it is valid, at least think it over. You do not have to agree with the people reviewing your paper, but you should respect their opinions. Give particular consideration to passages questioned by more than one reviewer. Such passages usually need extra attention even if you as the author do not immediately see the problem.

Your critics may not be able to pin down what is wrong, but the fact that they are questioning it probably means it could be improved. In revising these questionable passages, check for poor sentence location, sentence structure, word location, and word choice as well as for noun clusters, unclear comparisons, lack of parallel form, change in key terms, lack of transitions, and use of nominalizations.

➤ Check for grammatical and other errors before finalizing your manuscript

Remember, revision is the key to strong writing. Revising large sections in one sitting will make your document smoother. You may not be able to do this in the first few revisions, but the more you revise, the more you will be able to read through the document in one go.

Most authors need at least six to ten drafts to get a paper ready for submission. Be aware, however, that the process of revision can be endless. Inevitably, you will always find something to change, but you should not spend forever writing one paper. At some point, you have to stop revising. Keep in mind that the writing does not need to be perfect, just clear.

Last, check for grammatical and other errors, verify references with citations, confirm that figure numbers are sequential, and validate the information in the legend/caption. The final draft should be approved by all authors, although the final version will not read exactly as each individual coauthor would like it to read.

17.3   REVIEWING A MANUSCRIPT PRE-SUBMISSION

➤ **When reviewing someone else's manuscript, be as specific as possible and point out both strengths and weaknesses**

➤ **Always treat the author with respect**

Evaluating the work of others is one of the best ways to reinforce familiarity with revising strategies. Such peer review can also give you a deeper understanding of how writing affects different readers.

How you review a peer's work depends on when in the writing process you are doing the reviewing. Early drafts should be evaluated primarily with respect to major components of the paper such as the research question, the answer of the question, and the logical overall organization of the paper. Subsequent drafts should be evaluated particularly for style and composition as well as for flow, whereas a final draft should be reviewed for every aspect of the paper, including its topic, question, logical approach, format, style, composition, impact, grammar, and spelling.

The least helpful comment to receive from a peer reviewer is "It looks okay to me." To be an effective reviewer, you need to be as specific as possible and point out both strengths and weaknesses. Point out particular places in the paper where revision will be helpful. Do not hesitate to note when something is unclear to you, whether scientifically or in terms of the writing. If you disagree with the comments of another peer reviewer, say so. Not all readers react the same way, and divergent points of view can help writers see options for revising.

Always treat the author with respect. Avoid snippy comments such as "So what?" Instead, make suggestions and recommendations on how to improve and strengthen certain passages or on what else to add or omit from the document. If a passage reads well, point out this strength. If an argument is difficult to follow logically or does not make sense, raise objections politely or ask for explanations to clarify the argument. Write your comments either between the text lines or on the margins of the draft. Better yet, use the "Track Changes" option of MS Word (see also Appendix B), OpenOffice, Google Docs, or Pages as this will clearly show the author where and how to revise a document.

Read the paper you have been asked to review at least twice, once to get an overview of the paper and subsequent times to provide constructive criticism for the author. If you stumble across a phrase or sentence, this often indicates a spot that needs correcting. Mark such potentially faulty passages for the author. If possible, make suggestions on how to improve them. Use the following checklist as well as those provided at the end of Chapters 11 through 16 to ensure all important components have been included. When you have finished reviewing the document, you can then write up your overall impression of the paper by summarizing its strengths and weaknesses and listing specific areas of concern.

CHECKLIST FOR PRE-SUBMISSION PEER REVIEW

CONTENT
Purpose and interpretation

- ☐ Is the overall purpose of the paper and/or central question clear?
- ☐ Does the interpretation of the findings answer the overall question of the paper?

Support

- ☐ Is there sufficient evidence to support the answer?
- ☐ Is every paragraph and sentence in the paper relevant to the overall question?
- ☐ Are there portions of the text that could be omitted?

_____

Overall

- ☐ Does the paper advance the field?
- ☐ Does it provide interesting and important insights into the topic of interest?
- ☐ Have power positions been considered (especially in the Introduction, Results, and Discussion)?

Individual sections
Title:
- ☐ Is the title strong?

Abstract:
- ☐ Does the Abstract adequately summarize the paper?
- ☐ Have all necessary elements been included (Question, Experimental approach, Results, Conclusion)
- ☐ Is the Abstract concise?

Introduction:
- ☐ Does the Introduction clearly state the overall question of the paper?
- ☐ Does the question follow the unknown?
- ☐ Are all elements (known, unknown, question, experimental approach) clearly signaled?
- ☐ Has the topic been reviewed? _____

Materials and Methods:
- ☐ Have all experiments been described adequately?

Results:
- ☐ Is the main finding presented?
- ☐ Are there errors in factual information, logic, analysis, statistics, or mathematics?
- ☐ Are all figures and tables explained sufficiently?

Discussion:
- ☐ Is the overall interpretation of the results clearly stated?
- ☐ Did the writer adequately summarize and discuss the topic?
- ☐ Has a clear conclusion been provided?

References

- ☐ Have references been cited where needed?
- ☐ Are sources cited adequately and appropriately?
- ☐ Are all the citations in the text listed in the References section?

FORMAT AND ORGANIZATION

Overall format and organization

- ☐ Is the overall format and organization of the paper clear and effective?
- ☐ Are there unclear portions? _____
- ☐ Could the clarity be improved by changes in the order of the paper? _____

- ☐ Does the language seem appropriate for its intended audience?

Individual sections

Introduction:
- ☐ Does the Introduction follow a funnel structure?

Materials and Methods:
- ☐ Are experiments organized logically?

Results:
- ☐ Is the main finding presented in the first paragraph?
- ☐ Are all figures and tables labeled properly?

Discussion:
- ☐ Is the overall interpretation of the results clearly stated in the first paragraph?
- ☐ Is the significance clearly stated in the last/concluding paragraph?
- ☐ Is the discussion ordered in a way that is logical, clear, and easy to follow?

STYLE AND COMPOSITION

- ☐ Are the transitions between sections and paragraphs logical?
- ☐ Are key words repeated exactly?

☐ Are the paragraphs and sentences cohesive?
☐ Has word location been considered?
☐ Are there any grammar, punctuation, or spelling problems?
    _____

☐ Is the style concise?
☐ Are there any wordy passages? _____
☐ What other problems exist? _____

OVERALL QUALITY

☐ What are the paper's main strengths? _____
☐ What are the paper's main weaknesses? _____
☐ What specific recommendations can you make concerning the revision of this paper?_____

SUMMARY

FIRST REVISION

- Check the first draft for content and content location.
- Check logical organization of sections and subsections.
- Revise for style only after you are satisfied with the content and organization.
- Condense where possible.
- Proofread your manuscript.

SUBSEQUENT REVISIONS

- Let some time elapse between revisions. Count on ≥6 drafts.
- Recheck the first draft for content, logical organization, and style, and revise if needed.
- Give complete copies of revised manuscript to your coauthors.
- Show your manuscript also to a colleague in a related field and a friend in a different discipline.
- Ask for comments and constructive criticism in writing.
- Be prepared to accept the criticism.
- Check for grammatical and other errors before finalizing your manuscript.

REVIEWING A MANUSCRIPT

- When reviewing someone else's manuscript, be as specific as possible and point out both strengths and weaknesses.
- Always treat the author with respect.

CHAPTER 18

Final Version, Submission, and Peer Review

A final version of your manuscript moves you to the next stages in the publication process: (re)submission and peer review.

THIS CHAPTER PROVIDES AN OVERVIEW OF:

- Preparing the final version
- Submitting your manuscript
- Corresponding with the editor
- The journal peer review process
- Resubmission

18.1 GENERAL ADVICE ON THE FINAL VERSION

➤ Remember that first impressions are important

When you have finished revising the paper, make sure that the final versions of the text, tables, and figures are neat and professional. Double-check that you have followed the *Instructions to Authors* exactly (see also Chapter 7, Section 7.2). Pay attention to details such as font (Times Roman, Arial, or Helvetica size 10 or 12 are preferred), margins (usually 1 inch on all sides), line spacing (usually double-spaced), word count, page numbering, and line numbering if needed. Ensure that your sections are in the right order. Last, make sure that the version you submit is really the final version and that it is complete.

18.2 SUBMITTING THE MANUSCRIPT

➤ Submit to only one journal

Send your manuscript to only one journal at a time. Your paper will be considered for publication only if not submitted elsewhere. For advice on how to select a journal for submission of your article, see Chapter 7, Section 7.2 on Audience and Journal Choice.

➤ Follow guidelines, and ask for help if needed

Electronic submissions are standard these days. When submitted electronically, an article can be edited, stored, and distributed easily by the publisher. Electronic submission is also the fastest way to send your manuscript to an editorial office. Instructions for electronic submissions differ from journal to journal. They are usually very detailed and specific and should be followed carefully. If you are unsure how to submit your paper electronically, ask for help from a colleague or administrator who has experience with electronic submissions, particularly when it comes to submitting figures in a specific format and resolution.

When you submit your files, ensure that you are submitting the latest version of the manuscript. Name the file clearly and according to the submission instructions. Some journals require you to submit your manuscript as separate sections—cover letter, abstract, text, figures, tables, and a list of potential reviewers or of those you consider to have competing interests and would like to block from reviewing your manuscript. Check online instructions to determine whether this is the case, and prepare your final version accordingly.

The biggest problem in submitting papers electronically is getting figures into the correct format and resolution. Here, it is especially crucial to follow guidelines and suggestions explicitly from the start so that figures and tables do not have to be remade at the last minute to fit journal specifications. Usually when you submit figures you will receive electronic notification that the figures are in the correct format and have been accepted for submission.

18.3 WRITING A COVER LETTER

➤ Send a well-prepared cover letter or email with the manuscript

If possible, send a brief, well-prepared cover letter or email to the editor to go with the manuscript. In most scientific fields, the primary goal of a cover letter is to let the editor know why your research is novel and why it belongs in that particular journal. Make your case convincingly and quickly. That is, keep the cover letter short and simple while at the same time stating your objectives clearly. Do not list all your achievements to date, and do not list any complicated background information or details of your study. Also, do not ask a well-known scientist to contact the editor on your behalf. Note, however, that some journals have also instructions on what to include in the cover letter, such as the number of words in the text of the paper, the number of figures and tables, and so forth. Follow these instructions.

Prepare your cover letter/email to the editor carefully. It is typically submitted electronically with the manuscript. Find out the name of the editor if you can. Addressing the editor by name will make a good first impression.

Start the body of your letter with an introductory sentence stating the title of your manuscript, and mention that you would like to submit it for publication to that particular scientific journal. Tell the editor why your research makes an important contribution in the second paragraph. You may add a sentence that describes the results and their importance. You may also state why your work will be of interest to the target journal. Keep this paragraph short and at a high reading level. Do not include a complete summary of your research, nor many details. If you are submitting to a high-profile journal covering a wide variety of fields, such as *Science* or *Nature*, then you should argue why your paper would be interesting and relevant for a wide variety of disciplines.

If applicable, mention any special features of your paper. Such features may include prior related publications or conference abstracts, permissions to cite others' work or figures, or large tables and graphs. Be prepared to send files of related publications or conference abstracts. Where relevant, assure the editor that informed consent was obtained in accordance with ethical guidelines or that experimental animals were well treated and cared for. Attach signed copies of any required copyright assignment forms of your target journal.

Some journals allow you to suggest names of possible reviewers and to mention anyone to whom the editor could turn for more information. If allowed to do so, list only names of people who are not close colleagues, students, family, or friends. Editors will automatically exclude these people as reviewers. Include all relevant contact information for each suggested reviewer. You may either list reviewers within the letter itself or you can refer the editor to a separate section you have submitted. You may also be allowed a request for certain people not to review your manuscript. Explain why you make a particular request if possible. For example, explain that it might lead to a conflict of interest if any direct competitors review your work. Although most editors will try to accommodate you, be aware that your request may be ignored. Some journals also limit the number of reviewers that you are allowed to block.

Check the *Instructions to Authors* of your target journal to see if you have to send any additional documentation in your letter to the editor. This documentation may include signed statements of all authors or copyright assignment forms.

Two sample cover letters are shown in Examples 18-1 and 18-2. The first letter is very brief but contains all important information, including the title of the manuscript and the purpose of the letter. It also mentions the fact that the manuscript is an extension of a previous publication, which indirectly indicates that the manuscript advances the field, providing significance to the paper. The letter concludes with a general, positive closing statement.

Example 18-1	Sample cover letter
(omit information in brackets if sending an email)	[Letter head, if available] [Your name] [Name and address of editor] [your address] [Date]
Editor is addressed by name	Dear Dr. Riccardo:
Introductory paragraph states title and purpose	We would like to submit an article titled "Characterization of two Herpes simplex virus type I transcripts" by A. Wagner and J. Klein for consideration of publication in the *Journal of Virology*.
Special feature is mentioned as appropriate	The work reported in this article extends the work described in our earlier article, "Isolation and localization of HSV-1 Mrna abundant prior to viral DNA synthesis" (*J Virol* 2002;10:45–51). For your reference, a copy of this work has been included with the supplemental material.
Concluding sentence	We look forward to hearing whether you can accept this article for publication.

Yours sincerely,

Alfred Wagner, PhD

The letter in Example 18-2 is a bit more detailed. Its first paragraph provides the purpose of the letter as well as the name of the article. The second paragraph is a succinct statement about the contribution the work makes to the field and its expected impact. This letter also shows a sample request on who not to consider as reviewers (note again that requests for whom to include or exclude as a reviewer usually should be done by invitation only; check the *Instructions to Authors*).

Example 18-2	Sample cover letter
(omit information in brackets if sending an email)	[Letter head, if available] [Your name] [Name and address of editor] [your address] [Date]
Introductory paragraph states title and purpose	Dear Dr. Moore:

We would like you to consider the enclosed manuscript "XXX" by A. Piroletti. J. Klein, and U. Wettla for publication in your journal. The manuscript has not been submitted elsewhere, and all authors agree on its content. |

Overview of manuscript and impact are kept at high level	Our manuscript describes the development and testing of a novel vaccine against *Vibrio cholerae*. The study describes an innovative approach to vaccine development and would be of great interest in the medical field. Vaccines for this organism, which is the principle agent of cholera, have been available only on a limited basis. A new vaccine would therefore add much value to the treatment of this disease.
Special request	We would like to ask that Drs. D. Fuller and F. Gruppe, both from ABC University, not be selected as reviewers for this work, as they are in direct competition to our laboratory.
Concluding sentences end on a positive note	Thank you for considering our manuscript for publication in your journal. We are looking forward to hearing from you soon.

With best regards,

William Smith, PhD

18.4   THE REVIEW PROCESS

➤ **Understand the review process and when to expect a decision on your paper**

➤ **Be courteous and professional at all times**

Most journals will send an acknowledgment that the paper has been received. The *Instructions to Authors* or the acknowledgment you receive may say how long it will be until you can expect a decision on the paper. If you have heard nothing two to four weeks after the promised time, inquire whether a decision has been made. If you still do not hear back from the editor when another two weeks have passed after your inquiry, a phone call is not out of place. Be courteous and professional at all times, and do not submit the paper to any other journal until you get a letter of rejection from your target journal.

When the managing editor has received your manuscript, he or she will assign it to two to three anonymous, qualified reviewers. These reviewers will get back to the editor with specific comments and recommendations on the article. The editor then reads your paper and the comments of the reviewers. The editor may also comment on your paper, summarize the major comments of the reviewers, and state which comments should be taken most seriously. Most important, the editor decides whether to accept, accept pending revision, reject with encouragement to resubmit, or outright reject your manuscript.

This peer review, in which experts in the same or a related field evaluate scholarly articles or grants for scientific value, importance, and accuracy, ensures that publications adhere to strict scientific quality standards. For authors, such peer review lends respectability to their work. For editors, peer review informs their publication decision and adds to the reputation of the journal. For readers, peer review provides a measure of protection and confidence about the validity and respectability of the data and its interpretation.

Journals typically have a list of peer reviewers who provide high-quality reviews. Alternatively, editors may use the reference list of a manuscript to identify possible reviewers. Editors may also reach out to researchers they met through networking at conferences who have the necessary subject matter expertise to review a given manuscript.

Some journals allow authors to recommend preferred reviewers and/or to request exclusion of competitors. If allowed to do so, take advantage of this option. It can expedite the review process and, more often than not, works in your favor in terms of outcome for publication (see also Section 18.3 and Example 18-2). Usually, scientific peer review is single blind, that is, the names of reviewers are not known to the authors although those of the authors are revealed to the reviewers.

18.5 INFORMAL PEER REVIEW

➤ Evaluate major components, style, and composition of the manuscript

➤ Provide a constructive critique in writing

How you yourself review a peer's work depends on when in the writing process you are doing the reviewing. Early, pre-submission drafts should be evaluated primarily with respect to major components of the paper, such as the research purpose, the main findings/answer, and the logical overall content and organization of the paper. Drafts should also be reviewed for style and composition as well as for flow. Provide your comments in writing. Verbal comments are easily forgotten or confused. Write your comments either between the text lines or on the margins of the draft. Better yet, use the "Track Changes" option of MS Word because this function will allow you to suggest wording and clearly show the author where and how to revise a document.

The least helpful comment to receive from a peer reviewing your work is "It looks OK to me." To be an effective reviewer, you need to be as specific as possible. Point out particular places in the paper where revision will be helpful. Do not hesitate to note when something is unclear to you, scientifically or in terms of the writing. If you disagree with the author, say so. Not all readers react the same way, and divergent points of view can provide writers with options for revising.

➤ Always treat the author with respect

Always treat the author with respect. Avoid snippy comments such as "So what?" Instead, make constructive suggestions and recommendations on how to improve and strengthen certain passages or on what else to add or omit from the document. If a passage reads well, point out this strength. If an argument is difficult to follow logically or does not make sense, raise objections politely or ask for additional experiments if appropriate.

18.6   FORMAL PEER REVIEW

➤ Point out strengths and weaknesses

➤ Provide feedback on major and minor concerns

If you are asked to review a submitted manuscript for a journal, read over the work carefully. Point out both strengths and weaknesses, and distinguish between major and minor concerns. Provide a list of your impressions and comments for the editor (and author) in a thoughtful, detailed, and constructive written critique. In your critique, comment on the originality of the work. Moreover, remark on the validity of the science, pinpoint scientific errors, and assess the design and methodology used. In addition, comment on the importance of the work and the overall importance of the findings. As appropriate, point out missing or inaccurate references. Last, state whether you recommend the manuscript for publication or not in its current state and after recommended revisions.

18.7 CHECKLIST FOR PEER REVIEW

Content
Purpose and Interpretation

- ☐ Is the overall purpose of the paper and/or central question clear?
- ☐ Does the interpretation of the findings answer the overall question of the paper?
- ☐ Is there sufficient evidence to support the answer?

Overall

- ☐ Does the paper advance the field?
- ☐ Does it provide interesting and important insights into the topic of interest?

<u>**Individual sections**</u>

Title:

☐   Is the title strong?

Abstract:

☐   Does the Abstract adequately summarize the paper?

☐   Have all necessary elements been included (question, experimental/ study approach, results, conclusion)?

Introduction:

☐   Does the Introduction clearly state the overall question of the paper?

☐   Is the topic not reviewed exhaustively?

Materials and Methods:

☐   Have all experiments and observations been described adequately?

☐   Are methods detailed enough that the study can be repeated by another trained scientist?

Results:

☐   Have the main findings been clearly presented?

☐   Are there errors in factual information, logic, analysis, statistics, or mathematics?

☐   Are all figures and tables explained sufficiently?

Discussion:

☐   Is the overall interpretation of the results clearly stated?

☐   Did the writer adequately summarize and discuss the topic?

☐   Is a clear conclusion provided?

<u>**References**</u>

☐   Have references been cited where needed?

☐   Are sources cited adequately, appropriately, and accurately?

Overall Quality

☐   What are the paper's main strengths?

☐   What are the paper's main weaknesses?

☐   What specific recommendations can you make concerning the re

☐   vision of this paper?

18.8   LETTER FROM THE EDITOR

➤ Reply to any letters from the editor as quickly as possible

The answer from the journal of your choice usually includes a letter from the editor and the evaluation and comments from the reviewers. In the letter, the editor will tell you whether your manuscript has been accepted. Reply to any letters from the editor as quickly as possible.

A typical letter from the editor is shown in the following example:

Example 18-3   **Sample letter from the editor**

October 31, 2015
[Name and address]
RE: Journal and reference number: Title of manuscript

Dear Dr. Chu:

Your manuscript has been carefully evaluated by three external reviewers and the editor as an article in *Journal X*. We regret to inform you that the paper is not acceptable for publication in *Journal X* in its present form.

As you will gather from the reviews, the referees identified a number of substantive methodological problems. The editor concurs. Major issues include the absence of sufficient mechanistic insight into the role of ABC, and . . .

The editors would be willing to evaluate a revised version if you feel that you can effectively address the reviewers' concerns and are willing to perform the new experiments required. The paper would be reviewed again, with no assurance of acceptance.

As detailed in the reviewers' critiques, a responsive revision would require XYZ. Our current guidelines allow authors 90 days to complete the revision. Please ensure that your revised manuscript adheres to the Instructions to Authors as outlined at http://xxxxx.shtml. Please include a detailed response to each of the referees' and the editor's comments, providing each comment verbatim followed by your response.

We wish to thank you for having submitted this manuscript to *Journal X*.

Sincerely,

XXX
Editor-in-Chief

COMMENTS TO THE AUTHORS
Reviewer #1
The study by XX and co-workers examines the role of ABC. The authors report that . . .

General comment:
The study extends the group's recently published work on ABC. The findings are novel and of significant interest. The mechanistic direction of the manuscript needs to be strengthened by. . . . The following major concerns need to be addressed:

Major comments:
1. What are the cellular targets of ABC?
2. What is the cellular origin of ABC *in vivo*?. Is this supported by experimental evidence?
3. Mechanistic dissection of the pathway implicated ABC would significantly strengthen the study.

Minor comments:
Figure 6: Are the scale bars correct?

Reviewer #2

Previous work from the group identified ZZZ. The objective of the current study to determine why ABC is also interesting.

The authors begin by characterizing Y. The authors describe that XXX. However, it remains unclear how ABC is regulated, and the precise source remains obscure. Staining for ABC could help to clarify this point.

The overall strength of the study includes XX. Yet it is unclear if ABC is unique. Tests of ABC *in vivo* would increase confidence in the results reported.

Paper Accepted

➤ Make corrections by the requested date

Papers are seldom accepted without revision. If your paper is one of these rarities, thank the editor briefly. If the paper is returned to you for minor changes and corrections, make the corrections carefully and return the manuscript by the requested date.

Accepted Pending Revision

➤ Be prepared to make editorial changes in the event of acceptance

If the editor requests revisions before the paper is accepted, and if the requested changes are few or minor ones, make the necessary changes if you agree. Make sure that your coauthors approve of the changes. Many journals expect revised manuscripts to be returned within a certain time frame; otherwise, they will be considered new submissions. It is not in your interest to pass the set time frame when returning a manuscript for publication.

In the cover letter you send with the revised version, thank the editor and reviewers for their advice and enclose a list of the changes you have made in response to their suggestions. If you have rejected one or more of the recommendations, justify in the letter to the editor why, but keep rejections of recommendations to an absolute minimum. The editor will then decide whether you have addressed all concerns satisfactorily. The paper may be sent to one or both reviewers again. See also Section 18.6 on Resubmission.

Rejected with Encouragement to Resubmit

➤ Decide if you can and want to make revisions

If the editor rejects your manuscript but encourages you to resubmit after revisions, the revisions requested are probably major ones. Now you have to decide whether you can and will make the changes requested by the reviewers. If you decide to resubmit your manuscript to the same journal, you need to include another letter to the editor explaining how you have addressed the concerns and recommendations of the reviewers (see Section 18.6 on Resubmission). When you resubmit your rejected manuscript, the editor will most likely have it reviewed again, and this review may include the same reviewers as the first one. Sometimes the authors are asked to revise a paper a second time before it is accepted for publication.

If the editor offers nothing more than "further consideration" and requests major changes in his or her answer letter, decide whether the effort of making the changes is worthwhile. If you disagree, submit your manuscript to another journal. A second editor may give you a different opinion.

Rejection

➤ Consider how best to respond to a rejection

If the editor rejects your paper, read the reasons for rejection carefully. Various reasons for rejection can exist. Although the editor may not state them as such, the most common reasons include:

- Lack of originality/too similar to other studies
- Insufficient and unconvincing data
- Impact/significance not compelling
- Poorly organized document
- Buried or missing key points (problem/gap in knowledge unclear; purpose of study not clear, etc.)
- Style problems: too much jargon, too long, too technical
- Credibility killers: spelling errors, grammatical issues, incorrect technical terms, inconsistent format, language issues, etc.
- Incompatibility with the target journal
- Reviewer bias

See also Chapter 15, Section 15.7.

There are different ways you can respond to a rejection: If the editor says the article is outside the scope of the journal for whatever reason, send the article to another journal. If the editor says the article is too long and needs changes, decide whether to make the suggested changes—but, again, submit the revised article to a different journal. The editor would have included an offer to reconsider the article after revision if he or she had wanted to.

If the editor says the reviewers have found serious flaws in the paper, you should probably not resubmit the manuscript to the same journal. You should even consider obtaining more and better information before submitting it elsewhere. If you are sure that the editor and reviewers are wrong, send the paper to another journal or write a short but polite letter saying why you think the paper should be reconsidered. Reserve such appeal letters for extreme cases only, however, such as when a review is flawed. Do not phone the editor.

➤ Do not waste time trying to figure out who your reviewers were

Realize you are not alone. Up to 50% of articles submitted receive an initial rejection. More prestigious journals such as *Science* and *Nature* have even higher rejection rates (80%–90%). Relax. Then, go to work on a revised version for a new journal as soon as possible. The longer you wait, the harder it will be to get started again. Follow the suggestions of the reviewers when you revise your paper. You may get the same reviewers again even when you submit elsewhere.

Overall, do not react too fiercely if you receive a rejection letter from the editor or reviewers. Reviewers and editors can be very helpful, and your paper will usually improve if you take their advice. Writing a furious letter or—worse—arguing on the phone will not get your paper published in that particular journal. It will be better to consider the recommendations you received and to make constructive use of them. Do not give up. Perseverance usually pays off.

Reviewers are often, but not always, anonymous. Their anonymity may annoy you, but do not waste time trying to guess who they are. Your guess is likely to be wrong, and you may feel resentful toward the wrong person for the rest of your career. Rather, spend your energy on revising or re-writing your paper.

18.9   RESUBMISSION

➤ For resubmission, address every comment raised by the editor or by the reviewers

If you have been invited to resubmit your paper after revisions and have decided to do so, ensure that your revised manuscript undergoes the same rigorous editing and revising as the original one (see Chapter 17 and checklists in Chapters 11–16). When you have finished revising the manuscript, write a second letter to the editor and include any responses made to the criticism and comments of the reviewers. This letter must be written very carefully and clearly and must answer *all* of the concerns of the editor and reviewers, although you do not need to make all the changes. Your letter may be the last chance to get your manuscript accepted to this particular journal. You may or may not have the same reviewers as in your original submission. In any case, assume that the editor and the reviewers do not recall your paper in every detail, if at all.

Your response will consist of two parts: a letter to the editor and accompanying responses to the reviewers. Address the second letter to the same editor unless you know that the original editor is no longer there. In the first paragraph, thank the editor and the reviewers for their helpful comments and suggestions and reintroduce the title of your manuscript. Mention any manuscript number or identification number that has been assigned to your paper. In subsequent paragraphs, respond directly to the comments from the editor. In the last paragraph, introduce the more detailed list of your responses to the reviewers' comments, and shortly and politely state that you hope all concerns have been addressed and that you look forward to hearing about a final decision soon.

   Example 18-4   **Sample response letter for resubmission**

[Letter head, if available]

Dear Mr. Moore:

Introductory paragraph; includes title and number as well as a statement of thanks

We are submitting a revised version of our manuscript "ABC" (03-7062). In this revised version, we have addressed the concerns of the editor and the reviewers. We thank you for the helpful comments and suggestions.

Response to editor comments

In response to comments from the editor, in your letter dated August 10, 2008, you suggested to redraw Figure 4 such that the intron of gene A is more easily identified. In the revised version of our paper, Figure 4 has been redrawn according to your recommendations. You also suggested increasing the number of patients tested for the vaccine by at least a factor of 10. We now have analyzed our data with more than 50 times the original number of patients. We have included this new set of results in our revised manuscript (pp. 11–12 and Fig. 4).

Mention detailed responses to reviewers

We have revised the manuscript based on the suggestions and advice of the reviewers. An item-by-item response to their comments is enclosed. We hope that these revisions successfully address their concerns and requirements and that this manuscript will be accepted for publication.

Concluding sentences— letter ends on a positive note

We look forward to hearing from you soon.

With best regards,

William Smith, PhD
Principal investigator

Attach the accompanying responses to the comments of the reviewers to this letter. Your responses should address *every* comment the reviewers have raised. For each of their concerns, explain what specific changes you have made and why. If you disagree with a reviewer's suggestion and do not make a change, it is particularly important to explain why. You usually will have to have a solid reason to disagree and should do so very sparingly. Note that accompanying responses may be quite long.

Example 18-5   Sample response to reviewer comments

Responses to Reviewer 1:

We thank Reviewer 1 for the critical comments and helpful suggestions. We have taken all these comments and suggestions into account, and they have improved our manuscript considerably.

1. Reviewer 1 requested to map the cysteine residues involved in the disulfide linkages and to study the mechanistic significance. We have performed these experiments, and the results are now shown in Fig. 5. As mentioned in the manuscript, the full-length protein A contains 22 extracellular cysteines (p. 13, line 12). We did not attempt to do similar experiments with protein A because of the large number of possible disulfide linkages.

2. The reviewer suggested the citation of Norris and Manley with respect to the C-terminal truncation results. We have added this citation and compared the results (p. 12, line 6ff).

3. The reviewer was concerned about the clarity of describing the deletion series. We have improved this description and added the exact amino acid residues of the deletions in the Methods section (p. 5, line 5).

4. . . .

After you have addressed each comment and suggestion of Reviewer 1, you need to address those of Reviewer 2 in a similar fashion, starting with a polite introduction and then moving on to each individual comment and response. Address every concern of Reviewer 2, even if some of the concerns are identical to those of Reviewer 1. Note that the style of the response letter in Example 18-5 has been changed to show another way to write such responses. Here, the comments of the reviewer are shown in italics, and the corresponding answers of the authors are shown following each reviewer comment. Answers clearly reference where in the document the corresponding change has been made.

👍 **Example 18-6** **Sample response to reviewer comments**

Reviewer 2 also made many helpful comments and suggestions, and we thank this reviewer for them. We have taken all these comments and suggestions into account as follows:

Reviewer's comment

1. *p. 3, line 37. The GenBank accession numbers should be matched with the "isolate" designation in Table 2.*

Author's answer

We have revised the designation of the GenBank accession numbers in the table in the manuscript. Please refer to **Table 2**.

Reviewer's comment

2. *The pathogenic role of X in acute gastroenteritis is uncertain, particularly when rotavirus and norovirus were detected in 18 out of 20 X positive samples. What was the prevalence of rotavirus infection in this sample pool?*

Author's answer

The prevalence of rotavirus infection in this sample pool was 61.5%. We have included this information in the text. Please refer to **The Study** section **page 7, lines 98–99**.

3. . . .

Before you resubmit your manuscript, recheck it as carefully as you did for its original version that was submitted. Make sure the *Instructions to Authors* have been followed meticulously. Recheck especially page numbers, number of words, numbering of figures and tables, and the reference order and list as they may have changed during revision. You may also want to ask someone to read over your rebuttal letter before resubmission. Proofreading by others can help shave off the sharper edges of the document.

18.10 PAPER ACCEPTED

➤ When your manuscript has been accepted, celebrate!

When your manuscript is accepted, read the acceptance letter carefully for instructions on necessary forms and files, format of figures, structuring of highlights, submission of supplementary information, and so forth. Follow these instructions carefully. Additional final editing and copyediting may also be needed before publication (see Chapter 17), and you may be asked to assign copyright to the journal and to order any reprints.

After several weeks or months, the proof (most of the time a PDF file) is sent out to the corresponding author. This proof will look very much the way the paper will appear in the journal. Examine the proof closely and carefully. Look for errors, and make corrections and last-minute edits.

At this point, you should not introduce any new material or make any substantial changes. Also, do not make minor changes of wording or emphasis. Usually, only factual errors such as incorrect data in a table should be corrected now. If you need to make substantial changes, discuss this with the editor first. Return any future proofs by the requested date as quickly as possible, usually within 48 hours. The next time you will see your paper will be in print.

SUMMARY

SUBMISSION GUIDELINES

- Remember that first impressions are important.
- Submit to only one journal.
- Follow guidelines, and ask for help if needed.
- Send a well-prepared cover letter or email with the manuscript.
- Understand the review process and when to expect a decision on your paper.
- Be courteous and professional at all times.
- Reply to any letters from the editor as quickly as possible.
- Make corrections by the requested date.
- Be prepared to make editorial changes in the event of acceptance.
- Decide if you can and want to make revisions.
- Consider how best to respond to a rejection.
- Do not waste time trying to figure out who your reviewers were.
- For resubmission, address every comment raised by the editor or by the reviewers.
- When your manuscript has been accepted, celebrate!

PEER REVIEW GUIDELINES

INFORMAL REVIEW

- Evaluate major components, style, and composition of the manuscript.
- Provide a constructive critique in writing.
- Always treat the author with respect.

FORMAL REVIEW

- Point out strengths and weaknesses.
- Provide feedback on major and minor concerns.

PROBLEMS

PROBLEM 18-1 Cover Letter
Write a cover letter for a manuscript you would like to submit.

PROBLEM 18-2 Resubmission Letter
Write a resubmission letter for your manuscript after receiving pre-submission peer review.

B. REVIEW ARTICLES

CHAPTER 19

Review Articles

Review articles provide an overview of the current state of knowledge, literature, and research on a particular topic. As review articles usually reach a wider scientific audience, they also have a greater impact than peer-reviewed research papers.

THIS CHAPTER DESCRIBES:

- The difference between research papers and review articles
- Different types of review articles
- Selection of source material
- The overall format of a review article and the organization of each of its sections
- Reference management for review articles
- Signals in review articles
- Common problems to avoid in review articles
- Revising a review article

19.1 OVERALL

Review articles summarize what has been published or researched in a specific field. Some review papers also evaluate methods and results. Thus, review articles are not original articles with new data but secondary sources representing a well-balanced summary of a timely subject with reference to the literature. Although literature reviews can also be integral parts of dissertations or master's theses, they are typically stand-alone articles in journals. These articles are usually invited by an editor, although they can also be submitted independently. Most review articles, even if invited, are peer reviewed.

19.2   TYPES OF REVIEWS AND GENERAL CONTENT

Main Review Types

➤ Distinguish between different review articles

Different types of review articles exist. The most common reviews include:

- Literature/narrative review
- Systematic review
- Meta-analysis
- Integrative review

Literature/narrative reviews—the most basic reviews in the scientific literature—examine previously published findings on a subject in the current literature. This type of review may cover a topic comprehensively or more narrowly. The time range covered may also be broad or narrow although most popular are current matters. Topics may include clinical material, case studies, and research findings and are often focused on an area chosen based on availability or author selection. Thus, narrative reviews may contain some author bias.

Systematic reviews examine all the relevant literature on a topic to determine trends in a given research field. They aim to reduce bias by identifying how many studies have used certain research approaches, how they were carried out, and so on. These type of reviews may include a meta-analysis component.

Meta-analyses combine the findings and data from systematic reviews, and the researcher conducts secondary statistical analysis on these to test a hypothesis. A meta-analysis can also determine whether a publication bias exists by evaluating positive and negative studies. Note, though, that not all topics have enough data to permit a meta-analysis.

Integrative reviews summarize prior research findings, draw conclusions, and offer a more comprehensive understanding on a particular topic based on the literature. They assess the strength of scientific evidence, identify gaps, and generate new knowledge about the topic reviewed. These types of reviews can also identify the need for future research.

Aside from these types of review articles, shorter documents such as "perspectives" and/or letters to the editor are some of the most impactful journal pieces published by opinion leaders in the field. Such opinions, visions, and interpretation are also an important part of the scientific literature.

➤ Compare and contrast information to present an argument

Regardless of the type of article, all reviews analyze and evaluate current knowledge in a field. They do not just present a one-sided point of view or argument but compare and contrast the most important findings on a particular topic. They thus provide an overall picture on the specific subject matter or theory to others. Such evaluations provide scientists with the most up-to-date information as well as the history and a critical evaluation of the topic. Review articles in turn, then, are also extremely useful to writers who are composing new research articles or other scientific texts as they provide important overviews and background information on the topic across a field, including any controversies that may exist among different studies. As the author of such an article, you can suggest which side of the conflict seems to be presenting the better arguments. You can also suggest possible next steps or propose a new model.

➤ Make the information understandable to scientists in related fields

It is particularly important that the information in a review article is understandable to scientists in other related fields. Thus, reviews should use simple words and avoid excessive jargon and technical detail. Figures and boxed material should also be of a more general nature, summarizing and generalizing primary source data or highlighting new ideas.

Selecting Source Material

➤ Select a topic of appeal and importance

In writing a review article, it is best to select a subject of appeal and importance. Typically, issues that are of interest to the scientific community and for which sufficient source material exists are easier to cover than subjects that do not hold much interest or for which very little information can be gathered. Thus, issues of well-defined and well-studied areas of research generally give more fruitful topics. Stay away from outdated topics and those of little interest.

Particularly when you are new to writing review articles, it is often helpful to read samples from a variety of journals to find out what topics are "hot," those that generate the most interest from readers. Recent research articles might give you an idea of what is currently of interest in the field. Other ways to identify hot areas of research are through reading editorials and letters to the editor.

There are two main approaches to choosing an area of research to write about in a review article. One approach is to choose a point that you want to make and then select your primary studies based on this area of interest. Another approach is to familiarize yourself with a topic and arrange the material by theme or idea. That is, research the topic starting at general sources and work your way to specific sources.

➤ Look at tertiary sources first to get an idea, secondary to work on an outline, and primary to fill in the outline

Begin by looking at textbooks or Internet sources that are vetted through peer review or have refereed references that back up the statements. These are generally considered to be tertiary sources and can provide a good overview of a topic. Next, narrow your search by reading up on topics that have been summarized in secondary sources such as review articles. For both approaches, use secondary sources to come up with a general skeleton for your review paper. Then, use gathered, specific information from primary sources, which are first-hand accounts of investigations and include journal articles, theses, and reports, to fill in the skeleton of your outline.

For all your sources, keep good records from the beginning, particularly citation information. Remember, you have to cite information and ideas taken from research and publications of others (see also Section 19.9 and Chapter 8). There is nothing more aggravating than having to rediscover where the source material originated from.

19.3 FORMAT

➤ Use the following overall structure:

- Title
- Abstract (indicative)—not always required
- Introduction
- Main Analysis section
- Conclusion and/or recommendations
- Acknowledgments
- References

The standard organization of review articles does not follow the IMRAD (Introduction, Materials and Methods, Results, and Discussion) format. Instead, the organization of review articles usually includes an introduction to the topic, a main section with headings and subheadings, a conclusion with recommendations for further research, and a lengthy reference section. Some review articles also contain an abstract. Check the *Instructions to Authors* of your target journal. Depending on what type of review article you write, your paper could follow a slightly different organization. Note that many journals do not elaborate on how to write a review. In these cases, you may also want to examine several reviews published in your target journal to get an idea of how to organize the material.

When you write a review article, provide focus and direction. Focus your review on any trends, solutions, or unsolved issues. Delineate new models, explain novel perceptions, and suggest new lines of research that may provide or contribute to important answers. Information you select from your references should be able to let you answer not only what is known about the topic but also why the topic is important, what gaps exist, and how they can be filled.

➤ Create an outline and subsections based on gathered information

To compose a logically structured review article, create an outline. If you decide to compose your article without one, at least check the overall organization using a reverse outline during the revision stage. To create a clear outline for your review paper, create subsections based on the information you have gathered from the literature. Give these subsections individual headings and subheadings, and then sort the information you have collected into the various subsections. Use bullet points or whole sentences under each heading or subheading (see also Chapter 7, Section 7.5 for more guidelines on how to construct effective outlines). Subsequently, sort the information under each heading or subheading by similarities, contrasts, gaps in knowledge, and so forth.

In Example 19-1, an outline of a review paper is presented.

Example 19-1   Outline of review article

Title: Aftereffects of Ebola

Abstract

    <u>Background</u>: Ebola is a major health concern—overview sentence
    <u>Problem</u>: survivors continue to experience medical problems
    <u>Topic</u>: medical issues reported
    <u>Overview of report/article content</u>

Introduction

    Ebola—overview of disease outbreak
    Overview of disease progression
    Control measures and fatality rate

Main Analysis

    Statistics on survivors (age, gender, percentage)
    Aftereffects of surviving the disease
        Presence of virus
        General ongoing physical symptoms
        Effect on pregnancy
        Male impotency
        Depression
    Infectivity of survivors
    Treatment options

Conclusion

    Summary of main findings: viral presence, physical symptoms,
        treatment options
    Recommended future disease control measures
    Infectivity projection

Acknowledgments
References

The outline in Example 19-1 funnels from the 2014 Ebola outbreak and background on the disease progression to control measures and fatality rate in the Introduction. For the Main Analysis section, the article then logically connects by discussing statistics of survivors, followed by information on the diverse aftereffects found among survivors, their infectivity, and potential treatment options. The review article concludes by summarizing the main findings, recommending future control measures, and projecting future infectivity.

Example 19-2 presents another outline for a review article.

 Example 19-2   Outline of review article

Title: Species-specific cell recognition: Gamete adhesion of sea urchins

Abstract
> <u>Background</u>: Importance of gamete adhesion in sea urchins—overview sentence
> <u>Problem</u>: Lack of understanding of species specificity
> <u>Topic</u>: Fill in gap for species-specific interactions
> <u>Overview of report/article content</u>: **Novel hypothesis/model of species-specific gamete interaction**

Introduction
> <u>Background</u>:
>> Cell adhesion—overview
>> Cell adhesion and interaction during fertilization—sea urchins
>> Advantages of sea urchin model system
> <u>Problem</u>: Lack of understanding of gamete interactions in different species
> <u>Topic</u>: Fill in gap for species specific interactions
> <u>Overview of content</u>: **New model for gamete interaction**

Main Analysis
> Known macromolecular interactions in sea urchin fertilization
> Reported species specificity in fertilization for diverse sea urchin species
> Bindin protein and sperm adhesion
>> Support for bindin's function
>> Evidence for molecular interactions
>> Bindin's receptor
> **Deducted novel hypothesis: model of interaction between bindin and its receptor**
>> Evidence through deletion mutants
>> Evidence through sequence analysis

Conclusion
> Brief summary of main findings
> **Summary of novel hypothesis**
> Projections to other species

Acknowledgments
References

In this outline, the authors funnel from a general overview of cell adhesion to specific gamete interactions in sea urchin fertilization in the Introduction. For the Main Analysis section, the article is logically organized into discussing macromolecular interactions in sea urchin fertilization, the species specificity of these interactions, and the molecular interactions of sperm adhesion. Importantly, the Main Analysis section ends with a proposed new model for the interaction of the specific sperm and egg adhesive molecules based on the collective evidence from previous studies presented in prior subsections (in bold in Example 19-2). This hypothesis fills the gap in knowledge of the exact molecular interactions during sea urchin fertilization. The review article concludes by summarizing the main findings, restating the new resulting hypothesis, and projecting the possibility of similar interactions during fertilization in other species.

➤ Write iteratively

As you are filling in your outline, reread the source articles to ensure that you have not missed anything. Check also that the information you gathered is accurate, and ensure that you cite all sources (see also Section 19.9 and Chapter 8). Identify additional papers if needed, and re-sort your material again if necessary. Writing a review article is an iterative process. When you are satisfied with your outline, start writing the review article by linking all the ideas under each subheading (see Chapter 7, Section 7.5 on how to compose a document). The following sections provide more details on how to logically organize the Introduction, Main Analysis section, and Conclusion of a review article.

19.4 TITLE

➤ Make the title short but informative

The title of a review paper should be short and clear. Short titles are preferred by most readers as they are more memorable (see Chapter 16). This preference is particularly pronounced for review articles, which are geared toward a wider audience. Ideally, you should keep your title to 30 to 50 characters.

As for a research paper, the title of a review paper should also be informative to let the reader know immediately what the review is about and what main ideas will be discussed. It should contain the top three or four key ideas/key terms of the paper. One option to write a clear title for a review article is to start with the main overall topic followed by a colon and then one or more subtopics that best describe the contents of the review. Three examples of such titles are shown following.

Example 19-3   **Title of review article**

    a   Surviving Ebola: Aftereffects of a deadly disease

   b   Species-specific cell recognition: Gamete adhesion of
          sea urchins

   c   Proteins crystallization: Successful approaches

19.5   ABSTRACT OF A REVIEW ARTICLE

➤ **If an abstract is required, use an indicative abstract**

➤ **Structure the indicative abstract as follows:**

- Background (optional)
- Problem statement (optional)
- Purpose/topic of review
- Overview of content

Not all review articles require an abstract. If they do, their abstracts are usually written in the form of indicative abstracts, providing the reader with a general idea of the contents of the paper. Indicative abstracts differ from informative abstracts. Unlike abstracts for research papers, indicative abstracts are essentially tables of contents in paragraph form. They usually are not self-contained and need to be read together with the text of the article. They may contain some background information and/or a problem statement and should state the purpose or topic of the review. They usually include little if any methods or results. Whereas some review article abstracts end with a statement of significance, interpreting the main findings of a topic for the readers, most end with an overview sentence, listing what will occur in the document as shown in the bolded sentence of Example 19-4.

   Example 19-4   **Indicative abstract**

	This paper describes how plants adapt to desert environments in differing locales. **We outline various ways that allow plants to collect and store water as well as a variety of characteristics that allow them to reduce water loss. We also discuss tolerance, evasion, and water storage/succulence in detail.**
Overview of content	

The next example shows a complete, longer abstract for a review article. This example contains all of the elements of an indicative abstract: background, problem statement, statement of topic, and overview of content.

Example 19-5   **Indicative abstract**

	Aerosols serve as cloud condensation nuclei (CCN) and thus have a substantial effect on cloud properties and the initiation of precipitation. Large concentrations of
Background	

| Unknown/problem | human-made aerosols have been reported to both decrease and increase rainfall as a result of their radiative and CCN activities. At one extreme, pristine tropical clouds with low CCN concentrations rain out too quickly to mature into long-lived clouds. |
| Purpose/topic statement
Overview of content | On the other hand, heavily polluted clouds evaporate much of their water before precipitation can occur, if they can form at all given the reduced surface heating resulting from the aerosol haze layer. We propose a conceptual model that explains this apparent dichotomy. |

(With permission from the American Association for the Advancement of Science)

Following are another two indicative abstracts: the first (Example 19-6) for the outline on Ebola aftereffects presented in Example 19-1, and the second (Example 19-7) on the sea urchin fertilization example outlined in Example 19-2.

Example 19-6   **Indicative abstract**

| Background

Unknown/problem

Purpose/topic statement

Overview of content | Ebola has been a major public health concern, particularly after its outbreak in West Africa in 2014. Although the disease outbreak has largely been brought under control, its fatality rate averaged around 55%. The over 13,000 survivors of the disease, however, appear to continue to carry the virus, which can manifest itself in specific tissues. Survivors also continue to experience medical symptoms ranging from general malaise to depression, miscarriages, and impotency. Survivors may even infect other individuals. Here, we detail these ongoing medical issues and project future treatment options and necessary control measures. |

Example 19-7   **Indicative abstract**

| Background
Unknown/problem

Topic statement

Overview of content | Cell adhesion is essential for fertilization in most eukaryotes. Sea urchin gametes serve as an important model system of cell-cell adhesion in fertilization, particularly as only one adhesion protein, bindin, and its receptor appear to be involved. Although multiple reports have been published in the last few decades on specific interspecies gamete adhesions and fertilizations, a detailed understanding of the interspecies-specificity of sea urchin fertilization is lacking. This review analyzes the reported observations of interspecies cell adhesion and fertilizations and presents a hypothesis and novel model for the arising of species specificity among sea urchins. |

19.6   INTRODUCTION OF A REVIEW ARTICLE

➤ **Organize the Introduction:**

- Background
- Unknown or problem
- Purpose/topic of review
- Overview of content

The Introduction of a review article should provide the big picture of the topic and grab the readers' attention. It should present some general background and state the central purpose/topic of the review. It should also make clear why the topic warrants a review.

After discussing the general background and aspects of existing research, present any recent developments and describe what the problems with the existing research are and/or what is unknown. Subsequently, explain the overall purpose of the review article. This statement may be followed by a description of the organizational pattern of the review article. Do not make the introduction longer than one-fifth of the review article.

➤ **Phrase your topic statement carefully**

The statement of purpose/topic of a review article is similar to the question or purpose of a research paper. Your statement of purpose/topic will not necessarily argue for a position or an opinion; rather, it will argue for a particular perspective on the topic. The statement of purpose/topic sets the tone for the rest of the paper and makes the importance of the research area clear. Thus, these sentences present the topic of the entire article, and all paragraphs and sentences in your article relate to them. Sample statements of purpose/topic for literature reviews include those in Example 19-8.

   Example 19-8   **Purpose/topic statement for review papers**

    a   Various practices have been applied to counteract the spread of Africanized bees among European bees.

    b   The development of new antibiotics has become the main focus of several biotech companies as more and more resistance develops.

    c   Mathematical modeling of disease transmission is important to maximize the utility of limited resources.

In the Introduction of your review article, you may state the overall significance of the article. See Chapter 11 for a more detailed discussion of various components of introductions.

Following is an example of an Introduction of a review article.

Example 19-9   **Introduction of a review paper**

	Global climate change is altering the geographic ranges, behaviors, and phenologies of terrestrial, freshwater, and marine species. A warming climate, therefore, appears destined to change the composition and function of marine communities in ways that are complex and not entirely predictable (1–5). Higher temperatures are expected to increase the introduction and establishment of exotic species, thereby changing trophic relationships and homogenizing biotas (6). Because organisms in polar regions are
Background	adapted to the coldest temperatures and most intense seasonality of resource supply on Earth (7), polar species and the communities they comprise are especially at risk from
Unknown/ problem	global warming and the concomitant invasion of species from lower latitudes (8–10).
	Shallow-water, benthic communities in Antarctica (<100-m depth) are unique. Nowhere else do giant pycnogonids, nemerteans, and isopods occur in shallow marine environments, cohabiting with fish that have antifreeze glycoproteins in their blood. An emphasis on brooding and lecithotrophic reproductive strategies (11, 12) and a trend toward gigantism (13) are among the unusual features of the invertebrate fauna. Ecological and evolutionary responses to cold temperature underlie these peculiarities, making the Antarctic bottom fauna particularly vulnerable to climate change. The Antarctic benthos, living at the lower thermal limit to marine life, serves as a natural laboratory for understanding the impacts of climate change on
Background	marine systems in general.
	Recent advances in the physiology, ecology, and evolutionary paleobiology of marine life in Antarctica make it possible to predict the nature of biological invasions facilitated
Purpose/topic statement	by global warming and the likely responses of benthic communities to such invasions. This review draws on paleontology, biogeography, oceanography, physiology, molecular ecology, and community ecology. We explore the climatically driven origin of the peculiar community structure of modern benthic communities in Antarctica and the macroecological consequences of present and future
Overview	global warming.

(With permission from Annual Reviews)

19.7   MAIN ANALYSIS SECTION OF A REVIEW ARTICLE

➤ **Logically organize information within the Main Analysis subsections (similarities, contrasts, gaps in knowledge, etc.)**

It is difficult but critically important to organize the Main Analysis section logically. You will need to summarize, generalize, and organize findings reported in primary sources while also comparing and contrasting them. You may present key original data or your extrapolated conclusions based on the results from the source articles. Avoid introducing new data, however, unless absolutely necessary to make your point. Feel free to show data in figures and/or tables, but do not provide experimental details.

The overall organization of the Main Analysis section should sequentially unfold ideas in a logical order. This order may not be immediately apparent but usually becomes evident as you are organizing the gathered information, composing subsections and paragraphs, and revising your drafts (see also Sections 19.2 and 19.3 for information on selecting sources and deciding on which information to use).

➤ **Organize the Main Analysis section logically into subsections either**

- Chronologically
- Thematically
- Methodologically

Generally, this section of a review article can be organized chronologically, thematically, or methodologically. If you are organizing your article chronologically, you follow a logical timeline based on when relevant source articles were posted. Alternatively, you could examine the sources under the history of the topic. Such an organization would call for subsections according to eras within this history.

In contrast to the chronological presentation of topics, thematic reviews are organized around a topic or issue rather than around the progression of time. For example, as you deal with various levels of evidence pertaining to a question, your Main Analysis could move steadily downward in the level of inquiry from the organism, to the organ, to the cell, to the molecular mechanisms within the cell. Note that your discussion may still follow a chronological timeline, however.

If you are presenting your article using a methodological approach, you need to focus on the methodology presented in your source articles rather than on their scientific content. Accordingly, topics are organized by techniques or by methods or approaches.

You need to arrange the Main Analysis subsections of your review article based on its organizational structure. For a chronological review, subsections should relate to key time periods. For a thematic review, subsections should correspond to relevant themes or ideas, whereas for a methodological review, subsections should be based on description of different methods and approaches.

➤ Consider including other subsections in the Main Analysis as needed

Some reviews contain additional subsections outside those of your organizational structure such as subsections on "Current Status" or "Future Directions." In some instances, you may also merge a section with another or omit one altogether. Similar to the overall structure, what subsections you include in the body may become clear only as the review evolves. Be creative if needed.

In the next example, a partial Main Analysis subsection of a review article is shown.

Example 19-10   **Main Analysis section of a review paper**

Subheading | **The Opposing Effects of Aerosols on Clouds and Precipitation**

Context | With the advent of satellite measurements, it became possible to observe the larger picture of aerosol effects on clouds and precipitation. (We exclude the impacts of ice nuclei aerosols, which are much less understood than the effects of CCN aerosols.) Urban and industrial air pollution plumes were observed to completely suppress precipitation from 2.5-km-deep clouds over Australia (20). Heavy smoke from forest fires was observed to suppress rainfall from 5-km-deep tropical clouds (21, 22). The clouds appeared to regain their precipitation capability when ingesting giant (>1 μm diameter) CCN salt particles from sea spray (23) and salt playas (24). These observations were the impetus for the World Meteorological Organization and the International Union of Geodesy and Geophysics

Problem/ unknown | to mandate an assessment of aerosol impact on precipitation (19). This report concluded that "it is difficult to establish clear causal relationships between aerosols and precipitation and to determine the sign of the precipitation change in a climatological sense. Based on many observations and model simulations the effects of aerosols on clouds are more clearly understood (particularly in ice-free clouds); the effects on precipitation are less clear."

<table>
<tr><td>Proposed
solution
from literature</td><td>A recent National Research Council report that reviewed "radiative forcing of climate change" (25) concluded that the concept of radiative forcing "needs to be extended to account for (1) the vertical structure of radiative forcing, (2) regional variability in radiative forcing, and (3) nonradiative forcing." It recommended "to move beyond simple climate models based entirely on global mean top of the atmosphere radiative forcing and incorporate new global and regional radiative and nonradiative forcing metrics as they become available." We propose such a new metric below.</td></tr>
</table>

(With permission from the American Association for the Advancement of Science)

This subsection has its own heading ("The Opposing Effects of Aerosols on Clouds and Precipitation"), indicating its subject. The section is organized around the topic of aerosol effect on clouds and precipitation. Under the subheading, the authors provide context and state the problem. The problem statement is followed by (a) a proposed solution from the literature and (b) an overview sentence leading into the authors' own opinion on a solution (to follow), completing the logical structure of the subsection.

Following is yet another example of a Main Analysis section of a review paper. This analysis section has been composed in chronological order.

Example 19-11   **Main analysis section of a review paper: Chronological order**

Subheading | **The Evolution of Coined Money**

Context | In economics and social science, numerous theories have been published about the supposed reasons for the invention and consequent use of early money and coinage (Laum 1924, Schaps 2004, Carrier 2005).

. . . The earliest coinage, made of electrum (a naturally occurring alloy of gold and silver), emerged in the seventh century BCE in western Asia Minor. Although abundant literary sources of later date refer to the invention and use of coinage in the Aegean (e.g., Testart 2001, von Reden 2010), archaeological data are limited to a few well-documented finds and hoards. For early electrum coinage, discussion still rests heavily on the hoard excavated in 1904–1905 beneath the temple of Artemis at Ephesos, Turkey, and recent archaeological reassessments of the site (Karwiese 1991, Williams 1993, Muss 2008). This discovery remains the largest and best-stratified assemblage of weighed electrum pieces stamped with

<table>
<tr><td>Historical account</td><td>simple designs, together with unstamped silver nuggets. These finds imply the acceptance of coins and bullion equally as objects of value and suggest the overlapping monetary character of early coinage and valuable metals. The adoption and subsequent rapid spread of coinage did not mark the immediate end of noncoined money; in various parts of the Greek mainland weighed silver continued to be employed as currency, attested both by the written sources (e.g., the laws of Solon) and archaeologically through mixed hoards (Kroll 2008).</td></tr>
</table>

Subheading | **The Diffusion of Coinage in the Greco-Roman World**

| Historical account | Following its adoption in western Asia Minor, the concept of coinage was swiftly transmitted and elaborated through and beyond the Aegean. Thanks to the increasingly extensive data from well-documented excavations, it is now possible to examine how the economic, technological, and symbolic dimensions of coined money were transformed by successive encounters with different ideologies, value systems, and nonmonetary currencies (see, for example, papers in García-Bellido et al. 2011). However, the fact that key Mediterranean communities, notably the Phoenicians, Egyptians, and Romans, did not see fit to adopt coined money until the fourth century BCE demonstrates that this medium did not necessarily fulfill the economic, social, or political requirements of ancient societies, but instead became important within particular networks (von Reden 2010, p. 71). |

(With permission from Annual Reviews)

The subsections of the Main Analysis section shown in Example 19-11 are organized based on the progression of time, that is, chronologically. The authors first provide an overview sentence for context. They then relay chronologically how coinage evolved based on prior findings; how coinage was first adopted, when, and where; and how it eventually started spreading, as well as where it initially did not take root. Note that in this example both the topics of subheadings, as well as the information within the subsections, are organized chronologically. Such organization is easy to arrange for the author and equally easy to follow by the reader.

19.8   CONCLUSION OF A REVIEW ARTICLE

➤ **In the Conclusion section, summarize your topic, generalize any interpretations, and provide some significance**

The Conclusion section is one of the main highlights of a review paper. It recaps your review and your main conclusions, recommendations, and/or speculations.

You need to phrase this section with special care, summarizing and generalizing main lines of arguments and key findings. Discuss what conclusions you have drawn from reviewing the literature, and restate your interpretations. In addition, provide some general significance of the topic and results, and discuss the questions that remain in the area. Although this section is often longer than the Conclusion section of a research paper, try to keep it brief.

Example 19-12 provides an example of a well-written Conclusion section for a review article. It first provides a summary of the main findings in the field, generalizing these findings for the readers ("All of these effects appear destined to . . ."). In the second paragraph of this Conclusion section, the authors state their overall opinion/interpretation of the findings in the field and provide a general recommendation on how to solve the problem ("Global environmental policy must immediately be directed to . . .").

Example 19-12   Conclusion of a review paper

Beginning in the late Eocene, global cooling reduced durophagous predation in Antarctica. Despite some climatic reversals, the post-Eocene cooling trend drove shallow-water, benthic communities to the retrograde, Paleozoic-type structure and function we see today. Now, global warming is facilitating the return of durophagous predators, which are poised to eliminate that anachronistic character and remodernize the Antarctic benthos in shallow-water habitats. Rising sea temperatures should in general act to reduce the mismatch between the development times of invasive larvae and the length of the growing season. Increased survivability of planktotrophic larvae will decrease the selective advantage of brooding and lecithotrophy, increasing the pool of potentially invasive species. Warming temperatures will also increase the scope for more rapid metabolism and should ultimately obviate the adaptive value of gigantism. All of these effects appear destined to amplify the ongoing, worldwide homogenization of marine biotas by reducing the endemic character of the Antarctic fauna.

Summary

The fact that benthic predators are already beginning to invade the Antarctic Peninsula should be taken as an urgent warning. Controlling the discharge of ballast water from ships will be difficult but not impossible. Whether or not humans are the proximal vectors, however, the long-term threat of invasion in Antarctica has its roots in climate change. The Antarctic Treaty cannot control global warming. Global environmental policy must immediately be directed to reducing and reversing anthropogenic emissions of greenhouse gases into the atmosphere if marine life in Antarctica is to survive in something resembling its present form.

Interpretations and recommendations

(With permission from Annual Reviews)

19.9   REFERENCES

➤ Paraphrase and cite ideas and information from others accurately

Like in academic research papers, your interpretation of the available sources must be backed up with evidence. Therefore, keep track of and cite primary and secondary sources where needed (see Chapter 8). Use software such as ReferenceManager or EndNote that can be integrated with Microsoft Word to facilitate formatting and numbering of references. This management is especially important for reviews that can include hundreds of citations.

The type of information you choose should relate directly to the review's focus. That is, you need to select information known about the topic, what problems and questions exist, how they might be addressed, and why the overall topic is important.

In general, you will use many more references for a review article than for a research paper as you will need to cover a topic much more broadly. Consider citing other review articles where possible to keep your references to a minimum. Be aware that review articles sometimes misrepresent information. You should therefore verify information against the original article. Ensure also that every reference in the article is in the Reference List and that every reference in the Reference List is also cited in your article.

Paraphrase ideas, but ensure that you do not misrepresent them (see also Chapter 8, Sections 8.5 and 8.6). Place citations after the idea you are referring to or after the names of the authors rather than in the middle of an idea. In review articles, it is particularly important to be professional and courteous when referring to the ideas of other, even when you disagree with them.

19.10   SIGNALS FOR THE READER

➤ Signal all the necessary elements in a review article

As for research papers (investigative or descriptive), all the parts of the review article should be signaled to the reader. Signals are given at various levels and may consist of subheadings, topic sentences, as well as specific phrases and terms. Examples of such signals are listed in Chapter 11 (Introduction), Chapter 14 (Discussion), and Chapter 15 (Abstract).

19.11   COHERENCE

➤ Use topic sentences and techniques of continuity to tell the story

To ensure that the overall story is clear, use topic sentences and consider the following:

- Word location (Chapter 3 and Chapter 6, Section 6.3)
- Key terms (Chapter 6, Section 6.3)
- Transitions (Chapter 6, Section 6.3)
- Sentence location (Chapter 6, Sections 6.1 and 6.2)

19.12   COMMON PROBLEMS OF REVIEW ARTICLES

The most common problems of review articles include:

- Lack of analysis and commentary—writing the review as a simple list of facts and dates without interpretation and inference
- Not stating the unknown or problem—this leaves the reader hanging and wondering why the review is of interest
- Lack of logical organization of subtopics (Section 19.7)
- Lack of objectivity—article does not show conflicts between research "camps"
- Referencing errors (incorrect or missing citations, incorrect dates or volume; see also Section 19.9 and Chapter 8)

19.13   REVISING THE REVIEW ARTICLE

When you have finished writing the review article (or if you are asked to edit one for a colleague), you can use the following checklist to "dissect" the article systematically:

Overall
☐   1. Does the topic present something of interest to the field?

Individual sections
Title:
☐   2. Is the title short (ideally, 30–50 characters)?
☐   3. Does the title contain the main three or four key terms?
Abstract:
☐   4. Is the topic stated precisely?
☐   5. Is an overview of the article provided?
Introduction:
☐   6. Does the Introduction have the following components?
    ☐   a. Background
    ☐   b. Problem or unknown
    ☐   c. Topic or review
    ☐   d. Overview of content
Main Analysis Section:
☐   7. Is the section logically organized?
☐   8. Did you analyze and interpret all information (rather than simply list facts and dates)?
☐   9. Does your paper present information objectively, including contradictory data and ambiguities?
☐   10. Are all figures and tables explained and labeled sufficiently?
☐   11. Do all the components logically follow each other?
☐   12. Are the transitions between sections and paragraphs logical?
Conclusion and/or Recommendations:
☐   13. Is the topic summarized and interpreted in the Conclusion section?
☐   14. Is the result of your analysis clear?

References:

☐ 15. Have references been cited where needed?

☐ 16. Are sources cited adequately and appropriately?

☐ 17. Are all the citations in the text listed in the References section?

General:

☐ 18. Are all the necessary elements in the review article signaled clearly?

☐ 19. Revise for style and composition based on the basic rules and guidelines of the book:

    ☐ a. Are paragraphs consistent? (Chapter 6, Section 6.2)

    ☐ b. Are paragraphs cohesive? (Chapter 6, Section 6.3)

    ☐ c. Are key terms consistent? (Chapter 6, Section 6.3)

    ☐ d. Are key terms linked? (Chapter 6, Section 6.3)

    ☐ e. Are transitions used and do they make sense? (Chapter 6, Section 6.3)

    ☐ f. Is the action in the verbs? Are nominalizations avoided? (Chapter 4, Section 4.6)

    ☐ g. Did you vary sentence length and use one idea per sentence? (Chapter 4, Section 4.5)

    ☐ h. Are lists parallel? (Chapter 4, Section 4.9)

    ☐ i. Are comparisons written correctly? (Chapter 4, Sections 4.9 and 4.10)

    ☐ j. Have noun clusters been resolved? (Chapter 4, Section 4.7)

    ☐ k. Has word location been considered? (Verb follows subject immediately? Is old, short information at the beginning of the sentence? Is new, long information at the end of the sentence?) (Chapter 3, Section 3.1)

    ☐ l. Have grammar and technical style been considered (person, voice, tense, pronouns, prepositions, articles)? (Chapter 4, Sections 4.1–4.4)

    ☐ m. Has correct tense been used? (Chapter 4, Section 4.4)

    ☐ n. Are words and phrases precise? (Chapter 2, Sections 2.2 and 2.3)

    ☐ o. Are nontechnical words and phrases simple? (Chapter 2, Section 2.2)

    ☐ p. Have unnecessary terms (redundancies, jargon) been reduced? (Chapter 2, Section 2.4)

    ☐ q. Have spelling and punctuation been checked? (Chapter 4, Section 4.11)

SUMMARY

> REVIEW ARTICLE GUIDELINES
> - Distinguish between different review articles.
> - Compare and contrast information to present an argument.
> - Make the information understandable to scientists in related fields.

- Select a topic of appeal and importance.
- Look at tertiary sources first to get an idea, secondary to work on an outline, and primary to fill in the outline.
- Use the following overall structure:
  - Title
  - Abstract (usually optional and indicative)
  - Introduction
  - Main Analysis section
  - Conclusion and/or recommendations
  - Acknowledgments
  - References
- Create an outline and subsections based on gathered information.
- Write iteratively.
- Make the title short but informative.
- If an abstract is required, use an indicative abstract.
- Structure the indicative abstract as follows:
  - Background (optional)
  - Problem statement (optional)
  - Purpose/topic of review
  - Overview of content
- Organize the Introduction:
  - Background
  - Unknown or problem
  - Purpose/topic of review
  - Overview of content
- Phrase your topic statement carefully.
- Logically organize information within the Main Analysis subsections (similarities, contrasts, gaps in knowledge, etc.).
- Organize the Main Analysis section logically into subsections either
  - Chronologically
  - Thematically
  - Methodologically
- Consider including other subsections in the Main Analysis as needed.
- In the Conclusion section, summarize your topic, generalize any interpretations, and provide some significance.
- Paraphrase and cite ideas and information from others accurately.
- Signal all the necessary elements in a review article.
- Use topic sentences and techniques of continuity to tell the story.

PROBLEMS

PROBLEM 19-1 Review Introduction
Identify all the elements of the following review Introduction. Is the Introduction complete?

Mitochondrial genomes differ greatly in size, structural organization, and expression both within and between the kingdoms of eukaryotic organisms. The mitochondrial genomes of higher plants are much larger (200–2,400 kb) and more complex than those of animals (14–42 kb), fungi (18–176 kb), and plastids (120–200 kb) (Refs. 1–4). Although there has been less molecular analysis of the plant mitochondrial genome structure in comparison with the equivalent animal or fungal genomes, the use of a variety of approaches—such as pulsed-field gel electrophoresis (PFGE), moving pictures (movies) during electrophoresis, restriction digestion by rare-cutting enzymes, two-dimensional gel electrophoresis (2DE), and electron microscopy (EM)—has led to substantial recent progress. Here, the implication of these new studies on the understanding of *in vivo* organization and replication of plant mitochondrial genomes is assessed.

(With permission from Elsevier)

PROBLEM 19-2 Abstract
Is the following Abstract that of a research article or that of a review article? Explain why.

Interleukin 1 (IL-1), a cytokine produced by macrophages and various other cell types, plays a major role in the immune response and in inflammatory reactions. IL-1 has been shown to be cytotoxic for tumor cells (Ruggerio and Bagliono, 1987; Smith et al., 1990). To determine the effect of macrophage-derived factors on epithelial tumor cells, we cultured human colon carcinoma cells (T84) and an intestinal epithelial cell line (IEC 18) with purified human IL-1. Microscopic and photometric analysis indicated that IL-1 has a cytotoxic effect on colon cancer cells as well as cytotoxic and growth inhibitory effects on intestinal epithelial cells. Because IL-1 is known to be released during inflammatory reactions, this factor may not only kill tumor cells but also affect normal intestinal cells and may play a role in inflammatory intestinal diseases.

PROBLEM 19-3 Abstract
Is the following Abstract that of a research article or that of a review article? Explain why.

In the last few decades, Africanized honey bees have been spreading throughout South and Central America into the southern states of the US. During this spread, they have Africanized European bees largely through crossbreeding during mating flights, which almost always leads to the more aggressive Africanized bees. Various practices have been applied to counteract this trend. This paper analyzes different practices used to counteract this trend and recommends the optimal approach to ensure pure European honey bees.

PROBLEM 19-4 Review Title

Evaluate the following titles. Are they suitable as a good title for a review paper? Why or why not?

    a) Flying fish: Locomotion in water and air
    b) Spread of Africanized honey bees
    c) Renaturation and crystallization of the recombinant sea urchin protein bindin

PROBLEM 19-5 Conclusion

Is the following Conclusion section appropriate for a review paper? Explain why or why not.

Sea urchin gametes exhibit a high degree of species specificity. However, some interspecific inseminations have been observed among certain sea urchin species. Species specificity has been shown to arise through interactions of external surfaces macromolecules on both eggs and sperm. The sperm cell's primary adhesive substance, bindin, is found in the acrosome, and the cognate ligand on the egg is found in the egg's vitelline layer. Interactions between these macromolecules determine successful gamete recognition, adhesion, and fusion. While the egg ligand has not been studied in detail, analysis of the species specificity of bindin deletion mutants and a comparison of the protein sequences of these mutants among several sea urchin species suggests that the macromolecules function like a lock and key system whereby certain repeated sequences on the bindin molecule determine species specificity.

PROBLEM 19-6 Conclusion

Is the following Conclusion section appropriate for a review paper? Explain why or why not.

Many ectotherm animals hibernate in winter. Here, we reviewed the hibernation of land snails. Their hibernation is triggered by decreasing day length (from 16 to 12 hr), low temperatures, and a decline of the water. During hibernation, snails form an epiphragm to seal off their shell, stop feeding, reduce their oxygen consumption by half, and their heart rate by 90% or more, pending the temperature. We compared the evolution of these characteristics and the underlying differences to the hibernation of insects and vertebrates.

PROBLEM 19-7 Main Analysis Section

For the Conclusion section of (a) Problem 19-5 or (b) Problem 19-6, construct a potential outline of the Main Analysis section you would expect to see in the review.

PART FOUR

Grant Proposals

CHAPTER 20

Proposal Writing

G rant writing is as central to a scientific career, as are writing papers and doing research. Thus, careful targeting of a funder and careful preparation of a proposal can determine success or failure.

THIS CHAPTER DISCUSSES:

- Different types of proposals
- Different sponsoring agencies, including crowdfunding
- Basic sections of a grant proposal
- Preliminary steps to writing a proposal, including letters of inquiry and tips for postdoctoral fellows and junior faculty
- The importance of the first page
- The general proposal format (for detailed sections, see Chapters 22–27)
- Valuable online resources
- Outside help
- Interacting with a funder

20.1 GENERAL

➤ Know how to navigate and compose grant proposals

One of the greatest challenges for a scientist is finding sources of money to allow for research, salaries, supplies, travel, innovation, and technology. Applying for funding has become important for people at all levels of academia and industry, from graduate students and postdoctoral fellows to professors. To be successful and able to run experiments, scientists require funds, and they have to find sources for this money. Without sufficient funding, scientists cannot conduct research and even the best ideas cannot be tested. Thus, grant writing is as central to a scientific career as is writing papers and doing research at the bench. However, many candidates struggle with grant writing and make needless mistakes that impair potentially successful applications.

Common weaknesses include:

- Insufficient preliminary data (i.e., "proof of concept")
- Lack of a clear timeline with measurable milestones
- Overambitious research plans
- Incoherent or unfocused experimental designs
- Poor write-up and presentation of the proposed study

Knowing how to navigate and compose grant proposals increases your chances of positive reviews and of obtaining funding. The best way to learn about what constitutes a successful proposal aside from practicing to write one is to read grant applications that were funded, noting how they are structured, how much detail was included, and which ideas sold. Consider also applying for internal grants first, such as for travel grants or seed grants to build the skills necessary for becoming a successful grant writer. Such applications are particularly valuable if done as part of a grant team. Rejection of a grant proposal application is generally a high probability (sometimes more than 95% will not get funded), but the experience of writing a proposal is worthwhile and will often ultimately result in acceptance with perseverance and persistence.

➤ Familiarize yourself with various funding agencies

Another major stumbling block associated with attaining grants is locating funders. Do your homework. Read up on the various funding agencies, and carefully review individual grant programs. Then, decide which funders and programs align most closely with your research objectives. Funding agencies sponsor grant programs for various reasons.

➤ Understand that funders have goals and priorities

Before developing a grant proposal, it is vitally important to understand the goals of the funder and particular grant programs. Familiarize yourself with proposals the sponsor has funded previously to get to know the type of research and writing style that is expected.

If your proposed study does not fall within the grant maker's interest areas, the grant maker may not fund your work. Discuss your ideas with colleagues and with program officers at the funding agencies. Be prepared to adjust your original idea to also meet the funder's goals and to give your application a chance at being funded.

Aside from identifying the agenda and priorities of the various funders, you also need to find out the level of desired innovation and the respective risk tolerance (is the funder likely to fund riskier research in which outcome is uncertain?). Again, read previous proposals that the sponsor has funded. Take advantage of seminars or online guidance the agency offers, and if possible, talk to a program officer at the funding agency. Most are more than happy to give feedback and advice. (See also Section 20.7, subsection "Research.")

The success of grant proposals depends on six factors:

1. The sponsoring organization
2. The innovative nature or critical importance of the proposed project
3. The competition level
4. The skills of the grant writer in building a compelling case
5. Good luck (with reviewers, availability of funding, etc.)
6. Good timing (in terms of whether the funder and/or society are ready for the idea or approach)

Aside from these factors, in the past few years interdisciplinary proposals have become of more interest to many funders. Thus, view your research in a broader context and consider such collaborations.

Before you submit a proposal, or start writing one, find out whether there is an internal competition for this application. Many funders accept only a limited number of applicants per institution and year. In these cases, organizations usually decide which projects have priority through internal competitions.

➤ Be aware of the main questions of funders. Address them clearly

Because reviewers are busy people who are faced with many more requests than they can grant, make it easy for them to find the answers to the following main questions:

- What do you propose to do?
- Why is it important?
- Who will do this work?
- How long will it take?
- What does it cost?

20.2   TYPES OF PROPOSALS

There are various types of proposals:

- Solicited proposals—in response to a specific request for proposals (RFP) by a funder
- Unsolicited proposals—no solicitation but you believe the funder will likely be interested
- Preproposals—based on an abbreviated version of your proposal, the funder decides if a full proposal is warranted

- Continuation or noncompeting proposal—continued funding is contingent on whether initial progress is satisfactory
- Renewal (usually competitive)—request for continued support of an existing project

20.3   CHOOSING A SPONSORING AGENCY

➤ Keep a list of funders

When you are looking for funding for a project, there are various options. Broadly, funding sources can be divided into governmental and private sectors. The obvious funding agencies are federal agencies—private funders might not be so evident.

To have potential funders available at a glance, make a list and add to it throughout your career. When looking for private funders, start with networking. Check the Acknowledgments section of all those related research articles piling up on your desk to see who is funding those projects. Colleagues whose work resembles yours might also be able to advise you about potential sources. Contact your institution's program officers at the Office of Sponsored Programs (also called Office of Development, Corporation and Foundation Relations, or Office of Research) to ask for a list of funding sources. Program officers and/or proposal coordinators may be able to provide a list of private foundations and corporations that support scientific research.

Recently, another funding source has arisen through the use of the Internet: crowdfunding. Crowdfunders such as Kickstarter, GoFundMe, and Crowdrise have been successful in garnering support of scientific research. This emerging tool is becoming effective in supporting innovative projects and is employed by many institutions, including research universities.

20.4   FEDERAL AGENCIES

➤ Understand federal funding agencies and their instructions

The most obvious sources for scientists to obtain funding are federal agencies. If you are a medical researcher, you usually apply to the National Institutes of Health (NIH). If you are a nonmedical scientist or are looking for funding for education in nonmedical fields, you apply to the National Science Foundation (NSF). Other federal agencies for engineering, for example, include the Department of Energy (DOE) and the National Aeronautics and Space Administration (NASA). One of the largest funding agencies in Europe is the European Research Council (ERC), but individual countries also have federal and state agencies that distribute funding. Similar support exists in other countries and regions. If you expect to get tenure, you most likely will need to secure funding from one of the larger, more prominent agencies.

The NIH consists of several institutes with different missions, whereas the NSF organizes its research and education support through several directorates and offices (Table 20.1).

TABLE 20.1

NIH INSTITUTES

- National Cancer Institute
- National Eye Institute
- National Heart, Lung, and Blood Institute
- National Human Genome Research Institute
- National Institute on Aging
- National Institute on Alcohol Abuse and Alcoholism
- National Institute of Allergy and Infectious Diseases
- National Institute of Arthritis and Musculoskeletal and Skin Diseases
- National Institute of Biomedical Imaging and Bioengineering
- National Institute of Child Health and Human Development
- National Institute on Deafness and Other Communication Disorders
- National Institute of Dental and Craniofacial Research
- National Institute of Diabetes and Digestive and Kidney Diseases
- National Institute on Drug Abuse
- National Institute of Environmental Health Sciences
- National Institute of General Medical Sciences
- National Institute of Mental Health
- National Institute of Neurological Disorders and Stroke
- National Institute of Nursing Research

NSF DIRECTORATES AND OFFICES

- Biological Sciences
- Computer and Information Science and Engineering
- Cyberinfrastructure
- Education and Human Resources
- Engineering
- Geosciences
- Integrative Activities
- International Science and Engineering
- Mathematical and Physical Sciences
- Polar Programs
- Social, Behavioral and Economic Sciences

Note that the NSF places substantial emphasis on education, women, underrepresented minorities, and persons with disabilities in their consideration process.

Detailed instructions and guidelines for applications to the NIH are described in the application for a Public Health Service Grant PHS 398. For NSF solicitors, guidelines are found in the Grant Proposal Guide (see also https://www.fastlane.nsf.gov/fastlane.jsp for the NSF website). Check these guidelines frequently, as they are subject to change.

Federal grant applications tend to be fairly long (12 single-spaced pages for the narrative plus a one-page write-up for the Specific Aims) and have very specific, lengthy instructions. Therefore, when tackling a government submission, begin preparing well in advance of the deadline, break the application into manageable sections, develop a timeline for completing each section, delegate some of the preparation work to colleagues or staff if possible, and carefully review your draft against the selection criteria to ensure that no detail is overlooked. Above all, plan to revise your draft multiple times before submission.

➤ Do not be afraid to contact funding agencies

Feel free to contact these funding agencies. The NIH, for example, has program officers who are there to help. Tell them what you are interested in, or send a one- to two-page white paper, an educational overview that addresses the problems of your topic and how to solve them. Do not view the NIH as a black box. Follow up with the designated federal contact person to determine the estimated ratio of applicants to grantees based on prior rounds, planned deadlines for future grant cycles, the constituents who will serve on the review panel, and the precise definitions of funding criteria. Many federal agencies also conduct workshops for applicants, and online sources and examples are available as well.

➤ Be aware of potential policy changes

Securing any type of funding is difficult, time consuming, and unpredictable, and if you are considering federal agencies, you need to be aware of potential changes, especially in an election year. Presidential policies can impact the budgets of the NIH, NSF, and other federal funders. Government grant making can be political, so obtain support from elected officials and recognize that geographic considerations might limit your competitiveness. Remember also that the application process and bureaucratic obstacles are far more time consuming and costly for government grants than for private foundations.

20.5   PRIVATE FOUNDATIONS

➤ Write a letter of inquiry to a private foundation or corporation before preparing or mailing out a proposal

Although many scientists apply primarily to government agencies, private foundations and corporations can be equally good choices, both in the United States as well as abroad. In fact, in times of low federal support for research, private funders may be preferable. The notion that private sources fund at a much lower level or are more competitive is not correct. In terms of the success rate, you may be much more likely to be funded by a private funder than by government agencies at certain times and for particular projects, and this success rate is independent of the ask amount and length of the proposal.

Although grant proposals for some private funders are just as involved and lengthy as those for federal agencies, for most private funders applications are much less labor intensive, are shorter, and encompass much less red tape, if any. Note that for some private funders, particularly corporate ones, you may also have to provide an official agreement to be signed by both institutions once funding is approved.

Most foundations award funds based on their mission and area of focus. You should apply only to those foundations whose mission aligns with your research objectives. Foundations rarely support anything outside their funding interests. (See also Section 20.7, subsection "Research.")

Note, though, that most foundations do not respond well to unsolicited proposals unless you first send them an informal letter of inquiry (LOI) to which they can respond. The LOI usually is no more than two pages long and presents your proposed study without any ask amount to gauge the interest of the foundation. To make your LOI appealing to a foundation, explain why your project may be of interest to them.

Foundations typically like to fund projects that fall into one of the following categories:

- Urgent local, national, or international issues
- Novel educational approaches
- Proof of principle or low-risk pilot projects with promise for further independent funding
- Sustainable projects
- Projects of potentially high impact
- Innovative projects of importance or with promise of success but not eligible for federal or institutional funding

In your LOI, you may suggest a follow-up call or visit, should the foundation be interested in further discussion. Do not send a full proposal until you know that the foundation is receptive and interested in your work.

➤ Get help from your institute's administration

Note that when you are considering approaching a private funder (foundation or corporation), it is important to get help and guidance from your university's administration, such as the Corporations and Foundations Relations staff or the Office of Research. In some cases, you may even be *required* to work through the university when approaching a potential funder. (See also Section 20.7.)

20.6   CORPORATIONS AND OTHER FUNDERS

➤ Be aware that corporations often give to receive something in return

Aside from federal agencies and private foundations, corporations, individuals, family foundations, and public charities also may sponsor projects. Such funders exist worldwide, and many give internationally.

Corporations often give through a corporate foundation. Corporate foundations derive their grant-making funds primarily from the contributions of a profit-making business. These foundations often maintain close ties with the donor company but are legally separate organizations. Although some board members may be from the corporation, employees are not. In addition, the corporate foundation administers funding independently. These foundations are subject to the same rules and regulations as other private foundations.

Although corporations also award grants directly, they often give to get. What they hope to receive in return is exposure, publicity, community respect, market share, and intellectual property rights or parts thereof. Corporate funding can be a good source of support for new initiatives, special programs, and special events, but often, it may result in publication restrictions as corporations are more interested in producing and protecting the rights to a product rather than in publishing research results.

➤ Keep in mind other potential sponsors, which may include individuals, family foundations, and public charities; often, a direct tie and overlapping interests are needed

Most gifts to nonprofit organizations do not come through corporations or foundations, however. Rather, they come from individuals. To receive funding from an individual, you usually must have direct ties and overlapping interests, such as alumni giving to their alma mater.

Family foundations receive assets from individuals or families. Usually, at least one officer or board member of the foundation is a family member. The family member plays a significant role in governing and/or managing the foundation, often on a voluntary basis and receiving no compensation. Family foundations often fund in their immediate geographic area, with out-of-state exceptions made only for the donor's alma mater or a hospital where a relative received medical treatment, for example.

Public charities include churches or associates of churches, certain educational organizations, hospital and medical research organizations, endowment funds organized and operated in connection with state and municipal colleges and universities, and publicly supported organizations that normally receive a substantial part of their support from a governmental unit or from direct or indirect contributions from the general public. Some public charities make grants, although most provide other services. Note that private foundations usually derive their assets from a single source—such as an individual, family, or corporation—and not from the public. In the sciences and medicine, certain public charities, such as the American Chemical Society or the Polycystic Kidney Disease Foundation, can be important sources of funding, often providing scholarships, fellowships, and grant awards.

There are many useful fundraising resources available on the Internet, including program guidelines, application materials, and reports issued by private foundations. In addition to those Internet sites mentioned under Section 20.8, you may also consult the following:

> **http://foundationcenter.org/findfunders/topfunders/top50giving .html**—lists the 50 largest corporate foundations ranked by total giving (membership is required)
> **http://www.guidestar.org/**—the best source of information on grant-making public charities
> **https://www.devex.com/en/news/top-10-philanthropic-foundations-a-primer-75508**—provides a list of the top 10 international foundations

**https://en.wikipedia.org/wiki/List_of_wealthiest_charitable_
foundations**—contains a list of the wealthiest private foundations
**http://foundationcenter.org/findfunders/topfunders/top100assets
.html**—provides the top 100 US foundations by asset size

20.7   PRELIMINARY STEPS TO WRITING A PROPOSAL

Proposal Format

➤ Know the general proposal format

Note that some foundations and corporations have very few or even no
guidelines on how to write a proposal when you apply there. In these cases,
it will be up to you to decide which sections to include. As a general guideline, know that almost all grants will contain the following basic sections in
some form or other:

1. Abstract
2. Specific Aims
3. Background/Introduction/Statement of Need
4. Significance/Impact
5. Preliminary Results/Innovation
6. Research Design/Research Plan/Methodology/Implementation/Proposed
   Methods and Procedures/Approach/Strategy
7. Alternate Strategy/Technical Problems and Alternate Routes
8. References
9. Budget
10. Personnel/Biographical Sketches

Depending on the funder, your presentation preference, and the topic, some
of these sections may be combined; others may be split into two or more
subsections or may have differently named headings. All these basic sections are described in more detail in Chapters 21–27. Note that some funders
will ask for additional special sections, which are discussed in Chapter 27.

In general, federal agencies will require you to provide much more detailed proposals than private foundations or corporations. However, proposal details required by some private foundations may even exceed those
requested by federal agencies.

➤ Obtain and strictly follow the proposal guidelines

Do your homework. Most funders provide detailed instructions or guidelines regarding the format, content, length, and font size of a proposal.
These guidelines can be downloaded and should be studied carefully before
you begin writing the draft. Follow the guidelines! Funders post these
guidelines for a variety of reasons; sometimes it is out of respect for the time
of their reviewers who commonly review proposals without pay and out of
commitment to their field of study. A uniform format also makes it easier to

compare proposals and provides an easier means for program officers to gauge whether a proposal contains the necessary information and detail needed for evaluation. Enforcement of the rules is commonly Draconian, and many proposals are discarded without review because they fail to adhere to guidelines.

If you are given guidelines, provide the required information in the order listed, and answer all questions completely. Also, plan to start early— submission deadlines are usually *absolute*.

Format requirements are typically

- 8½" by 11" paper
- Single-spaced with all margins measuring at least 1"
- Times New Roman 12-point font

Most proposals are judged on content and presentation. Therefore, do not just pay attention to the content of your proposal but also to the presentation. Make it neat, professional, and organized.

➤ Consult administrators (department chair or dean and proposal coordinators)

Some universities have extensive resources you can tap into when looking for funding, such as postings of requests for proposals (RFPs), writing assistance, budget preparation, and application advice. In other universities, you are expected to do much, if not most, on your own.

When you are preparing a proposal, get all the professional help you can. Knowing about potential pitfalls and problems early on will help you move through the process much more smoothly and painlessly. Aside from letting your department chair (or dean) know about your plans, involve other professionals at your university early on. Development officers may be familiar with the agency you are planning to approach and may be able to advise on the proposal, the application, and navigation through the university system. If the proposal includes a budget, you may also be required to contact the office responsible for administering grants and contracts for your institution before you send out your proposal—the best route is usually through your department/program's business office.

Discuss your proposal plans with the chair or dean of the department. This is particularly important when the proposed work will affect your professional duties or administrative work. Such plans need to be approved, sometimes even at the provostial level.

Development officers and/or proposal coordinators are a general source of help for the whole process of planning and writing the proposal, as are professional grant writers and editors. Proposal coordinators can review the proposal for completeness, ensure compliance with all requirements, and raise pertinent questions that must be resolved before the proposal will be approved for submission such as human subjects review, the use of animals, potential conflicts of interest, equipment purchase, biological hazards, proprietary material, cost sharing, intellectual property, and many other matters. If you are a student or postdoctoral fellow, work closely with faculty and advisors in developing proposals.

Research

➤ Research the funder

In-depth research of targeted funders is essential for success. Aside from preparing a list of potential funders and obtaining the instructions for writing the proposal, you need to discover other important details about the funder. Every hour invested in research increases your chances of success.

What should you look for when trying to identify potential funding agencies? Most important, you need to ensure a match between your interests and those of the agency. Study the grant maker's priorities to decide on the most competitive project for a given grant maker. Ensure that you are applying for a program that is of interest to the funder or that falls under the funder's mission or geographic focus. Find out what kinds of projects the funding agency does and does not fund. Websites, annual reports, press releases, and listings of past funded projects all provide a great starting point to determine the type of projects an agency goes for.

Other details you need to know include the following:

- Do not over- or underestimate the required funding for your project, and know the amount you can ask for from the sponsor. If you need more than the sponsor offers on average, check to see whether the amount you are asking for will automatically eliminate you from consideration.
- Find out what the proposal deadline is to ensure that you can meet it. Estimate that it will take you about two to three months to prepare a well-written proposal. Also, take into account the start date of your project. It will take from three to six months for you to find out whether your proposal will be funded.
- Know who will review your proposal. Knowing who your reviewers will be, particularly their training and background, will help you gear your writing level to that specific audience.
- Inquire about indirect costs (or "overhead"). Institutionally administered grants more often than not require overhead or indirect costs to pay the institution back for the services it offers to the researcher (space, electricity, parking, facilities, etc.). The amount funding agencies allow for indirect costs varies from one agency to another (from 0% to more than 70%). The amount you request from the granting agency should be the sum of the direct and indirect costs.
- Find out the name and contact information for a program officer at the agency. Consider contacting this person if you have questions

about the content or format of the proposal you are considering to prepare. Usually, program officers are happy to advise on these and other questions. Do not make a nuisance of yourself, however, by calling the program officer on a daily basis.

Tips for Postdoctoral Fellows and Junior Faculty Members

Start building your personal list of potential funding sources early in your career. Pay particular attention to sources that specifically fund postdoctoral fellows or junior faculty. One such example is the Faculty Early Career Development program (CAREER) offered by the NSF. Funding under this program ranges from $200,000 to $500,000 payable over four to five years.

Take advantage of pilot projects, which are projects that are planned as a test or trial. You may be able to land a smaller amount of funding to test out projects. Although these projects do not come with large amounts of money, they may provide prestige and trigger further funding from other agencies down the road. In addition, many institutions also offer research support to junior faculty. Determine how your institution manages grant proposal applications and their administration. Find out whether you are eligible and what the deadlines are.

Above all, give yourself plenty of time to search for funding sources and to write and revise the proposal. Estimate how long it will take you to write a proposal, then double or triple the time to come close to the actual time you will need to spend. Count on weeks and months; writing and revising a proposal properly within a few days is unrealistic.

20.8   ONLINE RESOURCES

In addition to listings found using general Web search engines, the following specialized philanthropic websites contain profiles, articles, and links related to funders:

> **http://www.grants.gov**—lists federal funding opportunities
>
> **http://grants.library.wisc.edu**—contains information on grant program databases for foundations, corporations, US government departments, agencies, and other federal entities
>
> **http://foundationcenter.org**—a list of private foundations in the United States
>
> **http://www.philanthropy.com**—the *Chronicle of Philanthropy*, the leading fundraising news source
>
> **http://sciencecareers.sciencemag.org/funding**—the journal *Science* and the American Association for the Advancement of Science; lists various funding search engines, including those for international grants and fellowships
>
> **https://www.guidestar.org** or
>
> **http://www.grantsmart.com**—federal 990 tax forms are required of foundations and charities with gross receipts over $25,000

annually. The 990 tax forms provide information on funder assets, expenditures, grant making, and board members. Guidestar, Grantsmart, the Foundation Center, and many state government websites have begun developing online archives of 990 forms.

Additional sources include the following:

- Newsletters, annual reports, and websites of nonprofit organizations similar to your own
- Personal contacts who can help your organization approach a grant maker
- RFP announcements from large private or federal agencies that detail competition instructions and funding guidelines
- Cultivated relationships between your institution and a funder. Many foundations are loyal to grantees; a modest initial gift can lead to an ongoing partnership, with larger grants awarded each year
- Email discussion groups for grant writers. Professionals frequently share successful proposals and exchange insight about funders through these email lists.

20.9   STARTING TO WRITE A GRANT

➤ Make sure the first page is perfect

There is no best way to write a successful proposal, but successful proposals share similar characteristics. Know that your proposal will, with luck, be read by one or two experts in your field. The program officer and many members of the panel or board that judges your proposal against others likely will not be experts. You have to write your proposal for their benefit as well.

When you are ready to write, sketch out a research "mission statement" or overall objective/goal. Then, identify the specific aims that define how you will accomplish this goal. Draft expected project outcomes in measurable terms, and set a realistic timeline. If you have not done so already, conduct appropriate preliminary studies. For young scientists it is particularly important to see if and how this piece of research fits into their overall plans.

The first page of any proposal is the only page that may get read by the funder or reviewers. Thus, pay particular attention when outlining, writing, and revising this page. This page usually contains the Executive Summary or the Abstract and Specific Aims. These have to be perfect. The Summary should act as a stand-alone summary of the entire proposal. Every word in it will count and should therefore be weighed carefully.

Grant Writing

➤ Educate yourself in proposal preparation

The actual writing part of a proposal is not easy. It requires skill and experience, especially in style and composition of technical documents. If you do not have these skills, educate yourself or existing staff in proposal preparation. You can also consider hiring a grant-writing specialist or a fundraising consultant. If you hire a writer or consultant, look for a scientific grant writer or editor. General grant writers might be able to provide input in style and grammar but often lack the analytical mind and the scientific background to aid in overall composition, content, and structure of scientific grants.

Attending a grant-writing class, using this book, or researching online Web sources can help beginners learn the basics of grant writing and allow more seasoned researchers to improve their writing styles. Of course, any designated staff person should already know the fundamentals of clear writing, have an analytical mind and a scientific background, and be able to pay excruciating attention to detail. Above all, you need to be concerned that your proposal is well organized.

Online resources that discuss the standard components of proposals or offer courses on grant writing include the following:

> http://foundationcenter.org/getstarted/learnabout/proposalwriting
> .html—the Foundation Center site offers online advice and training courses on grant writing
>
> http://www.mcf.org/mcf/grant/writing.htm—Writing a Successful Proposal: tips from the Minnesota Council on Foundations offering answers to common questions for proposal writing and submission
>
> http://www.nsf.gov/pubs/1998/nsf9891/nsf9891.htm—the National Science Foundation's valuable guide on how to write proposals for the NSF

20.10 INTERACTING WITH THE FUNDER

➤ Establish a close relationship with funders

Just as it is imperative to create a strong case to sell programs, it is equally essential that you establish and maximize a close relationship with funders and potential funders. The cultivation process often begins when you call a grant maker to request current guidelines. Culture this relationship by making phone calls if warranted and sending progress reports, Christmas cards, and invitations to visit or for special events. The cultivation of a funder should never stop—not even after you have received funding. Be aware that it may take years of cultivation before some funding agencies make a significant contribution to your work.

Maintaining good relations involves good stewardship. Writing personalized cover letters and thank-you letters or emails is an absolute must.

Even if you have been rejected for funding, keep a courteous relationship. There may be other funding opportunities in the future to which you may want to apply, so do not burn any bridges. It may also be advantageous to ask a funder for feedback on a rejected proposal and to use such criticism constructively for new applications.

Aside from good stewardship, report key results and provide timely progress reports as these are often used to justify grant programs or to obtain renewal grants, particularly for federal funders. In addition, if you are an established researcher, you may also volunteer some time for consultation or for reviewing of other people's grants for a foundation.

When you are invited to apply to a funding agency, try to visit or call a program officer to get advice on timelines, proposal format, review process, and budget items. Pending your relationship with the officer and the interest level of the foundation in your project, you may even get feedback on preliminary drafts of your proposal.

SUMMARY

> GRANT WRITING GUIDELINES
> - Know how to navigate and compose grant proposals.
> - Familiarize yourself with various funding agencies.
> - Understand that funders have goals and priorities.
> - Be aware of the main questions of funders. Address them clearly.
> - Keep a list of funders.
> - Understand federal funding agencies and their instructions.
> - Do not be afraid to contact funding agencies.
> - Be aware of potential policy changes.
> - Write a letter of inquiry to a private foundation or corporation before preparing or mailing out a proposal.
> - Get help from your institute's administration.
> - Be aware that corporations often give to receive something in return.
> - Keep in mind other potential sponsors, which may include individuals, family foundations, and public charities; often, a direct tie and overlapping interests are needed.
> - Know the general proposal format.
> - Obtain and strictly follow proposal guidelines.
> - Consult administrators (department chair or dean and proposal coordinators).
> - Research the funder.
> - Make sure the first page is perfect.
> - Educate yourself in proposal preparation.
> - Establish a close relationship with funders.

CHAPTER 21

Letters of Inquiry and Preproposals

A letter of inquiry (LOI; sometimes also called letter of intent) is a short proposal in letter form that may be used instead of full grant submissions by some funders and as a preproposal or screening devices by others. An LOI can make or break your relationship with a foundation.

THIS CHAPTER OUTLINES:

- The importance of LOIs
- The different elements of an LOI
- General outline for writing an LOI
- Composing a cover letter for an LOI
- Revising an LOI or preproposal
- Verbal proposals

21.1 GENERAL

➤ Scrutinize every sentence for detail, clarity, and conciseness

The letter of inquiry is crucially important to securing funding for your project. For foundations, LOIs are a quick way to screen potential candidates for funding. For you, an LOI is a way to get an invitation from the foundation to submit a complete proposal. It is the most critical step to getting one foot in the door.

Before composing an LOI, you must research the foundation's priorities. Your letter must establish a connection between your project's goals and the foundation's philanthropic interests. Respect the funding source's stated preferences for geographic region, type of grant, and program areas. If the funding program is a good fit for your application, you should carefully review the application instructions to find out if an LOI or preproposal is required and what form it should take.

An LOI is much shorter than a full proposal—no more than two to three pages plus the budget. However, it is often more challenging to write a good LOI than it is to write a full proposal. Although the LOI is a miniproposal, do not just chop down your proposal to fit onto three pages. Every single word in your LOI or preproposal needs to be weighed and should be important. There are two secrets to a successful letter of inquiry: condense, condense, condense and edit, edit, edit. With only two to three pages of text, each sentence must be scrutinized when editing. Focus on detail, clarity, and conciseness. The LOI must succinctly but thoroughly present the need or problem, the proposed solution, and your and your organization's qualifications for implementing that solution.

When you need to prepare an LOI or preproposal, consider involving the help of a university representative who maintains liaisons with the funder, particularly someone trained in dealing with foundations and corporations. These professionals are well qualified to provide advice in approaching and coordinating activities with these organizations. They can also assist with budgetary concerns and with the actual writing and editing of proposals and letters.

Check submission requirements. Be aware that many funders ask for electronic submission; others require you to send your LOI or preproposal by regular mail. To submit your LOI by regular mail, address it to the appropriate contact person at the funding agency. Know that a rejection notification is not always sent out to inform you that your project is unsuitable.

21.2 COMPONENTS AND FORMAT

➤ Follow the funder's guidelines *exactly*

➤ Adjust the level of writing to the review board

If the foundation has published guidelines for an LOI or preproposal, follow them *exactly*. Sometimes LOIs or proposals are rejected simply because they do not conform to the required format specified by the funder's guidelines. Although the content and organization of an LOI can vary considerably from funder to funder, the primary challenge is to write clearly and concisely.

If an outline for the LOI or preproposal is not provided by the foundation, the following structural elements offer a starting place:

- Abstract
- Introduction/background
- Statement of need
- Objective and specific aims
- Strategy and goals
- Leadership and organization

- Budget
- Significance/impact

Note that each of these elements does not contain more than one or two paragraphs, as the LOI has to be very short. Also, the sequence of these elements is somewhat flexible depending on your project and organization. For example, instead of a summary or abstract, you may also consider stating your objective or hypothesis first. Follow the objective with specific aims (optional), and then provide a rationale, the approach, and an impact statement. Note that you often do not have to include citations or a list of references.

You may write the LOI separately from an accompanying cover letter or as part of the letter. If you are writing a preproposal, it should always be separate from the cover letter. Label the LOI and preproposal as such, and include a title. It is helpful if you make it plain that you are submitting an LOI or preproposal right from the start so the funder cannot miss it. In addition, in the header for each page, indicate your institution, your last name, and the date.

Adjust the level of writing to the foundation and the review board. Most of the time, you will have to write with the nonexpert in mind. Include facts, concrete verbs, and sentences that show action. Define terms if needed, but do not lecture the reader. Include an explanation of the issue you are addressing and how you will do it. Keep the foundation's interests in mind. You need to address these to sell your idea. Above all, convey confidence in the project and in your ability to carry out the work.

Pay special attention to verbs and their tense. Use the future tense (*will*) to describe anticipated project outcomes.

Consider Example 21-1.

 Example 21-1   If we receive the grant, then we <u>could</u> double our research space.

The conditional sentence structure of Example 21-1 is less authoritative and less certain than the definitive statement shown in the revised example.

 Revised Example 21-1   An X Foundation grant **will** fund 5,000 square feet of a new DNA sequencing center.

Note that using first person (*I* or *we*) and referring to your program as "our program" or your aims as "our aims" rather than "the program" or "the aims" often seems more natural and compassionate.

Generally, your LOI should contain the answers to the following questions:

- Why this project?
- Why you?

- Why at your institution?
- Why this sponsor?
- Why now?

Format your LOI carefully. Rather than filling each page with long blocks of text, aim for section breaks that catch the eye. Consider offering at least one format break per page and perhaps a bold heading or a short list of bulleted items. However, a LOI should be recognizable as a letter—do not include any figures, tables, photographs, or diagrams. Also, do not use color printing, and do not include a cover sheet. Preproposals, on the other hand, may contain an illustration or two.

21.3   ABSTRACT/OVERVIEW

➤ Include the following elements in the abstract:

- Background/general context
- Statement of need/problem
- Objective (and specific aims—optional)
- Approach
- (Funding request)
- Impact/significance

Summarize the proposal in the first paragraph. In the abstract, first provide context through broad background information. Then, state the problem or need followed by your objective, general approach, and overall significance. The abstract may also mention your institution and contain the qualifications of project staff, and possibly a timetable and the amount requested. Such a summary is sometimes also called an *overview* and may be combined with the introduction.

Pay special attention to power positions in your abstract. Put the most effort into writing the first sentence. Write and rewrite it. Start with an important word, and provide general context in a brief sentence. Then, state the problem or need, followed by the overall objective of the project, the specific aims (optional), and your proposed experimental approach or strategy. You may request a specific dollar amount for the proposed project and justify it. Conclude by talking about the expected outcomes and the impact of the proposed work. Remember to focus on philanthropic needs rather than on your or on institutional needs. Even if you seek funding for equipment or laboratory space, emphasize how these will impact services for people in need or allow you to gain new knowledge in the field more rapidly.

A well-written abstract is shown in Example 21-2. This abstract contains all required elements and starts with a strong first sentence. It is logically constructed and presents the individual elements in the order most reviewers expect to find them.

Example 21-2   **Abstract/overview of an LOI**

Background
Problem

> Feelings of connectedness to the fetus and newborn are the foundation of the maternal–infant bonding. Being separated from their babies can be devastating to mothers and can affect the health and development of infants.

Objective

> Here, we propose to evaluate perceptions of maternal–infant bonding and attachment in postpartum women within the context of forced separation of incarcerated women who are separated from their babies at birth. We will interview postpartum women in a hospital prison during the first 3 postpartum days and conduct semistructured interviews on these women over a 7-month period after delivery and forced separation. In addition, we will evaluate the health and development of the infants over the same period. Content analysis and constant comparison methods will be used to analyze the data. Knowledge about the emotional status of incarcerated postpartum mothers who underwent forced separation from their babies and knowledge on their babies' health and well-being will provide policy makers and health care providers with information necessary to make decisions regarding routine forced separation of mothers and newborns.

Strategy

Significance

21.4   INTRODUCTION/BACKGROUND

➤ Focus on highlights in the introduction/background portion

Follow the abstract/overview paragraph with one to two paragraphs of background information, which may also include important preliminary results. However, remember that a successful LOI or preproposal results in an invitation to submit a longer proposal. Save the extensive narrative history for that document, and just focus on the highlights here, as shown in the next example. This example starts with general background to provide context and then lists the problem and offers the hint of a possible solution by generally reporting on some preliminary results.

Example 21-3   **Introduction of an LOI**

Background
Overall problem

> Multi-drug resistant (MDR) TB is of increasing concern in countries with a high burden of HIV (1). Recent reports of MDR TB isolates resistant to several second-line drugs have magnified concerns about the continued transmission and effectiveness of treatment of MDR TB. Furthermore, the worldwide emergence of TB resistant to several second-line drugs (2), coined "extensively" drug-resistant (XDR) TB (3) has hampered advances in HIV treatment. Together, MDR TB and XDR TB account for an increasing proportion of TB cases globally (3). To decrease morbidity and

Preliminary
data for a
solution

Specific
problem

mortality from drug-resistant TB, the evaluation of new laboratory methods and clinical protocols has been priori-
tized (2,3). However, epidemic models of TB transmission
have yet to be integrated with operations research methods to determine the most effective control strategies for
drug-resistant TB in resource-limited settings.

(Alison Galvani, proposal to a federal agency, modified)

21.5   STATEMENT OF NEED

➤ Do not overuse statistics

The statement of need is a key component of an LOI. If it is not included, the
LOI will not be convincing to readers. Readers will not find any important
reason(s) for the proposed project as the need or problem that has to be met
is not laid out clearly.

Example 21-4   Statement of need

Salamander limb deformities detected in California are of
concern to public health experts. It is important to explore
the underlying causes of these deformities to establish preventive measures. An urgently needed next step in the protection of public health is the examination of the X
hypothesis. Investigations of this hypothesis may not only
determine the causative agent of the observed limb deformities but also forgo the danger of potential developmental problems in humans.

If appropriate, the statement of need may provide some statistical data and
examples. However, if you provide statistics to indicate the need for funding, do not overuse them. Most funders are well aware of such numbers if
they have a particular interest in your topic. For foundations that fund projects throughout the United States, focus on how local statistics compare to
national statistics.

21.6   OBJECTIVE AND SPECIFIC AIMS

➤ Clearly identify the overall objective

The overall objective of your LOI explains clearly what your long-term goal
or mission is to accomplish your proposed work. It is the most important
statement of your LOI. Ensure that funders can immediately identify this
statement by either putting it in italics or boldface or by placing it into a
separate section labeled, for example, "Objective" or "Goal." Signal this
statement ("Our overall objective is . . ."; "The goal of this study is to . . .").
The overall objective can be subdivided into specific (short-term) aims,

which can be listed within the abstract, with the objective, or as a separate section. The objective should follow logically from the statement of need or the problem. Two sample objective statements are shown in Example 21-5.

Example 21-5	**Objective**

    a The objective of this proposal is to develop an entirely new approach to X through Y and Z.

    b We propose to use a novel class of experiments to test nearly every theoretical prediction of T-violation that can account for the matter–antimatter asymmetry in the universe.

The specific aims link the overall objective to your research plan. Between two to five specific aims are standard, and these are often listed just after the objective, as shown in Example 21-6.

Example 21-6	**Objective and specific aims**

The proposed study aims to develop treatment strategies to minimize the emergence, amplification, and transmission of drug-resistant tuberculosis in the high-burden setting of Uganda. Specifically, we will (a) perform a meta-analysis of treatment success from five hospitals in the capital and (b) predict best treatment options through mathematical modeling.

21.7   STRATEGY AND GOALS

➤ Avoid excessive details in the experimental approach portion

The strategy/approach paragraph should describe briefly how you plan to address your objective. It should lay out a sound and attainable approach to address the stated problem or need. Do not provide much detail. Instead, provide an overview summary of your proposed methodology. Leave the details for a full proposal if invited. A brief strategy/approach section is shown in Example 21-7.

Example 21-7	**Strategy/approach**

Experimental Methodology

To block parasite transmission in mosquitoes, we will express *Plasmodium* products in the *Anopheles* commensal symbionts using expression system A. This system enables secretion of the expressed effector molecules to interfere with parasite viability in mosquitoes. We will also raise antibodies against this protein to demonstrate increased transmission in its absence by providing it in the infectious blood meal of mosquitoes.

In addition, we will evaluate the role of *Anopheles* symbionts on host reproductive physiology. We will focus specifically on oocyte and larval development. Aside from microscopic observations of the status of progeny development in the fertilized normal and aposymbiotic females, we will evaluate the expression of molecular markers associated with important genes in these processes. Our goal is to understand at the molecular level the basis of reproductive sterility that arises in the absence of the symbiotic flora. If the basis of this sterility is nutritional, supplementation of the female diet can rescue fertility and allow for the paratransgenic mosquitoes developed in the first year of the project to be fertile.

21.8   LEADERSHIP AND ORGANIZATION

➤ Describe any expertise briefly, including location if needed

Briefly describe the project's leadership and their expertise as well as that of other key personnel. The description of your team should be concise and focus on the ability to meet the stated need. Depending on the nature of your proposed project, you might provide a few biographical sentences about the principal investigator(s), research scientists, postdoctoral fellows, graduate students, or key board members to add credibility. You may also want to provide a very brief description of your organization and explain why your institution is the perfect place to conduct the proposed experiments. In addition, consider listing partners and collaborators if needed.

Following are some examples of personnel descriptions.

Example 21-8   Short bio/expertise of investigator

Dr. Paul Doe, Arthus Foundation Investigator and Chairman of the Department of Pathology, is a renowned expert on infectious diseases and has been instrumental in advancing genomics at Albert Einstein University. By coupling characterization of hundreds of pathogens from around the world with genomics, his group has mapped over 50 genes and has identified functional mutations in 28 of these. His findings have provided new insight into the mechanisms underlying HIV, malaria, West Nile virus, and the flu. These studies have identified new targets and pathways for development of novel therapeutic approaches to these diseases.

Example 21-9   Description of key personnel

a   **Key personnel:** The proposed project will be led by Martin Brown (PhD, Associate Professor of Geology and Geophysics) and Ron Robinson (PhD, Professor of

Applied Physics). Martin Brown's contributions to the project will build on his ground-breaking work in XYZ. Ron Robinson brings expertise in the design of Y. Both principal investigators have been collaborating on Z for a number of years.

b   **Key personnel:** Key personnel on the project include John Smith, a postdoctoral fellow working with Dr. Meng. Dr. Smith received his PhD and postdoctoral training in the study of the molecular systematics of medical microbes. He has expertise in a number of analytical methods including bioinformatics tools and will apply his skills to epidemiological modeling in the proposed research.

21.9   BUDGET

➤ State how the funding will be used

Briefly summarize the proposed use of grant monies. Request a specific dollar amount for the proposed project, and justify what the funds will be used for and how they will be distributed. Be explicit about committed funding and pending proposals if needed. Remember that foundations sometimes consult other philanthropic entities before making grant decisions. You may also want to point out the likelihood or difficulty of obtaining funding from other sources. Do not send a detailed budget; you can include a general, high-level one in the LOI even though you may need to prepare a detailed one for internal purposes of determining the funding request necessary to run a project.

The next three examples provide sample wording for a budget section of an LOI or preproposal.

Example 21-10   Funding request

The funding requested for this study will cover the salary of a full-time research scientist who will perform all the proposed experiments under the guidance of the Principal Investigator. The amount requested will also pay for all the reagents such as monoclonal antibodies, serum, plasticware for tissue culture, use of the flow cytometry facility, and chemical reagents.

Example 21-11   Funding request

We are requesting $150,000 to carry out the proposed study. This amount will cover the salary of a postdoctoral fellow, equipment, and travel to South America to collect data and specimens.

Example 21-12   Budget

Budget: The total project cost is $xxx,xxx, of which
$yyy,yyy is requested from the ABC Foundation and
$zzz,zzz is institutional support. The Foundation funds
requested are allocated as follows: Personnel, xx%; Equipment, xx%; and Operations, xx%.

21.10 IMPACT AND SIGNIFICANCE

➤ End the LOI or preproposal with a broad impact statement

In the last paragraph of your LOI, state the impact and significance of the
project if funding is provided as requested. Focus on how society will benefit rather than on construction, staff, or institutional issues. You may also
mention evaluation plans such as measurable objectives and reporting to
stakeholders. Examples 21-13 and 21-14 show how typical impact statements are worded. Pay particular attention to the last sentence. It should
relate the importance of your study as broadly as possible.

Example 21-13   Impact statement

Understanding the interaction between HCV virions and
the extracellular milieu is critical for combating this virus.
These interactions can be illuminated by defining the biochemical composition of the virus particles, elucidating the
mechanism of their assembly and maturation, and determining the differences in the entry of high-infectivity, low-
density particles to those of low-infectivity, high-density
particles. Insights into these mechanisms will contribute to
developing new drugs and vaccines to combat this major
human pathogen.

Example 21-14   Impact statement

Novel, bioengineered crops will allow farmers to grow
varieties of plants with enhanced productivity, quality, and
improved ability to adapt and survive when faced with
adverse environmental conditions. This need is particularly
urgent in light of the global food crisis, which has put
almost a billion people at risk of hunger and malnutrition.
By designing new technology and developing varieties
that allow more sustainable farming, the project aims to
provide innovative and sustainable solutions to the problems faced by crop growers worldwide.

21.11 COVER LETTER

➤ Send a cover letter together with your LOI or preproposal

In a separate cover letter to your LOI or preproposal, indicate that you are sending an LOI. Do so in the first paragraph. In the second paragraph, describe your project very briefly—no more than one paragraph—indicating the need, objective, strategy, and impact. In the final paragraph, propose to follow up, provide contact information, and reiterate the project's significance for the foundation.

Unless the foundation's guidelines state that attachments should be included with inquiry letters, resist the temptation to slip newsletters and other promotional material into the envelope. Such material is more likely to be appreciated during the full proposal phase or for a site visit. Trust that the presentation of your project and the clarity of your letter will entice the foundation into requesting additional information. Above all, be courteous and professional.

Example 21-15   Cover letter

Dear Dr. Miller:

We would like to submit the accompanying proposal on *Speciation Progression due to Global Warming* for the ABC Foundation's consideration. The proposed project is highly innovative and very ambitious, and we are very confident that it will bring about far-reaching advances in our understanding of and approach to climate changes.

Under the leadership of Professor X, the project will be overseen by a 10-member scientific advisory board, which will. . . . Although Y University will provide expertise and allocate state-of-the-art equipment and laboratory facilities, additional support is vitally needed, and the boldness of this project precludes funding from government sources. We are therefore turning to the ABC Foundation and its distinguished record of investing in research that expands the frontiers of knowledge.

We look forward to hearing the results of your deliberations and to future opportunities to collaborate in pursuit of our mutual goals.

With best regards,

21.12 VERBAL PROPOSALS

➤ Give verbal proposals a similar outline as LOIs or preproposals

For certain funders, it is not uncommon to give a verbal proposal in the form of a PowerPoint presentation during a meeting or site visit. Such presentations provide great opportunities for questions and answers with the potential sponsor. Often, a concept paper or LOI is prepared together with the verbal proposal, but a formal proposal may only be submitted if the sponsor is interested. Know that an invitation for a formal proposal does not guarantee funding.

Construct any verbal proposal using the same overall outline of an LOI or preproposal. Do not overcrowd your slides. See also Chapter 30 for more information on how to prepare and deliver a clear presentation.

21.13 LOI OUTLINES

Following is a general overall outline for an LOI or preproposal. Use this outline only if the funder does not provide any guidelines.

Outline:

Abstract/Overview (1 paragraph)
- Context/background (optional)—(1 to 4 sentences)
- Long-term mission/objective (and specific aims [optional])—(1 sentence)
- Statement of need (1/2 to 1 sentence)
- [Immediate, short-term goals/specific aims—1 to 3 sentences; instead of here, these may also be listed following the abstract in separate subsection]
- Strategy—(1/2 to 1 sentence)
- Significance/impact of initiative—(1 to 2 sentences)

Introduction/Background (1 to 2 paragraphs)
- Context and problem
- Selected accomplishments/findings/activities/events to date

Statement of Need (may also be folded into Background)
- Problem or unknown

Objective and Specific Aims/Goals
- Planned activities

Strategy/Approach
- Methodology
- Timetable (optional)

Leadership and Organization (as required)
- Key personnel profiles
- Description of your organization (optional)

Budget (Operations, Personnel, Equipment, Travel, etc.)
- Requests
- Other commitments (other funding or institutional commitments)

Impact and Significance
- Expected outcomes
- Significance of outcomes within your field, outside your field, and to humanity

---

21.14 REVISING AN LOI/PREPROPOSAL

When you have finished writing the LOI/preproposal (or if you are asked to edit these sections for a colleague), you can use the following checklist to "dissect" the sections systematically:

☐ 1. Is the proposed topic original?
☐ 2. Are all the components there? To ensure that all necessary components are present, in the margins of the LOI clearly mark the following:

	Abstract/overview
	Introduction/background
	Statement of need
	Objective and specific aims
	Strategy and goals
	Leadership and organization
	Budget/fundraising
	Impact/significance

☐ 3. Is the abstract kept short and within the set limits?
☐ 4. Does the abstract contain the following components?
  • Context/background (optional)
  • Statement of need
  • Objective
  • Strategy
  • Significance/impact of initiative
☐ 5. Is the overall objective stated precisely?
☐ 6. Does the introduction contain the following elements?
  • Context
  • Selected accomplishments/findings/activities/preliminary studies
  • Organization (optional)
☐ 7. Did you follow the funder's instructions on how to write the LOI?
☐ 8. Is your proposed work doable in the time frame given?
☐ 9. Revise for style and composition using the basic rules and guidelines of this book:
  ☐ a. Are paragraphs consistent? (Chapter 6, Section 6.2)
  ☐ b. Are paragraphs cohesive? (Chapter 6, Section 6.3)
  ☐ c. Are key terms consistent? (Chapter 6, Section 6.3)
  ☐ d. Are key terms linked? (Chapter 6, Section 6.3)
  ☐ e. Are transitions used, and do they make sense? (Chapter 6, Section 6.3)
  ☐ f. Is the action in the verbs? Are nominalizations avoided? (Chapter 4, Section 4.6)

☐ g. Did you vary sentence length and use one idea per sentence? (Chapter 4, Section 4.5)

☐ h. Are lists parallel? (Chapter 4, Section 4.9)

☐ i. Are comparisons written correctly? (Chapter 4, Sections 4.9 and 4.10)

☐ j. Have noun clusters been resolved? (Chapter 4, Section 4.7)

☐ k. Has word location been considered? (Verb follows subject immediately? Old, short information is at the beginning of the sentence? New, long information is at the end of the sentence?) (Chapter 3, Section 3.1)

☐ l. Have grammar and technical style been considered (person, voice, tense, pronouns, prepositions, articles)? (Chapter 4, Sections 4.1–4.4)

☐ m. Are words and phrases precise? (Chapter 2, Sections 2.2 and 2.3)

☐ n. Are nontechnical words and phrases simple? (Chapter 2, Section 2.2)

☐ o. Have unnecessary terms (redundancies, jargon) been reduced? (Chapter 2, Section 2.4)

☐ p. Have spelling and punctuation been checked? (Chapter 4, Section 4.11)

SUMMARY

LOI GUIDELINES
- Scrutinize every sentence for detail, clarity, and conciseness.
- Follow the funder's guidelines *exactly*.
- Adjust the level of writing to the review board.
- Include the following elements in the abstract:
  ○ Background/general context
  ○ Statement of need/problem
  ○ Objective (and specific aims—optional)
  ○ Approach
  ○ (Funding request)
  ○ Impact/significance
- Focus on highlights in the introduction/background portion.
- Do not overuse statistics.
- Clearly identify the overall objective.
- Avoid excessive details in the experimental approach portion.
- Describe any expertise briefly, including location if needed.
- State how the funding will be used.
- End the LOI or preproposal with a broad impact statement.
- Send a cover letter together with your LOI or preproposal.
- Give verbal proposals a similar outline as LOIs or preproposals.

PROBLEMS

PROBLEM 21-1 Abstract for an LOI
Write an abstract for an LOI on comprehensive measurements of CO_2 levels in ponds, lakes, rivers, and the ocean in the American Northeast. In this abstract, open with an important sentence or two and then state the problem/need, your overall objective, how you are proposing to solve it specifically, and the impact/significance of the proposed study. Feel free to invent, look up, or search for information on the Internet if needed.

PROBLEM 21-2 Components of LOI Abstract
Identify all components of the following LOI abstract/overview (background, need/problem, objective, specific aims, approach, impact):

The role that soils play in mediating global biogeochemical processes is a significant area of uncertainty in ecosystem ecology. One of the main reasons for this uncertainty is that we have a limited understanding of belowground microbial community structure and how this structure is linked to soil processes. Building upon established theory in soil microbial ecology and ecosystem ecology, we predict that the structure of belowground microbial communities will be a key driver of carbon and nutrient dynamics in terrestrial ecosystems. We propose to test and develop the established theories by combining state-of-the-art DNA-based techniques for microbial community analysis together with stable isotope tracer techniques. By doing so, we expect to advance our conceptual and practical understanding of the fundamental linkages between soil microbial community structure and ecosystem-level carbon and nutrient dynamics.

(Mark Bradford and Noah Frierer, proposal to private foundation)

PROBLEM 21-3 Abstract for an LOI
Rewrite the following structured abstract to an abstract that would fit for an LOI:

Background: Cardiovirus is a common cause of gastroenteritis. For routine vaccination of Chinese infants, a new cardiovirus vaccine has been recommended.

Objective: To evaluate the impact and cost-effectiveness of the Chinese cardiovirus vaccine program using a dynamic model of cardiovirus transmission.

Expected outcome: Our analysis will indicate the impact and effectiveness of a rotavirus vaccination program.

Significance: Findings can be used to inform policy makers to prevent rotavirus infection.

PROBLEM 21-4 Impact Statement

Point out what the problem is with the following impact statement.

Funding from the Foundation will not only provide for my postdoctoral fellow but also will result in the publication of two papers over the funding period.

PROBLEM 21-5 Personnel Section

Write a paragraph on personnel in which you will be the principal investigator leading the project.

CHAPTER 22

Proposal Abstracts and Specific Aims

When it comes to grant proposals, the Abstract and Specific Aims section might be the only part some reviewers read. It is therefore absolutely essential that this section is compelling and technically flawless.

THIS CHAPTER COVERS:

- The Abstract and Specific Aims section of grant proposals
- The difference between technical and lay abstracts
- Signals and wording for the different elements of a proposal abstract
- Common mistakes to avoid in the Abstract and Specific Aims section
- Reasons for proposal rejection based on the Abstract
- Revising the Abstract and Specific Aims section
- Several technical and lay proposal abstract examples

22.1   OVERALL

Most proposals open with an Abstract and Specific Aims section. The section may be written as one paragraph or as two separate parts, which together are about one page. The section(s) should be self-contained, provide a broad overview of the proposal, and be general in nature. Most reviewers will read your proposal only if your Abstract and Specific Aims section interests them. This section is therefore the single most important section of a grant proposal.

22.2 PROPOSAL ABSTRACTS

Components

> ➤ **For the Abstract and Specific Aims section include:**

- Abstract
  - ○ Brief background
  - ○ Unknown or problem
  - ○ Objective
  - ○ Preliminary results—if relevant
  - ○ General strategy (may include specific aims if not listed in separate part)
  - ○ Expected outcomes—optional
  - ○ Significance/impact
- Specific Aims

The Abstract articulates the highlights from each section of the proposal. It includes all of the main information covered in the proposal (background, unknown or problem, overall objective, general strategy, and significance/impact) in a single paragraph. Open the Abstract with a short portion of background information that provides broad context for the reviewers. Follow this information with a brief statement of what is unknown or a problem and then with the overall objective of the proposal. After the overall objective, consider providing some preliminary results if needed, and then describe your strategy in general terms. End the Abstract by adding a sentence or two about the overall impact of the proposed research. If you have not included the specific aims within your Abstract, add a lead-in sentence to introduce the specific aims after the Abstract, and then list the specific aims with brief explanations. Ideally, do not make the Abstract longer than one to two paragraphs (100–250 words), and when Specific Aims are included, no more than one page in length.

> ➤ **The first sentences should be informative, short, interesting, and provide broad background**

The first sentence of the Abstract deserves particular attention. Next to the title, it will be the first sentence the reviewers will read. Therefore, the first sentence needs to be perfect. It should provide general background, be informative and catchy, and be interesting to the reader at the same time. Do not state a cliché, however. Write and rewrite this sentence, and remember that shorter sentences are more powerful than long sentences.

The Abstract must be concise, informative, and complete. Write the section with nonspecialists in mind; that is, provide a broad context for the reviewers. Avoid abbreviations, unfamiliar terms, and citations. Do not include or refer to tables or figures. Do not include any references, but be sure to include all the important key terms found in the title because the

Abstract and the title have to correspond to each other. Include your specific aims in the Abstract itself unless they are repeated in a separate section shortly after the Abstract. In all cases, highlight your overall objective or goal of the project and address how your project furthers the goals of the sponsor.

The Central Point

➤ State the objective of your proposal precisely

The most important statement in your proposal is the overall objective or goal of the work. To ensure that reviewers and funders immediately find this statement, consider highlighting it using boldface and/or italics. This objective provides an overview of the entire proposal, that is, the overall goal you plan to achieve in the time frame of the proposal. Every paragraph and sentence in your proposal needs to relate to this objective. The overall objective can be subdivided into specific aims or steps toward the objective. Specific aims are included either within the Abstract, listed directly following it, or placed into an entirely separate section.

The overall objective and specific aims should follow logically from the previous statements of what is known or believed and what is still unknown or problematic. Thus, the objective and specific aims should state the purpose of the proposal one would expect after reading about what is unknown or problematic.

Following are some sample objective statements. Note also the clear signals (in boldface) that introduce the overall objective.

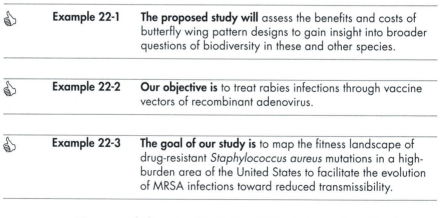

👍	**Example 22-1**	**The proposed study will** assess the benefits and costs of butterfly wing pattern designs to gain insight into broader questions of biodiversity in these and other species.
👍	**Example 22-2**	**Our objective is** to treat rabies infections through vaccine vectors of recombinant adenovirus.
👍	**Example 22-3**	**The goal of our study is** to map the fitness landscape of drug-resistant *Staphylococcus aureus* mutations in a high-burden area of the United States to facilitate the evolution of MRSA infections toward reduced transmissibility.

Many people have trouble distinguishing between the objective, specific aims, and overall impact/significance of a proposal. The schematic in Figure 22.1 lays out the key differences between these entities.

While the objective refers to the three- to five-year goal of a project described in a proposal, specific aims address the two to four steps required to achieve the objective. The impact/significance statement relates to the overall goal of the research over the entirety of a research undertaking,

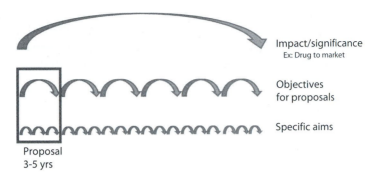

Impact/significance
Ex: Drug to market

Objectives
for proposals

Specific aims

Proposal
3-5 yrs

Figure 22.1 Timeline of research.

Types of Abstracts for Proposals

➤ Distinguish between technical abstracts and those written for a lay audience

The length and format of the Abstract are generally specified by the organization to which you are applying. Usually, abstracts contain between 100 and 250 words. You should not exceed the maximum allowed, but you may certainly summarize your proposal in fewer words.

In terms of format, proposals may contain two types of abstracts: technical abstracts and those written for a lay audience. Some funding institutions may require you to provide both. The components of the Abstract are the same for technical and lay abstracts. It is the level of sophistication of the writing that distinguishes the two. In other words, one is written at the level of a technical or peer audience and the other at the level of an educated lay audience. Accordingly, abstracts used in typical proposals to federal agencies are technical abstracts, as peer reviewers are the primary audience. Abstracts for private foundations and corporations can also be technical, but most of these abstracts need to be geared more toward an educated lay audience, especially if the decision makers or members of the board are nonscientists or scientists outside the field of expertise. As a point of reference, you will need to gear the level of your lay abstract to the person on the board who is least familiar with the topic of your proposal. (Note that proposal abstracts are not the same as abstracts for research papers or review articles; the latter are discussed in Chapters 15 and 19, respectively.)

As proposal lay abstracts may not only have to be presented to board members but may also be posted on websites, these types of abstracts need to be understandable by a wide audience. An example of a technical versus a lay abstract is shown in Examples 22-4a and 22-4b.

Example 22-4a   Technical abstract with integrated specific aims

Background

Problem

Objective

Preliminary
results and
hypothesis

Strategy/
specific aims

Significance

Most people over the age of 35 years exhibit emphysema, a major manifestation of chronic obstructive pulmonary disease (COPD). Cigarette smoke, pollutants, and gender are thought to be important determinants of the severity of the disorder. Curative therapies or reliable diagnostic bio-markers do not yet exist for emphysema/COPD. **Our objective is to identify new diagnostic or therapeutic targets for emphysema by applying our recent discovery of novel molecules in mouse models to humans.** Aging or cigarette smoke-exposed mice exhibit lung changes that partially mimic human emphysema, and mice deficient in toll-like receptor Z, a canonical receptor for lipopolysaccharides, exhibit an accelerated form of spontaneous, age-induced emphysema. We hypothesize that the synergistic or additive effects of age and smoking on Z function in suscep-tible individuals may explain the pathogenesis and temporal characteristics of emphysema. We have identi-fied two novel molecules regulated by Z, an oxidant-generating enzyme (X) and a protease (Y), and implicated both in the pathogenesis of emphysema in mice. This pro-posal will directly build on and expand our pilot findings. Specifically, we will first confirm the role of Z, X, and Y in the pathogenesis of age-induced and cigarette smoke-induced emphysema and validate their roles as therapeutic targets. Subsequently, we will analyze molecular interactions of these molecules in young and aged people in relation to cigarette smoke exposure, gender, and emphysema/COPD. These studies will provide important insights into the pathophysiologic mechanisms of emphysema, ulti-mately leading to the identification of novel targets for diagnostic or therapeutic interventions.

(Patty Lee, proposal to private foundation, modified)

Example 22-4b   Lay abstract with integrated specific aims

Background

Problem

Objective

Emphysema is a major subset of chronic obstructive lung disease, predicted to reach epidemic proportions by 2020. The condition develops in most people over the age of 35 and can lead to the loss of oxygen exchange, lung enlargement, and if severe, complete respiratory failure. Cigarette smoke, pollutants, and gender are thought to affect the severity of the disorder. Disease-altering treat-ments or reliable diagnostic features that can be used to measure the progress of the disease have not yet been determined. Therefore, **we propose to identify new diag-nostic or therapeutic targets for emphysema by exploring its underlying mechanisms.** Using genetically altered

Preliminary results | mouse models, we have recently discovered two novel molecules involved in the development of lung emphysema, X and Y. We found that a substantial increase in these molecules destroys lung tissue, resulting in emphysema. Interestingly, both molecules are controlled by a specific cell wall structure (receptor Z). We believe that the synergistic or additive effects of age and cigarette smoke on Z's function may explain disease development and characteristics. Analysis of the role of receptor Z, as well

Strategy | as those of X and Y, in age-induced and cigarette smoke-induced emphysema will provide insights into the underlying mechanisms of the disorder and may ultimately lead to the identification of novel targets for diagnostic or thera-

Significance | peutic interventions.

(Patty Lee, proposal to private foundation, modified)

Note the much simpler word choices and more extensive definition of terms in the lay version of the abstract. In contrast, in the technical abstract, technical terms and field-specific terminology are used, and a basic understanding of the field is assumed. In addition, in the technical abstract shown in Example 22-4a, specific aims have been integrated ("Specifically, we will . . ."), providing much more details on the technical approach to the problem at hand than described in the lay abstract. Although the lay abstract is generally easier to follow by nonexperts, it is often harder to write this type of abstract for a scientist than it is to write a technical abstract, where more familiar terminology is used. Writing a lay abstract takes time. Often, authors find it helpful to write a typical technical abstract first and then change it—largely through simpler and nontechnical word choices, as well as shorter sentences—to a lay abstract. A good level to aim for in a lay abstract is to write it such that your parents would understand it.

Following are additional examples of technical and lay proposal abstracts with and without integrated specific aims.

Example 22-5   **Technical abstract—with integrated specific aims**

Background | Plate tectonics distinguishes Earth from other terrestrial planets. Plate tectonics arise due to the convection in the mantle of the Earth, and for plate tectonics to occur, some mechanism must exist to compensate temperature-dependent viscosity. However, it is not well understood

Problem | what this mechanism is. In this proposal, we aim **to approach this long-standing mystery through employing an MCMC algorithm that will (1) systematically explore various sampling strategies and (2) seek the fastest forward**

Objective and specific aims | **model calculation by benchmarking competing Stokes flow solvers, including the pseudo-compressibility method.** To lay a foundation for realistic 3-D applications, all computations will be conducted in the 2-D formulation. Microsoft

Strategy	Windows OS will be the main platform for developing the Monte Carlo code and the flow solver as well as for analyzing and visualizing the results of MCMC simulations. Stokes flow calculations will be done with the PI's ABC server. Understanding the physics of plate-tectonic convection in the Earth's mantle will have profound implications for our understanding of the habitability of a terrestrial planet and the evolution of life.
Significance	

(Jun Korenaga, proposal to private foundation, modified)

The next example also presents a technical abstract, but this time the specific aims portion is set apart from the abstract, directly following the latter. In this combined abstract/specific aims section, the objective and specific aims should be highlighted by setting them off or by boldfacing them as they are the most important portion of this section. Other passages whose signals are often placed in boldface or italics include the strategy of the proposal and the specific aims, as shown in Example 22-6.

Example 22-6   Combined technical abstract and aims

Background	Hepatitis B virus (HBV) causes acute and chronic hepatitis as well as hepatocellular carcinoma in infected individuals. Current therapies for chronic HBV infection are only moderately effective and are limited by severe side effects
Problem	and viral resistance. Thus, there remains a need for new therapies for this serious disease. The most promising approach for a new therapy, and **the objective of this proposal,** is to treat HBV infections through vaccine vectors of
Objective	recombinant vesicular stomatitis virus (VSV). These vectors induce strong protective CD8 T cell and antibody responses to a variety of pathogens and are showing great
Background	promise as therapeutic vaccines. We will examine the hypothesis that recombinant VSV vectors expressing the HBV structural proteins will make effective vaccines for prophylactic and therapeutic vaccination against HBV. **Our general strategy** is to develop an effective therapeutic vaccine and/or an improved prophylactic vaccine that pro-
Strategy	vides long-term immunity in a single dose. Such a vaccine would have the potential to prevent millions of cases of
Significance	HBV-associated carcinoma.
Lead-in sentence	To evaluate this hypothesis, we will carry out three specific aims:

Aim 1. Generation of VSV vaccine vectors. We have previously produced ZZZ. . . . We will generate. . . .
Aim 2. Characterization of the immune response to VSV/ HBV vectors. Our preliminary data indicates that. . . . We will now comprehensively measure. . . . We will use . . . and expect to achieve. . . .

Aim 3. Determination of the efficacy of VSV. We will determine if . . .

(Michael Robek, proposal to federal agency, modified)

The two lay abstracts that follow do not contain detailed specific aims but rather present these as a general strategy.

Example 22-7   Lay abstract

Background	Infections caused by multidrug-resistant bacteria have increased markedly. To control this growing number, the search for new and better antibiotics is more important than ever before. However, only two new classes of chemical antibiotics have been approved in the past 30 years.
Objective	**We propose to develop a new family of designer ABC [antibiotics] that is effective against drug-resistant bacteria found in community and hospital settings.** We will design such antibiotics using ribosomes isolated from wild-type *Staphylococcus aureus* and applying structural analysis,
Strategy	crystallography, as well as computational chemistry techniques. This new family of ABC [antibiotics] will offer a broader spectrum of application in the field of antibiotics
Significance	and reduce ABC [antibiotics] resistance.

Example 22-8   Combined lay abstract and specific aims

Background	Dyslexia is the most common learning disability in children. Children with dyslexia have difficulty translating print into the sounds of spoken language. Otherwise, such children possess normal vision and intelligence and often demonstrate strengths in creative, visual, reasoning, and problem-solving abilities. Yet the needs and strengths of
Problem	these students far too often remain unrecognized, underappreciated, and unsupported throughout their educational training.
Objective	**We propose to organize a conference specifically to educate educators about dyslexia.** The conference will not only offer insights into the latest cutting-edge scientific research findings but also provide a forum
Strategy	for discussion and presentations of successful teaching practices for students who are dyslexic. Our goal is for new knowledge and perspectives disseminated through this conference to transform educational practices for students who are dyslexic, ultimately benefitting individuals
Significance	with dyslexia at all educational levels.

The proposed conference will
• Present the latest, cutting-edge scientific research findings
• Provide a discussion forum in which specific concerns and necessary accommodations for dyslexic students are addressed

Expected
outcomes
(optional)

- Demonstrate optimal strategies for teaching dyslexic students
- Introduce a film on the critical role of accommodations necessary for dyslexic students
- Present a panel of successful dyslexic college students discussing their individual experiences and successful strategies

22.3   SPECIFIC AIMS

➤ List your Specific Aims in active, precise language

If your specific aims are not included in the text portion of the Abstract, they should be listed in a separate section immediately following the proposal Abstract. Such a section would be similar to that shown in Example 22-6, but include the separate heading "Specific Aims."

In the Specific Aims section, you need to identify clearly and concisely what you plan to accomplish. You need to describe the specific questions you intend to answer. The questions or specific aims should be thematically related and fit together to clarify the problem raised. Reviewers will read this section very carefully as it serves as an orientation for the rest of the proposal and provides a detailed approach to your overall goal.

The specific questions or aims link the overall objective to your research plan. Between two to five specific aims are standard. They should be written in a list rather than in paragraph form, and they should be placed in boldface to highlight them. It is best to formulate your specific aims in active, precise language ("To quantify . . ."). Each specific aim may be followed by a brief (one paragraph) description of the proposed approach. This description typically summarizes the preliminary results or rationale, the hypothesis, and the approach to be taken. If a hypothesis statement is included, consider placing it in italics to set it apart. As a final sentence in this narrative, you may also state expected outcomes.

Example 22-9   **Specific Aims**

SPECIFIC AIMS:

We hypothesize that solar variability in energy output affects global temperature and thus climate. We plan to achieve our goal with two distinct specific aims:

Specific Aim 1. Model the interaction between turbulence and magnetic field. This interaction is a crucial element of the models and is currently represented through reasonable but arbitrary assumptions. The 3D code will be thoroughly tested, and solar features required for the complex and time-consuming 3D approach will be determined.

Specific Aim 2. Optimize operating modes for the satellite and develop software for on-board calculations and data analysis. Findings obtained in Aim 1 will be incorporated in modeling the properties of solar variability. To extrapolate implications of this work for the problem of global warming, we will collaborate with climate modelers. We expect to gain further insights into . . .

22.4   SIGNIFICANCE AND IMPACT

➤ State the significance of the study at the end of the Abstract

Stating an impact is particularly important at the end of the Abstract as most readers will scan only the first page and expect to find a statement of overall significance. To add significance to your study, you should state why your proposal and the expected outcomes are important. Do so by expressing what you expect to be the significance or impact of the study. This statement should be broad, explaining why your study is important for your field, for the scientific community, and for society at large. However, do not overstate the significance of the study as this would be as big a vote against your proposal as omitting that information.

Example 22-10   These studies are important because they may result in new approaches to prevent and/or treat a disease that causes high mortality worldwide.

Example 22-11   New knowledge and perspectives gained through this work will provide educators with the critical information to implement needed teaching techniques, ultimately transforming the lives of the next generation.

Example 22-12   The envisioned biocomputing system would dramatically expand the frontiers of computation and multidimensional architectures as well as bioelectronic interfaces and could have wide-ranging applications in diagnostics, chemical and biological sensors, and security.

Although typically only brief impact statements are used at the end of your Abstract and Specific Aims section (or at the end of the overall proposal), a longer impact or significance statement is required for some proposals, such as for many federal proposals (NIH or NSF, for example). some, such as for certain NSF applications, you may also need to indicate educational components and/or outreach activities or even the inclusion of underrepresented minorities in the research or its impact. An example of a longer impact statement containing an educational component is shown in Example 22-13.

Example 22-13   **Impact statement with educational component**

Intellectual Merit: Harnessing and controlling optical force on a chip will lead to the convergence of two important fields—nanophotonics and nanomechanics. Building silicon optomechanics and exploiting optical force on a silicon platform will also bring transformative advances in both photonics and nanoelectromechanical systems (NEMS).

Broader Impact: Translating this seemingly small yet fundamentally important phenomenon into an engineering reality will significantly impact not only the sciences and engineering but also society. Optically driven nanoscale machineries, as described in this proposal, could provide a prime example for the application of fundamental science in today's highly developed, technology-driven world, leading to diverse new applications of technology in a variety of fields.

   The educational component of this career proposal will address new curriculum development in nanotechnology by creating hands-on nanoscience modules. The knowledge gained through this project will be disseminated through a plurality of widely accepted multimedia platforms designed for outreach to the local high schools, area community colleges, and further to the general public.

(Hong Tang, proposal to federal agency [NSF])

22.5   APPLYING BASIC WRITING RULES

➤ Follow basic scientific writing rules and guidelines

To write your Abstract and Specific Aims, follow the basic rules (see Chapters 2–6). Pay particular attention to using simple words and avoiding jargon. Also, avoid noun clusters and abbreviations unless a long term occurs repeatedly in the proposal. Consider waiting to introduce an abbreviation until the Background section. If you choose to use an abbreviation that is not standard, show that you are doing so by including it first as a parenthetical.

   In the Abstract, it is even more important than in the rest of the paper to keep sentences short, dealing with just one topic each and excluding irrelevant points. To provide clear continuity, repeat key terms, use consistent order for details, keep the same point of view in the question and the answer, and use parallel form.

➤ Use past tense for observations and specific conclusions. Use present tense for general rules and established knowledge

The basic guideline is that if a statement is still true, use present tense. For completed actions and observations, use past tense. For anything that has not been done yet and you are proposing to do in the future, use future tense ("will"). Prefer active verbs. (See also Chapter 4, Sections 4.4 and 4.6.)

22.6 SIGNALS FOR THE READER

➤ Signal the unknown, the objective, the aims, the strategy, and the significance

Because Abstracts are usually written as one paragraph, it helps the reader if you signal the different parts of the Abstract. Examples of signals for the Abstract are shown in Table 22.1.

TABLE 22.1 Signals of the Abstract and Specific Aims section

PROBLEM OR UNKNOWN	OBJECTIVE	STRATEGY	SIGNIFICANCE OR IMPACT
. . . has not been determined	Our objective is . . .	We will achieve this goal by . . .	X will . . .
. . . is unclear	Specifically, we will . . .	Specifically, we will . . .	. . . is important for . . .
X is limited by . . .	Our objective is to . . .	. . . by . . .	These results may play a role in . . .
The question remains whether . . .	We propose to . . .	We will . . .	Y can be used to . . . . . . ultimately . . . resulting in . . .
	We will examine the hypothesis that . . .	Our general strategy is to . . .	. . . will provide insights into . . . This study is important because . . .

22.7 COMMON PROBLEMS

The most common problems of Abstracts include:

- Omission of parts
- Excessive length
- Unrealistic aims
- Excessive interdependence of aims

➤ Do not omit any parts of the Abstract

If any of the parts (known, unknown, objective, strategy, or impact) are missing or obscured in the Abstract, the reader may have to reread the Abstract several times because it is difficult to understand. The same problem arises when parts are not signaled.

Consider the following example.

Example 22-14   Technical abstract—separate Specific Aims section

> X is a major human pathogen, which infects over 100 million people per year, leading to high morbidity and mortality. Current therapies for X are expensive, poorly tolerated, and only partially effective in controlling the

pathogen and in limiting disease. Recently, we and others succeeded in establishing a system to grow X in cell culture. These systems will allow us to completely dissect the life cycle of X. Our initial characterization of cell culture–produced X indicates unusual physical properties. Understanding of X's life cycle will aid in the development of improved pharmaceuticals.

Here, the most essential part of an abstract is not stated: the objective. This omission lets the reader come away wondering what the author is proposing to do. Including all essential components of the Abstract (with their signals) is a must. Otherwise, you stand little chance of being funded as reviewers will not have the time to search for missing answers. The revised version of the same Abstract makes clear what the author is proposing. Here, all essential components have been included. The objective is stated, and a more specific problem statement has also been added.

 Revised Example 22-14   **Technical abstract—separate Specific Aims section**

Background	X is a major human pathogen, which infects over 100 million people per year, leading to Y and Z. Current therapies for X are expensive, poorly tolerated, and only partially effective in controlling the pathogen and in limiting disease. <u>The development of improved pharmaceuticals requires a thorough understanding of X's life cycle.</u>
Specific problem	<u>Yet, until now, only limited aspects of its life cycle have been studied in the laboratory because culture systems for X did not exist.</u> Recently, we and others succeeded in establishing systems to grow X in cell culture. These systems will allow
Preliminary results	us to completely dissect the life cycle of X. Our initial characterization of cell culture–produced X indicates unusual
Objective	physical properties. **Here, we propose to determine the exact life cycle of X and to study its unusual physical properties in more detail.** Understanding of X's life cycle will
Significance	aid in the development of improved pharmaceuticals.

➤ Keep the Abstract short

One of the most common problems for abstracts is excessive length. When you write the Abstract, consider every word carefully. If you can write your Abstract in fewer words than the maximum allowed, do so. If you find yourself in the situation in which you have to condense your Abstract, follow these suggestions:

To condense a long Abstract (see also Chapter 6, Section 6.4):

- Omit unnecessary words and combine sentences
- Condense background
- Omit or subordinate less important information (definitions, experimental preparations, details of methods, exact data, confirmatory results, and comparisons with previous results)

➤ Be realistic when listing aims

Your aims should be specific. They should not be so broad as to be unsustainable. Be realistic in what you propose to achieve. Remember, to get your next grant you will have to prove you followed through with the plan set forth in your specific aims.

Aims should also be feasible given your technical expertise, and it should be clear that the aims are relevant and important in the field. The aims should not appear to be merely a list of experiments but rather a targeted approach.

➤ Do not make aims too interdependent

If an aim depends on the success of an earlier aim, it becomes dubious to the reviewers and may be a justified reason for rejection. Aims that are too interdependent appear risky because if just one aim fails, the success of the overall objective is in jeopardy.

22.8   REASONS FOR REJECTION

You need to spike the interest of the reviewers with your Abstract and Specific Aims section(s). You are more likely to impress reviewers if you master the skill of writing simply, clearly, and concisely.

Besides the reasons listed in the previous section, several other common reasons for quick rejection of a proposal exist:

- **Lack of originality**—you must propose something new or better than what has been proposed by others.
- **Lack of context**—you need to provide a background of your work as well as its impact. Never assume that the reviewers will be sufficiently familiar.
- **Limited sample size**—few samples in a study may not convince a reviewer of the significance of the proposal.
- **Proposed work is too ambitious**—you leave the reviewers with the sense that you are overstating what can be achieved within a given time frame.
- **Lack of conformity**—you should follow the proposal instructions very carefully so as not to antagonize a reviewer.
- **Too many abbreviations**—this will turn off most reviewers.

22.9   REVISING THE ABSTRACT AND SPECIFIC AIMS

When you have finished writing the Abstract and Specific Aims (or if you are asked to edit these sections for a colleague), you can use the following checklist to "dissect" the sections systematically:

☐   1. Is the proposed topic original?
☐   2. Are all the components there? To ensure that all necessary components are present, on the margins of the Abstract and Specific Aims clearly mark:

Background
Unknown
Objective
Strategy
(Expected results)
Significance
Specific aims

☐ 3. Is the Abstract kept short and within the set limits (less than one page for Abstract and Specific Aims together)?

☐ 4. Is the objective stated precisely?

☐ 5. Do all the components logically follow each other? (Is the unknown what one would expect to hear after reading about what is known? Is the objective really the goal one would expect to read after reading the unknown? Does the significance of the project really come across?)

☐ 6. Did you follow instructions on how to write the proposal?

☐ 7. Is your proposed work doable in the time frame given? Are aims not too interdependent?

☐ 8. Revise for style and composition using the rules and guidelines of this book:

  ☐ a. Are paragraphs consistent? (Chapter 6, Section 6.2)

  ☐ b. Are paragraphs cohesive? (Chapter 6, Section 6.3)

  ☐ c. Are key terms consistent? (Chapter 6, Section 6.3)

  ☐ d. Are key terms linked? (Chapter 6, Section 6.3)

  ☐ e. Are transitions used, and do they make sense? (Chapter 6, Section 6.3)

  ☐ f. Is the action in the verbs? Are nominalizations avoided? (Chapter 4, Section 4.6)

  ☐ g. Did you vary sentence length and use one idea per sentence? (Chapter 4, Section 4.5)

  ☐ h. Are lists parallel? (Chapter 4, Section 4.9)

  ☐ i. Are comparisons written correctly? (Chapter 4, Sections 4.9 and 4.10)

  ☐ j. Have noun clusters been resolved? (Chapter 4, Section 4.7)

  ☐ k. Has word location been considered? (Verb follows subject immediately? Old, short information is at the beginning of the sentence? New, long information is at the end of the sentence?) (Chapter 3, Section 3.1)

  ☐ l. Have grammar and technical style been considered (person, voice, tense, pronouns, prepositions, articles)? (Chapter 4, Sections 4.1–4.4)

  ☐ m. Are words and phrases precise? (Chapter 2, Sections 2.2 and 2.3)

  ☐ n. Are nontechnical words and phrases simple? (Chapter 2, Section 2.2)

☐ o. Have unnecessary terms (redundancies, jargon) been reduced? (Chapter 2, Section 2.4)

☐ p. Have spelling and punctuation been checked? (Chapter 4, Section 4.11)

SUMMARY

> PROPOSAL GUIDELINES
> - For the Abstract and Specific Aims section, include:
>   - Abstract
>   - Brief background
>   - Unknown or problem
>   - Objective
>   - Preliminary results
>   - General strategy (may include specific aims if not listed in separate part)
>   - Expected outcomes—optional
>   - Significance/impact
>   - Specific Aims
> - The first sentences should be informative, short, interesting, and provide broad background.
> - State the objective of your proposal precisely.
> - Distinguish between technical abstracts and those written for a lay audience.
> - List your Specific Aims in active, precise language.
> - State the significance of the study at the end of the Abstract.
> - Follow basic scientific writing rules and guidelines.
> - Use past tense for observations and specific conclusions. Use present tense for general rules and established knowledge.
> - Signal the unknown, the objective, the strategy, the specific aims, and the significance.
> - Do not omit any parts of the Abstract.
> - Keep the Abstract short.
> - Be realistic when listing aims.
> - Do not make aims too interdependent.

PROBLEMS

PROBLEM 22-1 Proposal Abstract

For the following Abstract, ensure that the necessary elements (background, unknown, objective, strategy, significance) are present and clearly signaled.

Plants are our oldest source of medicines. Yet much of Earth's rich plant life remains unexplored. In recent decades, natural drug discovery has concentrated on tropical plants due to their great diversity. However, there is equally much diversity for the plants of our oceans, and this plant life has remained untapped. The overall goal of this proposal is to identify and purify natural chemicals of oceanic plants and to test their activity as potential medicines. We will apply new chemical fingerprinting technology for our screens of plant life in the Florida Keys and assess them for potential medicinal use using microbial techniques. Identification of plants with compounds active against important human diseases, and subsequent characterization of such compounds, will lay the foundation for new drug development and lead to novel treatment therapies and better outcomes for patients.

PROBLEM 22-2 Elements of Proposal Abstract

For the following Abstract, ensure that the necessary elements (background, unknown, objective, strategy, significance) are present and clearly signaled.

Global warming is arguably one of the most pressing concerns of our time. However, we lack an effective model to predict precisely by how much the temperature will rise as a consequence of the increased levels of CO_2 and other factors. The width of this range is due to several uncertainties in different elements of the climate models, including the variability in the Sun's rate of energy output. To gain greater insight into the relationship between solar energy output and global temperature, we propose to launch the internationally led ABC satellite in April 2012. Our aim is to collect for 2 years data on the solar diameter and shape, oscillations, and photospheric temperature variation. We will assess these data to model solar variability. Our findings will dramatically advance our understanding of solar activity and its climate effects.

PROBLEM 22-3 Proposal Abstract

The following paragraph is an Abstract that has far exceeded its permissible length of 100 words. Shorten the abstract to 100 words or less by establishing importance. Omit unimportant information, and subordinate less important information.

Tourette syndrome (TS) is an inherited neurological disorder. Patients with the syndrome have frequent tics—repetitive, uncontrollable movements and sounds. It has been reported recently that one way to treat this disorder may be a behavioral treatment called habit reversal therapy (HRT) (Deckersbach, 2006). The objective of this proposal is to compare the effectiveness of HRT with that of supportive psychotherapy (SP) in TS patients. We will assess 100 adult outpatients with TS and determine how effectively HR and SP reduce motor and vocal tic frequency, improve quality of life, and increase psychosocial performance upon treatment of patients with the syndrome. The HRT as well as the SP group will show which conditions and characteristics of the disorder improve, and to what extent, when patients are treated with either therapy. Improvement will be defined as a significant drop in tic frequency and an increase in patient-perceived quality of life and psychosocial performance for a minimum of 12 months. Insights into these treatment options and their effectiveness may lay the foundation for future expanded therapies in the field.

(177 words)

PROBLEM 22-4 Different Abstracts

1. **The following Abstract is written in the format of one for a research paper. Rewrite this Abstract to follow the format of a proposal abstract.**
2. **Identify the individual elements of the proposal abstract (background, unknown, objective, strategy, significance).**

Background: The role of nurses continues to expand and shift in response to high societal demand for health care services. Within this landscape, it is important to more fully explore and understand the affect on employment status of the profession.

Methodology: This is a prospective, blinded descriptive study designed to collect lifestyle, employment, and demographic data on a sampling of nurses in the UK. A validated survey will be mailed to representative cohorts of nurses to determine gender, age, medical field, work hours per week, total time, reasons for leave of absence from the workforce, and total leave time from the workforce that is planned. Standard descriptive statistics and multivariate analysis will be used to identify the impact of family and social responsibilities on work hours based on age and gender in the setting of different fields of medicine.

Results: Family and social responsibilities are expected to affect the number of work hours per week as well as the time and reason for work leaves. These responsibilities are anticipated to differ based on age and gender and may be more prevalent in some medical fields than others.

Conclusion: We study the effect of family and other social responsibilities on gender and age in diverse medical fields to better accommodate nurse professionals in relation to the health care delivery demands of our society. Identifying significant differences of employment patterns will enable policy makers to consider the effects of family and social responsibilities for the nursing profession.

PROBLEM 22-5 Intellectual Merit and Broader Impact Statements
Evaluate the following Intellectual Merit and Broader Impact statements written for a NSF proposal. Is the significance and impact of the proposed study clearly stated? Does the Broader Impact contain an educational component?

Intellectual Merit

The entire research plan is composed of the following seven sub-themes: (1) the ambient state of stress in oceanic lithosphere, (2) the energetics of slab rollback, (3) the onset of convection with internal heating, (4) scaling laws for stagnant-lid convection with mantle melting, (5) the initiation of plate tectonics, (6) the history of ocean volume and global water cycle, and (7) the nature of core-mantle interaction. Collectively, they constitute a major step toward a better understanding of the long-term behavior of Earth, by tackling unresolved first-order issues all together. Moreover, each of them is designed to address a stand-alone, basic physics problem with potential applications beyond the scope of this proposal, reaching out to earthquake seismology, regional tectonics, planetary sciences, Precambrian geology, igneous petrology, plume dynamics, and geomagnetism.

Broader Impacts

This proposal includes the education of one female PhD student in theoretical geodynamics. The PI will also assimilate research results into three existing undergraduate/graduate courses he regularly teaches. In addition, a new 200-level undergraduate course will be developed on the physics of Earth's evolution, with hands-on experience in scientific computing. The PI will also conduct community outreach to assist K-12 teachers in developing public school curricula by showing how geophysical topics can be used as friendly examples in science classes.

(Jun Korenaga, proposal to federal agency, modified)

CHAPTER 23

Background and Significance

When you compose the Background and Significance section for a proposal, consider, above all, the specific aspects of the topic that are of interest to the funder. This section has to convince reviewers that specific aims, once achieved, will have significant impact on the topic in question.

THIS CHAPTER DISCUSSES:

- The different elements of the Background and Significance section
- The format of the Background and Significance section
- Signaling the different elements
- Common mistakes to avoid in the Background and Significance section
- Revising your Background and Significance section
- A complete sample Significance section

23.1 OVERALL

➤ Aim to awaken interest and to relay importance

The Background and Significance section of a proposal has three main purposes: to awaken the readers' interest, to provide readers with relevant background information to understand the proposal independently of other publications on the topic, and to relay the importance of the proposed work relative to the current knowledge and state of the topic in the field.

23.2 EMPHASIS, FORMAT, AND LENGTH

➤ Know what to emphasize

The Background and Significance section provides context for the readers by describing the current state-of-the art, or what is presently known about the topic. The section also describes what is unknown, problematic, or needed and delineates why it will be important to fill this gap. The section should include both your findings and theories as well as those of others. Do not simply provide only background information, but explain what gaps need to be filled in order to gain sufficient insights to find a solution or to progress to the next logical stage in the research. Only if you identify the gaps and problems will the proposed research make sense to the reader and the importance of the work become clear. This section has to convince reviewers that specific aims, once achieved, will have significant impact on the topic in question.

The length and format of this section can vary and depend largely on where the proposal will be submitted. Short proposals to private foundations may have this section divided into two separate sections: one for background or rationale (usually, one to three paragraphs) and another for impact/significance (typically, one paragraph). If a separate Significance/Impact section is required, add this section in form of a summary at the very end of your proposal (see Example 23-9). Within this summary, you should state the unknown and the objective(s)/hypothesis, you may indicate your expected outcomes, and you should state the overall significance or impact. This impact should be very broad, addressing issues and topics of interest to the funder, the wider scientific field, or even humanity as a whole.

In federal grant applications such as for NIH Research Project Grant Programs (R01s), the Background and Significance section, which was formerly labeled as such, is now only called the "Significance" section (Grant Application Instructions are denoted SF424 (R&R) at http://grants.nih.gov/grants/funding/424/). For these proposals, you are expected to provide one to two pages of rationale, including background, highlighting the significance and importance of your proposal. For many private foundations and corporations, this section may be called only the "Background" section or might be labeled "Introduction and Overview" or "State of the Art." Whatever the label, do not provide too much background information or a full literature review, but be sure to include all relevant information on background as needed to explain why your research is important (see Section 23.4 and Example 23-13). Overall, this section should be no longer than roughly one-third of your entire proposal.

If your Background and Significance section is lengthy or contains different subtopics, consider dividing it into subsections according to topic. Such divisions make it easier for the reader and reviewer to find key points at a glance and to recall information. Subsections may be arranged according to specific aims, but they may also follow some other logical arrangement under various headings.

23.3   REFERENCES

➤ Be as objective as possible

When you write this section, be as objective as possible. Do not criticize the work of other investigators or possibly alienate reviewers with an opposing point of view. (Remember that you do not know who your reviewers might be!) Instead, provide clearly established facts but acknowledge controversy. You need to convince reviewers that you are thinking objectively about the topic. To do so, you need to present the pros and cons on the subject.

Know that for federal grants you should always provide citations and indicate sources and references (see also Chapter 8). Be clear and generous as to what others have researched and achieved on the topic. Address competing approaches, and explain why yours is different or better. Cite their work. In particular, do not overlook the work of any reviewer on the review panel. Go to the trouble of looking up who will be on the review panel—this is often possible for private foundations and federal agencies. Familiarize yourself with the expertise of the reviewers.

Unlike for federal agencies, for many private funders sources often do not need to be indicated. This option is particularly important to know when your proposal length is very restricted (two to four pages total), as cutting out all references and citations will allow you more space for text and figures. Do not worry about leaving out citations and references if needed in these cases. Your proposal is not intended for publication. Very few people will see it, and when they do, it will be solely for considering funding.

23.4   ELEMENTS OF THE SECTION

➤ Follow a "funnel" structure

- Summary/Overview (optional)
- Background:
  - ○ Within subsections:
    - ○ Background/known
    - ○ Unknown/problem/need
    - ○ Aim/hypothesis (optional)
- Summary (optional)
- Significance/Impact

To illustrate the different elements of the Background and Significance section, each of these components is explained in more detail in the following subsections. Note that, for many federal agencies, the section is written as one unit of the proposal that also is scored. For many private foundations, the elements of the Background and Significance section outlined here may also appear in separate individual subsections.

Background

➤ Provide pertinent background information, but do not review the literature

The background component of your Background and Significance section should provide context relevant to your proposed topic of interest. The amount of background information needed depends on how much the intended audience can be expected to know about the topic as well as on the guidelines of the organization to which you are applying.

Present the overall scope of the problem to demonstrate your knowledge of the research on the subject, but concentrate on aspects the proposal will address. Do not include an exhaustive literature review, and do not spend a lot of space on statistics if the topic is well known. After you provide some general context of your work, write about the existing research in the area, and discuss established scholarship. For shorter Background sections, a summary pertinent to the research you are presenting in the proposal should suffice.

Generally, readers expect the parts of the Background and Significance section to be arranged in a standard structure: a "funnel," starting broadly with background information and then narrowing to the unknown and a specific aim or hypothesis of the proposal, followed by a statement of significance. When you compose this section, consider, above all, the specific aspects of the topic that are of interest to the funder.

Following are some examples of Background sections that funnel from background to unknown/problem and in some case to the specific aim. Further examples for wording of the unknown and aim, as well as examples for the significance and impact portion of the Background and Significance section follow that.

    Example 23-1   **Background funneling to unknown**

	Global warming is arguably one of the most pressing concerns of our time. It has been linked to the rapidity of observed climate change—the fact that the Earth's temperature rose by approximately 0.7°C over the last century (the most dramatic increase documented in historic times) and the attendant threat posed by melting polar icecaps, rising sea levels, and potentially, more severe weather patterns.
Context/ background	
	We do not know yet what proportion of this global warming is due to human activity and what is due to natural variations. More important, we lack an effective model to predict precisely by how much the temperature will rise as a consequence of the increase of the levels of CO_2 and other greenhouse gases in the atmosphere of the Earth.
Unknown/ specific problem/need	
Specific aim	In this proposal we aim to model temperature changes in response to CO_2 levels using . . .

    Example 23-2   **Background funneling to unknown**

| | Stripes on the wings of butterflies are a common element and have long been regarded as a defensive strategy. They often indicate that a butterfly is unpalatable or toxic and are recognized by avian predators (1–3). An additional, but not necessarily exclusive, hypothesis is that striped wing pattern elements are important in butterfly |
| Context/ background | |

Unknown/
specific
problem/need

Aim

mate recognition. This role has been documented in some butterfly species, specifically in the family Danainae (4 and references therein). However, the role stripes play in mate recognition has not been evaluated in a butterfly species in which the stripes are thought to serve a defensive function and warrants further investigation (5). We propose to test this hypothesis using the butterfly *Ituna ilione* (Danainae).

Example 23-3   **Background funneling to unknown**

Context/
background

Unknown/
specific
problem

The long-term sustainability of forest productivity depends on the interactions between plants and soil microbes, and their effects on resource availability. Plants rely on microbes to transform nutrients to available forms, and microbes rely on plants to provide reduced carbon (C) for metabolism. The strong interdependence of this interaction has led to concerns that human-induced increases in atmospheric CO_2 and nitrogen (N) deposition may be decoupling the C and N cycles in forest ecosystems (Asner et al., 1997), resulting in unpredictable feedbacks to long-term forest productivity (Zak et al., 2003; Reich et al., 2006). Forest productivity is generally increased by elevated CO_2 (Ceulemans, 1999), and the magnitude of this growth enhancement is strongly regulated by soil N availability (Oren et al., 2001; Magnani et al., 2007). Thus, the stimulatory effects of elevated CO_2 on productivity have been predicted to decrease over time (i.e., a negative feedback) as pools of available N in soil become depleted (Strain & Bazzaz, 1983). However, empirical support for such progressive N limitation (PNL) in forests has been lacking (Johnson, 2006), suggesting that our understanding of the mechanisms by which trees influence soil N cycling needs further refinement.

(Richard Phillips, proposal to federal agency, modified)

Unknown and Aim

➤ State the unknown, problem, or need

After providing general context and specific aspects of existing research, describe shortcomings of or gaps in the existing research or unanswered questions. The unknown/problem/need is clearest if you signal it; for example, write "is unknown" or "is unclear." You can also use other phrases to state the unknown, problem, or need for funding request: ". . . is the underlying problem" or ". . . is needed." Alternatively, you can imply rather than state the unknown by using a suggestion or a possibility ("Previous findings suggest that . . ."). Remember to use an objective tone when criticizing any previous work. Avoid antagonistic or judgmental phrases.

Unknown/need statements should allude to the specific aims or objective of the proposal and set the stage for the subsequent section on research design and methods (also called the "Approach"). Ideally, unknown statements are placed toward the end of subsections within the Background section and often are immediately followed by one to two sentences of a corresponding aim of the proposal or by a hypothesis. Examples 23-1 through 23-3 have already shown examples of unknown/problem statements. Here are some additional examples of stating an unknown/need followed by the corresponding aim or hypothesis.

☝	**Example 23-4**	The depth to which microbial life exists in the Antarctic ice core is not known. We hypothesize that microbial life forms can survive even in the deepest ice cores due to interconnected liquid veins in which bacteria can move and obtain energy and carbon from ions in solution.

☝	**Example 23-5**	Because the consequences of global warming for life on Earth greatly depend on where the actual warming lies within this predicted range, it is critical to narrow the range by improving our understanding of the uncertain components of the climate models with utmost urgency. We will investigate these key issues in this proposal by . . .

☝	**Example 23-6**	These treatments are only moderately effective and are often accompanied by severe side effects and viral resistance. Thus, there remains a need for new therapies for this serious disease. We propose to use a novel class of experiments to test . . .

Significance and Impact

➤ State the significance of your study clearly

Highlight why your proposal and the expected outcomes are important by stating what you think the significance or impact of the study is. Stating the significance or impact is particularly important at the end or the beginning of the Background and Significance section (especially in the Significance section of federal grant applications). Note that for some proposals, rather than a combined Background and Significance section, separate sections on Background and on Significance/Impact may be requested. (See also Chapter 22, Section 22.4.)

Describe the significance of the problem by relating the problem to one or more of the following criteria, but do not overstate its significance. Instead, place your work into the proper context:

- Timeliness
- Practical solution to a problem
- Important to a wide population

- Fills a gap in knowledge
- Provides a key apparatus for future observation or analysis of data
- May improve current or prevent future wide-ranging problem
- Provides possibility for a fruitful exploration with known techniques
- May decrease costs (e.g., cost-effectiveness, equity of resource allocation across populations, etc.)
- May improve quality of life
- May bridge theoretical and practical knowledge
- May provide a sustainable solution to a problem

Three examples of a statement of significance or impact are shown next. Note that in these examples, the impact of the proposed study is not limited to just the advancement of the topic or the field but is stated much more broadly in terms of society. While Examples 23-7 and 23-8 are sample statements that may be found in combination with background portion, Example 23-9 is an example of an independent Significance/Impact section.

Example 23-7   **Significance and Impact statement**

Presently, we have a time-sensitive opportunity to model and gain greater insight into the relationship between global temperature and tidal cycles: the internationally led TOPEX/Poseidon Jason-2 satellite scheduled for launch in summer 2011. The TOPEX/Poseidon Jason-2 mission was developed in close consultation with the National Center for Meteorological Studies to provide a better under-standing of variability in tidal cycles. It will gather data over three or more years on the solar diameter and shape, oscillations, and photospheric temperature variations. When these data are modeled, they will yield information on the magnitude, depth, and shape of the internal mag-netic field in the Sun (variations that contribute to pressure, internal energy, and energy transfer) and thus allow us to model the engine of tidal variability. This satellite is the best opportunity in decades to advance our understanding of the solar activity engine and its climate effects.

In Example 23-7, the significance of the study is related to timeliness, whereas in Example 23-8, the impact is related to a possible solution to a major problem.

Example 23-8   **Significance and Impact statement**

Optically driven nanoscale machineries, as described in this proposal, could provide a prime example for the appli-cation of fundamental science in today's highly developed, technology-driven world. Harnessing light force in inte-grated silicon photonics will not only allow us to bring transformative impact in the field of NEMS but will also have significant impacts in the sciences and engineering as well as for society as a whole.

(Hong Tang, proposal to federal agency [NSF])

Example 23-9   **Independent Significance/Impact section**

Impact and Significance.

We have proposed a new approach to vehicular fuel in a post-oil future by proposing production of ethanol and virtual ethanol storage. Our proposed approach avoids production of undesirable byproducts, does not compete with food production, and is viable on a global scale. Our findings would thus generate a valuable alternate vehicular transport fuel. The investigations proposed herein will provide the first insights into this fuel development. The understanding that will be gained from these investigations will be extremely far reaching. Therefore, this proposal is uniquely relevant to the goals of the Trust for Research of Alternate Fuels.

Summary

➤ For lengthy sections, add a summary/overview highlighting why your proposal is important

If your Background section is long or if a statement of significance is not requested, consider adding a one-paragraph summary or impact statement at the end of either the Background section or the entire proposal to remind the reviewers once again what the most important aspects of this section are. An example of one such summary is shown next.

Example 23-10   **Summary**

Summary. Hepatocellular carcinoma is a common cancer found throughout the world, and a large proportion of this disease is associated with chronic HBV infection. New approaches to prevent or treat HBV infection therefore have the potential to prevent the development of this cancer in a large number of individuals. We will examine the hypothesis that VSV-based vaccine vectors represent a powerful tool for improved prophylactic and novel therapeutic vaccinations. These studies are important because they may result in new approaches to prevent and/or treat a disease that causes high mortality worldwide.

(Michael Robek, proposal to federal agency)

23.5   SAMPLE SIGNIFICANCE SECTION FOR FEDERAL GRANTS

The role of the Significance section, particularly within federal grant applications, is crucial. It is the first subsection in the overall Strategy section, which follows the Abstract and Specific Aims section. The Significance section will therefore likely be read/glanced at by most reviewers. Although this section has only the heading "Significance" for federal grants, the section contains background information as well and is, accordingly, longer than a simple, individual one-paragraph Significance section you may encounter in many grants written for private foundations (see Example 23-7 and 23-9). Note that there is no set length for Significance sections, but generally 1 to 1½ pages will do. You want to leave plenty of room for the largest section of your proposal, your Approach section (see Chapter 26).

Significance sections of federal grants generally follow the outline provided for Background and Significance sections in Section 23.4:

- Summary/Overview
- Background/Rationale:
  - Provide context
  - Introduce the problem
  - Aim to solve problem
- Significance and Impact

The emphasis of the Significance section is more on significance than on background information. Therefore, highlight the significance and importance of the proposal and concentrate only on immediately relevant background information when providing the rationale of the proposed study. You will need to emphasize the significance of the idea(s) as well as the significance in a broader context. Point out *why* your study is important and *how* it is going to change the field by addressing the following questions: Why is your idea important or different? How will it advance the field? Why is it important for society? (For NIH also: How will it improve public health?)

The significance and importance of this section may be further highlighted by adding a brief summary/overview paragraph at the very beginning of this section or at the very end of it. Such a summary will also make it easy for reviewers to find relevant information immediately. Signal this information verbally ("This work is important because . . ."; "The proposed work is significance because . . .") and/or visually (by placing it into boldface or italics.) Such highlighting will allow reviewers to simply cut and paste it into their review.

Two examples of such overviews/summaries are shown next, followed by an example of a complete Significance section for a federal grant proposal.

Example 23-11   Significance

> *The proposed research is important because it harnesses new and original engineering (synthetic biology) approaches that hold exceptional promise to get to the heart of long-standing and fundamental questions about the central energy-producing organelle, the mitochondria.*

Example 23-12   Significance overview

The proposed work is significant because it addresses an important unmet clinical need: Alzheimer's disease. The fundamental information generated from this proposal will contribute to novel treatments and improved disease outcomes for affected patients. Such information is particularly important as our older population is starting to outgrow our younger population within the next three decades (7). Accordingly, the number of Alzheimer's patients can be expected to increase. Yet we understand little about how to treat people effectively once afflicted with the disease.

Example 23-13   Complete Significance section

Significance

Within the next forty years, the number of older people will exceed the number of young people for the first time in history.[1] As the number of older people grows, their health care needs will increasingly burden the health system. Altered immune responses in the elderly are responsible for many diseases, as well as for increased susceptibility to infections and cancer. For example, influenza viral lung infections are a growing public health concern in older people because of their predisposition for community-acquired pneumonia and the ability of viral lung infections to exacerbate chronic heart and lung diseases.[2] Influenza viral lung infections induce higher morbidity and mortality in older people than in younger people.[3] Furthermore, these infections are increasing in prevalence, as exemplified by a 20% increase in hospitalization rates due to community acquired pneumonia from 1988 to 2002, for patients aged 65 to 84 years.[4]

Efforts to protect older people from influenza viral infection have only been partially successful. The trivalent seasonal influenza vaccine reduced hospital rates for pneumonia by 48–57% and mortality by 36–54% in community dwelling older people.[5,6] However, these vaccines do not protect older people as efficiently as they protect younger people, who typically mount immunity to influenza vaccination with greater than 90% efficacy.[6,7] Similar to the increased susceptibility of older people to microbial infection, reduced efficacy of vaccines in older people is likely caused by age-induced alterations in the immune system. Hence, discerning how aging affects the immune system is critical for development of more effective vaccination strategies and therapies to protect older people from viral infection.

Although the current paradigm indicates that declining immune responses underlie increased susceptibility to viral infections and reduced efficacy of vaccines in the older population,[8] *our preliminary work suggests that exaggerated immune responses actually render older people **more** susceptible to viral infections and less responsive to vaccines.* We propose to use murine models and a translational human observation study to determine if and how exaggerated immune responses underlie the aged phenotype.

Our proposal is important as it will provide new insights into the molecular interactions and pathways of the aging immune system. The results of these studies may lead to novel anti-inflammatory therapies that could improve outcomes in older people infected with viruses and increase vaccine efficacy in older people. Such therapies would differ vastly from the mainstream therapeutic pipeline of drugs aimed at boosting immune responses in aging individuals.[8,9] **Thus, this study could ultimately transform the treatment of respiratory viruses and potentially other vectors in older people.**

(Daniel Goldstein, Significance section of a federal proposal, with minor modifications)

23.6   SIGNALS FOR THE READER

➤ Signal all the elements of the section

Generally, all the parts of the Background and Significance section should be signaled so the reader does not have to guess about the information provided. The signals vary depending on how the known/background, unknown/need, objective/aim, and significance/impact were phrased. Numerous variations on these signals are possible. Some examples of such signals are shown in Table 23.1.

TABLE 23.1 Potential signals for the Background and Significance section

BACKGROUND	UNKNOWN/NEED	OBJECTIVE/AIM	SIGNIFICANCE/IMPACT
X is . . .	. . . is unknown	We propose to . . .	. . . may result in . . .
X affects . . .	. . . is unclear	Our objective is to . . .	. . . will contribute to . . .
X is a component of Y	. . . has not been determined	We will examine the hypothesis that . . .	. . . may provide insight into . . .
	. . . is needed	Our overall goal is . . .	. . . are important for . . .
	. . . is necessary		. . . may be used to . . .

23.7   COHERENCE

➤ Use topic sentences and techniques of continuity to tell the story

In long Background and Significance sections, the storyline can be difficult to follow. To ensure that the overall story is clear when ministories are placed into the section, use topic sentences and all of the techniques of continuity presented in this book:

- Word location—Chapter 3, Section 3.3
- Key terms—Chapter 6, Section 6.3
- Transitions—Chapter 6, Section 6.3
- Sentence location—Chapter 6, Section 6.3

23.8   COMMON PROBLEMS

The most common problems of the Background and Significance section include:

- Poor organization
- Lack of objectivity
- Amount of detail
- Equating significance with a problem
- Equating significance with innovation

➤ Organize your background section logically

The section may also be divided into different subsections, which may be listed from most to least important or chronologically. In each subsection, organize your topic sentences and paragraph. Each subheading and topic sentence within each subsection should make it easy for the reviewers to absorb the key points at a glance. Follow a logical order.

➤ Present evidence objectively

Do not omit opposing viewpoints when presenting background information. Only if you discuss both pros and cons will you sound convincing to reviewers. Also, remember that no reviewer wants to see himself or herself left out should they be working on the same topic.

➤ Provide the necessary amount of detail

Adjust your writing to the needs of the reviewers. Too much or too little detail can work against you. If you are writing to a panel of reviewers within your field, you may not have to provide much detail on the topic. If you are writing to an educated lay audience, however, your background section will need to start quite broadly and will have to explain more technical details or use simpler expressions.

➤ Distinguish between significance and a problem

Beginning writers easily confuse the severity of a problem with the significance of their proposed work. When you write your Significance section, you should not argue that a particular problem or illness is significant. Rather, state what you will do to alleviate the problem or illness and how your proposed work will advance the field/public health.

👍	**Example 23-14**	Heart disease continues to be the number one killer among white and black adults. Every year in the United States, around 600,000 deaths result from heart disease. . . .
👍	**Revised Example 23-14**	Our work will shed important insights into coronary heart disease by . . . thus laying the foundation for new treatment options and ultimately a reduction in the number of deaths associated with this devastating disease.

➤ Distinguish between significance and innovation

Do not confuse innovation with significance. Your study is not significant because you employ a new approach, a new combination of expertise, or undertake scientific research that has not been done before. Reviewers evaluate your study within the context of a research field. They are looking to see how your study will advance that field, hasten translation, and improve lives. For a discussion on the Innovation section of a proposal, see Chapter 24.

23.9 REVISING THE BACKGROUND AND SIGNIFICANCE SECTION

When you have finished writing the Background and Significance section (or if you are asked to edit these sections for a colleague), you can use the following checklist to "dissect" the sections systematically:

☐ 1. Are all the components there? To ensure that all necessary components are present, on the margins of the sections clearly mark:

Background
Unknown
Aim/goal
Significance

☐ 2. Is the section written objectively?

☐ 3. Did you adjust your writing to the needs of the reviewers? Not too much detail nor too little?

☐ 4. Do all the components logically follow each other? (Is the unknown what one would expect to hear after reading about what is known? Is the aim really the goal one would expect to read after reading the unknown? Does the significance of the project really come across?)

☐ 5. Revise for style and composition using the rules and guidelines of
       this book.
   ☐ a. Are paragraphs consistent? (Chapter 6, Section 6.2)
   ☐ b. Are paragraphs cohesive? (Chapter 6, Section 6.3)
   ☐ c. Are key terms consistent? (Chapter 6, Section 6.3)
   ☐ d. Are key terms linked? (Chapter 6, Section 6.3)
   ☐ e. Are transitions used, and do they make sense? (Chapter 6,
         Section 6.3)
   ☐ f. Is the action in the verbs? Are nominalizations avoided?
         (Chapter 4, Section 4.6)
   ☐ g. Did you vary sentence length and use one idea per sentence?
         (Chapter 4, Section 4.5)
   ☐ h. Are lists parallel? (Chapter 4, Section 4.9)
   ☐ i. Are comparisons written correctly? (Chapter 4, Sections 4.9
         and 4.10)
   ☐ j. Have noun clusters been resolved? (Chapter 4, Section 4.7)
   ☐ k. Has word location been considered? (Verb follows subject im-
         mediately? Old, short information is at the beginning of the
         sentence? New, long information is at the end of the sentence?)
         (Chapter 3, Section 3.1)
   ☐ l. Have grammar and technical style been considered (person,
         voice, tense, pronouns, prepositions, articles)? (Chapter 4, Sec-
         tions 4.1–4.4)
   ☐ m. Are words and phrases precise? (Chapter 2, Sections 2.2 and 2.3)
   ☐ n. Are nontechnical words and phrases simple? (Chapter 2,
         Section 2.2)
   ☐ o. Have unnecessary terms (redundancies, jargon) been reduced?
         (Chapter 2, Section 2.4)
   ☐ p. Have spelling and punctuation been checked? (Chapter 4,
         Section 4.11)

SUMMARY

BACKGROUND AND SIGNIFICANCE GUIDELINES
- Aim to awaken interest and to relay importance.
- Know what to emphasize.
- Be as objective as possible.
- Follow a "funnel" structure.
  - Summary/Overview (optional)
  - Background:
    - Within subsections:
      - Background/Known
      - Unknown/Problem/Need
      - Aim/Hypothesis (optional)

- Summary (optional)
- Significance/Impact
- Provide pertinent background information, but do not review the literature.
- State the unknown, problem, or need.
- State the significance of your study clearly.
- For lengthy sections, add a summary/overview highlighting why your proposal is important.
- Signal all the elements of the section.
- Use topic sentences and techniques of continuity to tell the story.
- Organize your background section logically.
- Present evidence objectively.
- Provide the necessary amount of detail.
- Distinguish between significance and a problem.
- Distinguish between significance and innovation.

PROBLEMS

PROBLEM 23-1 Background and Significance Section of a Grant
For the following Background section, ensure that the necessary elements (background, unknown, objective) are present and clearly signaled.

Healthy older adults often experience mild decline in some areas of cognition. The most prominent cognitive deficits of normal aging include forgetfulness, vulnerability to distraction and other types of interference, as well as impairment in multitasking and mental flexibility (Albert, 1997; Bimonte, 2003). These cognitive functions are the domain of the most evolved part of the human brain known as the prefrontal cortex, the brain region that is the last to fully mature in children and the first to decline as we age. Indeed, prefrontal cortical cognitive abilities begin to weaken already in middle age and are especially impaired when we are stressed. Loss of these organizational abilities is a particular liability in this Information Age when our demanding lives require that we multitask and navigate through endless interferences. Thus, understanding how the prefrontal cortex changes with age is a top priority for rescuing the memory and attention functions we need to survive in our fast-paced, complex world.

PROBLEM 23-2 Background and Significance Section of a Grant
For the following Background section, ensure that the necessary elements (background, unknown, objective) are present and clearly signaled.

Age-induced emphysema ("senile emphysema") is an under-recognized and poorly understood phenomenon that occurs in the lungs of most people over the age of 35 years. It leads to the loss of effective oxygen exchange in the lungs, resulting in progressive shortness of breath and, if severe, complete respiratory failure. We postulate that cigarette smoke-induced emphysema represents a form of accelerated lung aging, akin to what is observed in the skin and cardiovascular system of smokers. Toll-like receptor (TLR) deficiency predisposes people to emphysema. In mice deficient in TLR4, the canonical receptor for lipopolysaccharide (a component of specific bacteria) is an accelerated form of age-induced lung enlargement that resembles human emphysema both histologically and functionally (1). In people, TLR4 function declines with age (2, 3), which may in part contribute to "senile emphysema" even in the absence of significant exposures. Human studies have also found that chronic smoking leads to decreased TLR4 function in the lung and that the level of TLR4 depression is correlated with the severity of COPD (4). The synergistic or additive effects of age and smoking on TLR4 function in susceptible individuals may explain the pathogenesis and temporal characteristics of smoke-induced emphysema, but this synergy has not been explored in detail. We aim to explore how TLR4 is affected by age and cigarette smoke.

(Patty Lee, proposal to private funding agency, modified)

PROBLEM 23-3 Background and Significance Section of a Grant
For the following Background section, ensure that the necessary elements (background, unknown, objective) are present and clearly signaled.

One potential therapeutic approach for treating chronic HBV infection is through therapeutic vaccination to disrupt the immunological tolerance and to induce an immune response that is capable of controlling the virus. Despite the promise of therapeutic vaccination for treating chronic HBV, progress in this area has been limited (16, 29, 34, 64, 73, 83). A successful therapeutic vaccination strategy must accomplish two goals. First, immunological tolerance must be broken, and virus-specific T cells must be generated. Second, these T cells must efficiently perform their effector functions, including killing target cells and producing the antiviral cytokines such as IFN-γ and TNF-α, which can noncytopathically inhibit virus replication. We will test the hypothesis that recombinant VSV vectors are ideally suited for therapeutic vaccination for chronic HBV infection.

(Michael Robek, proposal to federal agency)

PROBLEM 23-4 Background and Significance Section of a Grant
For the following Background section, ensure that the necessary elements (background, unknown, objective) are present and clearly signaled.

The Pliocene climate was significantly different from our contemporary one. Most important, it was significantly warmer—high-latitude temperatures, for instance, were 4–6°C higher than today (4, 5). Surface temperatures in polar regions were in fact so much higher that continental glaciers were absent from the Northern Hemisphere, and the sea level was approximately 25 m higher than today. A number of numerical simulations have previously been conducted using coupled and atmospheric GCM and the data from the PRISM projects (6). However, many of these studies have assumed either that the Pliocene tropical climate was similar to the modern one or that the models themselves produced climate conditions in the tropics not far different from the modern one. Only recently, a wealth of new evidence has accumulated indicating that not only high latitudes but also the tropics and the subtropics had a very different climate in the early Pliocene. In particular, the tropical Pacific was characterized by what is often referred to as "a permanent El Niño-like state" (7). This phenomenon implies that the mean state of the Pacific had a significantly reduced or absent zonal SST gradient along the equator. In addition, a number of other dramatic climatic changes occurred throughout the tropics and the subtropics, resulting in very different patterns of sea surface temperatures from what we typically observe today. Because recently discovered changes in the tropical climate of the early Pliocene are so significant, we propose to undertake a systematic study to understand the physical mechanisms responsible for such a different climate state.

(Alexey Federov, proposal to federal agency, modified)

CHAPTER 24

Innovation

The Innovation section is a key peer review criterion to assess the importance of a grant application at certain private and federal funding agencies, particularly for the National Institutes of Health (NIH). Many researchers struggle with composing this section. Given this challenging section and the importance of this section for these funding agencies, this chapter is dedicated to this particular section of a grant proposal. The section is also critical to understanding just how innovative your research should be to warrant funding.

THIS CHAPTER DISCUSSES:

- The elements of the Innovation section
- The format of the Innovation section
- Signals in the Innovation section
- Common mistakes of the Innovation section
- Revising the Innovation section
- A complete Innovation section written for a federal agency

24.1 GENERAL REMARKS ON PROPOSAL SECTIONS

Proposals from almost all private and federal agencies follow similar outlines, albeit with different headings and diverse distribution of elements (see also Chapter 20, Section 20.7 on Proposal Format). For example, the NIH expects the Research Plan to be structured as Specific Aims, Significance, Innovation, and Approach. The USDA-NIFA Project Narrative format is Introduction, Rationale and Significance, and Approach, with instructions for specific subsections within each of those. The NSF format is more free-wheeling and has no specified format except that a "Broader Impacts of the

Proposed Work" must be included (http://www.nsf.gov/pubs/policydocs/ pappguide/nsf15001/gpg_2.jsp#IIC2d). For NSF-IOS preproposals, the use of the following subsections is recommended: Conceptual Framework **or** Objectives **or** Specific Aims, followed by Rationale and Significance **or** Background, followed by Hypotheses **or** Research Question(s), then Research Approach **or** Experimental Plan **or** Research Design, and last, Broader Impacts of the Proposed Work.

Unlike the NIH, most funders do not explicitly require an Innovation section. Thus, at first glance such a section may seem unusual. Yet it is important in all proposals to show how the proposed work is innovative, either under the heading of "Innovation" or another proposal heading. Although this chapter follows the required outline for basic research project grants of the NIH, such as the R01, R03, or R21, most of the information about innovation statements in the chapter applies to other grants as well.

24.2   COMPONENTS

➤ Make your Innovation section concise

In the Innovation section, you will need to show that your project is original and point out why your proposal is innovative. Although not very long (on average, one-half to one page), the Innovation section is very important as it is often scored by reviewers along with other key sections, including Abstract and Specific Aims, Significance, and Approach. It is therefore essential to compose a winning, concise Innovation section.

In the Innovation section, emphasize any novel, original approach you will be using. Aim to answer the following questions:

1. What is your innovation?
2. Why is your approach better than what has been done before?
3. What makes it novel?

To highlight why a proposal is innovative, you can emphasize:

- Innovative methods
- Refinements
- Improvements
- A new specific aim
- New applications of theoretical concepts
- New approaches or methodologies
- New combination of expertise
- Novel equipment
- An innovative way of looking at a problem
- A paradigm shift, that is, a change in the basic view of a theory of science (with caution—see note below)

Do not just describe why your proposed work is innovative; explain also any advantage over existing methodologies, instrumentation, or intervention(s).

In short, write an innovation statement and provide context, such as an explanation why your approach is novel compared to that of others.

☞ **Example 24-1** **Innovation statement and explanation**

Innovation statement	Experiments outlined in this application will employ a novel goal in deciphering eukaryotic ribosome structure. Previous studies have only been carried out without inhibitors. Our studies will look at eukaryotic ribosome structure in the presence of inhibitors.
Context/ explanation	

☞ **Example 24-2** **Innovation statement and explanation**

Innovation statement	Combining our respective expertise in microbiology and plant genetics is a novel approach that will allow us to obtain critical insights into chloroplast structure and function. Only through this combination will it be possible to assess how chloroplasts are related to certain primitive life forms. . . .
Context/ explanation	

24.3 FORMAT

➤ **Include the following elements in your Innovation section:**

- Background/general context/preliminary results
- Statement of need/problem
- Innovation statement(s)
- Overview of approach (optional)
- Impact statement

In your Innovation section, precede or follow your innovation statement with sufficient background information, including general key preliminary results, to provide context. Keep the background information as concise as possible. Your Innovation section should be no longer than one page. You may also add a very brief overview of your approach (one-half to one sentence) if necessary to explain your innovation(s). Round off the Innovation section with a one-sentence impact statement, but do not expand on impact/ significance as this discussion is covered in the Significance or Background and Significance section.

➤ **For longer sections (more than half a page), provide an overview or a summary listing your innovations**

For Innovation sections that are longer than one to two paragraphs, add a brief summary/overview paragraph at the very beginning or at the very end of the section. This overview will aid reviewers to find relevant information immediately.

Example 24-3 is a sample overview/summary. This overview contains several innovation statements, listed under (a) through (d), that set the stage for any subsequent expansion by the author on each of these topics.

Example 24-3 **Innovation overview/summary**

Signal for innovation statements

List of innovations

> Our proposal is innovative in several ways: (a) it combines epidemiology, network theory, economics, psychology, and game theory to meld important characteristics of an epidemic in a single multilevel modeling framework; (b) it extends behavioral modeling in epidemiology to allow for multiple interacting behavioral factors, and specifically considers the potential impact of prosocial preferences, imitative behavior, behavioral clustering within social groups, and habitual behavior; (c) it evaluates cross-national scale differences in vaccination decisions; and (d) it yields probabilistic predictions based on novel data and informative models, guiding how policy could be developed to account for and leverage prosociality, imitation, and habitual behavior in the design of public health education.

(Alison Galvani, proposal to a federal agency)

A complete Innovation section to a federal agency follows, in which innovations have been highlighted clearly. This section starts by providing context (known/background, unknown/problem, and recent study as evidence), then lists several innovations before ending with a brief impact statement.

Example 24-4 **Complete Innovation section**

Context (background, problem, evidence)

Signal for innovation statements

List of innovations

> The controlled release of immunosuppressants can greatly enhance drug efficacy while decreasing toxicity. Nanoparticles are a well-established methodology to deliver a variety of compounds in a controlled fashion. However, to date, administration of nanoparticles encapsulated with two immune suppressants simultaneously has not been attempted in experimental solid-organ transplantation. Our recent study documents our experience with this platform for organ transplantation with one agent, MPA (Shirali et al., 2011).
>
> Our proposal is innovative as we aim to: (i) assess dual encapsulation of immune suppressants within nanoparticles to avoid drug toxicities induced by soluble administration of agents; (ii) combine and deliver two drugs via nanoparticles whose effects may join by complementary and independent pathways in the cells of interest via safe delivery platforms; (iii) enhance a pro-tolerogenic immune phenotype in dendritic cells by dual encapsulation of MPA + RAPA in nanoparticles that promotes transplant tolerance inducing effects of co-stimulatory blockade without

Impact
statement
<u></u>

drug toxicity. The combined administration and delivery of two immunosuppressants in nanoparticles may yield a novel drug delivery platform that can effectively avoid drug toxicity, suppress alloimmunity, and enhance transplant tolerance.

(Daniel Goldstein, proposal to a federal agency)

24.4   SIGNALS FOR THE READER

➤ Signal the innovation statement

To highlight your innovation statement(s), signal it explicitly so the reader does not have to guess about the information provided. If you provide an overview or summary paragraph, consider making this paragraph stand out visually by styling the overview paragraph in boldface or italics. Highlighting this paragraph makes it easier for reviewers to find it. The overview/summary can then simply be cut by reviewers and pasted into the review document that they usually have to compose.

Some examples of signals include those shown in Example 24-5.

Example 24-5   **Signals for innovation statements**

a   The proposed research employs a novel method that we developed . . .

b   These aims are original in that . . .

c   Experiments outlined in this application will employ a novel approach to generate . . .

d   In this study, we will employ innovative methods to . . .

e   Our study is innovative because . . .

f   X represents a novel approach that will . . .

g   My work provides a novel contribution to science by . . .

24.5   COMMON PROBLEMS

The most common problems of the Innovation section include:

- Your innovation is too far out of the box
- Equating innovation with significance (see also Chapter 23, Section 23.8)
- Innovation is not feasible and/or credible
- Omission of important points

➤ Be cautious about seeming too innovative

Innovation can be tricky. In the Innovation section, you certainly have to show what is novel or innovative about your proposed work. However, be cautious about seeming too innovative or sounding too boastful. Proposing a paradigm shift can be dangerous. If your work is controversial or too far out of the box, it may not be convincing to reviewers. Challenging the status quo of something may be perceived by reviewers as challenging their world view. Reviewers tend to be particularly critical if you are a novice to the field.

Remember that you will have to get your application past your reviewers and that the primary objective of your application is getting funded. So do not ignore the reviewers' perspective or present them with an innovative idea they are unlikely to appreciate. Instead, propose work they will view as important. Explain the knowledge your proposed work will add to the field in a different, innovative, and distinct way. Describe a new approach or model, test an innovative idea, apply an old idea to a new area, or explore new scientific avenues. Show the gap your proposed project may fill or explain how it will bring a problem closer to a solution. In other words, think of innovation as making significant incremental progress, not a giant leap forward. Sometimes, though, your ideas may simply be ahead of their time, and reviewers are just not ready to deal with them.

➤ Distinguish between innovation and significance

Do not confuse innovation with significance. Do not describe how your study will advance that field, hasten translation, or improve lives. These items describe significance. Your study is innovative because you use a new approach, a new combination of expertise, or investigate new science (see also Chapter 23, Section 23.8).

➤ Balance innovation with feasibility and credibility

Many Innovation sections receive a bad score because they do not seem feasible or credible. To prove that your idea is feasible and that you can get the job done, show that you have the resources, preliminary data, and expertise to conduct the proposed research. That is, ensure that your preliminary data indicate a likely chance of success for the proposed research, that your training has prepared you well, and that you have solid relevant collaborations and resources. Any previous publication can back up your credibility (see Example 24-4), as can training and collaborators.

Many projects require more than one person's expertise. If needed, reach out to other scientists for collaborations or ask them to become a co-investigator. Having experienced people on your team will help build credibility and make reviewers trust in the success of your project.

24.6   REVISING THE INNOVATION SECTION

When you have finished writing the Innovation (or if you are asked to edit this section for a colleague), you can use the following checklist to "dissect" the section systematically:

☐   1. Have you clearly stated what is new about the proposed study (new hypothesis, new methods or approaches)? (Section 24.2)

☐ 2. Do you have sufficient expertise to perform the proposed work? Will the reviewers view your expertise as credible? (Section 24.2)

☐ 3. Are your preliminary data and scope of work feasible? (Section 24.2)

☐ 4. Did you clearly signal what is innovative? (Section 24.4)

☐ 5. Did you distinguish between innovation and significance? (Chapter 23, Section 23.8)

☐ 6. Did you provide a clear overview/summary for a longer Innovation section?

☐ 7. Revise for style and composition using the basic rules and guidelines of this book:

    ☐ a. Are paragraphs consistent? (Chapter 6, Section 6.2)

    ☐ b. Are paragraphs cohesive? (Chapter 6, Section 6.3)

    ☐ c. Are key terms consistent? (Chapter 6, Section 6.3)

    ☐ d. Are key terms linked? (Chapter 6, Section 6.3)

    ☐ e. Are transitions used and do they make sense? (Chapter 6, Section 6.3)

    ☐ f. Is the action in the verbs? Are nominalizations avoided? (Chapter 4, Section 4.6)

    ☐ g. Did you vary sentence length and use one idea per sentence? (Chapter 4, Section 4.5)

    ☐ h. Are lists parallel? (Chapter 4, Section 4.9)

    ☐ i. Are comparisons written correctly? (Chapter 4, Sections 4.9 and 4.10)

    ☐ j. Have noun clusters been resolved? (Chapter 4, Section 4.7)

    ☐ k. Has word location been considered? (Verb follows subject immediately? Old, short information is at the beginning of the sentence? New, long information is at the end of the sentence?) (Chapter 3, Section 3.1)

    ☐ l. Have grammar and technical style been considered (person, voice, tense, pronouns, prepositions, articles)? (Chapter 4, Sections 4.1–4.4)

    ☐ m. Are words and phrases precise? (Chapter 2, Sections 2.2 and 2.3)

    ☐ n. Are nontechnical words and phrases simple? (Chapter 2, Section 2.2)

    ☐ o. Have unnecessary terms (redundancies, jargon) been reduced? (Chapter 2, Section 2.4)

    ☐ p. Have spelling and punctuation been checked? (Chapter 4, Section 4.11)

SUMMARY

INNOVATION GUIDELINES

- Make your Innovation section concise.
- Include the following elements in your Innovation section:
  - Background/general context/preliminary results
  - Statement of need/problem
  - Innovation statement(s)
  - Overview of approach (optional)
  - Impact statement
- For longer sections (more than half a page), provide an overview or a summary listing your innovations.
- Signal the innovation statement.
- Be cautious about seeming too innovative.
- Distinguish between innovation and significance.
- Balance innovation with feasibility and credibility.

CHAPTER 25

Preliminary Results

When describing preliminary results, you need to prove your expertise in the topic area and establish credibility in your ability to perform the work.

THIS CHAPTER DISCUSSES:

- The function of the Preliminary Results section
- The components of the Preliminary Results section
- The format of the Preliminary Results section
- Signals of the different elements
- Common mistakes of the Preliminary Results section
- Revising the Preliminary Results section

25.1 FUNCTION

➤ Establish credibility

Many proposals have an independent section on preliminary results, but note that certain federal proposals require you to integrate preliminary results into other sections, such as into the Approach section (see also Chapter 20, Section 20.7 and Chapter 24, Section 24.1). Regardless of placement or heading, the main purpose of a Preliminary Results section is to establish credibility in your ability to carry out the specific aims of your proposal. The section should therefore show how your work to date has revolved around the topic that is directly pertinent to the funding agency's goal and to the objective of the proposal. You need to convince reviewers that your work is likely to succeed:

- Under your leadership
- At your institution

- Using your methods
- As supported by your preliminary data

For established, independent investigators, this section carries more weight than for first-time applicants or for training grants. If you do not yet have many publications, related prior work or training and knowledge of the work of others can help demonstrate your competence.

25.2 CONTENT

➤ **Report important previous findings relevant to the topic and to the specific aims**

➤ **Indicate logical next steps or what is unknown or problematic**

➤ **Use figures and tables if necessary to enhance findings**
The Preliminary Results section should describe the major scientific contributions of your work to date as relevant to the proposed objective. The section should also specify the unknown or problematic areas and set the stage for the Approach/Research Design section. That is, it needs to indicate logical next steps. The section should present the results of initial experiments carried out by you or your mentor's laboratory and point the reader to the data shown in figures and tables.

➤ **Do not overstate your competence or capabilities**
For grants that emphasize successful research, you especially need to prove your expertise in the area and establish credibility in your ability to perform the work. You will have to prove not only that you have some data on which to base your future aims but also that you and your team are sufficiently trained and equipped (laboratory, skills, resources) to accomplish the goals set forth in the proposal. Some proposal formats ask you to provide a detailed overview of expertise and resources (see Chapter 27, Section 27.2). Do not overstate your competence or capabilities. Good reviewers will know if you do.

➤ **If needed, use subsections to organize the Preliminary Results section**
Organize the Preliminary Results section much like the Results section of a paper, with subsections pertaining to specific findings and observations. Use meaningful headings for these subsections. Report only results that are pertinent to the topic of the proposal. Exclude results that are not relevant. If needed, explain the purpose of an experiment briefly. Include results whether they support your hypothesis or not, and explain any contradicting results if necessary. Consider ending each subsection with a sentence that summarizes the observations and conclusions.

➤ Adjust the level of writing according to the topic and to your potential readers

If you are writing for a group of scientists, give sufficient scientific and technical details and results. Particularly for federal grants, especially R01s, provide detailed data and technical feasibility. Include figures and tables to support your findings and hypothesis and to show your ability to perform the proposed experiments. Realize, however, that novel and unusual techniques and methods typically require more detailed descriptions and explanations.

If you are writing for a lay audience, you probably need to provide more background and generalizations and discuss much broader implications. Here, too, include figures and tables if needed. Consider your readers knowledgeable in the general research area but not experts on your specific topic.

Data and Their Interpretation

➤ Interpret your results for the reader

Do not simply describe the conclusions of prior publications or work. Instead, include relevant key findings and their interpretation. Adjust the level of describing data and results according to your reviewers' background (technical or lay). Young investigators can show competence and capability despite having few publications by, for example, reporting results of preliminary studies or pilot studies, unreported data, or data presented at conferences. You may also include results determined by others, but you should clearly identify them as such. In addition, if you are applying for an R01, distinguish yourself from your mentor. You need to establish your role as an independent leader.

➤ Distinguish between data and results

Differentiate between data and results. Data are values derived from scientific experiments (concentrations, absorbance, mean, percent increase). Results are interpretations of data (e.g., "Growth rate **decreased** when samples were incubated at 15°C instead of 25°C"). In the Preliminary Results section, do not just present data but summarize and interpret their meaning for the reader by presenting them as results. (See also Chapter 13, Section 13.2.)

   Example 25-1   The plants grew 2 cm in 24 hrs when auxin was added.

Unless your readers are plant hormone specialists, they may not be able to put "2 cm in 24 hrs" into any relation, especially if no comparative value is given. Is "2 cm in 24 hrs" more growth or less growth than normal? The data are not interpreted for the reader. The revised example is much more meaningful.

| Revised Example 25-1 | Plant growth **increased 50% in 24 hrs** when auxin was added. |

In the revised example, the data have been interpreted and are presented as a result, which has been given a comparative value to put it into relation for nonspecialists in the field.

25.3   FORMAT

Overall Format

➤ Summarize and generalize relevant previous results

Rather than providing much details on rationale and experimental set-ups and other information from previous experiments, simply summarize and generalize relevant earlier results. This will allow you to get to the important points more quickly and help keep your readers focused on key findings.

➤ Organize shorter or integrated Preliminary Results sections by including (in this order):

- Background and methods to make the study understandable
- Enough results to be convincing
- The conclusion (emphasize the interpretation of results)

If you are applying to a federal agency, you will likely have to place your preliminary results into the Approach/Research Design section as many of these proposals no longer contain a separate Preliminary Results section. For these proposals, you may either include preliminary results as one subsection under the Approach or as separate subsections under each aim. In many federal applications, the shift has been away from volume toward relevance/quality of the preliminary data you include.

➤ Organize longer Preliminary Results sections by subtopics and under each subheading chronologically or from most to least important. Consider providing an overview upfront

For more extensive preliminary results—whether within the Approach or as a separate section—divide your Preliminary Results into manageable subsections to provide the reader with a better overview. The overall structure within these subsections is normally either chronological or from most to least important. You may also consider providing an overview, rationale, or introduction of your preliminary results before going into the specifics of your subsections. The next two examples show two such overviews.

Example 25-2   **Overview of Preliminary Results subsections**

Preliminary studies

To understand the multitude of potential functions of short tandem repeats in the human genome, we are employing a battery of molecular, cellular, genetic, functional genomic, and bioinformatic approaches to analyze short tandem repeat-binding sites. We summarize our findings to date below.

Example 25-3   **Overview of Preliminary Results subsections**

Introduction. We are initially focusing our VSV vaccine production on two viral antigens HBMS and HBC. We will produce VSV vectors that encode HBMS. Expression of HBMS has the advantage of encoding the largest number of possible cell epitopes without potential toxicity. In addition to producing these vectors, we will produce recombinant VSV vaccine vectors that express the viral capsid structural protein HBV Core. The use of these two particular proteins has the technical advantage that (a) they are highly expressed, (b) the immune response to these antigens is very well characterized both in humans and in mice, and (c) a variety of reagents (antibodies, ELISA assays, etc.) are readily available for analyses. We have initiated studies in these directions as described below.

(Michael Robek, proposal to federal agency)

Organization Within Subsections

➤ Organize each Preliminary Results subsection

- Purpose or background
- Experimental approach
- Results
- Interpretation of results
- Optional: problem statement/unknown/logical next steps/important results

After the overview (if provided), describe details of subtopics in the different subsections. Start each subsection by presenting context first. You can provide context by giving a short background or writing an overview of the main findings on this topic or by stating the purpose of the particular set of experiments right after the subheading, as shown in Example 25-4.

Example 25-4   **Subsection of Preliminary Results**

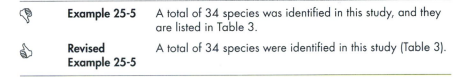

Subsection heading

Mechanism of Regulation of Cell Cycle Synchronization. Numerous studies in *Drosophila melanogaster* have focused on determining the underlying mechanism(s) of cell cycle synchronization over the last decade. These studies have increased our understanding of cell cycle synchrony revealing . . .

Topic sentence

After providing context, briefly state the experimental approach and the major findings on this topic in the field. Then, give an interpretation of the results to make them meaningful for the reader.

When describing preliminary results, be sure to refer to each related figure and table in the text, but do not repeat all the data shown in figures and tables. Instead, describe your results, and point the reader to a figure or table by citing this figure or table in parentheses after the description of the results.

Example 25-5   A total of 34 species was identified in this study, and they are listed in Table 3.

Revised Example 25-5   A total of 34 species were identified in this study (Table 3).

If appropriate, and if the subsection's topic directly precedes a specific aim of the proposal, end the relevant Preliminary Results subsection by stating what has not yet been determined or what logical next steps would be. Such statements will lead the reviewers into the next section describing your proposed research on this specific aim. You may also reemphasize important findings on which your specific aims are based. Consider highlighting such last thoughts or problem statements by placing them in italics or by underlining them. Examples of ending sentences within the Preliminary Results section are shown in the next example; these sentences serve as a setup of the corresponding specific aim, which will be described in the Approach/Experimental Design section.

Example 25-6   **Ending sentences of Preliminary Results subsections**

a   *This dramatic change indicates that further research into X may yield important results. However, this factor has only been investigated in one small observational study conducted in the absence of Y (19).*

b   At present, it is unclear what the function of A is.

c   *Together, these observations strongly suggest that X may increase Y by promoting Z.*

> **d**   However, many educators are unaware of these find-
> ings and of the special accommodations required by
> children and adults who are dyslexic.
>
> **e**   *Therefore, we confirmed that the specificity of the im-*
> *munodominant A cell response in XX-immunized mice*
> *is directed to known epitopes in this antigen.*

If necessary, compare your data to that of other studies, speculate on possible mechanisms, or draw general conclusions, as shown in Example 25-7.

Example 25-7   **Specific conclusion at the end of a Preliminary Data section**

Specific conclusion	The practicality of this approach is shown by the fact that Cooper and Pez at Air Products Corp have produced a practical H_2 storage material, N-ethyl carbazole, which holds ~5.5% w/w H_2 and is fully recyclable.

Following are three examples of subsections containing all necessary
components.

Example 25-8   **Preliminary Data subsection**

Subsection heading	
Topic sentence	**The current Y vaccine.** Currently, a vaccine is available that prevents Y infection. This vaccine is a recombinant protein
Details of preliminary results	preparation consisting of the Y envelope protein expressed in yeast. . . . <u>Despite its success, the Y vaccine has a</u>
Problem	<u>number of characteristics that are suboptimal</u>:

1) . . .

2) . . .

Example 25-9   **Preliminary Data subsection**

Subsection heading	**Downstream targets of X**
Topic sentence	Malignant gliomas are the most common primary malignant brain tumors in the US with more than 15,000 new cases each year (Fisher et al., 2005). The five-year survival rate for malignant glioma is only 3.3% (Scott et al., 1998), and recurrence is almost universal (Sathornsumetee and Rich, 2006). Recurrence is largely due to a small fraction of glioma stem cells that have tumor-initiating functions (Fan et al., 2007) and drive tumor formation and tumor cell migration (Clarke, 2004).
Details of preliminary results	Long-term remissions and cures require eradication of glioma stem cells. However, glioma stem cells are resistant to standard radiotherapy (Bao et al., 2006). In addition, current chemotherapy drugs, such as cisplatin, vincristine,

Problem/gap

methotrexate, and etoposide, are not able to eliminate glioma stem cells effectively (Dean et al., 2005). The reason why chemotherapy does not work well for these tumors is likely that conventional drug delivery approaches do not effectively reach glioma stem cells due to the blood-brain barrier (Kreuter, 2001).

Example 25-10   **Summary of preliminary data**

Overview

To search for slab-related heterogeneities beneath Central America we have performed P-wave seismic migration analysis using several intermediate and deep earthquakes in South America and several Californian seismic networks. Both conventional migration and the high-resolution migration method, Slowness Back azimuth Weighted Migration (SBWM), were tested.

Details of preliminary results

Prior to migration, all of standard seismic arrivals (such as pP, sP, PcP, PP, etc.) were masked out to enhance weak but coherent energy from small-scale heterogeneities. A 3-D volume of migration grid points was defined in the study area (Figure 5a) from the surface to the lowermost mantle, with the vertical and horizontal grid intervals of 50 km. Theoretical travel time, back azimuth, and slowness were computed for each grid point and for each earthquake and receiver pair, using the IASP91 reference Earth model (25). We found that the migrated energy is concentrated at the topside of the high-velocity anomaly imaged by tomography. Migrated energy is weaker in the SBWM image than that by conventional migration, but its distribution along the slab-like velocity anomaly remains similar, suggesting its robust nature. What these scatterers actually represent is of course an open question at present, and it will become clearer when we estimate the physical properties of these scatterers and compare them with mineral physics predictions for subducted oceanic crust. . . . These

Specific conclusion/ applicability for aim

preliminary results support that our plan to investigate the fine-scale structure of the mantle beneath Central America is promising given the available source-receiver pairs for this region and the power of new-generation migration methods.

(Jun Korenaga, proposal to federal agency)

➤ For lengthy Preliminary Results sections, add a summary at the end

If your Preliminary Results section is long, consider adding a one-paragraph summary at the end of it to remind the reviewers about the most important aspects of this section. Two examples of such summaries for a Preliminary Results section are shown next.

Example 25-11a   Summary of preliminary data

> Through our preliminary data we started to unravel how IkappaB kinase alpha (Ikkα) regulates diverse functions. We found that there is a new functional isoform of Ikkα that has a broad expression pattern. In addition, we generated a polyclonal Ikkα antibody and successfully used this antibody for chromatin immunoprecipitation and for subsequent cloning to identify direct Ikkα targets. Our data also show that in addition to other phenotypes, reducing Ikkα signals affects cell survival. We will further examine the molecular relationship between Ikkα and cell survival, particularly examining how this affects progression of cancer.

Example 25-11b   Summary of preliminary data

> We produced recombinant WT VSV vaccine vectors that express the HBMS or HBC proteins and have preliminarily characterized the magnitude and specificity of the CD8 T cell response to HBMS. We now propose to (1) produce additional highly attenuated and boosting vaccine vectors, (2) test the ability of these vectors to induce an HBV-specific antibody and T cell response, and (3) determine if the VSV vaccine vectors can function as therapeutic vaccines using two mouse models for chronic HBV replication. These studies have the potential to result in new approaches to prevent and/or treat a chronic disease that is a major cause of cancer worldwide.
>
> *(Michael Robek, proposal to federal agency)*

25.4   IMPORTANT WRITING RULES

Word Choice

➤ Pay attention to word choice

Words in the Preliminary Results section should be chosen carefully. Select the most exact and descriptive wording that reflects what you want to say. Use simple, precise words, and avoid jargon and repetition. (See Chapter 2 for more details on clear word choice.)

Pay particular attention to the following specific words and phrases in the section. These words are often used carelessly by authors, especially ESL authors, but they should be distinguished because of their implied meaning (see also Appendix A).

ESL advice

Clearly/It Is Clear/Obvious

Omit *clearly* and similarly subjective terms and phrases. *Clearly* makes authors seem arrogant and is an overt, but not necessarily persuasive, attempt to influence the reader.

Example 25-12   Figure 6 *clearly* shows that Grizzly bears prefer south-facing slopes for hibernation.

 Revised   Figure 6 shows that Grizzly bears prefer south-facing
Example 25-12   slopes for hibernation.

Significant

Significant in science refers to "statistically significant." If you write, for example, "Flow rate decreased significantly," the reader expects statistical details to follow this phrase. If you report results of statistical significance, specify the significance level.

If you do not plan to provide statistical details, use *markedly* or *substantially* instead of *significantly*. Also, remember you should quantify these qualitative words by using precise values or referring to data or figures such as in the following example.

 Example 25-13   Flow rate decreased substantially (23%).

Tense

➤ Use past tense for results but present tense for general conclusions

Results are usually reported in past tense because they are events and observations that occurred in the past.

 Example 25-14   a   Imidazole **inhibited** the increase in arterial pressure.

    b   Once nectar **was depleted** from the *D. wrightii* flower, all of the moths **switched** to feeding from the *A. palmeri* flowers (Riffell, JA, 2008).

Exceptions are results that are considered general knowledge and are still true. These findings and generalizations should be reported in present tense.

 Example 25-15   a   The *fgk* gene **has** several different introns.

    b   These results **suggest** that learning the association between nectar reward and flower type is primarily olfactory mediated (Riffell, JA, 2008).

See also Chapter 4, Section 4.4 for more information and examples on verb tense.

TABLE 25.1 SIGNALS FOR THE PRELIMINARY RESULTS SECTION

BACKGROUND/ PURPOSE	EXPERIMENTAL APPROACH	RESULTS	INTERPRETATION OF RESULTS	PROBLEM/ UNKNOWN
X is a member of . . .	. . . when X was subjected to . . .	Gyl et al. found . . .	. . ., indicating that . . .	. . . is unclear.
Z was tested . . .				X needs to be improved
For the purpose of XYZ . . .	X was subjected to . . .	Experiment X showed . . .	. . ., consistent with . . .	. . .
	ABC was performed.	Ho reported that . . .	Thus, A is specific for . . .	. . . do not include . . .

25.5   SIGNALS FOR PRELIMINARY RESULTS

➤ **Signal the elements of your Preliminary Results section**

To emphasize different portions of the Preliminary Results section, consider using signals, such as those shown in Table 25.1.

ESL advice

These examples provide great starting points and may be particularly useful for those authors who have writer's block or ESL authors.

25.6   COMMON PROBLEMS OF PRELIMINARY RESULTS

The most common problems of the Preliminary Results section include:

- Findings not directly related to proposed work
- Experience of investigator not adequate
- Omission of important points or procedures
- Poor organization

➤ **Provide only findings directly relevant to your proposed work**

All findings and information in this section should be directly relevant to your proposed work. If you provide other, unnecessary information, it may confuse reviewers. Additional information will also add unnecessarily to the overall length of the proposal.

➤ **Demonstrate your expertise**

You need to convince reviewers that you have sufficient expertise on the topic and/or preliminary data and collaborations to carry out the proposed Specific Aims. Expertise may be achieved through credentials, prior publications, preliminary data, training, or collaborations. In describing your preliminary data, you will also show indirectly that you have the necessary means and resources (laboratory, equipment, and team).

➤ **Do not omit important ideas or techniques**

If you neglect to describe an important idea or technique, you may not convince the reviewers that your project is feasible or that your expertise is sufficient. Having a peer outside your field review your proposal will help you identify missing ideas and techniques.

➤ Provide logical transitions

If you organize your subsections and the information contained within them and provide logical transitions, the Preliminary Results section will be more interesting for the reviewers. Reviewers are particularly bothered by information that is not logically organized and clearly presented. An outline or reverse outline may help you identify poor organization.

25.7   REVISING THE PRELIMINARY RESULTS

When you have finished writing the Preliminary Results (or if you are asked to edit this section for a colleague), you can use the following checklist to "dissect" the section systematically:

☐ 1. Are your findings directly related to the proposed work?
☐ 2. Are all important results and procedures described?
☐ 3. Did you indicate logical next steps or what is unknown or problematic?
☐ 4. Did you interpret your results for the reader and distinguish between data and results?
☐ 5. Is the Preliminary Results sections organized chronologically or from most to least important within subsections?
☐ 6. Do subsections state:

	Purpose/background
	Experimental approach
	Results
	Interpretation of results
	Problem statement/unknown/logical next steps

☐ 7. Is the section organized well?
☐ 8. Do all subsections logically follow each other?
☐ 9. Did you follow instructions on how to write the section and what to include?
☐ 10. Revise for style and composition using the basic rules and guidelines of this book:
    ☐ a. Are paragraphs consistent? (Chapter 6, Section 6.2)
    ☐ b. Are paragraphs cohesive? (Chapter 6, Section 6.3)
    ☐ c. Are key terms consistent? (Chapter 6, Section 6.3)
    ☐ d. Are key terms linked? (Chapter 6, Section 6.3)
    ☐ e. Are transitions used and do they make sense? (Chapter 6, Section 6.3)
    ☐ f. Is the action in the verbs? Are nominalizations avoided? (Chapter 4, Section 4.6)
    ☐ g. Did you vary sentence length and use one idea per sentence? (Chapter 4, Section 4.5)

☐ h. Are lists parallel? (Chapter 4, Section 4.9)

☐ i. Are comparisons written correctly? (Chapter 4, Sections 4.9 and 4.10)

☐ j. Have noun clusters been resolved? (Chapter 4, Section 4.7)

☐ k. Has word location been considered? (Verb follows subject immediately? Old, short information is at the beginning of the sentence? New, long information is at the end of the sentence?) (Chapter 3, Section 3.1)

☐ l. Have grammar and technical style been considered (person, voice, tense, pronouns, prepositions, articles)? (Chapter 4, Sections 4.1–4.4)

☐ m. Are words and phrases precise? (Chapter 2, Sections 2.2 and 2.3)

☐ n. Are nontechnical words and phrases simple? (Chapter 2, Section 2.2)

☐ o. Have unnecessary terms (redundancies, jargon) been reduced? (Chapter 2, Section 2.4)

☐ p. Have spelling and punctuation been checked? (Chapter 4, Section 4.11)

SUMMARY

PRELIMINARY RESULTS GUIDELINES

- Establish credibility.
- Report important previous findings relevant to the topic and to the specific aims.
- Indicate logical next steps or what is unknown or problematic.
- Use figures and tables if necessary to enhance findings.
- Do not overstate your competence or capabilities.
- If needed, use subsections to organize the Preliminary Results section.
- Adjust the level of writing according to the topic and to your potential readers.
- Interpret your results for the reader.
- Distinguish between data and results.
- Summarize and generalize relevant previous results.
- Organize shorter or integrated Preliminary Results sections by including (in this order):
  ○ Background and methods to make the study understandable
  ○ Enough results to be convincing
  ○ A conclusion (emphasize the interpretation of results)
- Organize longer Preliminary Results sections by subtopics and under each subheading chronologically or from most to least important. Consider providing an overview upfront.

- Organize each Preliminary Results subsection:
  - ° Purpose or background
  - ° Experimental approach
  - ° Results
  - ° Interpretation of results
  - ° Optional: problem statement/unknown/logical next steps/ important results
- For lengthy Preliminary Results sections, add a summary at the end.
- Pay attention to word choice.
- Use past tense for results but present tense for general conclusions.
- Signal the elements of your Preliminary Results section.
- Provide only findings directly relevant to your proposed work.
- Demonstrate your expertise.
- Do not omit important ideas or techniques.
- Provide logical transitions.

PROBLEMS

PROBLEM 25-1 Preliminary Results Section

Assess the following partial Preliminary Results section. Ensure that all the parts of a paragraph for the Preliminary Results section are provided and signaled (purpose or background of the experiment, experimental approach, results interpretation of the results, problem statement/ unknown/logical next steps).

ELISA for secreted HBMS. Because HBMS is a secreted protein (46), we determined whether the HBMS protein produced in VSV-infected cells is secreted into the culture media. We detected HBMS in the media of VSV-HBMS-infected cells but not in uninfected or recombinant WT VSV-infected cells by both Western blot (data not shown) and qualitative ELISA (Table 2). We found by quantitative ELISA that secreted HBMS levels reached 35 ng/ml in the media of VSV-HBMS-infected BHK cells by 24 h post infection. <u>Therefore, the HBMS produced in VSV-infected cells is correctly processed for secretion.</u>

(Michael Robek, proposal to federal agency)

PROBLEM 25-2 Preliminary Results Section

Assess the following partial Preliminary Results section. Ensure that all the parts of a paragraph for the Preliminary Results section are provided and signaled (purpose or background of the experiment, experimental approach, results interpretation of the results, problem statement/ unknown/logical next steps).

There have been few measurements of root exudation in situ in forest ecosystems largely due to methodological challenges. As a first test to examine the role of CO_2 on exudation in loblolly pine, we grew seedlings for 150 days in controlled chambers under conditions designed to approximate the CO_2 and N gradients at the Duke FACTS-1 site. We found that pine seedlings increased mass specific exudation rates in response to elevated CO_2, and that the magnitude of the CO_2 effect was greatest in our lowest N treatment (Figure 5a). In contrast, higher N supply reduced exudation. Our preliminary data suggest that exudation may be increased by 60% in the elevated CO_2 plots, representing an increased flux of ~10 g C m^{-2} yr^{-1} to soil.

. . .

Our preliminary data also suggest that elevated CO_2 may increase the importance of rhizosphere processes by increasing the flux of soluble exudates from roots to soil. In addition, our results indicate that CO_2-induced changes in the chemical composition of the exudates released can affect the magnitude of the microbial response and the quantity of N mineralized in the rhizosphere. Given the paucity of root exudation studies from trees in situ, and the absence of studies of the exudation response of field grown trees to elevated CO_2, there is a need to understand the effects of elevated CO_2 on both the quantity and chemical composition of the exudates released from roots.

(Richard Phillips, proposal to federal agency, modified)

PROBLEM 25-3 Preliminary Results Section

Assess the following partial Preliminary Results section. Ensure that all the parts of a paragraph for the Preliminary Results section are provided and signaled (purpose or background of the experiment, experimental approach, results, interpretation of the results, problem statement/ unknown/logical next steps).

Calculations with a General Circulation Model of the atmosphere (6) indicate that during idealized permanent El Niño-like conditions, the warming of the Eastern equatorial Pacific reduces the area covered by stratus clouds, thus decreasing the albedo of the planet. At the same time, the atmospheric concentration of the powerful greenhouse gas, water vapor, increases. This scenario may have happened during the early Pliocene, amplifying the warm conditions at that time. Consequently, projections of the effect of increasing concentration of greenhouse gases on climate should include a thorough consideration of the tropical climate conditions.

(Alexey Federov, proposal to federal agency, modified)

CHAPTER 26

Approach/Research Design

In the Approach/Research Design section, you will need to convince the reviewers that your plans can achieve the stated specific aims in the proposed time frame.

THIS CHAPTER COVERS:

- The different elements of the Approach section
- The format of the Approach section
- The closing paragraph
- Signals for the reader in the Approach section
- Common mistakes in the Approach section
- Revising the Approach section

26.1 OVERALL

The Approach/Research Design section, often also referred to simply as Approach or Methods, is the largest section and the heart of your proposal; it will receive much scrutiny during review. Many proposals are rejected by reviewers because of a bad Approach section. Your goal is to convince the reviewers that your plans are sound and you can achieve the stated specific aims.

26.2   COMPONENTS

➤ Address the problems stated in the Background and Preliminary Studies sections

The main function of the Approach is to propose in detail what experiments you will perform to address each specific aim. Each aim will need to be achievable during the time frame set for the proposed project. The section not only needs to clearly show the reviewers your plans to go about achieving the proposed aims, it also should explain alternate routes if you hit roadblocks.

To delineate your plans convincingly, the Approach should elucidate the rationale and, potentially, preliminary findings for your proposed specific aims, what you expect to add to the topic, the specific experimental approaches, the expected outcomes of your experiments, and what alternate approaches you could take. It should also summarize expected outcomes and the impact of your findings.

Other items that may be included in this section include (a) potential limitations, such as possible technical problems that may arise and alternate plans you may implement; (b) a timeline, which shows that your experiments are doable in the proposed time and that you have a logical plan for carrying them out (see Chapter 27, Section 27.2); (c) a description of your research team and what each member will contribute (see Chapter 27, Section 27.2); and (d) a list of facilities and infrastructure you have available to you (see Chapter 27, Section 27.2).

➤ Dedicate about two-thirds of your proposal to the Approach/Research Design section

Depending on the funder's requirements, the length and content of this section may vary. Typically, however, count on roughly two-thirds of your entire proposal being dedicated to the Approach.

➤ Adjust your level of writing to your audience

Even though the preliminary data may be valid and interesting, the logical continuation and next steps may be obscure. Good style and clear, logical presentation are especially important for this section. To reach all of your reviewers, be sure to adjust your level of writing to your audience. Overall, you do not have to include exact technical specifications or detailed compositions of solutions, but you should describe what experiments you propose, why, and what you expect them to show.

Tone and Tense

➤ Convey confidence—use future tense

The tone of your writing is important. Make your writing convey confidence and authority. Show that you are knowledgeable about the subject and take responsibility for your propositions. Do not be afraid to take a stand. The most important way to indicate that you are confident is to use future tense ("We will . . ."). Do not use "could," "should," "might," "may," or "we see an opportunity to."

26.3   FORMAT

➤ Organize the Approach into subsections according to your specific aims

➤ Within the subsections, cover

- Rationale, including hypothesis and preliminary studies (optional)
- Experimental design
- Analysis
- Expected results (outcomes and significance)
- Alternative strategies (not always required)

Begin the Approach with a brief overview. Consider restating your overall rationale, objective, and approach as a reminder and transition for the reader. You may also provide a progress report if you are applying for a renewal or an overview/summary of relevant preliminary results if they are not presented in a separate section (see Chapter 25) or integrated under the approach for each aim. Then, cover each specific aim in a separate subsection.

Give each specific aim subsection the exact heading as that of the corresponding specific aim listed in the Abstract and Specific Aims section. Subsections within the Approach may be divided into further subsections.

Under each specific aim, you may want to start by providing the rationale, preliminary studies, or hypothesis of the aim. Then, tell how you will go about finding the answer to the question of the specific aim. Explain your experimental design and your proposed analysis. The level of detail here depends on where you are planning to send the grant proposal. Generally, for federal grants this section will have to be detailed and technical, although buffer compositions and detailed protocols are generally not included. If you are applying to private foundations or corporations, read their individual instructions. Some require you to be as detailed as federal agencies. Others simply ask for a short general overview of your planned approach.

An example of a partial Approach is shown next. This example starts off by relisting the specific aims as a title of the section. It then provides a brief overview followed by a subsection on experimental design and another on Analysis. Such subsections may not always be labeled as such, but the components (experimental design and analysis) are usually provided. Note that the Analysis subsection is further divided into different components, which consist of experimental approach, expected outcomes, and significance. These components may also be described under their own subheading. Expected outcomes and significance should be written in overview form. These latter two sections give importance to your proposal.

☞ | **Example 26-1** | **Research Design and Methods subsection**

Heading of subsection | **Aim 1. Determining regulation of Y and Z by different XX isoforms**

overview | This aim focuses on examining how the different XX isoforms as well as XX-interacting proteins affect Y and Z.

Experimental design | **Experimental design**
We have already established a function for at least one new isoform. To continue this analysis, we will generate peptide antibodies. Although peptide antibodies are slightly more expensive than traditional fusion constructs, they allow for the precise specificity required to delineate each isoform. We will use sequence analysis protocols as outlined that are specific for each antibody (Fig. 11).

Experimental approach | **Analysis**
First, we will confirm the spatial and temporal expression patterns of the different isoforms to determine . . . We will examine where the particular isoforms are expressed and . . . Second, we will examine the expression of the different proteins by Western blot to see if we can detect distinct regulatory effects of the different isoforms. Finally,

Expected outcome | if the antibodies function well, we will use them for additional analyses involving interacting proteins (Section 5.1).

Significance | These experiments will ultimately define how the different isoforms are regulated.

Example 26-2 is another example of an Approach. This particular section starts by providing an overview of the experimental approach and methods that will be employed in the study. It does not contain a rationale. This is okay as the rationale portion is optional. However, the section could be strengthened by stating the expected outcome.

Example 26-2 | **Research Design and Methods subsection**

Methods

Overview | Research study components. The four watershed sites will be studied as follows: We will establish 20 m × 20 m plots in each of the four high elevation watersheds. Transects will be established along the contour of each plot. Random locations along each transect will be used for each soil sample collection. Soil samples will be collected quarterly for two years for the eight, 3-monthly soil samplings.

(1) At the first of the eight 3-monthly samplings, the sites will be characterized for vegetation composition, soil carbon and nitrogen fraction sizes, bulk density, and aspect and slope.

Experimental design

(2) Eight, 3-monthly samplings for temporally responsive soil microbial and abiotic variables will be taken.

(3) In the second year of sampling, the remaining four 3-monthly samplings will be complemented by short-term, ^{15}N pulse-chase field assays to follow the fate of inorganic and organic nitrogen inputs to the soils. In addition, ^{13}C-glucose will also be pulse-chase to follow its speed of movement through the soils.

Specific experimental approach

Soil sampling and processing: We will set up 20 m × 20 m plots in each of the four watersheds. Four transects will be established across each of these plots for selection of random sampling locations. Quarterly, ten 5 cm diameter and 10 cm deep cores will be taken and pooled, permitting microscale spatial heterogeneity in soil variables to be factored out of the watershed comparisons. Immediately following collection, composited soil samples will be sieved to <6 mm and extracted for NO_3 and NH_4-N determinations. Soils will be transported on dry ice (for later molecular analyses) or on wet ice and then transferred to −80°C or 5°C. At the first sampling date from each watershed, subsamples will be air dried and used for SOC fraction determinations. Additional cores will also be taken at the first sample date for bulk density determinations.

Expected outcome is not stated

(Mark Bradford lab, proposal to federal agency, modified)

Alternative Strategies

➤ Think of the Alternative Approach section as a decision tree

Often, the experimental design and analysis portion is followed by a subsection on alternative strategies, especially for federal grants and for proposals that are hypothesis based. As experiments frequently do not go as planned, reviewers often like to see that you have thought about potential pitfalls and alternatives strategies. If you do not foresee any pitfalls, consider stating that your experiments are straightforward and no major roadblocks are expected.

Including an Alternative Strategies subsection will indicate to the reviewers that you have thought through your experiments very carefully. It will also show them that the proposal has coherence even if some of the hypotheses or assumptions turn out different than anticipated. In this section, you may describe how you will continue your study if your outcome does not support a particular research hypothesis, that is, if one experiment determines the next. Think of the Alternative Approach section as a decision tree: if your experiment leads to A, you will do X; and if it leads to B, you will do Y; and explain how result A or B affects the interpretation of the second experiment or leads into a subsequent experiment. If you logically construct your Alternative Approach section, it shows reviewers that the study and experiments can be completed as proposed and will lead to useful results.

In the Alternative Approach section, you may also describe alternate routes to arrive at an expected finding in case one or the other experimental setup does not work. Know that all designs have limitations and that these limitations require either controls in the experiment or acknowledgment in interpretation and implications. Limitations may exist for interpretation of data. These limitations require a discussion of appropriate experimental controls. Other limitations arise due to generality of results and require qualifying interpretation of findings. Yet other limitations are due to a particular design. When you discuss limitations in your Research Design and Methods section, include only limitations of major concern to the purpose and potential conclusions. Do not list every possible item that "could go wrong" as this will not leave a positive impression on the reviewers.

An example of an Alternative Approach section is provided in Example 26-3.

Example 26-3   **Alternative Approach section**

Expected Results and Alternative Approaches

Expected outcome

> We expect to successfully generate the highly attenuated, single cycle and G-protein exchange vectors for use in the vaccination studies. We do not anticipate any significant problems with this aim based on the more than ten years experience using these techniques, and our successful recovery of recombinant WT VSV-HBMS and VSV-HBC in collaboration with the X lab. However, if we encounter unanticipated problems, other alternative approaches exist. For example, if expression of HBMS or HBC does not induce a strong immune response in subsequent experiments, we can express the proteins from the first position within the VSV genome to increase protein expression (70). We chose the fifth position because these recombinant viruses are better characterized and easier to recover (unpublished observations). However, expression from the first position in the genome typically leads to 3–4-fold greater protein expression and can therefore significantly enhance the immune response (63).

Alternative approach

(Michael Robek, proposal to federal agency)

If you organize your subsections as described, reviewers are sure not to miss any important information. The following example shows an Approach subsection that contains a subsection on expected results and alternative approaches.

Example 26-4   **Research Design and Methods subsection**

SPECIFIC AIM 3. Determination of the efficacy of VSV as a therapeutic vaccine for HBV.

Rationale

HBV vaccine vectors elicit very strong protective antibody responses to virus-expressed proteins in antigen-naive animals (Table 1). However, relatively little is known about the ability of these vectors to induce an immune response in a therapeutic, rather than prophylactic, setting. One such ineffective immune response is associated with HBV infection. We will test the hypothesis that . . . These experiments are important because they may identify therapeutic vaccination with VSV as a novel treatment for HBV infection.

Experimental Design and Methods

Mouse models of HBV replication. Studies on the immune response to HBV are limited by the relative lack of animal models in which to study virus infection. However, mouse models of HBV replication are available. . . . The use of mice for these experiments is advantageous in that . . . Each model has a number of distinct advantages inherent to the specific system (Table 4). For example, model 1 . . . and model 2. . . . The use of these two models is advantageous in that it allows us to perform our studies on therapeutic vaccinations in both contexts.

To measure virus replication and persistence, we . . . Virus replication and persistence will be measured by ELISA. Based on a previously published study (30), we expect that . . .

Analysis

Animals replicating HBV will be immunized to generate a HBV-specific immune response. We will use five different immunization permutations, and analyze how . . . Immunizations will be performed as described in Specific Aim 2a.

Expected Results and Alternative Approaches

If these experiments are successful as defined by the criteria outlined in 3c below, then we will test . . . However, if these immunization procedures do not induce . . . , then we will determine if . . .

(Michael Robek, proposal to federal agency)

To ensure continuity within the Approach, provide a chain of topic sentences. Use transitions and key terms, and consider word locations and power positions. Organize the topics by proceeding from most to least important unless there is a reason for putting one topic before another. The topic sentence of each paragraph must indicate not only the topic or message of the paragraph but also the relation of the paragraph to the previous paragraph(s) and thus to the corresponding specific aim. Reviewers who study your proposal more carefully will read the chain of topic sentences to identify subtopics quickly. Note again that the subtitles for individual specific aims within the Approach should match the aims stated in the Specific Aims section of the proposal word for word.

26.4   CLOSING PARAGRAPH

➤ **Provide a concluding summary for the overall section**

➤ **Indicate the significance of your work**

At the end of the proposal, consider providing some closure by writing a one-paragraph concluding summary. This last paragraph should summarize your specific aims, generalize the expected outcomes, and give importance to the proposal. Do not bring in new evidence for the summary. Rather, complete the "big picture" by restating your specific aims and the expected overall outcomes and significance.

The significance of the work can be provided by including far-reaching interpretations and conclusions. Try to generalize your expected findings to other, broader situations. Depending on your level of certainty, significance can range from the practical application to the theoretical proposition. Adding a practical application, giving advice, implying an action, or providing a proposition in the concluding paragraph give the proposal importance.

		Level of certainty	Example
Practical	**Application**		". . . can be used for . . ."
Theoretical	**Proposition**		". . . will provide an understanding of . . ."

You can highlight the significance of the problem by relating the problem to various general topics of urgency or importance, such as an important point in time, solution to a practical problem, applicability to a wide or critical population, or filling an essential research gap (see also Chapter 23).

Often, the importance of the expected outcomes is not discussed or not adequately explained, which leaves the reviewer wondering about the significance of the proposal and the overall understanding of the researcher. It is therefore important that you indicate the significance of your expected findings at the end of the Approach section. Ending this section with an Alternate Approach subsection is not nearly as convincing and powerful as ending with a concluding impact statement or summary.

Because the conclusions are the major message of your proposal, you should phrase them with great care. Possible ways to provide some closure of your work at the end of the Research Design section are shown in the next examples.

Example 26-5   **Summary of Research Design and Methods section**

Study question | How does the one protein coordinate various neural cell functions to ultimately regulate migration and maturation? This grant application has three specific aims that are independent but closely related, and are all aimed at determining how *DCX* functions to modulate several different regulatory processes. We will find how the different *DCX* isoforms regulate Doublecortin outputs (5.1), we will perform genome wide location analysis to define direct outputs of Doublecortin under different conditions (5.2), and we will determine how changes in genomic stability affect
Specific aims | Doublecortin-dependent phenotypes (5.3). At each step, we will examine how two *DCX* dependent processes are affected: neural migration and maturation. These findings will likely be relevant to cortical development and open up new areas to explore in the understanding of the connec-
Significance | tions between neural migration and maturation.

The previous summary opens with the overall question of the study. A question such as this can be used instead of stating the overall goal of the project. The summary then lists the three specific aims and provides a brief, general overview of the experimental approach before it provides a significance/impact statement.

Following are two other examples geared toward a private foundation.

Example 26-6   **Summary of Research Design and Methods section**

In summary, we propose to develop a new modeling approach to examine the intersection of behavioral, evolutionary, and ecological dynamics with the overall goal of developing novel theory on cooperation and conflict in real populations. This theory will be used to make specific predictions about cooperation and conflict during mating and parental care, allowing us to predict what factors determine when cooperation and conflict arise from social interactions. The ability to understand what factors shift an interaction from cooperative to conflict has broad relevance to our understanding of social behaviors in a wide variety of contexts, including human behavior.

✍

Example 26-7   Summary of Research Design and Methods section

General strategy

Objective

> Our research project will combine cutting-edge imaging technique, genetic manipulation, and state-of-the art sequencing facilities to elucidate how X controls brain cancer cell invasiveness. This innovative interdisciplinary approach will develop important new tools to examine how interactions with X affect neural migration and maturation. Our laboratory's ability to perform . . . coupled with our expertise in . . . provides us the unique opportunity to achieve these goals.

Specific aims

Expected outcome

Significance

> We believe that aberrant neural migration can be blocked in one of two ways: (1) by targeting . . . ; and/or (2) by inhibiting . . . Both strategies represent novel therapeutic approaches to attenuate brain tumor invasiveness. Our studies should identify the most critical targets for possible prophylactic treatment to block tumor cell invasiveness. The information we will gain from these studies will provide a solid foundation for better diagnosis and novel strategies to treat invasive brain cancer and reduce brain cancer-related mortality.

26.5   SIGNALS FOR THE READER

➤ Signal the elements of the Approach; use subheadings

Signal the different elements of the Research Design section so that the reviewers recognize immediately what they are reading. Possible signals for various elements include some of the phrases listed in Table 26.1, but may also be subheadings.

ESL advice

Table 26.1 may provide great starting and reference points when you put together your first draft and may be particularly useful for those authors whose native language is not English.

TABLE 26.1 Signals for the Approach Section

OBJECTIVE	EXPERIMENTAL DESIGN	EXPECTED OUTCOME	SUMMARY	SIGNIFICANCE
We will test the hypothesis that . . .	We will . . . Animals will be . . .	Findings will define . . .	In summary, . . .	Our findings can/ will serve to . . .
We aim to determine if . . .	. . . will be analyzed . . .	The proposed study will . . .	In conclusion, . . .	. . . can be used . . .
The study has three specific aims . . .	To determine . . .	We expect to find . . .	Overall, . . . Taken together . . .	. . . will provide insight into . . .

26.6   COMMON PROBLEMS

The most common problems of the Approach section include:

- Experimental plan is too ambitious
- Aims depend on previous aims excessively
- Amount of detail
- Expected results are not clear

➤ Ensure your experimental plan is feasible

Your experimental plan cannot be too ambitious in scope or volume. If you are looking to get the next round of research funded as well, you have to prove that you can follow through with this one. In addition, you have to convince the reviewers that you have the required expertise to complete the proposed work.

Example 26-8   **Sample research objective and aims that are too ambitious**

Research Objectives

The overall goal of this three-year research program is to identify the underlying genetic and mechanistic causes during neural development that lead to cerebral palsy. Our findings will revolutionize both diagnosis and treatment options for this disabling condition. The proposed program consists of three aims:

Aim 1: To elucidate the genetic sources of cerebral palsy.

Aim 2: To determine the mechanistic causes of cerebral palsy.

Aim 3: Using knowledge gained through Aims 1 and 2 to develop diagnostic screens, therapies, and/or treatments for the condition.

This research objective and aims are too ambitious for a three-year plan. Although Aims 1 and 2 may be achievable depending on preliminary results, previous findings, and complexity of the underlying mechanisms, Aim 3 would span many more years, and potentially even decades. The authors would stand a much better chance for a convincing proposal if they were to convert Aim 3 into a statement of significance or impact as shown in the revised example.

Revised Example 26-8   **Sample research objective and aims that are too ambitious**

Research Objectives

The overall goal of this three-year research program is to identify the underlying genetic and mechanistic causes

during neural development that lead to cerebral palsy.
Our findings will revolutionize both diagnosis and treatment options for this disabling condition. The proposed
program consists of two aims:
Aim 1: To elucidate the genetic sources of cerebral
    palsy.
Aim 2: To determine the mechanistic causes of cerebral
    palsy.
**Knowledge gained through our expected findings will lay
the foundation for future development of diagnostic
screens, therapies and/or treatments for the condition.**

➤ Do not use too many interrelated aims

If the research design depends too much or solely on the first specific aim,
failure of this aim will guarantee failure of any subsequent aims. As a consequence, your grant may not get funded unless it is a high-risk grant application
with a promising high payoff. Including a section on alternative approaches
may come in helpful to address failure issues as such a section shows that the
study will still result in interpretable and important findings.

Example 26-9   **Sample aims that are too interrelated**

Specific Aims

Spider silk is one of nature's strongest fibers and shows
great promise for use in biomedical applications. However, to date we have been unable to produce such silk
artificially. Here, we propose to
Aim 1: Study the conditions and components of spider silk
    production of the species *Nephilia clavipes*.
Aim 2: Use insights gained in Aim 1 in order to generate
    artificial spider silk using a recently developed spinning
    device.
Aim 3: Employ synthetic spider silk to suture wounds in
    mice and promote wound healing.

The aims in the previous example are too interrelated and too dependent on
the success of Aim 1. Any good reviewer will be concerned about what happens to the rest of the project if Aim 1 is not successful. Even if Aim 1 works
out, Aim 3 is still dependent on Aim 2, and similar concerns will arise for
this Aim. The proposed work should be revised to include aims that are less
interdependent. For example, Aim 2 could suggest to study mechanical
properties of the silk from *Nephilia clavipes* and possibly other species.

➤ Do not use too much or too little detail

Adjust your Research Design section according to who your potential readers will be. If you are writing for a very specific group of people, stay within

that group's area of interest. Assume that your readers are intelligent but not experts in your area. If you are writing for a lay audience, you probably need to discuss much broader strategy and provide more generalizations and background. Generally, you want to keep the level of detail directly proportional to the novelty of the techniques and methods. Do not exaggerate, however. Note that if you make it sound too easy, the reviewers will think you do not understand the problem. If it sounds nearly impossible, they may think the risk is not worth funding.

Example 26-10   **Sample Approach section that contains too much detail**

Degradation in the presence of aqueous-substance X. To produce aqueous substance X, PIPES (piperazine-N,N'-bis(2-ethanesulfonic acid); pKa (6.76 at 25°C)) buffer will be mixed with 10 g of X and the resulting mixture will be filtered with membranes (0.25 μm). The concentration of X in aqueous phase will then be determined by subtracting the amount/weight of insoluble X from the total amount of X. To determine the degradation of Y, 5 h of Y particles will be mixed with 2 L of PIPES buffer containing aqueous X, shaken for 24 hrs at 4°C, and Y degradation will be measured by spectrophotometry at A_{450}.

The Approach in Example 26-10 contains too many technical specifications that detract from the flow of the paragraph. Such specifications are particularly troublesome if used throughout the entire Approach section. They also add unnecessary bulk to the text and may not even be followed in actual procedures. Projects are perceived as more convincing—and the authors as more confident—if the Approach section is composed at a higher level (i.e., with less technical details, even for a peer audience).

Revised Example 26-10   **Sample Approach section that contains too much detail**

Degradation in the presence of aqueous-substance X. To produce aqueous substance X, PIPES (piperazine-N,N'-bis(2-ethanesulfonic acid)) buffer will be mixed with X and the resulting mixture will be filtered to remove insoluble X. The concentration of X in aqueous phase will then be calculated. To determine the degradation of Y, Y particles will be mixed with PIPES buffer containing aqueous X, and Y degradation will be measured by spectrophotometry.

➤ State expected results clearly, and interpret them

When your expected results are not clearly interpreted or when they are not listed, your reviewers may not follow your train of thought and may not understand how you will fulfill the proposed specific aims. Even if the results may seem obvious to you, the reviewers are not necessarily experts on this topic and will need to be told precisely what the expected outcomes are and what they mean; otherwise, they are left hanging.

26.7   REVISING THE RESEARCH DESIGN AND METHODS SECTION

When you have finished writing the Approach (or if you are asked to edit these sections for a colleague), you can use the following checklist to "dissect" the sections systematically:

☐   1. Is the proposed experimental plan doable in the time frame given?
☐   2. Are most aims not too dependent on each other?
☐   3. Did you convey confidence? (Use future tense?)
☐   4. Has each of the specific aims been addressed under different subsections with about the same detail?

	Rationale/purpose/preliminary results
	Experimental design
	Analysis
	Expected results
	(Alternative strategies)

☐   5. Did you follow the funding agency's specific instructions on how to write this section?
☐   6. Did you address the problems stated in the Background and Preliminary Studies sections?
☐   7. Did you adjust your level of writing to your audience?
☐   8. Did you organize the Approach into subsections according to your specific aims?
☐   9. Did you provide a concluding summary?
☐ 10. Did you provide alternative approaches if needed?
☐ 11. Revise for style and composition based on the basic rules and guidelines of the book:
    ☐ a. Are paragraphs consistent? (Chapter 6, Section 6.2)
    ☐ b. Are paragraphs cohesive? (Chapter 6, Section 6.3)
    ☐ c. Are key terms consistent? (Chapter 6, Section 6.3)
    ☐ d. Are key terms linked? (Chapter 6, Section 6.3)
    ☐ e. Are transitions used and do they make sense? (Chapter 6, Section 6.3)
    ☐ f. Is the action in the verbs? Are nominalizations avoided? (Chapter 4, Section 4.6)
    ☐ g. Did you vary sentence length and use one idea per sentence? (Chapter 4, Section 4.5)
    ☐ h. Are lists parallel? (Chapter 4, Section 4.9)

☐ i. Are comparisons written correctly? (Chapter 4, Sections 4.9 and 4.10)

☐ j. Have noun clusters been resolved? (Chapter 4, Section 4.7)

☐ k. Has word location been considered? (Verb follows subject immediately? Old, short information is at the beginning of the sentence? New, long information is at the end of the sentence?) (Chapter 3, Section 3.1)

☐ l. Has grammar and technical style been considered (person, voice, tense, pronouns, prepositions, articles?) (Chapter 4, Sections 4.1–4.4)

☐ m. Has tense been used correctly? (Chapter 4, Section 4.4)

☐ n. Are words and phrases precise? (Chapter 2, Sections 2.2 and 2.3)

☐ o. Are nontechnical words and phrases simple? (Chapter 2, Section 2.2)

☐ p. Have unnecessary terms (redundancies, jargon) been reduced? (Chapter 2, Section 2.4)

☐ q. Have spelling and punctuation been checked? (Chapter 4, Section 4.11)

SUMMARY

APPROACH/RESEARCH DESIGN GUIDELINES

- Address the problems stated in the Background and Preliminary Studies sections.
- Dedicate about two-thirds of your proposal to the Approach/Research Design section.
- Adjust your level of writing to your audience.
- Convey confidence—use future tense.
- Organize the Approach into subsections according to your specific aims.
- Within the subsections, cover
  ° Rationale, including hypothesis and preliminary studies (optional)
  ° Experimental design
  ° Analysis
  ° Expected results (outcomes and significance)
  ° Alternative strategies (not always required)
- Think of the Alternative Approach section as a decision tree.
- Provide a concluding summary for the overall section.
- Indicate the significance of your work.
- Signal the elements of the Approach; use subheadings.
- Ensure your experimental plan is feasible.
- Do not use too many interrelated aims.
- Do not too much or too little detail.
- State expected results clearly, and interpret them.

PROBLEMS

PROBLEM 26-1 Approach Section
On the left margin of this Approach section, identify the components (rationale, methods, analysis, expected outcomes, alternate strategies, and significance).

Specific Aim:

Optimization of triage and treatment in the context of HIV and XDR TB.

In this model, we will improve upon preliminary studies to evaluate more precisely whether factors that can be assessed within a few days of a patient's admission can be used to determine which patients are most likely to be infected with drug resistant strains. We will thereby determine an optimal empirical treatment regimen for the interim period until their drug sensitivity results are available. Key clinical factors to be incorporated into the model include TB treatment history, hospitalization history, sputum smear results, lack of response to first-line treatment, chest X-ray deterioration, and rapid rifampin, kanamycin, and ciprofloxacin resistance testing results. Parameterizing our model with the data collected, we will calculate the likelihoods that a patient has non-MDR, non-XDR MDR, or XDR TB, and will determine what further studies are required to enhance confidence in the likelihood calculations. The model will also account for the benefits and disadvantages of different treatment regimens in terms of treatment outcomes, side effects, costs, pill burdens, and the probability of acquired or amplified resistance.

(Alison Galvani, proposal to private foundation, modified)

PROBLEM 26-2 Approach Section
Assess the following, partial Approach section. Ensure that all the parts of an Approach section are provided and any unnecessary parts are omitted.

Here, we propose to study stress response behaviors in autistic children using both well-developed laboratory paradigms and dense array electro-encephalographic assessments of maturing cortical coherence in response to inhibitory demands on standard neuropsychological tasks under non-stressful and stressful conditions. We will study children's response to stressors in a well-characterized cohort of boys and girls age 5 to 12 years old. We will follow the children over one year with detailed assessments of their social and scholarly development as well as their overall health. Laboratory stress procedures include asking the children to have a conversation with a trained professional and to complete simple math and social tasks in a controlled test setting. These procedures have been found to reliably induce changes in physiological indices of stress (heart rate, salivary cortisol, blood pressure) and to be a reliable index of individual differences in stress response.

PROBLEM 26-3 Approach Section
Assess the following, partial Approach section. Ensure that all the parts for an Approach section are provided and any unnecessary parts are omitted.

I. Laboratory and Field Behavior Study

<u>Objective</u>: To determine the function of butterfly wing patterns by characterizing the detailed behavioral responses of insect and predators in response to the Monarch butterfly, *Danaus plexippus* L. Laboratory experiments will incorporate information observed in the field.

<u>Approach and Analysis</u>: Field experiments will include a) tethering of live butterflies to vegetation to observe who predates them; b) observation of interactions of insect predators, such as mantids, with Monarch butterflies; and c) observation of interactions of avian predators, such as Blue Jays, with the same butterflies. For our experiments, artificial habitats will have representative irradiance, background color, and complexity based on ecological information we collect in the Mojave Desert. Our model insect predator will be *Mantis religiosa*, a common mantid species. Using high-speed video, we will examine whether the width and number of stripes affect the target of mantid attacks.

CHAPTER 27

Budget and Other Special Proposal Sections

In all Proposals, you will have to include a budget. Depending on where you send your Proposal, you may also be asked to include Proposal sections other than the ones discussed in previous chapters. The budget and potential other Proposal sections are described in this chapter.

THIS CHAPTER COVERS:

- The budget and budget narrative for Proposals
- Other potential Proposal sections, including:
    - Biographical Sketch
    - Collaborative Arrangements
    - Competing Interest Statement
    - Description of Facilities
    - Description of Organization
    - Dissemination Plans/Data Sharing
    - Education and Outreach
    - Endorsement Letters and Letters of Reference
    - Executive Summary
    - Expected Outcomes
    - Expertise and Role
    - Evaluation Plan
    - Future Directions
    - Impact Statement

- ○ Importance of Funding
- ○ List of Publications
- ○ Organizational Chart
- ○ Personal Statement
- ○ Personnel
- ○ Project Management
- ○ Recognition Statement
- ○ References/Bibliography
- ○ Relevance to Foundation
- ○ Table of Contents
- ○ Timeline
- ○ Title Page/Cover Page

27.1   BUDGET

➤ Follow specific instructions on how to prepare the budget if provided by the funder. Ask your administration for help

Aside from the Abstract, Specific Aims, Background, Significance, Preliminary Results, Innovation, and Approach, every Proposal should contain a Budget section. The budget is a detailed breakdown of the financial support requested from the sponsoring agency. It lists the expenses required to successfully perform the proposed research. Your business office and/or the grants and contracts administration office in your institution usually helps in setting up a budget.

➤ Include direct and indirect costs

Proposal budgets should not just include direct costs but must also take into account indirect costs (also known as assessment or overhead). The latter costs refer to ongoing expenses of operation and usually cover administration, utilities, taxes, repairs, and so forth. Rates can vary widely depending on what your organization requests and what the funder is willing to pay (range can be from 0%–70%). Costs to also be included are fringe costs (insurance, pension plans, etc.). Together, all of these expenses equal the total costs.

To prepare the budget, you will also need to know (a) the sponsor's likely upper limit and (b) whether cost sharing or matching funds are required. Some funders supply budget forms to be filled out along with corresponding directions. Others only provide guidelines, and yet others do not supply you with any of these. If a funder requests a specific budget format and/or provides guidelines, follow the instructions carefully. If the funding agency to which you are applying does not have a set budget form, you may have to provide one yourself.

Depending on your proposed project as well as on the sponsor to which you are applying, the following items may be included as direct costs in your budget.

Personnel: If needed, include salaries and prorated fringe benefits (Social Security, retirement benefits, etc.). Also include an inflation factor and projected salary increases when applying for multiyear grants. For staff, identify the percentage of time that each individual will spend on this particular project and prorate the appropriate costs. If your case worker's salary is 100% covered by the County Department of Mental Health, then a supplemental 20% of this position's salary cannot be charged to a different funder. However, the 20% contribution to this project can be shown as a county contribution to the project. Calculate fringe benefits and payroll taxes on the exact cost percentage based on information from your human resources department.

Equipment: You may have to provide cost estimates for new equipment.

Operations:

Materials and supplies: List expenses for supplies, transportation, photocopying, or a similarly appropriate category.

Travel: Do not include nonessential travel in the budget for a project that only has local impact. If conference attendance is of vital importance for project effectiveness or results disseminations, identify the conference and provide ample justification in the budget narrative.

Animal costs: Include costs for animal housing and care.

Facilities: Only list those relevant to the project and no other facilities costs.

Consultation: List any contracts if needed.

Other: If you include a "miscellaneous" or "other" budget category, be sure to itemize what this category refers to. Other budget costs may include greenhouse space costs, land use costs for field experiments, user fees for equipment, and so forth

➤ Make sure costs are realistic and justified

Excel is a great program for setting up your budget because it will calculate as you go. When you prepare your budget table, use only whole numbers (no cents) with proper formatting: $50,450 (not 50450). Do not overestimate or underestimate costs. Ensure that you conduct the project within the proposed budget, that the costs are realistic and justified, and that the budget matches the proposed goals and methods. Describe any cost sharing (what your institute/university will contribute). Once you have all numbers in place, check for their consistency between the project description, budget narrative, and budget line items. Be prepared to justify and negotiate your budget with a potential funder.

A sample budget is shown in Example 27-1.

☞ **Example 27-1**

Table 27.1 Sample budget table for matching costs

CATEGORY	YEAR 1		YEAR 2		CUMULATIVE		TOTALS
	ABC Funding	Other Sources*	ABC Funding	Other Sources*	ABC Funding	Other Sources*	All Sources
Personnel (Salary + Fringe Benefits)							
Professor X, Principal Investigator (50%)	$ –	$xxxxxx	$ –	$xxxxxx	$xxxxxx	$xxxxxx	$xxxxxx
Associate Research Scientist (100%)	$xxxxxx	$xxxxxx	$xxxxxx	$xxxxxx	$xxxxxx	$xxxxxx	$xxxxxx
Personnel (Stipend + Fringe Benefits)							
Postdoctoral Fellow (100%)	$xxxxxx	$xxxxxx	$xxxxxx	$xxxxxx	$xxxxxx	$xxxxxx	$xxxxxx
Personnel Subtotal	$xxxxxx	$xxxxxx	$xxxxxx	$xxxxxx	$xxxxxx	$xxxxxx	$xxxxxx
Equipment		$xxxxxx				$xxxxxx	$xxxxxx
Equipment Subtotal	$ –	$xxxxxx	$ –	$ –	$xxxxxx	$xxxxxx	$xxxxxx
Operations							
Consumable supplies	$xxxxxx	$xxxxxx	$xxxxxx	$xxxxxx	$xxxxxx	$xxxxxx	$xxxxxx
Animal costs	$xxxxxx	$xxxxxx	$xxxxxx	$xxxxxx	$xxxxxx	$xxxxxx	$xxxxxx
Travel/symposiums	$xxxxxx	$ –	$xxxxxx	$ –	$xxxxxx	$xxxxxx	$xxxxxx
Consultation	$ –	$xxxxxx	$ –	$xxxxxx	$xxxxxx	$xxxxxx	$xxxxxx
Facilities/Overhead (y%)	$ –	$xxxxxx		$xxxxxx	$xxxxxx	$xxxxxx	$xxxxxx
Other*	$ –	$ –		$xxxxxx	$xxxxxx	$xxxxxx	$xxxxxx
Operations Subtotal	$xxxxxx	$xxxxxx	$xxxxxx	$xxxxxx	$xxxxxx	$xxxxxx	$xxxxxx
Totals	$xxxxxx	$xxxxxx	$xxxxxx	$xxxxxx	$xxxxxx	$xxxxxx	$xxxxxx

*Other includes institutional and external forms of support.

519

For shorter Proposals or if the full budget form is provided as an appendix, simpler tables may be shown within the text of the Proposal. Examples of such tables are shown next.

👍   **Example 27-2   Short sample budget table**

Use of Funds

The total project cost is $xx,xxx. The requested funds are allocated as follows:

Personnel	PI, salary for 2 months	$xx,xxx including benefits
	Postdoctoral fellow	$xx,xxx including benefits
Equipment	laptop computer	$x,xxx
Travel	domestic and foreign	$xx,xxx
Publication costs		$x,xxx

👍   **Example 27-3   Short sample budget table**

PERSONNEL

Personnel	Title	Salary	%Effort	Requested	Fringe	Total
Peter Mayer	PI	$xxx,xxx	5%	$x,xxx	$x,xxx	$x,xxxx
Todd Blewett	Graduate Student	$xx,xxx	90%	$xx,xxx	$x,xxx	$xx,xxx
TBN	Postdoctoral Fellow	$xx,xxx	100%	$xx,xxx	$x,xxx	$xx,xxx

Subtotal for Personnel = $xx,xxx

MATERIALS AND SUPPLIES

Tissue Culture Supplies: $x,xxx

Molecular Biology Reagents: $x,xxx

Subtotal for Materials and Supplies = $xx,xxx

Overhead (xx%) = $xx,xxx

TOTAL REQUESTED = $xxx,xxx

Budget Narrative or Budget Justification

In addition to the budget table described in the previous section, you may be required to provide a brief narrative or justification of each budget line item and to specify funding source(s). Such a section often consists of two parts: a budget justification/narrative and a list of other support. Examples on how to write such a section are shown following:

Example 27-4   **Sample budget narrative/justification**

Budget Justification

Peter Warneke, PhD, Associate Professor and Principal Investigator, will devote 5% of his effort to this research project. He will provide guidance to the research staff and be involved in the design of the experiments.

Erin Toddl, Graduate Research Associate, will devote 90% of her effort to this research project. She will examine the localization of . . .

Other Funding Sources

You may be required to list other active or pending funding sources:

Example 27-5   **Other funding sources**

Other support

There is currently no other support available for the proposed project.

Example 27-6   **Other funding sources**

<u>Current Support/Institutional Support</u>

Department of ABC, Y University, "Start-Up" Support (PI: John Doe) (9/1/2006–) $xxx,xxx

Roles of X in Z   Principal Investigator NIH/NINDS (R01) 02/01/04–01/31/08

$xxx,xxx direct costs/current year

Overlap: None

<u>PENDING</u>

Regulation of X   Principal Investigator Department of Defense (IDEA Award)

01/01/08–12/31/10

$xxx,xxx direct costs/year

Overlap: None

Investments of Organization

If your Proposal is one that is meant to be matched or one that is meant to establish a partnership, you may also have to include a budget subsection about the investment your organization is making and another requested from the collaborating organization. Here, you should list any commitments that will be made toward the match or partnership such as facilities and equipment, as shown in the next example.

 Example 27-7 **Investments of organization**

Investments by ABC University

ABC University's investments in the field of chemistry constitute $xx million over x years. In addition to building institutional strength in Chemistry via a program of long-term investments in faculty recruitment and research infrastructure, ABC University has made specific new commitments to essential core facilities over the next 5 years. These include commitments for a clean room and for an Ion Cyclotron Resonance Fourier Transform Mass Spectrometer. The resulting cores will enrich the proposed collaboration and will be fully available to all participants.

These investments will be accompanied by continuing faculty recruitment for Chemistry at ABC University and are superimposed on a background of $xx million of research grants and contracts annually for Chemistry at ABC University (2013 data).

Summary of ABC University contributions to the collaboration:

	Years 1 through 5
Clean room	$xx,xxx
Ion cyclotron	$xx,xxx
Total	$xx,xxx

27.2 OTHER SPECIAL PROPOSAL SECTIONS

Overall

➤ Check the Proposal instructions carefully to ensure that your Proposal contains all sections as required by the funder

The overall content of a Proposal varies depending on where you are planning to send it. Whereas many federal agencies have very defined requirements as to what sections to include in your Proposal, many private funders do not. Even the names of various sections and their content may be different for the latter. In addition, you may be asked to include other special sections in your Proposal. Check the Proposal instructions carefully to ensure that your sections contain what is required by the funder.

The following elements are a list of special components that you may be asked to provide:

- Biographical Sketch
- Collaborative Arrangements
- Competing Interest Statement
- Description of Facilities
- Description of Organization
- Dissemination Plans/Data Sharing
- Education and Outreach
- Endorsement Letters and Letters of Reference
- Evaluation Plan
- Executive Summary
- Expected Outcomes
- Expertise and Role
- Future Directions
- Impact Statement
- Importance of Funding
- List of Publications
- Organizational Chart
- Personal Statement
- Personnel
- Project Management
- Recognition Statement
- References/Bibliography
- Relevance to Foundation
- Table of Contents
- Timeline
- Title Page/Cover Page

Again, note that specific names for each section may vary, and that some of these sections may be combined and others may be split into two or more subsections depending on the funder.

Biographical Sketch

In your Biographical Sketch, include your professional information as well as selected publications. Typically, biographical sketches do not exceed one to two pages. Keep your Biographical Sketch short, listing only the most important details. Do not cram too much information into it.

Example 27-8   **Biographical Sketch section**

John Doe

Current Address

Dept. of Biochemistry, Y University, XLML 332

3445 Froh Street, P.O. Box 3454345, Cincinnati, Ohio, USA

Phone: ####### Fax: #######;
e-mail: john.doe@yuniversity.edu

Education

PhD in Molecular Biology, 1983, University of California, Los Angeles, CA

B.S. in Biology, 1973, Lafayette College, Easton, PA

Professional Experience

Professor, Dept. of Biochemistry, 1997–present, Yale University, New Haven, CT

Asst/Associate Professor, Dept. of Biochemistry, 1989–1997, Yale University, New Haven, CT

Other Appointments and Awards

- President, International Society of Biochemistry, 2001–present

Other Activities:

- Co-Editor-in-Chief, *XXX Journal,* Elsevier (1999–present)
- Associate Editor, *YYY Journal* (1991–present)
- Editorial Board, *Journal of ZZZ* (1997–present)

Selected Publications (5 of 122 total)

1. XXX, YYY, and **J. Doe,** *XXX Journal* 3749: 61–76, 2006.
2. AAA, BBB, and **J. Doe**. *Medical Imaging,* 13(9): 40–48, 2002.
3. CCC, DDD, EEE, and **J. Doe,** *Science,* 22(5): 86–100, 2002.

. . .

Collaborative Arrangements

Some Proposals require you to include a Collaborative Arrangement. Such a section should state with whom you are planning to collaborate, what this collaboration would involve, and how the collaboration would be maintained.

Example 27-9   **Collaborative Arrangements section**

Collaborative network

Our collaborative network will be maintained by visits, email, conference calls, and meetings. We will have a minimum of two conference calls per month, and an annual meeting, as well as site visits to both the United States and country X to facilitate greater collaboration between X scientists. The proposed research is based on a collaboration that has already been established and has worked well for several years. Our modeling study will be an expansion of the collaborative process that has already resulted in the detection and initial study of Z. Furthermore, several members of our collaborative network are instrumental in the implementation of Z treatment and control protocols. Thus, the proposed research will have a direct impact on public health policy designed to reduce Z.

Competing Interest Statement

Some Proposals ask for a statement about competing interests (sometimes also called a dual commitment or conflict of interest statement). A conflict of interest may arise if an author's or coauthor's judgment, action, and interpretation are affected due to financial and other gains or due to personal relationships and affiliations. A conflict of interest may also arise if the potential exists for their judgment to be affected. Often, people are not aware that their interests pose a conflict, which can range from negligible to conflicts involving a direct stake in the content or competition. To assess a potential conflict of interest, consider if something at work is associated with your private interests and may lead to a personal gain. If so, disclose this possible conflict of interest to avoid any appearance of bias in grant applications and to ensure their integrity. Conflicts may include being on an advisory committee, a spouse or relative in a related field, a recent collaborator or advisor, a former student, financial ties, board of directors, potential employer, and more. Generally, the greater the possible gain, the more serious the conflict of interest.

Example 27-10   Competing interest statements

1. "The authors declare no competing financial interests."
2. "No potential conflict of interest relevant to this application or line of work is reported."
3. "Dr. Abajaratna serves as a consultant to company X, and Dr. Barnes has a license for Y with Z hospital. We report no other conflict of interest relevant to this Proposal."
4. "A grant (AG0967) has previously been awarded to DB by Y for the study of ZZZ. No other author has any further financial link to Y."
5. "Y University owns a patent, xx/yyy,zzz, that uses the approach outlined in this application and which has been licensed to Z."
6. "The applicants have nothing to disclose."

Description of Facilities

If you are asked for details on your facilities, describe those available to you, particularly those at your institution, as well as those required for the completion of your study.

Example 27-11   **Description of Facilities section**

Professor X's laboratory occupies 3,900 ft² in the Center for Biochemistry. Professor Y's laboratory (1,000 ft²) is located in the immediately adjacent Center of Immunobiology. Professor Y also directs the BBB lab, which occupies 760 ft² of space in the basement level of the same building. Professor X's laboratory contains all the major equipment needed for modern biochemical research, including –70°C freezers, PCR machines, spectrophotometer, electrophoresis equipment, power supplies, protein sequencer, HPLC system, and three chemical fume hoods. The BBB lab is equipped with three advanced CCC setups, including DDD setups.

Description of Organization

It may be helpful to include the following passage or portions thereof on your organization in a Proposal in which it is important to highlight the significance of your institution. Two such sample passages are shown in Example 27-12.

Example 27-12   **Why X University—statement**

a   As an internationally recognized university, X attracts exceptional students and scholars. The University offers a remarkably open and flexible community that encourages working across disciplines and a transdisciplinary spirit of inquiry especially suited to broad, new, and innovative questions such as the one at the core of this Proposal. Furthermore, the X community fosters and supports healthy risk taking by its investment in individuals with nontraditional career trajectories and in questions and programs that cross and/or blur traditional academic disciplines and boundaries. The University supports structures to nurture creative thinking and innovative approaches even if these stand outside departmentally based organizational models.

b   X University is one of the world's foremost research institutions. Comprised of the Graduate School of Arts & Sciences and a total of eleven professional schools, the University is diverse in many research areas. Its expert faculty and bright students are engaged in rigorous interdisciplinary collaborations, making substantial contributions across the academic spectrum in such fields as biomedical sciences, law, public policy, and the humanities. As such, the University maintains a prominent national and international presence, attracting a diverse student body of over 11,000 from countries around the globe.

X's unique geographical location, together with its reputation for stellar academic scholarship and advancement, makes the University a leader among research institutions.

Dissemination Plan/Data Sharing

If you need to include any dissemination plans in your Proposal, describe how the findings of your study would be publicized. Include any meetings, conferences, publications, reports, newsletters, broadcasts, teaching materials, or websites.

Example 27-13 a   Dissemination Plan section

We plan to present our research findings at two major conferences: the Fall Meeting of Y and the General Assembly Z. In addition, we aim to submit our research articles to either the Journal of A or B Journal International. These two journals are the top-class journals in the field, thus securing the most critical reviews for our research.

Example 27-13 b   Dissemination Plan section

The Institute website will contain a regularly updated resource section on current understanding of X as well as a section on publications and materials from the Institute. We will make available information on the methods developed in phase one and also offer trainings in the use of Y. Results achieved from our studies will be published in peer-reviewed, high-profile journals and presented at national and international meetings. We will also develop instructional materials derived from X. These materials will be a critical part of our plan to disseminate the work of our community of scholars. These instructional materials are also a key step in developing prevention and intervention approaches for large groups of children and adults and are aimed to help more individuals access a capacity for healthy adaptation to X, a goal we anticipate will be a priority of the Institute in its later phases of development.

Education and Outreach

Some funders ask for a section on education and outreach. For others, this section may be included under a different heading, such as "Broader

Impacts" for the NSF. Subsections may also be included under this umbrella, including:

- Workshops
- Guided Scientific Tours
- Seminar Series
- School Programs
- Scientist Training

Following is an example of an Education and Outreach section. See also Chapter 22, Section 22.4 for a brief example of a "Broader Impact" section.

Example 27-14   Education and Outreach section

The funding requested in this Proposal will be used to support and train four minority graduate students at X and Y. These students will interact closely with each other and will be exposed to both simulations and experimental work. Their training will include regular and frequent interactions and discussions with their respective PI, as well as regular conference calls, presentations, and frequent visits to the laboratory of the other PI, facilitated by the close proximity of the two institutions. In this project, the graduate students will be introduced to a new, more integrated approach to molecular techniques and gain invaluable skills, such as DNA sequencing, PCR, and data analysis— skills they will be able to apply in future research.

Funding of the proposed project will also ensure that two undergraduate students will continue to have an opportunity to engage in cutting-edge research at the laboratories of the PIs. The student populations at the two participating institutions are diverse and the PIs have built a strong track record of engaging undergraduate students of diverse backgrounds in their research projects, either over the summer or through independent research courses. PI#1 has trained three female undergraduate research students, all of whom are now in PhD programs at different universities. PI#2 has supervised two undergraduate students, one Native American and one African American, the latter of which also entered a graduate program.

As part of the outreach planned for the proposed project, we will make any data we obtain available on the Web in an easy-to-use web format. Open sharing of these data through the ABC Project will enable a multi-pronged approach to research on Z as it allows others to use the information for their own research and approaches, including for data mining or for developing or verifying diverse theories. In addition to participating in a community outreach program, in which we will give three workshops on X, we will hold an annual seminar series to which the wider community will be invited. Key findings of our project will also be made available through publication and presentations at conferences.

Endorsement Letters and Letters of Reference

Many grant Proposals require an endorsement letter or a letter of reference. Such letters are often written by a department head, a dean, or even the president of a university. If such a letter is required, give plenty of time to the person composing this letter so your application is not delayed because of it. You may even consider drafting this letter to help the writer with information about you and your project.

 Example 27-15   Endorsement letter

Date

Dear Mr. Miller:

I am delighted to write this letter of support for Professors Laura South and Peter Plowe's application to the AAA Foundation. We are honored that these outstanding faculty have been invited to present a full Proposal on their program entitled "Moon Phases and Tidal Variability."

Professors South and Plowe aim to answer important questions on if and how moon phases influence tides throughout the planet. Funding for their proposed program will allow them to bring to light the forces that drive tidal variability. I am convinced that a collaboration between the AAA Foundation and X University on this project will be very productive and advance our shared institutional values and aims in this and related fields.

We are extremely grateful for the previous awards provided by the Foundation and thankful for the opportunity to submit this current request. We look forward to hearing the results of your deliberations and to future opportunities to collaborate in pursuit of our mutual goals.

Sincerely,

Dean of X University

For examples of Letters of Reference/Recommendation, see Chapter 31, Section 31.9.

Evaluation Plan

For certain funders, evaluation plans are very important and need to be addressed in your Proposal. These usually list annual or final reports, meetings, and conferences. An evaluation plan (performed by external evaluators) may also be an important component of a federal Proposal, especially for NSF grants. Such a plan is required particularly for large-scale, multi-investigator Proposals. In the plan, you have to show how you can prove that you have been successful in meeting your objectives.

Occasionally, program evaluations can be complex and may require you to hire a professional evaluator. Some funding agencies may even require you to use specific, trained specialists as evaluators. Such requirements are most often the case for larger programs that run for an extended time and for which outcomes are expected to be multifaceted and extensive.

Example 27-16 **Evaluation Plan section**

The project will be measured not only by publications but by its impact on the greater community. As an evaluation mechanism, we plan to a) establish a Scientific Advisory Board (SAB) composed of experts in A, B, and C and b) have two retreats to discuss and identify major scientific issues relevant to the project. The SAB will consist of Professors Peters, Falk, and Gusto. We will convene the scientific advisory board once a year to discuss and identify major scientific issues relevant to the project and to provide evaluation. We will also consult with the AAA foundation for the selection of external advisors. At the end of Years 1, 2, and 3, we will have an annual retreat where all involved parties will participate and present their work by posters and talks. The SAB will be present and will reconvene to advise and offer improvements. A final report will also be presented to the AAA Foundation.

Executive Summary

Some funders may require you to provide an Executive Summary. Usually the Executive Summary does not exceed two pages and should be written for a well-educated lay audience. The components of the summary can be split into different subsections.

Aside from providing an overview and objective, the Executive Summary should contain methodology. Briefly describe the methods that will be employed, highlighting what is unique or distinctive in one to two paragraphs. You may also have to include a rationale, a summarized timeline, and a section on personnel. Identify the principal investigators and other key personnel, and summarize their expertise and roles in the project. In addition, the summary may include a brief version of the budget, stating the total cost of the project, the amount requested from the funder, and the amount of other sources of support such as from your institution. Describe how the percentage of funds requested from the funder will be allocated.

Expected Outcomes

Although sometimes Expected Outcomes will be included in other subsections, such as in the Approach/Research Design section or in the Significance section, other times you will be asked to place your expected outcomes into a separate section. State how your proposed study will advance the field(s) involved, and mention also any broader applications or significance.

Example 27-17   Expected Outcomes section

> The proposed project represents a major step toward a self-consistent theory of X. Overall, the study will broaden our understanding of Y. For the field of physics, this project will produce a new model that can generate insights into X. For the computational field, the project will lay a knowledge base for the application of specific methods to large-scale Y.

Future Directions

Some sponsors ask for a section on future directions. Such a section should cover what will happen after funding is expended and should describe how the proposed project will continue.

Example 27-18   Future Directions section

> The proposed research aims to develop X methodology for the analysis of desertification. We expect that development of this methodology will find utility in a broad range of application areas and will advance the field of geology in general. An initial commitment from the AAA Foundation will serve to launch the project, but we hope to leverage AAA support to obtain additional funds from other granting agencies. With this goal in mind, we plan to submit one or more such Proposals by the end of the second year. Over time, we will seek to extend the X-based approach into complementary areas of geology with the addition of collaborators from our university and other institutions as appropriate.

Impact Statement

An impact statement is similar to a summary but expands more on the significance portion of it. This statement is often one of the most important paragraphs in a Proposal. As described in earlier chapters (Chapters 22 and 23), in the impact statement highlight the significance of the problem by relating it to one or more of the following criteria:

- Important point in time
- Solving a practical problem
- Applicability to a wide or critical population

- Filling a research gap
- Having many implications for a wide range of practical problems
- Advancing a specific field of research

Describe the impact your expected outcomes will have for future projects in your field, outside your field, and for society at large.

 Example 27-19   **Impact Statement section**

X represents perhaps our best hope for tackling many of the seemingly intractable problems facing the world: from engineering tissues to expanding computing capacity; from responding more effectively to global climate change and pollution to meeting the threats from bioterrorism. The preceding requests will have both immediate and long-term impact on advancing X and its applications. The enhanced Seed Project Funding will nearly double the number of new projects that can be started, benefiting not only University students and faculty but also the new fields and discoveries that will emerge. Y will have both immediate and long-lasting positive impact on the kinds of research that may be conducted on campus, thereby encouraging faculty from every field to extend their research horizons and to think more broadly about new opportunities for A, B, and C than would otherwise be possible.

On a long-term scale, support for a new building would have the greatest and most lasting impact on X. Such a dedicated facility will meet evolving, long-term educational and research needs. In recognition of such a transformative gift, the new building will provide a wide range of highly visible naming opportunities such as an appropriate recognition of the AAA Foundation. It is our hope that the AAA Foundation will be a partner in this important endeavor.

Importance of Funding

Some funders like to know why their funds are important for your project. Explain if your funding options are limited, and highlight the importance of your proposed research.

 Example 27-20   **Importance of Funding section**

The goal of this Proposal is to support an innovative program that brings together A, B, and C. An application to the NIH would not be appropriate at this time because this work is high risk as it will be conducted at the interfaces of three radically different disciplines. However, the potential payoff in terms of advances in technology, theory, and not least our understanding of XXX requires forward-looking investment. Thus, the information that we expect to obtain from this Proposal will be relevant to many different fields of biology and medicine. The methodology will be widely applicable beyond the field of XXX, for example, to studies of Q, O, and P. The AAA Foundation can provide the critical initial impetus to enable these advances.

List of Publications

Investigators are sometimes asked to provide a list of their publications. Provide this list using a scientific format in presenting the references. If your list is too extensive, consider listing only selected publications, and label the list as such.

Example 27-21   **List of publications**

Selected Publications

Cove, D.J., & Knight, C.D. The moss *Physcomitrella patens*, a model system with potential for the study of plant reproduction. Plant Cell, 5 (1993) 1483–1488.

Cove, D.J. Regulation of Development in the moss *Physcomitrella patens*. In: Russo, V.E.A., Brody, S., Cove, D. & Ottolenghi, S. Development. The molecular genetic approach. pp 179–193. Springer Verlag Berlin, 1992.

Cove, D.J., Kammerer, W., Knight, C.D., Leech, M.J., Martin, C.R., & Wang, T.L. Developmental genetic studies of the moss, *Physcomitrella patens*. Symp. Soc. Exp. Biol. 45 (1991) 31–43.

. . .

Personal Statement

When you are asked to provide a personal statement, such as in a bio sketch, briefly describe why your experience and qualifications make you particularly well-suited for your role in the project.

Example 27-22   **Personal statement**

The proposed project will test how X modifies Y through a transdisciplinary case study. My training and professional career paths have prepared me well for the proposed study. I have a broad background in A and B. As a postdoctoral research fellow, I worked on. . . . As an Assistant and Associate Professor at X University, I broadened my field of expertise in A and its integration with diverse disciplines, such as . . . In the combined years of experience and expertise, my work has applied interdisciplinary approaches to address public health challenges for a wide range of diseases, including . . . As in the proposed research, I have used these results to do A. My research has advanced both methodology and public health applications. . . . The proposed project builds logically on my prior work. During these investigations I will oversee and collaborate on A and on the application of B, as well as supervise a postdoctoral associate, who has previously worked at the interface of climate change and disease transmission.

Personnel

Sponsors usually want to know who the team is. Provide information on your team in the Personnel section. Allow insight into your team's background and respective areas of expertise. Sometimes you may even be asked to provide an organizational chart. This information may be combined with other sections under a Project Management section, but often funders request a separate section describing each member on the team. These descriptions can be done either in narrative form or by listing each team member's responsibilities and qualifications.

 Example 27-23   Personnel section

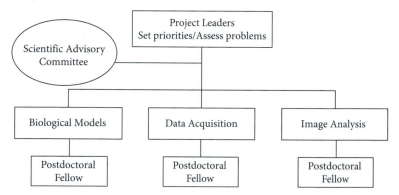

The team is in a unique position to carry out the proposed project due to the complementary expertise of its members in three essential areas. Professors X, Y, and Z will together be responsible for directing the overall project, including setting priorities and assessing problem areas. Professor X will guide the biological parts of the work, whereas Professors Y and Z will manage the technical and mathematical aspects. The organizational chart of the project personnel illustrates the roles and responsibilities of key personnel. Work in each project area will be carried out by 4 to 5 postdoctoral fellows, 3 of whom will be paid by this grant; fellowships for the others will be funded from external sources. The entire project group will convene in biweekly group meetings. A Scientific Advisory Committee will be established consisting of experts in relevant areas.

Example 27-24   Personnel section

The PI is among the few scientists who is versatile in both A and B. She has worked on a variety of X problems in the last 12 years, including XX [9,15], XXX [17], and XXXX [20]. She has also made various contributions to AA, such as B [18,19], C [10,11,13], and D [12,14]. She has been collaborating with experts of almost all disciplines, such as . . . Before her faculty appointment at Y University, the PI was a Z Research Fellow at the University of California, Irvine. In 2004, she was awarded a 5-year NSF CAREER grant for her study on ZZ. In 2007, she was awarded the ABC prize from the AG Union, which is the highest honor bestowed to young scientists. Most recently, the PI was elected as a Frontiers Fellow by the National Academy of Sciences.

Collaborators:

An ambitious project of this nature requires varied expertise. The following investigators will contribute:

Dr. Dana Ghys, Associate Professor of X: Dr. Ghys is an ENT surgeon who will help us with injection and sampling of X.

Dr. Peter Stern, Associate Research Professor, Neuroscience, X University: Dr. Stern is an expert on immunohistochemistry and will consult on how best to perform injections.

Example 27-25   Personnel section

The funding for this project will support Mary Holmes, a talented scientist with expertise in X who recently joined my group as a postdoctoral associate. After her arrival, she developed Y, which will be used in this project.

For other examples on how to describe personnel, see Chapter 21, Section 21.8.

Project Management

Often, a funder who asks for an evaluation plan also asks for a management plan in which you have to describe the scientific advisory board, the external advisory board, the directors of the program, personnel, and the organizational structure. Whereas for smaller Proposals these sections may be individual sections, for larger Proposals a separate overarching Project Management section may be presented, which includes all these sections. A possible outline for a Project Management section is shown next.

Example 27-26   Project Management

Composition and Governance

Overall Organizational Arrangement

Scientific Advisory Committee

Project Principal Investigator

Affiliated Investigators

Staff and Facilities

Budgeting and Accountability (funding, progress reports, etc.)

Communication (among participating laboratories)

Recognition Statement

You may have to include a recognition statement in your Proposal, which should describe how your institution will recognize any extraordinary funding from your sponsor. Possible recognitions include press releases, naming opportunities, acknowledgments in publications, or conferences to which the funder will be invited.

 Example 27-27   Recognition statements

1. Z University proposes to name the program in recognition of the AAA Foundation's support. This program will consist of the faculty, postdoctoral fellows, and students associated with the research program over the 3 years of the grant.

2. Press releases will be issued by Z University regarding noteworthy accomplishments resulting from the AAA foundation sponsored research. Approval of these texts will be sought prior to release.

3. Scientific papers and presentations that arise from the research conducted within this project will acknowledge the support of the AAA Foundation.

4. At the end of Years 1, 2, and 3, the program will host a retreat for all involved parties (students, postdoctoral fellows, investigators, Scientific Advisory Board members). These events will honor the AAA Foundation for its support, and representatives of the AAA Foundation will be invited to attend.

Bibliography

Proposals that will be peer-reviewed require you to use citations throughout the narrative and to provide a reference list at the end—both similar in format as for those for research articles and review papers. Note, though, that not all Proposals require citations and references. Particularly short Proposals to private funders often do not ask you to provide citations. Or, they may allow you to list references in much abbreviated format and in smaller font. These agencies are not concerned about crediting other sources as the applications are not intended for publication, and many of such applications are not peer-reviewed. Not listing references or listing them in abbreviated form affords you more space for text—so take advantage if you are provided with such an option.

Relevance to Foundation

Certain funders request a section that describes very directly in what way your Proposal is relevant to the vision and goals of the funder. The content of such a section depends largely on the goals of the sponsor. In general, you need to point out and highlight any alignments of your work with the vision and goals of the funding agency. A sample section is provided in Example 27-27.

Example 27-28   Relevance to Foundation section

> Our proposed project aligns extremely well with the vision and goals of the AAA Foundation. The significance of gaining insight into why and how X occurs is monumental. We intend to approach such studies by joining Y and Z in a multifaceted, cross-disciplinary approach to gain critical and transformative knowledge in the field of A. In addition to Y and Z, the implications of this research touch other core themes of the AAA Foundation such as D, E, and F. Our findings could be a tremendous potential benefit to society. Our work will include inquiries into ABC to gain insights into X, promoting an overall new line of inquiry that will encompass the understanding of these different and yet connected core themes.

Table of Contents

For Proposal narrations over five pages, a table of contents is sometimes requested. Even if the narration section is brief, a table of contents is helpful when multiple attachments or appendices are included. All of the attachments should be page numbered sequentially. Provide a title for appendices and other supplementary material. The table of contents and the sequence of attachments should follow the format specified by the funder.

Timeline

In many Proposals, you have to provide a timeline for the research project, outlining the unfolding of the proposed research in reasonable detail. The research duration usually ranges between one and five years. In a timeline, you need to show that your experiments are doable in the proposed time and that you have a logical plan for carrying them out. To show timing of activities, it is a good idea to use a chart or table displaying both start and end points of your proposed project and that of particular tasks and aims comprising it. If specific start dates are not indicated by the funder, suggest that your project begin 9 to 12 months after you submitted your Proposal.

Example 27-29   Timeline

> The PI will be involved in all theoretical and empirical research. The postdoctoral fellow will focus on field research of X, whereas the graduate student will conduct fieldwork on Y. Theory and analyses will be conducted by all involved scientists. The first two years will focus on A. In Years 3 and 4, field data from the first two years will be used to test the general models and inform the development of theoretical predictions for the two study parts. Field experiments will then examine B and whether these responses can be predicted from the specific models examining the interactions between A and B.

 Example 27-30   Timeline

> We plan to complete the proposed project within two years, starting August 1, 2011. Specific milestones are depicted in the following timeline:

AIM	Year 1	Year 2
Aim 1 Description of X	⟶	
Aim 2 Generation of Y		⟶

Title Page/Cover Page

If required, provide a title page for the Proposal. This page is the first page of the Proposal and contains the title and the names and affiliations of the authors. You may also include the seal of your institution, the name of the funder, and the date of submission. Certain foundations, corporations, or federal agencies may have their own specifications for the title page. Check their instructions.

ESL advice

If you are from a country where names are usually written by listing the last name before the first name, know that in English the names are written in Western style name order: first name, middle initial, then last name. Take this into consideration when preparing the title page.

SUMMARY

> **BUDGET AND PROPOSAL GUIDELINES**
> - Follow specific instructions on how to prepare the budget if provided by the funder. Ask your administration for help.
> - Include direct and indirect costs.
> - Make sure costs are realistic and justified.
> - Check the Proposal instructions carefully to ensure that your Proposal contains all sections as required by the funder.

CHAPTER 28

Revision and Submission of a Proposal

I mprove your chances of getting funded through repeated revisions and by ensuring that your first page is perfect.

THIS CHAPTER INSTRUCTS YOU ON:

- General and common federal proposal outlines
- Revising your proposal, including:
    - The importance of the first page
    - Strategic placement of key information
- Submitting the proposal, including writing a cover letter
- The review process (NIH, NSF, private funders)
- Site visits
- Common reasons for rejection
- Proposal resubmission and addressing reviewer's comments

28.1   GENERAL

Writing a convincing proposal is a problem of persuasion. Such persuasion needs to be specifically tailored toward each particular funder in terms of the content, format, and budget request. To tailor your proposal, you need to know the potential funder's priorities, funding budget, and expectations.

28.2   BEFORE SENDING OUT THE PROPOSAL

➤ Follow instructions; fulfill all requirements

Proposals range widely in their format and quality. It is important that you obtain instructions (if provided) and follow them. Double-check that you have fulfilled all requirements before sending out your final version.

28.3   REVISING THE PROPOSAL

➤ Be clear and concise

➤ Edit, edit, and edit

Ensure that the proposal is clear and concise. You can improve your chances for obtaining grant dollars enormously by ruthlessly writing and rewriting. Therefore, edit, edit, and edit; then, have some other people edit. (See also Chapters 2–6 and 17 for editing and revising your writing.)

➤ Make sure reviewers find the most important information

Know that program managers and board members are often inundated with proposals; therefore, you may have one minute or less to grab their attention. Make sure that they can find the most important information immediately (see also Chapter 6 and information on the first page and placement of information in the following proposal outlines).

Key ways to get their attention:

- Present a clear and concise proposal
- Make the proposal look neat and professional
- Follow all the requirements made by the funder

➤ Ask for help

Ask colleagues inside and outside your field to help you improve your proposal. Listen to what they say. Write and rewrite your proposal so the goals and intentions cannot be misunderstood and the proposal's significance is obvious. Be clear and concise.

➤ Make sure the first page is perfect

Assume that many readers will get no further than the first page. It should therefore act as a stand-alone summary of the entire proposal. See Chapter 20, Section 20.9 and Chapter 22 for details on content of the first page.

➤ Check your proposal outline

When you are ready to revise the text of the proposal, start by ensuring that all necessary elements are contained in each section (see Chapters 22–27 for details of each section). For ease of reference, double-check the overall structure against a proposal outline such as the following.

Example 28-1   Proposal Outline I

Abstract/Aims:
background, problem/need, overall objective (specific aims), approach, impact

Specific Aims: list aims
Under each aim: rationale, approach, expected outcome

Background:
1. Paragraph: background/introduction, problem, objective
Last paragraph—hypothesis/approach; expected outcome

Preliminary Results (and/or Review of Literature):
by specific aims, chronologically/most-to-least important
1. paragraph/sentence overview of literature
Last paragraph—summary and unknown

Approach/Research Design:
by specific aims and by subsections
within subsections: cover design, analysis, outcome, alternate strategies, significance
1. paragraph/sentence overview of aim/ experiment

Conclusion: Expected outcomes and significance

Example 28-2   Proposal Outline II (for certain federal proposals)

Abstract/Aims:
background, problem/need, overall objective (specific aims), approach, impact
Specific Aims: list aims
Under each aim: rationale, approach, expected outcome

Significance:
First or last paragraph: summary/overview paragraph highlighting all reasons why the proposed work is significant
Include background information

Innovation:
First or last paragraph: summary/overview paragraph highlighting all reasons why the proposed work is significant

Approach/Research Design:
by specific aims and by subsections

within subsections: preliminary results, cover design,
analysis, outcome, alternate strategies,
significance

1. **paragraph/sentence** overview of aim/
experiment
**Last paragraph—summary, expected out-
comes, and brief significance**

Note that in these overall outlines, key power positions are written in bold,
as they indicate the most crucial structural locations of a proposal. Pay particular attention to the content and location of these power positions
throughout your revisions. Note also that other similar outlines may be
used, depending on the funder's application requirements (see also Chapter
20, Section 20.7 and Chapter 24, Section 24.1). Follow instructions set by
the funding agency carefully. If instructions are not followed (e.g., page
length, budget total, abstract word count, etc., are not kept as requested),
your proposal could be returned without review.

➤ Revise in stages

When your structural components are in place, revise each section. Refer to
summaries and checklists in the previous chapters for the proposal Abstract and Specific Aims, Background, Significance, Preliminary Results,
Innovation, and Approach/Research Design (Chapters 22–26). After you
are satisfied with content, logical arrangement, and structure, revise for
style and composition using the basic scientific writing rules presented in
Chapters 2 through 6. Pay particular attention to word, sentence, and paragraph locations. Then, check for spelling and punctuation. Finally, review
your overall layout and spacing. First impressions are important. Ensure
that your proposal looks professional and that it does not appear too dense
and busy. Space out figures, and leave some white space to break up
text-heavy sections and pages (see also Example 28-3).

	Example 28-3	First impression	
		Text-heavy proposal	

Revised **First impression**
Example 28-3

Expect several revisions, and give yourself enough time to write and revise the proposal. Consider also time for colleagues and mentors to read and comment on any drafts, and do not forget that it will take you time to incorporate their suggestions into the proposal. Calculate several weeks of preparation and editing before sending out a proposal. For specific questions regarding the proposal, you may also consider contacting the program officer. Program officers can be key to getting a grant and to understanding the process. Usually, they are happy to talk with a new grant applicant. Many times, the officers can remind applicants of what to emphasize and provide other important tips (see also Chapter 20, Sections 20.4 and 20.7).

After you have revised the proposal a final time, recheck that you have followed the guidelines of the foundation if they were provided. Ensure that you have paid attention to detail such as font and margins and that your sections are in the right order. Double-check that you have included all required sections in the requested order. Make sure that you have contacted the Grants and Contracts Office of your university and that you have obtained all required signatures from your administration or business office. Start this process early in order to secure all the signatures required for approval (see also Chapter 20, Section 20.7).

If there are no more corrections, and if you are required to send hard copies, make the required number of copies of the proposal for the funder. Also, make one copy for yourself. For electronic submission, ensure that the version you send is really the final version and that it is complete.

28.4 SUBMITTING THE PROPOSAL

➤ Ensure that you have included all sections in the order requested

Depending on the funder, your proposal may have to be submitted electronically or by hard copy. Electronic submission is the fastest way to send your proposal to the funder.

Instructions for electronic submissions are usually very detailed and specific and should be followed carefully. If needed, ask for help from someone that has experience with electronic submissions.

When you submit hard copies, package your final copies with care. Realize that first impressions are important. Submit only a neat proposal that follows all the requirements provided by the funder. Check that the copies to be submitted are complete. Consider placing the proposal pages in a folder or binder. Ensure that the proposal will reach the funder in time. Use a dependable delivery service, check schedules and delivery times, and address the package clearly. Keep a complete spare copy of everything in case the package is lost or damaged in the mail.

➤ Include a cover letter

Include a brief cover letter to the sponsor (or an endorsement letter with the dean's or university president's signature) to go with the manuscript. Keep the cover letter short and simple. Do not list all your achievements to date, and do not list any complicated background information or details of your proposed study. Any cover letter or endorsement letter should be carefully written and well presented. Find out the name of the program officer if you can.

Start the letter with an introductory sentence stating the title of your project and that you would like to submit it for funding. Add a sentence that describes the expected outcomes and their importance. You may also state why your proposal would be of interest to the funder.

Some funders allow you to suggest names of possible reviewers. Make sure you include all relevant contact information for each suggested reviewer. You may list reviewers either within the letter itself or you can refer the editor to a separate sheet.

A sample cover letter is shown in Example 28-4.

Example 28-4   Cover letter

Name   Date

Address

Dear Dr. Brunner:

I am pleased to submit the enclosed proposal, *Gene targeting in the moss Pyscomitrella patens,* to the ABC Foundation. The project investigators, Professors Gary Jillter, who will lead the effort, as well as Professor Robert Johnson, are very enthusiastic about their project. The proposal's strong interdisciplinary approach and ambitious aims would seem to fit well within the Foundation's goals of supporting innovative research that lies outside the funding purview of conventional sources.

The proposed project aims to use recent breakthroughs in X to achieve Y. Support from the ABC Foundation for equipment and investigations will be critical to enable these scholars to address fundamental scientific questions, surmount a range of technical challenges to basic research, and point the way to the design of new technologies that promise to improve treatment methods in cancer therapy.

Thank you for your willingness to consider this proposal. We look forward to hearing the results of your deliberations and to a future opportunity to collaborate.

Yours sincerely,

Signature

28.5 BEING REVIEWED

➤ Understand the review process

Usually, when your proposal has passed the first screening, it will be discussed at panel meetings. Know that reviewers have many proposals to review but only a limited amount of time to do so. They also bring with them different experiences, which may affect the review process. Similarly, their level of knowledge in your area of expertise will influence their understanding and judgment.

If you are not familiar with the review process, do not hesitate to clarify the process with the program officer in charge of your proposal. A clear understanding of the process, such as who will be reading your proposal, for example, research staff, board of directors, and so forth, will also help you better tailor the technical details of the proposal to the appropriate audience. Understand the general review criteria (see NIH and NSF criteria outlined next), and know what may result in unscoring or rejection of a proposal (see Section 28.7 and also Chapters 22–26, Sections 22.8, 23.8, 24.5, 25.6, and 26.6).

NIH Review

The NIH is one of the biggest funding agencies in the United States. It receives more than 60,000 applications from the medical and scientific fields each year. If you are applying to the NIH or another federal agency for funding, your grant will be assigned to specific study sections for review (see Office of Extramural Research: Scientific Review Group Roster Index: http://era.nih.gov/roster/index.cfm). Look carefully at the composition of the study section rosters. People in your field should be on the panel. It will be important that your grant goes to the correct panel. The wrong panel may not have appropriate expertise and may misinterpret your ideas or findings.

If you are unsure to which panel to submit your proposal, call or email the Scientific Review Administrator to find out if your grant is suitable. A cover letter may also help guide your grant to the correct panel. In the cover letter you can also request someone on the panel not to score your grant due to a personal or scientific conflict. Make such requests very sparingly; doing so is a conspicuous step, and the program officials will want to know what the conflict is.

Once submitted and at the correct panel, only a few people will actually look at it closely many months later (first, second, or third reviewer). Reviewers have a heavy load: they are assigned five to eight grants for which they need to write reviews, and an additional two to five grants that they have to read. Reviewers give a preliminary impact score between 1 and 9 to each assigned and read grant proposal. The preliminary scores determine if the proposal receives a full review at the meeting and in which order.

The panel will try to eliminate at least 50% of the grants right away. Unscoring grants reduces the amount of work and the time people have to sit at the table. So there is peer pressure to unscore grants. When your name, grant title, and grant number are read, the first and second reviewers are asked if they want to unscore the grant. Unless they are in consensus, the grant will be discussed and scored. If the grant application is unscored, you receive the reviewers' comments but no summary of the discussion as none took place.

If a grant application is discussed, the first reviewer summarizes the goals of the grant and its strengths and weaknesses and evaluates the grant based on the review criteria. The second and third reviewer and any of the readers may or may not add to the overview provided by the first reviewer. Everyone at the table, including reviewers who have not even looked at your grant, score the grant based on the discussion they have heard (overall impact score). Although reviewers will weigh five core review criteria (see next section), their overall impact score is not derived from the scores of the core review criteria. Rather, the overall impact score is a measure of the likelihood of the proposed work making an impact on the corresponding research field(s). The overall impact score for each proposal is the mean of all impact scores multiplied by 10. This score is provided on the summary statement. (See also http://public.csr .nih.gov/aboutcsr/contactcsr/pages/contactorvisitcsrpages/nih-grant -review-process-youtube-videos.aspx and http://grants.nih.gov/grants/peer_ review_process.htm#scoring2.)

NIH Review Criteria

Each of the five criteria below are assigned a score between 1 and 9 (1 being best, 9 worst; see also http://grants.nih.gov/grants/peer/reviewer_ guidelines.htm). Note that in the following list, the order of review criteria reflects how criteria are weighted.

- **Significance**: Does the study address an important problem? How will successful completion of the aims change scientific knowledge, technical capability, and/or clinical practice in this field?
- **Investigators**: Are the principal investigators (PIs) well suited for the work? Do early-stage investigators have appropriate experience and training?

- **Innovation**: Is the project original? Does the application use novel theoretical concepts, approaches or methodologies, instrumentation, or interventions?
- **Approach**: Is the design/method appropriate to accomplish the specific aims? Are potential problems, alternative strategies, and benchmarks for success presented?
- **Environment**: Does the environment contribute to the likelihood of success? Is the level of institutional support appropriate (space, time, equipment, resources, assistance/guidance with project, etc.)?

Applications for the same project can be submitted a maximum of two times. The entire selection and review process from submission to notification takes about 6 to 10 months.

Typical NIH Notification

A typical NIH summary statement is shown next, along with a brief sample critique from one reviewer.

Example 28-5   Summary statement

SUMMARY STATEMENT
(Privileged Communication)   *Release Date:* 12/11/2015

Principal Investigators (Listed Alphabetically): ROSE, DORO . . .

Applicant Organization: ABC UNIVERSITY

Application Number: 1 R01 AL123456-01
Review Group: XXX1 YY-Z (33)

Center for Scientific Review Special Emphasis Panel
PPP12-234: Aging

Meeting Date: 11/23/2015   *RFA/PA:* PPP12-234
Council: JAN 2016   *PCC:* HHAATN
Requested Start: 04/01/2016

Dual IC(s): AG

Project Title: Disease X and Aging

SRG Action: Impact Score: 36   Percentile: 26 #
Next Steps: Visit http://grants.nih.gov/grants/next_steps.htm
Human Subjects: 10 - No human subjects involved
Animal Subjects: 30 - Vertebrate animals involved - no SRG concerns noted

Project Year	Direct Costs Requested	Estimated Total Cost
1	xxx,xxx	xxx,xxx
2	xxx,xxx	xxx,xxx
. . .	xxx,xxx	xxx,xxx
TOTAL	x,xxx,xxx	x,xxx,xxx

ADMINISTRATIVE BUDGET NOTE: The budget shown is the requested budget and has not been adjusted to reflect any recommendations made by reviewers. If an award is planned, the costs will be calculated by Institute grants management staff based on the recommendations outlined below in the BUDGET RECOMMENDATIONS section.

RESUME AND SUMMARY OF DISCUSSION: The applicants propose a novel mechanism for the progression of disease X with age due to reduced yyy in the aged cells. The significance is high because disease X is a major public health problem, with morbid consequences that are increasing with age. Our understanding of the underlying mechanisms is incomplete, and the current therapy options are not very effective. Strengths were the outstanding expertise and accomplishments of the investigators in this area of research; well-designed, highly comprehensive experimental approaches; good animal models for the in-vivo studies; and the innovative concepts. Weaknesses were the rationale for some study components, such as ZZZ; Aim 2 was not integrated well with the focus of the proposal. . . . Although the strengths were acknowledged, the overall enthusiasm was somewhat reduced because of the weaknesses.

DESCRIPTION (provided by applicant): . . .

PUBLIC HEALTH RELEVANCE: . . .

CRITIQUE 1:
Significance: 3
Investigator(s): 2
Innovation: 2
Approach: 3
Environment: 2

Overall Impact: The PI proposes to study the disease X with aging focusing on YY. Understanding the particular contribution of disease X is of relevance. Strengths include . . . Two major concerns are noted. The first, . . .

1. Significance:
Strengths
• The underlying mechanisms of disease X during aging are not clear. This proposal is aimed at determining a role of YY of disease X.
Weaknesses
• It will be difficult to tease out effects of aging vs. effects of Z. Most of the proposed comparisons will fall short of defining a specific role of aging.

2. Investigator(s):
Strengths
• The PI is a Professor at ABC University and has significant expertise related to disease X and aging. She is well qualified to direct these studies.
Weaknesses
• None noted.

3. Innovation:
Strengths
• The project contains novel elements such as. . .
Weaknesses

4. Approach:
Strengths
- The proposal is well organized and thought out. It included some discussions on . . .
- Appropriate emphasis on aging in Aim 1, includes . . .
- The plan is very comprehensive. The PI will look at . . .
- Pilot data are presented for each Aim and are supportive of . . .
Weaknesses
- Aim 2 is not well integrated in some of these experiments.
- It is unclear if . . .

5. Environment:
Strengths
- The environment is excellent.
Weaknesses
- None noted.

Protections for Human Subjects: Not Applicable
Vertebrate Animals: Acceptable

CRITIQUE 2: . . .

NSF Review

The NSF preferably funds high-risk science and engineering projects. If your proposal is one of the more than 50,000 yearly applications to the NSF, it will be assigned to a panel for confidential merit review. As the proposer, you have the possibility to suggest and/or block reviewers. Usually, three reviewers, selected based on their expertise under consideration of your input, review the proposal and discuss it on a panel. Reviewers provide their reviews for each proposal to a program officer, who in turn also reviews the proposal. Reviews may also be made through a panel, which then provides a review summary to the program officer, or through a site visit. Aside from the reviews, the program officer takes into consideration other criteria such as capacity building of a specific area, achievements of program objectives, previously funded proposals, and other current proposals. The program officer then makes a recommendation to the cognizant divisional director to award or decline funding for a given proposal.

NSF Review Criteria
- **Intellectual merit**—refers to the potential to advance knowledge. Specifically, reviewers evaluate the significance of the proposed work in its field and beyond, the qualifications of the principal investigator and those of the team, the innovation and creativity of the proposed work, its conception and proposed approach, as well as access to resources.
- **Broader impact**—refers to the potential to benefit society. Here, reviewers consider how well the proposed research combines with

education and training at academic and research institutions. In addition, they look at the integration of minorities and underrepresented groups, dissemination of findings to others, advancement of scientific and technical knowledge, and its importance to society.

As appropriate, the NSF will also use other criteria needed to evaluate specific aims of proposed activities. The entire review process takes about six months due to the volume of applications and the thoroughness of the process.

Reviews by Private Foundation and Corporations

Private foundations and corporations fund a large variety of projects. Each of these funders has its own review mechanism. Many foundations that do not have staff with scientific background will employ a scientific advisory committee to review the content of the grants. In this case, your proposal will have to appeal technically to them as well as to the foundation staff. For others it may be the head of the agency who makes the decision, or the agency may form its own internal review board. Often, these boards include nonspecialists and even nonscientists. Thus, you have to gear these proposals to an educated lay audience, using nontechnical language where appropriate and possible (see also Chapter 20, Sections 20.5–20.7). A good rule of thumb is to integrate a *Scientific American* style of writing with that of a news and review section in *Science* or *Nature*.

If possible, find out who the reviewers will be. Typically, they are listed on a funder's website. Gear your writing toward the person with the least science experience in your field.

28.6 SITE VISITS

➤ Offer to meet the funder, and put your strongest effort forward: yourself and your team

Be prepared for a site visit if requested. Many philanthropic organizations that make large grants will require a site visit along with proposal submission. When you get that call or email announcing a visit, do not panic. First, you must determine the expectations of the visitor. Knowing what they expect to get out of the site visit will be key to your success in garnering the grant. For example, some foundations use site visits only as a formality. This may be the case when you have already established a solid relationship through past grants. Other funders may look to validate that all the staff are well versed in the goals and expectations of the proposal or that research space and equipment is adequate for completion of the project. Some foundations look for steps for sustainability and backing by the research institution. Many foundations do not like to make grants in which they are the sole funder—they like to know that other funding organizations also think that the project is worth investing.

When potential donors come to visit, find out ahead of time who they are, what the goal of the meeting is, and with whom they would want or need to visit. Consider involving a professional at your organization who is used to dealing with site visits from potential funders. Chances are, these professionals will be happy to help out or provide advice. Depending on the number of visitors, the amount of money under consideration, and the length and involvement of the visit, you may want to hire or compensate staff for the time put into preparations for the visit. Some site visits, particularly from federal agencies, are highly time-consuming and require professional coordination skills from others. Be prepared to spend significant resources and time for the best chance to have a successful visit.

Do clean up your laboratory and office, but do not stress out your staff with your nervousness. When money is involved, know that business attire is appropriate. So dress accordingly, but also remember that you are not attending a wedding or gala reception.

Ensure that meeting times and places are clear. Consider preparing some handouts or note pages of slides for your visitors. The latter in particular are useful for visitors to take notes if you give a presentation. If needed, make reservations at an appropriate restaurant and know how to get there. Alternatively, have food catered.

During the site visit, be objective and try to answer any questions funders may have with respect to your project, your laboratory, and the organization. Do not get angry or complain. If you think they are wrong, stay objective and positive.

Most sponsors like to meet people they are funding in person. They want to learn more about you, your work, and your laboratory firsthand. They also love the opportunity to ask questions. Thus, it may not be a bad idea to make the offer to come by and meet the funder, or to invite them to visit with your organization and laboratory. In brief, put your strongest asset forward: yourself and your team.

28.7 REASONS FOR REJECTION

The following factors are considered the most common reasons for a proposal not getting funded:

1. Logical inconsistencies
2. Project feasibility not convincing/the proposal is overly ambitious (common for early career investigators)
3. Significance is not clear: Failure to focus on the overall problem or need and the resulting payoff provided through your project. No compelling potential impact is offered
4. Innovation: Proposal does not distinguish itself from work of others
5. No persuasive structure: poorly organized, power positions and location of information has not been considered. Key points are buried or not signaled

6. The aims are interdependent/do not stand alone
7. The investigator lacks sufficient expertise or publications in the area of study
8. Proposal is difficult to read: full of jargon, too long, or too technical
9. Oversimplifying the problem at hand
10. Incorrect nomenclature, sloppy formatting, spelling and grammatical errors, and so forth.

28.8   IF YOUR PROPOSAL IS REJECTED

➤ Be persistent. Try, try again

To fundraise successfully, many things have to come together: the idea, the team, the timing, the right place—to name just a few. In addition, every sponsor—whether federal, state, private foundation, or industry—has its own review and decision-making processes. Thus, success in funding often requires multiple submissions.

Seldom does a proposal get funded on first application (especially those sent to federal agencies), particularly if you are new to the field or a young investigator. Expect rejections, and take them as opportunities to get feedback and improve your proposal. Although receiving a rejection can be upsetting, do not get frustrated if your proposal does not get funded. Nearly all researchers have applications rejected. If possible, find out the reason for rejection. It may be something that you can adjust, giving you another chance to be considered for funding. Reapply if possible. Do not give up! Some funders even encourage resubmission of worthy but unfunded proposals on the basis of reviewer comments.

Even if you will not reapply to the same agency, reevaluate your proposal/project, and make any necessary changes. Do not throw out your proposal! You can use many parts again, possibly in other proposals.

28.9   RESUBMISSION OF A PROPOSAL

If you revise a rejected proposal based on reviewers' comments, you may have to decide whether to resubmit the application to the same funding program or to apply to a different one. If you have not received any reviewers' comments or other feedback from the funder, consider seeking feedback from the program officer. If the agency encourages you to resubmit, your chances of success on a second go-round are good. If the agency discourages you to resubmit, consider finding an alternative funding source or modifying your project idea or approach. Reworking a proposal and resubmitting it, either to the original agency or to a different one, often results in a funded project.

Most reviews are meant to be helpful and are intended to provide constructive criticism, pointing out weaknesses in any aspect of the application, including the scientific ideas presented, the research methods, and the clarity of presentation. If you plan to resubmit a proposal, you have to decide how and to what degree you will respond to reviewers' comments. It may be worth discussing any feedback and criticism with your collaborators, peers, and senior faculty members at your institution, especially difficult-to-address comments and suggestions. You may also want to contact a program officer for further insights into how best to tackle any revisions. Having to revise mainly for style and composition is usually easier and quicker than having to rework content, produce more preliminary results, or redesign an aim or approach. If possible, work closely with the program officer and be willing to rethink aspects of the project based on the agency's feedback.

If you are resubmitting a proposal to the same funding agency, clearly state that the proposal is a resubmission, and clearly identify revisions that have been made based on reviewers' comments in the proposal. If you are planning to resubmit your proposal to a different funding agency, do not make any revisions based on reviewers' comments from the first funding agency if those comments run contrary to the priorities of the new agency or program.

Some sponsors, such as the NIH or NSF, limit the number of resubmissions of a proposal to two. If resubmitted, these proposals have compulsory redress by the principal investigator. For the NIH, this is usually restricted to the first three pages of the proposal and is essential for increasing your chances of funding on resubmission. Here, you need to show the reviewers that you have made efforts to address their concerns and point out where the reviewers were mistaken in their assessment if applicable. It is common to get reviewers who do not know the proposal's field and make uninformed comments, which must be dealt with in the resubmission. However, do not antagonize reviewers by arguing about critiques. If you disagree with a reviewer's suggestion, explain why in a professional, objective manner. You will have to have a solid reason to disagree and should do so very sparingly.

Following is a sample redress of a resubmitted proposal. Note that responses to reviewer 2 are shown in a different style from those of reviewer 1 in order to highlight how you may condense subheadings if space is limited.

Example 28-6   **Resubmission—Addressing reviewers' comments**

Introduction

This is a resubmission application. We are pleased that the panel members have acknowledged the significance of X.

Our revised proposal has been strengthened by the incorporation of changes suggested by the reviewers and provides more detail about X. Major changes are pointed out below. Critiques of the reviewers are italicized.

Critique 1

For Aim 1, reviewer 1 asked how procedures for development and validation of X will be addressed.

We have explained in detail how our models will be developed and validated in Section D.2.

For Aim 2, the reviewer asked for a detailed description of Y.

We have now added these details in Section D.3.

For Aim 3, reviewer 1 requested more details of X. Specifically, the reviewer wanted to see . . .

We have added these specifics in Section D.5. We apologize for the lack of clarification concerning these details. We have added a new detailed description of the data that are being used in this study (Section G) . . .

Critique 2

R2: Please mention Z and to provide more details of its derivation. As suggested, we are now comparing. . . .

R2: Define all terms. We have now defined these terms in the text of the resubmission.

R2: It is unclear why it is necessary to develop a local database system to collect data instead of available standard database software. It is necessary to develop a local database because this permits dynamic collection of data as the disease progresses. The system is a simple Microsoft access-based system used to collect. . . .

28.10   IF YOUR PROPOSAL IS FUNDED

➤ Know who is negotiating your agreement

When funding is offered, you may have to negotiate with the sponsor. Involve your Grants and Contracts Office and/or Office of Research for assistance (see also Chapter 20, Section 20.7). The Grants and Contract Office has your interests at heart and will usually ensure that you are not being taken advantage of. Surprises can be fatal.

Know who is negotiating your contract. Stay involved, and keep others involved as well.

SUMMARY

REVISION GUIDELINES

- Follow instructions; fulfill all requirements.
- Be clear and concise.
- Edit, edit, and edit.
- Make sure reviewers find the most important information.
- Ask for help.
- Make sure the first page is perfect.
- Check your proposal outline.
- Revise in stages.
- Ensure that you have included all sections in the order requested.
- Include a cover letter.
- Understand the review process.
- Offer to meet the funder, and put your strongest effort forward: yourself and your team.
- Be persistent. Try, try again.
- Know who is negotiating your agreement.

PART FIVE

Posters and Presentations

CHAPTER 29

Posters and Conference Abstracts

Posters are intended for visual impact. Their text is meant to support the illustrations they display. Mastering excellent presentation skills is arguably the most effective and most rapid way to disseminate new findings in science.

THIS CHAPTER DISCUSSES:

- Conference abstracts
- Components of a poster and their location
- Poster formats
- Poster figures
- Preparing a poster
- Revising a poster
- Presenting a poster
- Sample posters

29.1 FUNCTION AND GENERAL OVERVIEW

A poster is a visual presentation of your work and can serve as a concise communication tool for small groups of people. Posters facilitate the rapid communication of scientific ideas. When presented well, they can be more effective than a talk in establishing a relationship with your audience because such presentations allow you to interact one on one with the people who are interested in your research and findings. Not only can a poster attract viewers and provoke curiosity, but it can also serve as an advertisement and as a summary of your work, which can be viewed in your absence.

If you are planning to present a poster at a conference, you usually have to write an abstract first. This abstract will be reviewed by a conference committee (see Section 29.2 for a more detailed discussion of conference abstracts, as well as Chapter 15 for more details on writing an abstract). If you are invited for the conference, you will be told whether your presentation is a poster or an oral presentation (see Chapter 30 for the latter).

29.2 CONFERENCE ABSTRACTS

➤ **Ensure that your conference abstract fits the topic of the conference and summarizes your work accurately**

Sending in an abstract paper to a conference is often a requirement to be considered for a poster or for a talk. As this abstract has to convince the conference committee to invite you for a poster or a talk and the conference audience to attend your poster session or your talk, it needs to take all these readers into consideration.

A conference abstract not only permits you to display your findings but also allows you to get initial feedback for work that you have not yet tried to publish. Thus, your abstract has to summarize your work accurately. The abstract has to fit the topic of the conference and should be interesting enough to attract an audience when displayed as a poster.

Your abstract is made available to the conference participants and will often be published as submitted. You may also be invited for a paper, which gets published in the conference proceedings. Your abstract (and poster) may present preliminary results or results close to publication.

A conference abstract is different from the abstract on your poster (if required) (see Section 29.5). Therefore, do not use your conference abstract as the abstract on your poster. The underlying format for a conference abstract is similar to that of a research paper but is usually longer (see also Chapter 15). The conference abstract may be 350 to 500 words long. It includes a title, a longer background/context portion, and sometimes a few references.

➤ **For a conference abstract, ensure your overall format consists of**

- Background/context
- Question/purpose
- Experimental approach
- Results
- Conclusion (answer)/implication
- (References)

See Chapter 15 for a more detailed description of these elements.

➤ **Follow the guidelines provided for a conference abstract**

It is important to make a good first impression when submitting your conference abstract. Follow guidelines exactly, or you may risk being eliminated from the start. Conference abstracts are usually considered a publication of a sort, and many are published as a supplement to the association's journal.

Submit your abstract on or before the due date and in the required way, usually electronically or by email. Ensure computer compatibility of documents, and include your name, title, organization, and contact details.

Abstracts are typically reviewed anonymously. A few conferences will send comments from reviewers about your abstract. This is very valuable information, and you should request it if available.

Following is an example of a conference abstract.

Example 29-1   **Conference abstract**

Features of sperm cell differentiation that alter the meiotic program in C. elegans

Shakes, D. C., Wu J., Sadler P.L., LaPrade K., Moore L.L., Noritake, A., Chu D.S.

The fundamental process of meiosis underlies two differentiation programs that occur at different rates and generate vastly different cell types, sperm and oocytes. There is a limited understanding in any organism, including *C. elegans*, regarding how sperm or oocyte specification either coordinates with or modifies meiosis to give rise to these disparate cell types. We have conducted an in-depth analysis of sperm cell formation to understand how gamete-specific features influence meiotic events and progression.

Our work has produced a detailed timeline of late meiotic prophase during spermatogenesis in *C. elegans*. This study is unique in that it defines a broad set of cytological and molecular landmarks that inter-relates changes in chromosome morphology and dynamics with germ cell cellularization, subcellular organelle disassembly, spindle formation, and meiotic cell cycle transitions to accurately stage nuclear progression. This analysis has uncovered differences in sperm meiotic chromatin composition and morphology compared to oocyte meiosis. Nuclei progressing past pachynema undergo distinct morphological changes after the incorporation of sperm nuclear basic proteins into sperm chromatin. Most strikingly, *C. elegans* spermatogenesis includes a previously undescribed extended stage when chromosomes form a constricted mass within an intact nuclear envelope. This karyosome stage, which is a common feature of meiosis in many organisms, follows desynapsis and is largely transcriptionally inactive. However, it is highly dynamic, as multiple cell signaling pathways are sequentially activated during this stage and in an order that is distinct from that of diakinetic oocytes. Also, in contrast to developing oocytes, spermatocytes exhibit centrosome-directed microtubule dynamics that are distinct in both their timing and morphology. Correspondingly, we observe that kinetochore structures that mediate chromosome segregation also exhibit sperm-specific features. Overall, several of these gamete-specific features effectively increase the efficiency and pace of meiotic progression during sperm formation. These studies identify specific features of the meiotic program that differ in sperm and oocyte meiosis, revealing that the underlying molecular machinery required for meiosis is differentially regulated in each sex.

(With permission from Diana Chu et al.)

29.3   POSTER COMPONENTS

➤ **Design the poster around your research question. Include:**
- Title
- (Abstract)
- Introduction
- Materials and Methods
- Results
- Conclusion
- (References and Acknowledgments)

Overall, posters follow the standard scientific format in that they contain all the sections found in a research paper except for the Discussion. Strict IMRAD style headings do not need to be used in all posters. Titles and subheadings can vary, and additional or different subheadings, such as "Study Participants" or "Study Site," may be warranted to make the poster more appealing and easier to navigate.

➤ Concentrate only on the main points in each section

The major difference between sections in a poster and those in a research article is that poster sections concentrate only on the main points and present these sections briefly and visually. As posters are not intended for publication, you do not need to include much detail about materials and methods or overwhelm the reader with too many figures and tables. If needed, you can fill in your audience verbally, for example, during the poster session, or you can prepare handouts with this information and leave them at your poster.

When your abstract has been accepted for poster submission, review the poster guidelines before starting to make the poster. Note especially what poster format the conference requires as well as the specific time and place to submit and display it.

Sample poster layouts are shown in Figures 29.1, 29.2, and 29.3. Poster layouts and designs can vary widely. Some have a symmetrical layout while others do not. Some are horizontally laid out; others are vertical posters. Many larger conferences are now also using digital posters.

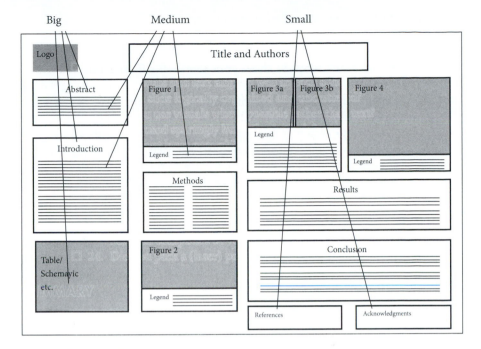

Figure 29.1 Horizontal poster layout, asymmetric.

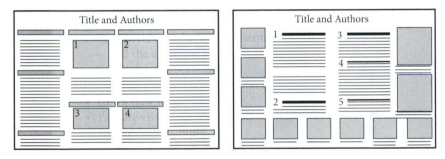

Figure 29.2 Additional sample poster layouts.

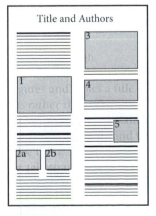

Figure 29.3 Vertical sample poster layout.

➤ Find visual ways to show your work—the illustrations tell the story

Sometimes it can be helpful to think of each part of the poster as a slide that you would show to an audience (see Chapter 30). As done for a slide presentation, try to keep text to a minimum by using key words and phrases throughout the poster. Do not simply paste your journal articles, conference abstracts, or other manuscript portions onto poster board. Instead, present a well-designed and engaging display of scientific information:

- Design the poster around your research question or discovery (see Chapter 11, Section 11.3). During your poster presentation time, you can expand further on this central theme, your approach, and more.
- Prepare a brief, distinct, and memorable take-home message. State implications and conclusions clearly, and direct them toward an educated audience. Depending on the conference, you may also have to gear your poster toward experts in your field.
- Add an Acknowledgments section, as needed (see also Section 29.5 and Chapter 8, Section 8.12). This section may be written in smaller font to avoid drawing too much attention to it. In the section, also acknowledge any sponsors or funders.

Remember that the clarity of the presentation stems from the proper arrangement of information and that graphical elegance is often the simplest design. Because the visual impact is most important for a poster, use illustrations, symbols, colors, and so forth wherever possible rather than text.

29.4   POSTER FORMAT

Overall Layout

➤ Use an easy-to-follow layout

➤ Use different sizes and arrangements for the various poster sections, but maintain a consistent style

➤ Aim for about 20% text, 40% graphics, and 40% blank space

Scientific posters are judged by their content as well as by their presentation. Effective poster design is therefore important. One of the first things you should do when you find out that you have to present a poster is to find out how much space you are allowed as this will determine the amount of detail you will be able to present. Unlike a manuscript, posters can adopt a variety of layouts (see Section 29.3).

Although layouts can vary, viewers expect to find certain sections in specific places. The most important text sections (Abstract and Conclusion) are usually placed in the top left and bottom right corner, respectively, as readers read from left to right and top to bottom (see also Chapter 6, Section 6.2 for a more detailed discussion of power positions). The most important visuals should be placed in the middle of the poster. The least desirable real estate on a poster is usually on the very bottom. References, acknowledgments, and logos are often placed there.

You can get creative, but at the same time must keep in mind the importance of where to place sections, as well as the need of guiding the reader through the logical sequence of the components while providing enough blank space to make your poster visually appealing.

For clarity:

- Present the information in a sequence that is easy to follow. You could, for example, number sections, use arrows, and/or choose headings to show the flow of the information.
- Arrange the material into columns—most posters allow for three to four columns (see Figures 29.1, 29.2, and 29.3).
- Use different sizes and arrangements for the various poster sections.
- For maximum visual layout and impact, aim for about 20% text, 40% graphics, and 40% empty space.
- Maintain a consistent style. If the style is not consistent, it distracts readers and may even disrupt the logical flow of your presentation. Consistent style means that you should use consistent color schemes, fonts, positioning of headers, figures and tables, and general layout.

Poster Background and Color

The background design is important in the presentation of your data. These days it is typical, but often expensive, to have your completed poster reproduced as a single large sheet of paper, which can then be rolled into a cylinder for transport. If your school or department does not have access to a printing facility or the funding for such a poster sheet, you can also use mat board to make a solid background for the entire poster, or just frame the individual elements of the poster onto smaller pieces of poster board. This latter format is easier to handle when you travel, although it may be more cumbersome to put up for display.

For well-designed poster backgrounds, consider the following:

Do

- *Use a colored background to unify your poster. The choice of a background color is up to you. You may also use a second background color to frame individual elements of the poster, or use colored text boxes on a white background to add visual interest.*
- *Use muted color(s) for the background. They are easier on the eye and offer the best contrast for text, graphic, and photographic elements.*
- *Use thoughtful contrast when displaying illustrations. For example, use a darker colored background for white or light-colored illustrations, and vice versa, a white or light-colored background for dark illustrations.*

- *Use colors in a consistent pattern. Otherwise, your viewers will spend their time wondering what the pattern is rather than reading your poster.*
- *Consider people who have problems differentiating colors, especially when designing graphics. One of the most common visual color weakness or blindness is an inability to distinguish green and red.*

Don't

- *Don't use more than two or three colors in the background—much more will overload and confuse viewers.*
- *Don't use overly bright colors. They will attract attention but then wear out readers' eyes. To add emphasis, you can use a more contrasting color for borders, but be conservative.*
- *Avoid using designs in the background.*

Text Format

Preparing a poster is very different from preparing a paper. Your main objective in preparing text for a poster presentation is to edit it down to very concise language (see Chapters 2–6). People are attracted to posters that have good graphics, a clear title, and few words. Conference poster sessions are high distraction environments—it is therefore particularly important to keep text simple.

Suggested Text Fonts and Sizes

- Use a san serif font such as Arial or a serif font such as Times New Roman.
- Font sizes should be large enough to be read from 6 feet away.
- Use font sizes proportional to importance.
  - Title: 90 point, boldface
  - Subtitles: 72 point
  - Section headings (Introduction, etc.): 32–36 point
  - Other text: ideally 22–28 point
  - Boldface with 1½ to double-spacing
- Pay attention to the text size in figures—it must also be large.
- Use lowercase lettering, with initial capitals where needed.
- Use italics to add emphasis.

Do

- *Use an interesting title in large font.*
- *Use only a small amount of text.*
- *Use bullets and numbers to break text visually and make text more readily available.*
- *Use double-spacing, and either left or left and right justify the text for ease of reading.*
- *Use active voice where possible.*
- *Omit unnecessary references, citations, and parenthetical information.*
- *Spell check and proof text carefully before your final print out.*

Don't

- *Don't use too much text or information.*
- *Don't make font sizes too small.*
- *Don't use your conference abstract as text on the poster.*
- *Don't use more than one or two fonts in the poster.*
- *Avoid underlining or exclamation marks, as they are not considered effective.*

29.5   SECTIONS OF A POSTER

➤ Minimize text and use images and graphs instead

As posters are primarily a visual medium, you should minimize text and use images and graphs as much as possible (see the section on photos, figures, and tables in this chapter; see also Chapter 30, Section 30.3 for a more in-depth discussion of visual aids). Use short sentences, simple words, and bullets to illustrate discrete points. Do not use long paragraphs. Instead, use simple statements. Also, consider basic writing rules and guidelines such as use of active voice and avoiding jargon and redundancies (see Chapters 2–5).

Title

➤ Make the title informative, attractive, and large enough

The title should convey the topic, the approach, and the system (organism). It needs to attract the reader and should be no longer than two lines. The title should be at least 2 inches (5 cm) tall. See also Chapter 16 on titles.

Below the title of the poster, include your name and those of the coauthors, indicate organization affiliations through superscript symbols, and add the name(s) of the organization(s) directly below the authors (see also Chapter 7, Section 7.3 on authorship). The font size for names of authors and associations should be smaller than that of the title (< 90 points) but larger than that of subtitles (> 72 points).

Abstract

➤ Distinguish between a poster abstract and a conference abstract

Some posters include a short poster abstract. Others do not. If you are including an abstract on your poster, keep the abstract to a minimum (50–100 words). Do not paste in your conference abstract, which is much longer (see Section 29.2 for conference abstracts). If your conference abstract is published, you usually do not need to include a poster abstract on your poster. Follow the guidelines of how to write a good abstract and ensure that all essential elements are present (see Chapter 15). To keep the poster abstract short, include only the research question/purpose (Chapter 11, Section 11.3) and the main finding(s) and conclusion(s).

Example 29-2

Question/
purpose
Main finding
and conclusion

Abstract

To examine the specificity of gene disruption in the haploid moss *Physcomitrella patens*, we have disrupted one member, *ZLAB1*, of the multigene *Cab* gene family. We found that the *ZLAB1* gene is specifically targeted in three out of nine integrative transformants, indicating that gene disruption in *P. patens* is highly specific.

Introduction
➤ Keep background information to a minimum in the Introduction

In the Introduction, you need to draw in your audience to interest them in the topic. Do not overwhelm the readers with background information, however. Provide only the minimum background information needed to capture their attention. Start broadly, then quickly place your research in the context of published, primary literature; state your research question clearly; and provide a brief description and justification of the general experimental approach. Keep background information to a minimum—you are at the poster session to fill in details if needed. The introduction for a poster should be short (ideally 50–100 words but no more than 200 words):

Example 29-3

Introduction

The landscape of the Middle East has been altered by human activity for most of the Holocene period. The rate of these modifications has accelerated in the last century, and today rapid population growth, political conflict, and water scarcity are common throughout the area. All of these factors increase the region's vulnerability to potentially negative impacts of climate change while decreasing the likelihood of successfully emerging region-wide adaptation strategies. In this study, we analyzed climate change in the Middle East during the 21st century as predicted by 18 Global Climate Models. The simulations were run as part of the Intergovernmental Panel on Climate Change Fourth Assessment Report (IPCC AR4) and used the Special Report on Emission Scenarios (SRES) A2 emission scenario, which is the scenario closest to a "business as usual" scenario in the SRES family.

(With permission from Roland Geerken, modified)

To reduce the length of a poster introduction, you may also present the introduction in bullet point form, as shown in Example 29-4:

Example 29-4

> **Introduction**
>
> • Sea urchins are model systems for fertilization studies
> • Fertilization among sea urchins is species specific
> • Interaction of surface proteins (bindin and its receptor) on the egg and sperm are largely responsible for specificity
> • Differences in success rates of fertilization exists for individual urchins, but the reasons for these differences are unknown
> • We analyzed fertilization specificity quantitatively for 10 male and 5 female *Strongylocentrotus purpuratus* urchins

Materials and Methods

➤ Keep Materials and Methods information to a minimum

Summarize your experimental approach only very briefly. Where possible, use illustrations rather than words. You can also use flow charts and schematics to summarize experimental procedures if needed. Include photographs or drawings, and use references where relevant—use only names and dates within the text for the latter. Again, stay well within a maximum of 200 words, but note that in Methods posters, this section will be longer.

A sample flowchart of a Materials and Methods section from a poster is shown in Example 29-5.

Example 29-5 Materials and Methods section of a poster

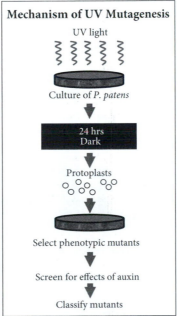

Results

➤ Present results in figures and tables

Describe your most important, overall results. Distinguish between results and data (see Chapter 13, Section 13.2). For basic research posters, this will be the largest portion of the poster. Most, if not all, of your findings should be presented in the form of figures and tables (see also Chapter 9 on figures and tables). Show your data analysis as it led up to your main findings. Provide engaging figure legends that can stand on their own, and interpret your findings, particularly if you do not have any separate text on results. On posters, place legends with tables as well. Ensure a consistent order between your results and the conclusions.

    Example 29-6

Results

Table 2 shows the domain average multimodel ensemble mean change in annual temperature and precipitation. There is high agreement amongst the Global Climate Models for the predicted temperature change and significant disagreement for the predicted precipitation change. This is reflected in the magnitude of the change and standard deviation shown in Table 2.

Table 2 Multimodel ensemble mean change in annual temperature and precipitation

	Temperature (K)		Precipitation (mm)	
	2050–2005	2095–2005	2050–2005	2095–2005
Mean Change	1.41	3.95	−8.42	−25.45
SD	0.32	0.73	16.08	28.66

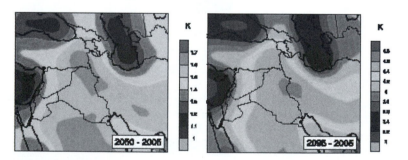

Figure 2 Multimodel ensemble mean change in annual temperature.

(With permission from Roland Geerken, modified)

Sometimes the Results and Conclusions are displayed on the same panel, often in bullet-point format as shown in Example 29-7.

Example 29-7

> **Results**
>
> - 30% of *S. purpuratus* gametes displayed twofold to ninefold differences in intraspecies fertilization efficacy
> - Surface components of both egg and sperm are involved in intraspecies fertilization success rates
>
> **Conclusions**
>
> - We hypothesize that different alleles of the surface proteins are responsible for variable fertilization success rates
> - These alleles may contribute to the process of reproductive isolation and, ultimately, speciation
> - Sequences of individual surface proteins from different individuals of the same species may shed light onto this hypothesis

Conclusions

➤ List your main findings and their meaning in the Conclusion

The last section of the poster is usually the Conclusion section. The discussion is left for you to present to your viewers and for publication. For the Conclusion section, assign importance to your results and summarize them accordingly. Use the same key terms you used consistently throughout the poster, and make use of bullets, arrows, italics, or colored text to emphasize major points. Usually, this section is very brief, as the poster is indicative of only a portion of the research. Concentrate on your main findings and their interpretation. Do not list all of your research findings. Instead, mention only two to four main points in your Conclusion or Summary. If written as bullet points, these findings will be more visually pleasing than a whole paragraph of text.

Example 29-8

> **Conclusions**
>
> - Mean annual temperatures will increase by ~4 K by the late 21st century.
> - Changes in precipitation are more variable; the largest change, however, is a precipitation decrease that occurs over an area covering the Eastern Mediterranean, Turkey, Syria, northern Iraq, northeastern Iran, and the Caucuses.
> - Changes in precipitation will have a significant impact on fresh water resources.

(With permission from Roland Geerken, modified)

As part of your Conclusion, you may also consider depicting a model or other figure to highlight a hypothesis or proposed model or mechanism, such as the one shown in the following example.

Example 29-9 Proposed model on a poster

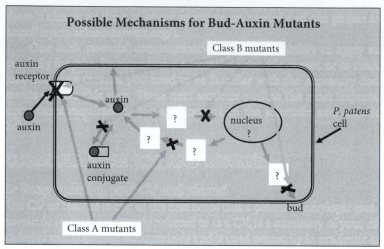

Sometimes a brief section titled "Future Research" follows the Conclusion section. In this section, explain briefly how you are planning to extend the presented work.

References

➤ Keep references to a minimum

References should be included if the technique is someone else's, but keep references to a minimum and keep them brief. Usually, names, dates, and journal information is enough for the Reference List. To cite information within poster text, use names and years only. If possible, do not use more than five citations. (See also Chapter 8 on References.)

Example 29-10   Text

To evaluate locomotion in the mice, we used the Basso-Beattie-Bresnahan (BBB) locomotor rating scale (Joshi & Fehlings, 2002) to score paralysis after SCI (Basso et al., 1996).

References

Basso, D. M., Beattie, M. S., & Bresnahan, J. C. (1996). *Exp Neurol* 139, 244–256.
Joshi, M., & Fehlings, M. G. (2002). *J Neurotrauma* 19, 175–190.

Acknowledgments

If applicable, thank individuals for specific contributions to the project (e.g., equipment donation, statistical advice, laboratory assistance, comments on earlier versions of the poster), and mention who has provided funding. Also, include in this section disclosures for any conflicts of interest and conflicts of commitment. (See also Section 29.3 and Chapter 16, Section 16.7.)

29.6   PHOTOS, FIGURES, AND TABLES FOR POSTERS

General Components and Format of Visual Aids

➤ **Use visuals where possible**

➤ **Prepare illustrations well ahead of time**

Clear visuals, such as figures, tables, photos, and schematics, are the most important pieces of a poster. Therefore, use such visuals where possible. Make illustrations well ahead of the meeting to give yourself time to check, replace, or improve them. Like for text, ensure that illustrations are large enough to be clearly discernable from two yards away. (See also Chapter 30, Section 30.3 and Chapter 9 for more details on visuals.)

Figures and Tables

➤ **Keep exhibits simple**

➤ **Make exhibits look attractive**

All illustrations and tables should be comprehensible on their own. Your exhibits should be attractive but professional. That is, the key information in them needs to be large enough to be recognizable from 6 feet away. Exhibits also should be kept simple and contain little text. They should not look cluttered. Individual figures should be recognizable as being part of a set (same colors, style of font, and emphasis techniques) and have sufficient resolution to avoid looking fuzzy. Furthermore, exhibits need to have a good balance of where information is shown within them across the entire display. The important information—the data—should be the main focus within exhibits.

Consider the following examples.

Example 29-11 Possible figures on a poster

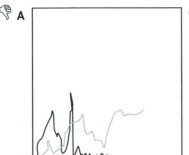

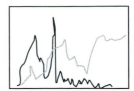

Example 29-11B is much more balanced than 29-11A, as the curves take up the entire frame in B versus only a fraction of it in A. Most readers find version B more attractive than version A because much unnecessary space has been eliminated.

Example 29-12 Possible figures on a poster

 A    B

The same source image has been used in Example 29-12A and B. However, version B has been cropped and enlarged to allow readers to focus on the berries more and eliminate unnecessary space. The cropped and enlarged image is perceived as more attractive by readers. It also allows them to recognize the key focus from farther away.

Most viewers look at the illustrations first. Not only do these need to be visually attractive, they also should be simpler than the ones in published papers such as the conference proceedings. In addition, you may need illustrations such as flow diagrams, which are typically not found in papers.

For the best possible presentation, follow these guidelines:

Do

- *Provide a title for each figure and table.*
- *Make the illustrations tell the story.*
- *Provide a legend for each figure and table.*
- *Use names for tables rather than numbers or letters.*
- *Choose graphs rather than tables.*
- *Make tables simple, if you use them.*
- *Emphasize key data through color and contrast, arrows, circles, pullouts, and other highlighting.*
- *Make annotations and lines in illustrations larger than normal and symbols easy to tell apart.*
- *Write all text horizontally.*

- *Explain each variable and its significance.*
- *Keep the scale consistent for all figures and graphs; keep axes consistent in all graphs.*
- *Write equations large enough to be read from 6 feet away.*
- *Eliminate all unnecessary material from illustrations.*
- *Label items directly in the figures rather than using a key or describing them in a legend.*
- *Use solid shades and colors rather than patterns in bar graphs.*

Don't

- *Don't use more than six bars in bar graphs and no more than three to four lines in a line graph.*
- *Don't use more than five columns and rows in tables, excluding the title and column headings.*
- *Avoid nonstandard colors, fonts, graphs, and abbreviations.*
- *Don't present unnecessary or unimportant equations.*
- *Don't use gridlines that distract from plotted graphs.*

Examples 29-13 and 29-14 show two poster panels with well-constructed illustrations.

Example 29-13 Sample poster panel

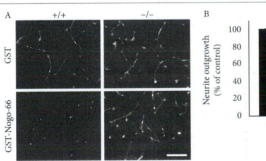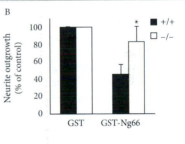

Neurite outgrowth assays using NgR KO P6 cerebellar neurons

(A) Dissociated cerebellar granule neurons isolated from *ngr +/+* or *ngr −/−* P6 pups were plated on dried spots of GST or GST-Nogo-66 (45 ng) and incubated for 12 hr. Scale bar equals 50 μm.

(B) Quantitation of neurite outgrowth shown as percentage of GST control for *ngr +/+* or *ngr −/−* P6 cerebellar neurons.

All data are represented as mean ± SEM. *, significantly different from wild type, $P < .05$ (Student's t test).

(With permission from Betty Lui)

Example 29-14 Sample poster panel

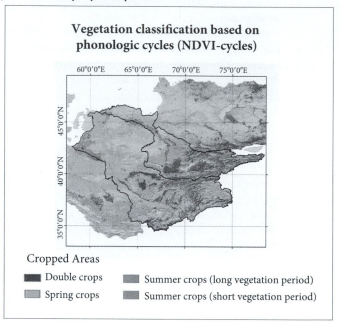

(With permission from Roland Geerken, modified)

29.7 RESOURCES FOR PREPARING AND PRESENTING A POSTER

On Campus

Your university or institution may have valuable resources that can help with designing and printing of a poster. Consult with these campus resources, which may include printing services or an instructional design center.

Software and Hardware Options

Although some posters are hand-created, posters generated electronically using a layout program usually look much more professional. Electronically generated posters can be printed as one large document using a variety of software packages such as Microsoft PowerPoint®, Adobe Photoshop®, Canvas®, Corel-Draw®, Illustrator®, PaintShopPro®, Adobe FrameMaker®, Adobe InDesign®, Keynote®, Pages®, or Sun/Solaris®. Large-format printers come in various sizes. Some department printers can handle posters up to 42 inches wide (length is flexible). For more detailed instructions or supplements on how to use diverse programs to create a poster, see also the following websites:

> **http://www.posterpresentations.com/html/free_poster_templates
> .html**
> Free research poster PowerPoint templates

http://www.emich.edu/apc/guides/apcposterpowerpoint2010.pdf
>   Creating a poster in PowerPoint
http://faculty.washington.edu/robinet/poster.html
>   Creating a poster using PowerPoint
http://blogs.ksbe.edu/ksedtech/download/Create%20a%20
>   Poster%20Using%20Pages.pdf
>   Making a poster using Pages on a Mac
https://wiki.wooster.edu/display/itdocumentation/Creating+a+
>   Poster+using+Pages
>   Creating a poster using Pages on a Mac

Other Useful Links

http://www.osti.gov/em52/workshop/tips-exhibits.html
>   Tips for effective poster presentations
http://www.bio.miami.edu/ktosney/file/PosterHome.html
>   How to create a poster that graphically communicates your message
http://www.ncsu.edu/project/posters/#Note0
>   Guidelines and additional links on designing and presenting a
>   scientific poster, including links to YouTube
http://www.aspb.org/EDUCATION/poster.cfm
>   How to make a great poster
http://colinpurrington.com/tips/poster-design
>   Tips on academic poster design
http://guides.nyu.edu/posters
>   Guidelines on poster basics

29.8   REVISING A POSTER

➤ Revise your poster repeatedly. Be ruthless when you edit

After you have drafted the poster sections, check them for mistakes, legibility, and inconsistency in style. Try different layout arrangements if necessary, and see if anything is missing that casts doubt on the content. Ensure
that the sequence of display panels is clear so that the viewer knows what to
look at and in what sequence. Make sure also that your layout emphasizes
important information and avoids visual distraction and that the graphics
speak for themselves together with their titles and captions.

Test the layout and content on other people such as friends, colleagues,
and your supervisor. Ensure they can glean the most important information from your poster. Ask them to write down the message they took away
from your poster so you can confirm that your poster is effective. Ask them
whether the title is engaging, informative, and appropriate for the target
audience and if the colors and layout are attractive and the lettering large
enough. Other questions to ask are: What is your favorite part of the poster?
Why? What would you do to make the poster stronger? Why? You may disagree with advice given, but at least listen to and think about the reactions

of other people to your poster, particularly if you get similar reactions from different viewers. They may have a point.

One of the most difficult tasks is to condense text, change layouts, adapt printed figures and tables for a poster, and eliminate information that is not absolutely necessary. Do not be afraid to omit unnecessary passages, figures, or table information, or information that might be interesting but not absolutely required to get your message across. Be ruthless when editing. Concentrate on the main points. Scrutinize your graphs and images. Do not intimidate the viewer with jargon and complicated graphics. Instead, simplify, omit, and summarize where possible.

29.9 PRESENTING A POSTER

➤ Be at your poster during the assigned poster session to answer questions and to tell viewers about your work

Although posters should look professional, the actual poster sessions are usually informal and interactive, unlike a talk in which people are generally more afraid to ask and answer questions.

For your poster session, arrive early at the display site. Unless you are confident the organizers will have proper supplies, bring a poster hanging kit with you. Hang your poster straight and neat, and do not encroach on your neighbor's space. You may consider bringing copies of a handout or miniature version of the poster for your readers. Such handouts can easily be created when the poster is designed using a layout program. Put handouts, business cards, and reprints nearby—on a table or in an envelope hung close to the poster so people can take them as they are passing by. Do not forget to restock supplies periodically if your poster is up for a long time.

Although the material you are presenting should convey the essence of your message, make sure you are at your poster during your assigned presentation time to be available for discussion. In addition, ensure that you have prepared a three- to five-minute talk highlighting the key points of your poster. Practice this talk (see Chapter 30, Sections 30.4–30.9). Your task as the presenter is also to answer questions and provide further details and to convince others that what you have done is excellent and worthwhile (see Chapter 30, Section 30.10). During the actual presentation, focus on your graphics. Use your poster as a visual aid—do not read it! Tell viewers the context of your research problem and why it is important, the objectives and how you achieved them, as well as the data and its significance.

29.10 SAMPLE POSTERS

The following figures show well designed posters.

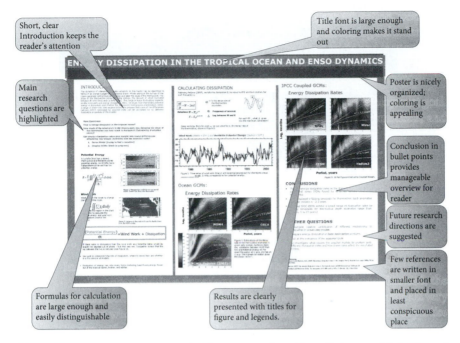

Short, clear Introduction keeps the reader's attention

Title font is large enough and coloring makes it stand out

Main research questions are highlighted

Poster is nicely organized; coloring is appealing

Conclusion in bullet points provides manageable overview for reader

Future research directions are suggested

Few references are written in smaller font and placed in least conspicuous place

Formulas for calculation are large enough and easily distinguishable

Results are clearly presented with titles for figure and legends.

Figure 29-4   (With permission from Alexey Federov and Jaclyn Brown)

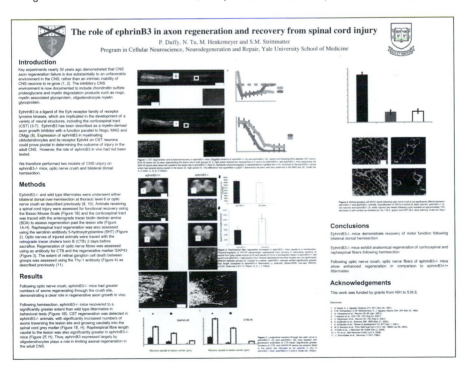

Figure 29-5   (With permission from Philip Duffy)

CHANGES IN THE LAND COVER AND LAND USE OF THE ITACAIÚNAS RIVER WATERSHED, ARC OF DEFORESTATION, CARAJÁS, SOUTHEASTERN AMAZON

P. W. M. Souza-Filho[a,b,*], W. R. Nascimento Jr.[b], B. R. Versiani de Mendonça[c], R. O. Silva Jr.[a], J. T. F. Guimarães[a,b], R. Dell'Agnol[a,b], J. O. Siqueira[a]

[a] Vale Institute of Technology, [b] Intelligence Planning and Environment, Vale S.A., [c] Geoscience Institute, Universidade Federal do Pará

Introduction

Long-term human induced impacts have significantly changed the Amazonian land cover and land use (LCLU) since early 1970s. The modifications were more intense in the arc of deforestation and were accelerated from 1980s **(Figure 1)** due to implantation of agricultural settlements, cattle ranching in large farms, and large development projects, such as Carajás mining, dams and ports for exportation. The studied area covers protected and non-protected domains, which correspond respectively to 28% and 72% of the Itacaiúnas river watershed.

The aims of this paper is to present a combined object-based classification and manual interpretation methodology for a quantitative assessment of LCLU changes in the Itacaiúnas River watershed in the Amazon Region from 1984 to 2013 **(Figure 2)**.

Fig. 1. Total deforestation in Brazilian Amazonia from PRODES data. Most of the forest clearing has taken place in the "arc of deforestation".

Fig. 2. 2013 Landsat-8 OLI mosaic images in 8R5G4B color composite showing the location of Indigenous land and protected areas. Green = forest; reddish = non-forest.

Dataset and methods

1984, 1994, 2004 Landsat-5 TM and 2013 Landsat-8 OLI; 1,060 GCPs collected along 2,400 km of roads;

LCLU classes mapped **(Figure 3)**;

Landsat-8 OLI images were converted to the Top of Atmosphere reflectance. Landsat-5 TM images were converted to ground reflectance. Furthermore, we derived the NDVI and band ratio 6/4 of Landsat-8 OLI and 5/3 of Landsat-5 TM;

The multi-resolution image classification of LCLU classes based on combined manual interpretation and automatic classification synergy from geographic object-based image analysis (GEOBIA) **(Figure 4)**;

LCLU "from-to" change detection approach to recognize the trajectories of classes based on GEOBIA from 1984 to 2013 **(Figure 5)**;

Accuracy assessment was undertaken using confusion matrices, Kappa index and Tau statistics.

LCLU System	Segmented Landsat-8	Fieldwork Photograp	Description
Land cover classes			*Observed (biophysical cover on the earth's surface.*
Forest			Ombrophilous forest canopy with more than 30 m in height.
Montane savannah			Deciduous open and shrub montane savannah occurring over ferruginous duricrusts in high altitude.
Water bodies			Rivers and small lakes discriminating whitewater and blackwater.
Land use classes			*Anthropogenic arrangements, activities and inputs undertaken in a certain land cover type.*
Pasturelands			Pasturelands in large farms. Croplands rarely occur.
Mining			Open pit mining with baresoil and rock outcrops.
Urban			Dense urban settlements with considerable area.

Fig. 3. Description of LCLU classification system used in this study.

Segmentation process based on combined manual interpretation and automatic classification synergy

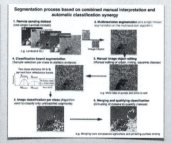

Fig. 4. Empiric GEOBIA workflow that illustrates the principles of the segmentation.

Results

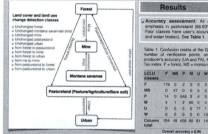

Figure 5. Conceptual model of trajectories of LCLU changes over time.

Accuracy assessment: All classes have producer's accuracy higher than 90%, with emphasis in pastureland (98.83%), montane savannah (97.92%) and urban areas (95.06%). Four classes have user's accuracy higher that 97% (forest, montane savannah, pastureland and water bodies). See **Table 1**.

Table 1. Confusion matrix of the GEOBIA 2013 Landsat-8 OLI classification. The matrix shows the number of verification points, omission's and commission's errors (OE and CE), user's and producer's accuracy (UA and PA), general Kappa index and Tau index per class, overall accuracy. F = forest, MS = montane savannah, P = pastureland, M = mining, U = urban, W = water.

LCLU classes	F	MS	P	M	U	W	Row total	OE (%)	PA	CE (%)	UA	Kappa per Class
F	176	0	2	3	0	0	181	9.28	90.72	2.76	97.24	0.97
MS	0	47	0	0	0	0	47	2.08	97.92	0	100	1
P	14	0	649	2	4	0	669	1.37	98.63	2.99	97.01	0.92
M	4	1	2	60	0	0	67	7.69	92.31	10.45	89.55	0.89
U	0	0	5	0	77	1	83	4.94	95.06	7.23	89.55	0.92
W	0	0	0	0	0	13	13	7.14	92.86	0	100	1
Column total	194	48	658	65	81	14	1060					

Overall accuracy = 0.96   Kappa index = 0.94   Tau index = 0.93

LCLU "from-to" change detection: Figure 6

From 1984 to 1994 unchanged forest - 2.8 millions ha (68% of the study area). Conversion from forest to pasture - ~800,000 ha. Unchanged pasture - ~300,000 ha **(Table 2)**;

Between 1994 and 2004, unchanged forest - 2 millions ha. Unchanged pasture - ~1 million ha. Forest to pasture - ~800,000 ha;

From 2004 to 2013 unchanged forest and unchanged pasture - 1.8 and 1.7 million ha, respectively. No change occurred in ~85% of the area;

Between 1984 and 2013, ~47% (1.9 million ha) of forest kept unchanged; almost ~41% (1.7 million ha) of changes correspond to conversion from forest to pasture.

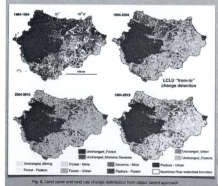

Fig. 6. Land cover and land use change distribution from object based approach.

Table 2. A summary of LCLU change trajectories from object based approach. Quantification of LCLUC changes between 1984-1994, 1994-2004, 2004-2013 and 1984-2013. Area in 1,000 ha.

LCLU change class	1984-1994 Area	%	1994-2004 Area	%	2004-2013 Area	%	1984-2013 Area	%
F - M	2.63	0.06	2.49	0.06	3.64	0.09	8.73	0.21
F - P	792	19.17	796	19.25	301	7.27	1,680	40.82
F - U	0.72	0.02	0.42	0.01	0.90	0.02	5.79	0.14
P - F	65.21	1.58	72.26	1.75	139	3.36	41.99	1.02
P - U	1.23	0.03	0.00	0	7.51	0.18	6.72	0.16
MS - M	0.39	0.01	0.56	0.01	0.49	0.01	1.43	0.03
UF	2.818	68.16	2.046	49.5	1,807	43.7	1,925	46.55
UM	1.36	0.03	3.40	0.08	5.49	0.13	1.12	0.03
UMS	10.23	0.25	9.93	0.24	9.42	0.23	9.12	0.22
UP	307	7.43	1,040	25.15	1,710	41.36	333	8.05
UU	1.13	0.03	2.74	0.07	5.10	0.12	1.18	0.03
Un	134	3.23	161	3.88	146	3.52	122	2.94

Conclusions

The synergy of visual interpretation to discriminate fine level objects with high contrast associated to urban, mining and montane savannah classes and automatic classification of coarse level objects related to forest and pastureland classes is most efficient than use these methods individually. In essence, this approach combines the advantages of the human quality interpretation with quantitative computing capacity.

LCLU changes are associated to strong negative relation between deforestation processes and formation of new landscape dominated by pasturelands in the Itacaiúnas River watershed.

Acknowledgement

We would like to thanks USGS that provides Landsat-5 TM and Landsat-8 OLI images used in this work.

Figure 29-6

(With permission from Pedro Walfir M. Souza Fihlo)

29.11 CHECKLIST FOR A POSTER

Use the following checklist to ensure that you have addressed all important elements for a poster:

- ☐ 1. Do the illustrations tell the story?
- ☐ 2. Does the layout emphasize important information and avoid visual distraction?
- ☐ 3. Is the purpose of the research or topic clear?
- ☐ 4. Did you avoid attaching the conference abstract?
- ☐ 5. Does the Introduction have the following components?
  - ☐ Background
  - ☐ Problem or unknown
  - ☐ Purpose/topic or review
  - ☐ Overview of content
- ☐ 6. Did you concentrate on the main points in each section?
- ☐ 7. Is the flow of the panels self-evident to the viewers?
- ☐ 8. Is the topic summarized and interpreted in the Conclusion section?
- ☐ 9. Do all figures and tables have a title and a legend?
- ☐ 10. Is your poster layout uncluttered?
- ☐ 11. Did you use visuals where possible rather than text?
- ☐ 12. Did you keep text to a minimum?
- ☐ 13. Is text written in consistent font, and is the font large enough?
- ☐ 14. Are exhibits kept simple?
- ☐ 15. Are exhibits attractive? Is color used well?
- ☐ 16. Did you use active voice in the text?
- ☐ 17. Have all jargon and redundancies been omitted?
- ☐ 18. Did you proofread your text?

SUMMARY

> **POSTER GUIDELINES**
> - Ensure that your conference abstract fits the topic of the conference and summarizes your work accurately.
> - For a conference abstract, ensure your overall format consists of
>   - Background/context
>   - Question/purpose
>   - Experimental approach
>   - Results
>   - Conclusion (answer)/implication
>   - (References)
> - Follow the guidelines provided for a conference abstract.

- Design the poster around your research question. Include:
  - Title
  - Abstract
  - Introduction
  - Materials and Methods
  - Results
  - Conclusion
  - (References)
  - (Acknowledgments)
- Concentrate only on the main points in each section.
- Find visual ways to show your work—let the illustrations tell the story.
- Use an easy-to-follow layout.
- Use different sizes and arrangements for the various poster sections, but maintain a consistent style.
- Aim for about 20% text, 40% graphics, and 40% empty space.
- Minimize text, and use images and graphs instead.
- Make the title informative, attractive, and large enough.
- Distinguish between a poster abstract and a conference abstract.
- Keep background information to a minimum in the Introduction.
- Keep Materials and Methods information to a minimum.
- Present results in figures and tables.
- List your main findings and their meaning in the Conclusion.
- Keep references to a minimum.
- Use visuals where possible.
- Prepare illustrations well ahead of time.
- Keep exhibits simple.
- Make exhibits look attractive.
- Revise your poster repeatedly. Be ruthless when you edit.
- Be at your poster during the assigned poster session to answer questions and tell viewers about your work.

CHAPTER 30

Oral Presentations

Every scientist should be able to prepare and deliver a good oral presentation. Although most scientists desire to present their work at international conferences, many of them also fear having to present a talk. Unfortunately, as with scientific writing, most scientists are not formally trained in this art. However, being a good presenter is something that can be learned. For many people, this art is much easier to master than the art of writing a paper.

THIS CHAPTER DISCUSSES:

- Components and format of a scientific talk
- Planning and preparing for a presentation
- Visual aids
- Giving a talk, including conquering nervousness
- Voice, delivery, body actions, and motions
- Vocabulary and style
- The question-and-answer period
- Other speech forms: introductions, impromptu talks, TED talks
- Resources for presentations

30.1 BEFORE THE TALK

➤ Prepare your talk well ahead of time

The key to becoming a good presenter is being prepared. As soon as you know you are going to speak, begin preparing your slides. Preparing your slides will make your subconscious mind work on the words for the actual talk.

➤ Get to know your audience

To plan the best possible talk, you have to know your audience. Find out to how many people you will be presenting as well as their level of expertise. Design your talk accordingly in terms of its direction and necessary background information. If your audience includes nonexperts in your field, provide a much broader background using nontechnical terms. You may also want to consider not going into as many technical details as in a talk geared toward your peers.

➤ Practice, practice, practice

To deliver your talk well, you *need* to practice and practice and practice. You should practice your talks with the slides or other visuals, and you should practice from beginning to end. Without practicing a talk, you will not know whether you stay within the given time limit, nor will you know if your flow of words is smooth or if your voice has the right pitch. Practicing will make you realize at least some of these potential problem areas. Consider videotaping yourself, or ask someone else to do this for you. You can use your smartphone to record audio and video, for example. Review the video and note any areas that may need improvement. In addition, take advantage of opportunities to give and practice presentations, such as in departmental talks or in a formal class.

Conference Talks and Abstracts

Getting invited to speak at a conference is an important recognition of your work and may not only help you to gather new ideas but also increase your visibility in the field. If you would like to get invited for a talk at a conference, you usually have to submit an abstract. (See Chapter 29, Section 29.2 for details on Conference Abstracts.)

30.2 COMPONENTS AND FORMAT OF A SCIENTIFIC TALK

Components

To present a good talk within a given timeframe, you need to

- Identify the most important information
- Illustrate it in well-organized and visually attractive slides
- Present the information clearly to your audience

The content of a talk is similar to that of a journal article with an introduction, results (combined with overview of experimental approach), discussion, and conclusion. However, an oral presentation needs to be structured and worded differently from a written paper. If you want to deliver your presentation well, do not just read it off. Understand the differences of how information is extracted by readers versus listeners. Whereas text can be read and reread at your own speed, listeners get a chance to listen to a piece of information only once and at the speed the presenter sets. Similarly, listeners have no control over the order or type of information they will see and hear, whereas readers can easily scan headings and subheadings and skip ahead when they want. In addition, when presenting, the emotions of a speaker can easily be conveyed; this is much harder to do in writing.

Generally, your talk should start broadly, providing information on the background of your research. See Figure 30.1 for a graphical representation of the structure of a scientific talk. For research presentations, the slides then narrow to what is unknown and to the overall purpose/question of your study. (In case of a verbal proposal, your slides would narrow to the objective of the proposal.) The main part of our presentation will remain more narrowly focused on your key results (or aims for a proposal). During the conclusion portion, the presentation usually broadens out again to help your audience understand the impact and significance.

If you have to present your talk in a language you do not know well, consider persuading a native speaker to listen to your talk and to comment on pronunciation. Alternatively, ask a native speaker to make a tape recording of the talk for you. Listen to the tapes a few times. Note the pronunciation of difficult words and the intonation of sentences. You should try to adjust the language of the wording on your slides to that of your target audience. If a translation service is provided for the talk, it's imperative to speak slowly so the translators do not get rushed in the translation—this cuts time off of your talk as well. Practice the presentation as often as possible.

ESL advice

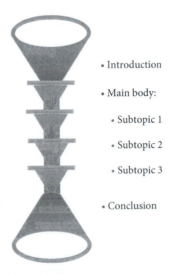

- Introduction

- Main body:

  - Subtopic 1

  - Subtopic 2

  - Subtopic 3

- Conclusion

Figure 30.1 Ideal flow of information during a lecture.

Layout of Talks of Various Lengths

Whether you are presenting a short, 10-minute conference talk or a full hour seminar, follow the overall format of introduction, results, and discussion for your talk. Depending on your target audience, different numbers of introductory slides may have to be included. For a nonscientific audience, it is particularly important to include more background slides, as most speakers lose their audience in the first few minutes by failing to give an appropriate introduction to the problem.

Shorter talks are usually more difficult to prepare and present than longer talks. For shorter talks, you need to be extremely selective regarding the most important information for each of the sections, and you need to prepare slides with this focus in mind. Generally, you will have to reduce the number of background slides for shorter talks and concentrate primarily on the main findings and their interpretations in the rest of the slides. In fact, depending on your topic and audience, you may only have time to present one or two main findings.

When your time is limited, consider skipping the title slide as well as the overview slide. Instead, just verbally inform the audience about your talk's overall title and about the outline for your talk. Do provide a concluding/summary slide in all cases.

General Format

➤ Organize your slides
Include the following:

Optional: Title slide	Title
First slide	Overview of talk
Second slide	Introduction and background
Subsequent slides	What you studied and how you studied it; your results
Final slide	Your conclusions and the main points that support it
Optional final slide:	A credit slide in which you acknowledge those who worked with you or financed your research

➤ As an overview for your presentation:
- Tell the audience what you are going to tell them.
- Tell them.
- Tell them what you have told them.

Use and explain the title slide if you are not introduced for your talk. If you are introduced, start your presentation with an overview of the talk by telling the audience what you will speak about: introduce the overall topic, and mention how you will present your findings and that you will then summarize what you have told them in the talk.

Follow this overview with a presentation on the background of the overall topic that funnels down to your work, the unknown, and then the overall purpose/question of your project. State your experimental approach briefly when you present your results, and discuss findings in the general scientific context. In the final slide, present your interpretations and conclusions. Concentrate on the broad picture and the importance of your findings to society. You may add an acknowledgement slide but, depending on the time limit, consider simply putting it up or skipping it altogether. When you are finished talking, thank the audience.

30.3   VISUAL AIDS

➤ Know how to use visual aids

Competent speakers must know how to use visual aids to clarify and reinforce their talk. Visual aids come in various shapes and forms. PowerPoint is currently the most powerful and impressive way of making presentations. PowerPoint slides make a professional and effective presentation and can easily be changed at the last minute if needed. In PowerPoint presentations you can also include short cartoons and jump back and forth between exhibits easily. You can either show the whole slide at once or build up an illustration using animations. Appropriate spots on the slide can be indicated using a laser pointer.

To present these slides, you need to know how to use a projector and be aware of potential problems that arise when using different computer programs and systems. Particularly, when your presentation has to be converted from a Macintosh to a PC or vice versa, you need to check that figures and font types convert properly. Also, check conversion between different programs or program versions on the same systems. Do not wait until shortly before your talk to check whether your slides will work correctly, however. Conversion problems may take time and expertise to resolve. Try out programs, equipment, and set-ups ahead of time, and bring your computer as a back-up. Be prepared to give the talk even if the slides fail.

➤ Prepare visual aids well ahead of your talk

Prepare visual aids well ahead of the meeting to give yourself time to check, replace, or improve them. Visual aids, regardless of whether you are using overheads, slides, or PowerPoint presentations, should be comprehensible on their own.

➤ Keep exhibits simple

Keep visual aids simple. Your slides must be clear, legible, and easy-to-understand. Avoid nonstandard colors, fonts, graphs, and abbreviations. You need to create exhibits that add to, not distract from, your work. (See also Chapter 29, Sections 29.4 and 29.6 and the following sections.)

➤ Think graphically

When it comes to oral presentations, you need to think graphically. That is, whenever possible, display information visually as figures, tables, photos, diagrams, and so forth rather than as text. Be creative. The adage "a picture is worth 1,000 words" holds true here. Generally, it is more powerful to use a picture or figure than a table or only text. See, for example, Example 30-1 and its revised versions.

➤ Make exhibits look attractive

Your exhibits should be visually pleasing but professional. Moreover, individual slides should be recognizable as being part of a set (same colors, style of fonts and emphasis techniques). (See also Chapter 29, Sections 29.4 and 29.6.)

➤ Communicate one main idea per slide

Before you can prepare your slides, you need to decide on what and how much to include in your talk. A good rule of thumb is to communicate one main idea per slide, and emphasize this central message when speaking. Prepare slides with the audience in mind. The total number of slides you will be able to display in your talk depends on the time you are allotted to speak. (See also Sections 30.2 and 30.4 for more information on content for various talks and time limits.)

➤ Use conservative colors and high contrast between background and writing or figures

When presenting a talk, there is no substitute for clear, graphical visual aids. To create visually attractive slides, pay attention to how other people present colors, fonts, and graphics on slides that are easy on your eyes. Avoid bright colors, and avoid red/green (or blue/orange) color contrasts as some people are color-blind to these. Look for high contrast between background and writing or figures. Choose dark text against a light background or vice versa. A medium to dark blue background with white or yellow writing, for example, is commonly used and easy to read because of the high contrast between the background and writing.

Not only do your slides need to be visually attractive, they also should be simple. Your slides should be informative, but discipline yourself to use as few words as possible to convey this information. Use your voice to fill in the rest.

To create effective slides using as few words as possible, it is essential to know how to write clear but brief bullet points. Slides that are text-heavy are usually constructed for the benefit of the presenter rather than that of the audience. Avoid such slides. To make your slides more attractive for your audience, use key words and phrases—they are more effective than whole sentences.

➤ Follow the 5 × 5 rule

Avoid numbering items in a list. Use bullet points instead. Try to follow the 5 × 5 rule: Use about five words per bullet point and five bullet points per slide. At the same time, keep punctuation to a minimum and start text at the left rather than in the center. If you absolutely have to include more than five bullets on a slide, consider grouping the bullets and creating a slide for each group.

Visual fatigue is the biggest enemy of presentations, especially if you ask your audience to sit in a dimly lit room for 30 to 60 minutes. With just five words per bullet point, you will have to fill in the rest of the information verbally, and the audience will have to pay attention to you, the speaker. This results in a more animated presentation, which is more interesting for the audience.

If your slides contain too much text, your audience will concentrate on reading the text and not listen to what you have to say, or they will be listening to you and not pay attention to what is on the slide. Neither case is what you as the presenter really desire.

Following is an example of a slide that contains too much text.

 Example 30-1   Text slide

> Overview of the Yale School of Medicine
>
> - Founded in 1810, the Yale School of Medicine is a world-renowned center for biomedical research, education and advanced health care.
> - The School is viewed internationally as a leader in biological and medical research.
> - The Yale School of Medicine has over 900 faculty members and consists of 9 basic science departments and 17 clinical departments.
> - The School of Medicine consistently ranks among the handful of leading recipients of research funding from the National Institutes of Health and other organizations supporting the biomedical sciences.

To improve a text-heavy slide such as the one shown in Example 30-1, decide on what information in each bullet point is really important and/or which bullet points can be omitted. Then, list the important information by itself in a bullet point and omit the rest. A possible revision of Example 30-1 follows.

 Revised Example 30-1a   **Text slide**

> Yale School of Medicine Overview
>
> - Founded in 1810
> - Leader in biomedical research
> - Over 900 faculty members
> - 9 basic science departments
> - 17 clinical departments
> - Top biomedical research funding

In the revised version, the bullet points have been reduced to their main piece of information and are visually distributed better on the slide, all of which is preferred by the audience.

If we further revise Example 30-1a by adding a different background as well as a picture and using different styled and shaded text for the heading, a more visually attractive slide results, as shown in a further revision.

 Revised Example 30-1b   **Text slide with background and graphic**

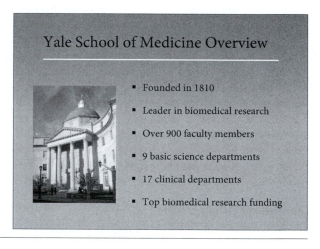

Revised Example 30-1b makes for a much more attractive and interesting slide than Example 30-1a. This is because Example 30-1b not only has a shaded background but also a pleasing picture that directly relates to the information on the slide.

➤ Make lettering large enough

Lettering should be large enough to read easily even from the back of the room. Ensure that the minimum font size is larger than or equal to 18 points for anything other than references or figure labels.

➤ Use textual highlights where appropriate

Use a sans serif font type such as Arial. For many people, sans serif font types are easier to read than serif ones. Serif fonts are also accepted, however, such as Times New Roman. Note that lowercase lettering, with initial capitals where needed, is easier to read than lettering that is all in capitals.

TABLE 30.1 Font sizes for slides

FONT SIZE	USE
32–40 point	slide heading
24–28 point	body of slide
18–24 points	sub-bullets
14 points	references, figure labels

 Example 30-2A **Text in capitals**

WORDS ARE HARDER
TO READ AND TAKE UP
MORE SPACE

 Example 30-2B **Text in capitals**

Words are harder to read and take up more space

Put the most important information in a larger print size. Italics can also be used to add emphasis, but avoid underlining or exclamation marks as they are not considered effective.

Use maximum contrast and boldface for maximum legibility.

 Example 30-3 **Text contrast**

This is not OK

Revised   **Text contrast**
Example 30-3

This is OK

Aside from textual highlights, remember that it is important to add emphasis by using colors and/or visuals such as photographs or figures where appropriate (see Example 30-1b).

➤ Give poster figures and tables a title

Figures and tables on each slide should have a title, placed at the top and separate from the rest of the material by extra space. Unlike for posters and in printed text, figures and tables on slides do not contain any legends. Therefore, it is particularly important that you use names for different groups rather than numbers or letters. Also, do not simply transfer published figures or tables as PDF files or similar onto your slides. Their lettering is often too small, they do not have a heading, and more often than not they contain additional information that is not needed for your slide.

The difference between printed, poster, and slide figures is apparent in the next example.

  Example 30-4a   **Printed figure**

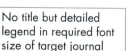

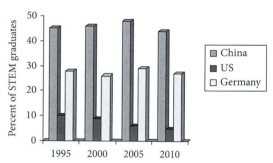

Figure X. Percent of STEM majors graduating in scientific disciplines in China, the US, and Germany over the past two decades. Although roughly twice as many students initially declare a major in a STEM field in US universities, less than half of them graduate in the field and graduation numbers are declining. STEM graduates include male and female students finishing a four-year degree from a public or private university in each country. The survey spanned 420 universities in the US, 382 in Germany, and 764 in China. Numbers were averaged by students that declared a stem major by the end of the first year of their studies.

Example 30-4b   **Poster figure**

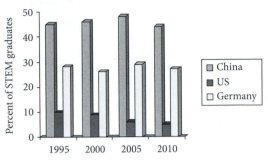

Title and brief legend in large font size

We surveyed male and female students for STEM majors at 420 universities in the US, 382 in Germany, and 764 in China finishing a four-year degree from a public or private university. Numbers are averaged by students that declared a STEM major by the end of the first year of their studies.

Example 30-4c   **Slide figure**

Title but no legend

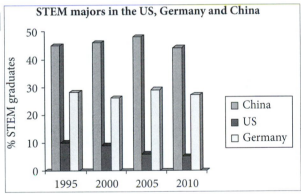

➤ Choose graphs rather than tables

Generally, you should choose graphs rather than tables. Bar graphs are often preferable because their message can be quickly understood. In bar graphs, keep the number of bars to a minimum. Write all text horizontally, including labels for vertical axes if they are not too long. For graphs, curves should be smooth and the lettering clear (i.e., Arial or Helvetica). Symbols must be easy to tell apart. Provide a key for figures if needed (see also Chapter 9 and Chapter 29, Sections 29.4 and 29.6 for advice on good layout of figures). Make sure that exhibits are aligned well within the slide and with respect to each other and possible text.

➤ **Follow the 3 × 5 rule for tables: maximum 3 columns and 5 rows**

If you have to use a table, follow the 3 × 5 rule: keep it to a maximum of three columns and five rows, excluding the title, row, and column headings.

 Example 30-5   Slide with table

Thermal Induction Cytokinin Overproduction

		Concentration of cytokinin in medium (nM)	
Mutant	cytokinin	15°C	25°C
thiAl	iP	0.16	0.23
	[9R]iP	0.09	0.16
OvesT25	iP	6.34	60.0
	[9R]iP	0.54	9.51

➤ **Construct well-balanced figure and table slides**

An example of a well-constructed slide containing figures is Example 30-6.

Example 30-6   Slide with figures

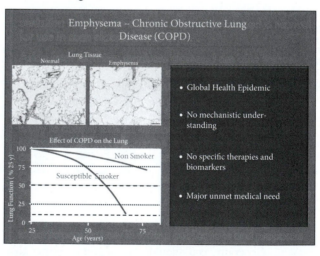

(With permission from Patty Lee, modified slide; and Robert Homer, EM images)

The figures and the text in Example 30-6 have been well placed. They are aligned nicely, visually pleasing, and easy to grasp. They are also clearly labeled, making for a very balanced slide.

Similarly, the schematic displayed in the slide for Example 30-7 is well balanced and easily graspable. The author could have used a text slide to explain the concept, but the schematic brings the message across much more effectively and memorably.

Example 30-7   Slide with schematic

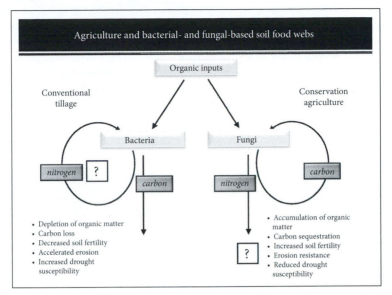

(With permission from Mark Bradford, modified)

Following is another slide in which a well-constructed bar graph is shown.

Example 30-8   Slide with bar graph

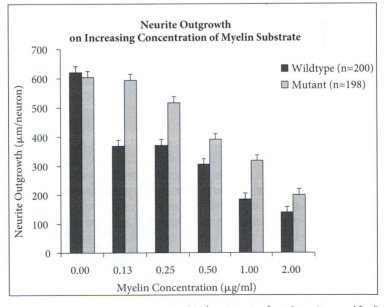

(With permission from Betty Liu, modified)

Note that in this slide, the title for the graph is at the same time the title for the slide. Depending on how your presentation is structured, it is possible to compose such titles if your slide contains a single figure. Note also that the slide contains no figure legend, but a key is shown for the graph. The presenter's words will fill in any additional necessary information and summarize the slide.

In other slides, the title may serve as a summary or overview of a finding depicted in the slide. This use of a title will reduce the number of words that need to be placed on the slide, as shown in the next example.

 Example 30-9   Slide with result in title

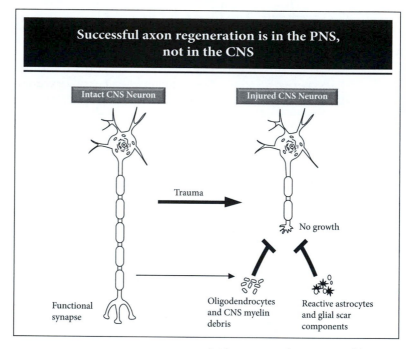

(with permission from SAGE publications)

30.4   PLANNING AND PREPARING FOR A TALK

Strategic Preparations Ahead of Your Presentation

➤ Stick to the time limit

It is customary to assign a limit to the length of a speech. Adhere to this time limit. Find out whether you tend to speed up or slow down your talk during delivery. Allow for it during rehearsal. On average, you will be going through one to two slides per minute. Note that this is an average number, which very much depends on your slides. For some slides, you may spend considerably more time. For other slides, you may be able to go through them in substantially less time than the average. The only way to know how long it will take you is to time yourself when practicing your talk.

One of the most annoying things for any audience is a talk that goes overtime. To help you better keep track of the time, place a timer on the lectern. If you find that you need to leave out some important slides to stay within the time limit, do so. Do not speed through your talk. You run the risk of losing your audience if you present your material too fast.

If the chairperson signals to you that your time is up, summarize any remaining material and give your conclusions immediately. You can prepare for this eventuality by inserting a hyperlink to the last slide strategically on some previous slides. This hyperlink may be disguised in a design feature or text of the slide. (See also Appendix D for the top 20 PowerPoint tips.)

➤ Structure your talk to maximize the attention span of your audience

On average, after a talk, most people remember about 10% of what they hear, 20% of what they read, and up to about 50% of what they hear and see. After about a week, your audience will remember only 10% of your presentation. Yet much of this retention depends also on when during a talk the message is delivered.

No audience listens to 100% of a speech. Typically, your audience's attention will only be high in the first 10 minutes and during the last 5 minutes of a 40-minute talk (Figure 30-2). They will be thinking about work to be done or last night's movie, or perhaps they will just be quietly slumbering after a heavy meal.

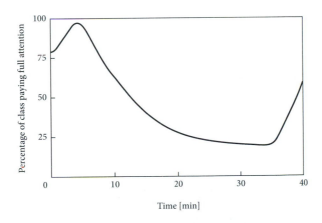

Figure 30-2 Attention span of average audience.
(Mills, H.R. (1977). Techniques of Technical Training, 3rd ed. Macmillan, London; Permission of Macmillan Press.)

To work against this pattern, you have to engage your audience by using an enthusiastic tone of voice, highlighting information verbally, and pointing out new topic areas. To maintain the audience's concentration on your talk, use attention-getting words to focus their minds on what you are saying ("Ladies and Gentlemen, I want to point out . . ."; "The two most important points of this talk are . . ."; "To the graduate students I say . . ."). Then, use word pictures, emotion, and impression in your voice. Keep your words simple and easy to understand, and vary the pitch and tone of your voice.

➤ Identify the purpose/question and the take-home message of the study

Similar to that of a research paper, the central theme of your talk needs to revolve around a specific question, that is, the purpose of your study. Identify this purpose/question clearly, and present it right after your introduction. (See also Chapter 11, Section 11-3.)

Next, pinpoint the take-home message of the talk. The take-home message is your interpretation of the collective significance of all conclusion points. Together with the purpose/question, the take-home message will serve as the filter for what to include in the talk. All other slides and data of your talk will be selected based on the question and take-home message. This take-home message should be presented at the end of your talk.

➤ Pick out the most important figure for your core slide

When you are preparing your talk, ensure that you address the most important information, no matter what the time limit. To help you select your most important figure or core slide, imagine you were given only two minutes to present your work. The core slide will be the single slide you would need to include to make your talk compelling.

The core slide needs to be clear, uncluttered, and immediately understandable. As it will be your most important selling point, you need to prepare the core slide with extra care. It should contain your most important figure, be highly visual, and be the slide(s) that everyone remembers when they leave. (See also Sections 30.2 and 30.3 for slide design.)

Example 30-10   Cytokinin overproducing mutant of the moss *Physcomitrella patens.* A. at 15°C, B. at 25°C

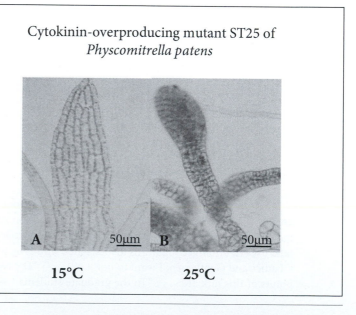

➤ Plot your talk by choosing supporting slides

While the core slide helps you center the talk, other slides will be supporting slides that highlight the work performed in the course of answering the overall question. To plot your talk, write down your main findings and then determine corresponding images to illustrate them in your supporting slides. Show only highlights to stay within the time limit.

➤ Prepare notes

Notes are the road maps for speakers. Too many notes can cause problems, however, and so can too few. To record notes, use index cards (3 in. × 5 in.) or write into the Notes section of your PowerPoint slides and display them using the presenter's view during your presentation.

Index cards are a convenient size to handle. They easily fit into a pocket and are not visible to the audience. Do not memorize a speech. Instead, strive to maintain an image of spontaneity. Most important, never read a speech unless you are forced to by legal requirements or time restrictions.

Your notes on the index cards or on the computer should be easily readable—use extra-large print if necessary. Notes should be in outline form and should not be written in sentences. There are, however, a few exceptions when notes should be written out in full and may even be memorized:

- Write out the opening of your speech—memorize the first few sentences.
- Write out the closing section of your speech.
- Write out transitions on note cards.
- Write out quotations in full.

Having these passages available in full will give you comfort in knowing where to find them if you need them.

Conquering Nervousness

➤ Write out your opening sentence in full. Deliver it firmly and accurately

No human has ever or will ever conquer nervousness. Only experience teaches a person to control his or her nerves and to appear confident to the audience.

Know that an audience usually will watch the body language of a speaker even more than listen to the speech. Thus, you need to use your body to communicate the message you want to present to your audience. Confidence is key. The most powerful way to appear confident is to look directly at the audience. You should also stand up straight and gesture, but not wildly. Check your appearance and dress. For more informal talks, such as departmental presentations, business casual wear is appropriate. You may also get away with business casual for conference presentations. For many job interviews or for formal presentations to wealthy funders, you should consider business formal. Do not accessorize excessively as accessories can be distracting to the audience.

Study your surroundings. The more familiar you are with your surroundings, the more comfortable you will feel. Take a number of deep breaths while being introduced. If you have prepared and rehearsed the speech adequately, you may feel more comfortable. If your notes are well written, you are assured that a glance at them will put you back on track.

In your notes on the cards, or in your notes on screen, have your opening sentence written out in full. Deliver the opening sentence firmly and accurately—such delivery will give you confidence to continue. The early sentences of a speech should have been practiced over and over again as confidence builds on itself. Nervousness normally abates when you realize that you are off to a good start.

Practice is the only way not to act in peculiar ways. Practicing in private helps only to a certain extent, however. Actually speaking before an audience is the only effective answer.

Self-Improvement Suggestions

➤ Add excitement to your voice, and pause for emphasis

To give yourself the opportunity to present the best possible talk, consider doing the following: Read aloud. Talk to yourself. Read to your children. Address your dog, and above all, practice in front of your colleagues. Hearing your voice is as important as finding the right intonation. Poetry makes the best material for practicing reading and speaking. Get away from the monotone. Learn to add excitement to your voice, and learn to add pauses for emphasis. Learn to read without gluing your eyes to the printed text.

Use a recording device to play back your performance. Check quality of voice, word flow, delivery speed, vocabulary, grammar, and body movement. Become aware if you are uttering appalling sounds such as "um," "ah," "uh," and "You know." If so, break this habit. Videotaping yourself or recording your voice and playing back the recording is usually the best remedy for this annoying habit. You could easily use your smartphone for such recordings.

I encourage ESL speakers also to listen to the radio, to channels such as NPR, CBC, BBC, and so on to get a feel for the rhythm of speech. Go over what you want to say often enough that the proper word will not be difficult to conjure. Moreover, actively seek speaking opportunities. In short: Practice, practice, practice.

ESL advice

30.5   GIVING THE TALK

Vocabulary to Know

Podium—low, raised platform on which a speaker stands (also known as a dais or rostrum)

Lectern—reading desk on which speakers place their notes; a lectern may also contain ports for laptops and controls for projectors, microphones, lights, and screens

Most speakers perform standing on a podium behind a lectern.

Set-up

➤ Check the set-up ahead of time

Before you present any talk, you should check the set-up. Familiarize yourself with the room as well as with the presentation equipment. Also, check lighting, plugs, and chalkboard.

If you are planning to give a PowerPoint presentation without using your own computer, make sure your PowerPoint file is compatible with the program of the computer and the projector that you are planning to use. Make sure your notes are in the proper order. Last but not least, get the correct pronunciation of the name of the person introducing you.

Going to the Lectern

Your movements should be unhurried and dignified. Do not begin your speech before you reach the lectern. Also, do not begin as soon as you reach the lectern. First, place your notes on the lectern. Place any other material you may need next to your notes, such as a pointer, a watch, or some water. Once you have your materials in place, compose yourself and look out at the audience. Then, thank your introducer and begin your speech.

30.6 VOICE AND DELIVERY

➤ Make sure you can be heard by the entire audience

Do not speak too softly. Soft speech signals that the speaker is uncertain. However, do not blast the audience out of its seat by the volume of your delivery either.

English is a language in which stress is crucial. Although pronunciation is important, it is less important than using the correct stress. If you have the stress right, you should not worry about having some kind of accent. If you have trouble pronouncing words correctly, it may help to put accent marks on syllables to be stressed in your notes and to mark places where your voice should pause. You may also want to underline phrases to emphasize.

➤ Speak neither too fast nor too slow

A good talk requires speech that is slower and clearer than in normal conversation while adhering to the time limit. Speakers who are nervous often speak too fast, as do speakers who realize that their talk is too long and that they will likely go over the time limit. Try using a deliberate pace in speaking. Practice your talk so you are able to judge your rate of speaking

ESL advice

compared for the assigned time limit. Above all, listen to yourself as you talk. Record your presentation if possible. Ensure that you talk only as fast as you can comprehend it.

➤ Avoid appalling sounds, and pay attention to pitch

Avoid any appalling sounds when you speak, such as "um," "ah," and "mmmh." These can be very distracting for your audience, and some people may even start counting the number of times you use these sounds rather than focusing on your presentation.

Also, pay attention to the pitch of your voice. Higher tones of pitch may lead the audience to assume that a speaker is less professional and more childlike. Speaking in deeper, fuller tones makes your voice more pleasant to listen to and can be achieved by using the diaphragm or lower throat to control the pitch of your voice rather than the upper throat or the nasal passages.

➤ Explain everything on a slide

When you present your talk, explain everything on your slides. Do not skip over figures, table values, or text. Walk the audience through each slide step-by-step. If you skip information, you will lose your audience. Skipping information shows that you have not sufficiently prepared for the presentation.

The Most Important Do's and Don'ts for an Oral Presentation

Do

- *Memorize the first few sentences.*
- *Look at the audience as much as possible.*
- *Use a deliberate pace in speaking.*
- *Simultaneously show on your slides the visual information you are providing verbally.*
- *Explain everything that is on a slide.*
- *Show all key points on the screen.*
- *Proofread your visuals for spelling.*
- *Use spoken English.*
- *Use simple words but technical terms.*
- *Use uncluttered, visually attractive slides.*
- *Use informative headings.*
- *Be yourself.*
- *Pay attention to your body, arms, hands, and legs.*
- *Dress appropriately.*

Don't

- *Don't look only at the screen.*
- *Don't look only at your notes.*
- *Don't read word for word from your PowerPoint slides or from your notes.*

- *Don't pace across the front of the stage or room.*
- *Don't utter appalling sounds, such as "ah" and "um."*
- *Don't speak in written English.*
- *Don't fiddle with objects or play with your hair.*
- *Don't read subheadings.*
- *Don't use uninformative headings or text.*
- *Don't skip over information on a slide.*
- *Don't argue with a questioner in the audience.*

Other important points

Be yourself. Being natural is the most valuable asset of a speaker.

30.7   VOCABULARY AND STYLE

For the most effective talk, you need to be aware of the vocabulary and style of your delivery in your presentation. Thus, when you practice and revise your talk, focus on the following:

Word Choice

➤ Use smoothers and transitions

Words used verbally are different from those used in written communications. When presenting a talk, you need the "smoothers" and soft transitions that are typically edited out in final drafts of published research articles. You need these soft transitions in your voice but not on the slides. Practice using these transitions ahead of your talk.

ESL speakers need to pay particular attention to the difference between written and spoken language. When you listen to native speakers talk, consider preparing a list for yourself that contains soft transitions. Do not overuse such smoothers, however. These transitions could include the following:

> . . . and, yes, the . . .
> . . . actually . . .
> . . . well, . . .
> . . . anyhow . . .
> . . . okay . . .
> . . . let's see— . . .
> . . . once again . . .
> . . . all right . . .
> . . . so . . .
> . . . for example . . .
> . . . just as an aside . . .
> . . . now . . .
> . . . please . . .
> . . . sorry . . .

ESL advice

As with soft transitions and smoothers, overview words and phrases are used commonly in talks but not in writing. Examples of overview words and phrases include those in the following list:

 I am going to present . . .
I would like to . . .
What I am going to talk about is . . .
To start, . . .
This talk is about . . .
My presentation deals with . . .
Then I am going to discuss . . .
At last, I would like to . . .
Now we move on to . . .
I want to spend some time on . . .
For the rest of the time . . .
On the next slide . . .

Unlike overview words and smoothers, jargon is not preferred or accepted in a talk. Thus, get rid of the jargon (see Chapter 2, Section 2.4). Do not use coarse words or profanity in a speech. Adjust the use of technical words to your audience, but never talk down to an audience and do not use sexism or racism.

Grammar and Sentence Structure
Make a conscious effort to speak in reasonably short sentences. Do not let occasional slips of the tongue bother you. Correct yourself calmly and keep going.

Anecdotes, Jokes, and Personal Experiences
Do not forget that the message is the important factor. A joke is purely supplemental. Therefore, do not feel that your talk absolutely has to have a joke or funny cartoon. Many successful, clear presentations do not. If you do include humor in your presentation, ensure that it is not offensive to anyone in the audience and that your audience will understand the joke.

Personal experiences enliven and reinforce points made—provided, of course, that they fit logically into the speech. If you feel like telling your audience about a personal experience, I encourage you to do so, but keep it relevant to your talk.

30.8 BODY ACTIONS AND MOTIONS

➤ Maintain eye contact
Be very conscious of head and eye movements. Eye contact with the audience is essential. Look at individuals in the audience, but do not exclude sections of listeners. Look at different sections of the audience at least 5 to

10 seconds at a time. If it distracts you to look at individual faces, you can look in between faces but at the level of faces for a large audience.

Do not look at your notes excessively. PowerPoint slides especially tempt speakers to glue their eyes on the computer screen. Resist this temptation; keep your eyes on the audience. Also, do not fix your eyes on a spot way beyond the EXIT sign at the back of the room.

➤ Face the audience

Keep the front of your body facing the audience as much as possible. Avoid hiding behind the lectern. Consider stepping out next to the lectern occasionally or moving closer to the audience at times, but avoid turning your back to the audience and speaking at the same time. When you show slides, turn halfway toward the screen rather than turning your back on the audience. Face the audience again after pointing out relevant parts of each slide.

➤ Stay within the presenter's triangle

As the speaker, take care not to obtrude the projection with your head, hands, or other parts of your body. In addition, ensure that you are not obstructing anyone's view. The best position for a presenter is on the left side of the room or screen as seen from the audience when facing the screen (Figure 30.3). This position will ensure that you do not block the projection of your slides and also that you do not block the view of the audience. In addition, it is better to work your bullet slides from the left side as text is written from left to right.

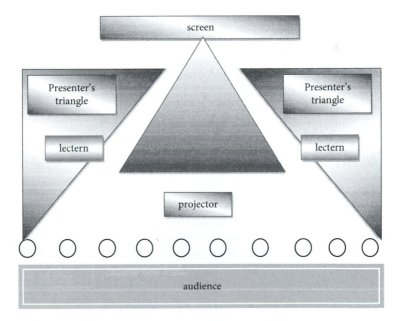

Figure 30.3. Ideal presenter's location.

➤ Use gestures

Use gestures to reinforce and complement your talk. Gestures should be smooth and not jerky and not a wild, windmill style. When you are not using your hands and arms, let them hang naturally. Do not stick your hands into your pockets. Do not grip the sides of the lectern. Do not cover the front of your body with your arms or fold your hands. Do not use your hands to straighten your clothing, rub your nose, explore your ears, smooth down your hair, or play with keys, bracelets, and so forth. Use a pen to calm busy hands if needed. Stand up straight but do not stand rigidly.

If you are using a laser pointer, learn how to use it correctly. Above all, learn where the "Off" button is. To use a laser pointer correctly, learn to employ a single, steady spot of light to show the audience where to look. Keep the light pointed at the spot two to three seconds and then turn it off. Pointing to items is most effective if you do not talk while you are pointing the laser to the point of interest. Talking will distract the audience from where to look. Exceptions exist such as when you are explaining a flow chart or comparing items on a slide. Hold the laser pointer in the hand closest to the slide presentation so you do not have to cross your body with the pointer or turn away from the audience when pointing out things on the screen. If your hands tend to shake while you are pointing, consider supporting the pointing hand with your other hand or resting it on the lectern.

➤ Be conscious of body movement

Your feet should be securely on the floor, each leg carrying an equal share of your body weight. Balance may be shifted occasionally but only for body comfort. Do not teeter back and forth or pace. In general, your feet should neither be seen nor heard. Stay at the lectern unless you need to point out data on a slide or exhibit. However, when you are engaging the audience directly, outside the formal talk (during the question-and-answer session, for example), step away from the lectern to its side and move closer to the audience.

30.9   AT THE END OF THE PRESENTATION

When your speech has ended, stand for a short moment doing nothing. If your words and the tone of your voice do not make it clear that you have finished, you can thank the chairperson, or just say "Thank you" and stop. Do not worry about ending a bit early. No one has ever been upset if a speaker ended early, but the audience is easily upset if a speaker ends late. If questions are to follow immediately, stay at the lectern. Do not ask for questions yourself—this is the chairperson's job. Acknowledge any applause or "thank yous." Gather your notes deliberately; also, gather the remainder of your items. Then, walk back to your seat dignified and unhurriedly.

30.10 QUESTIONS AND ANSWERS

➤ **Make sure that you are in charge**

➤ **Stay calm and polite**

ESL
advice

Often, an oral scientific presentation is followed by a brief question-and-answer period during which anyone in the audience can ask a question of the presenter. Many, if not most, presenters are nervous about this period, especially about being asked questions to which they might not know the answer. The question-and-answer period is particularly frightening for nonnative speakers. Follow the next few points given in this chapter to help ease your concerns.

Anticipate questions before the talk. Try to look at your talk from the audience's point of view, and envision what questions you would be asking. Practice receiving and answering questions with your peers, your PI, or a colleague. The best way to practice with a colleague is to go to a quiet conference room. Give the talk standing up, using your slides (even if not quite finished), a projector, and laser pointer; in short, pretend to be in front of your audience. When you practice in front of your colleagues, be prepared to accept criticism from them.

To deal with questions after a talk in the best possible way, follow these do's and don'ts.

Do

- *Be courteous in your answers at all times.*
- *Tell the audience that if there are questions you will be happy to answer them at the conclusion of your talk.*
- *Admit when you do not know the answer.*
- *Direct the answer to the entire audience.*

Don't

- *Don't give your audience an opportunity to interrupt your presentation.*
- *Don't maintain eye contact with the questioner when you give your answer; this can otherwise easily turn into a two-person conversation.*
- *Don't make up an answer if you do not know.*

If no questions are forthcoming during the question period

At many conference meetings, the chairperson is instructed to think up a question in case no one else from the audience asks one. Alternatively, before you begin your speech, plant the first question with the chairperson or with a friend in the audience. Once the ice is broken, other questions should flow spontaneously. Sometimes you can start the session by asking a question yourself: "Many of you may have been wondering how . . . ," and then go on to give the answer. Alternatively, you may ask the audience a question. Note, however, that the latter two options should be employed rarely as in many fields they may be considered inappropriate.

If there is no chairperson

You must exercise firm control so that not too many people compete for questions at the same time. You must also be prepared to deal with a member of the audience who wishes to make a speech of his or her own. If time is running out for the question period, announce that you will be able to accept only one more question.

If questions are not relevant

Sidestep these questions gracefully. Be especially gracious when you duck a question.

If a single individual in an audience digs in and will not give up trying to turn the question period into an argument

Handle this person politely but resolutely. Isolate the opposition. Smile at the person. Say, "It looks as if we do not agree on this point. Rather than take the time of the whole group, why don't we meet for coffee and discuss it further?" The odds are 10 to 1 that person will never show.

If a questioner is never satisfied with an answer but counters with a further question

Never look at such a questioner when you complete your answer. Look somewhere else and pick a new question as quickly as possible from a different sector of the audience.

Difficult questions

When you get a tough question, always repeat it, ostensibly to make sure all of the audience has heard it. This gives you a few seconds to think about the answer. You have one of several options:

- Ask the questioner to rephrase the question.
- Ask the questioner to come talk to you after the session.
- State that you do not understand the question and ask for an explanation.
- Admit that you do not have an answer and that you might have to look into that.
- Ask if someone in the audience will help you answer.

30.11   OTHER SPEECH FORMS

➤ **Know how to make a proper introduction**

➤ **Know how to give an impromptu talk**

➤ **Become familiar with TED talks**

If you want to become a well-rounded speaker, you will not only have to learn to master the art of preparing and delivering presentations, but you will also have to learn to handle other speaking assignments confidently.

You must learn to

- Make a proper introduction
- Talk intelligently when called on unexpectedly
- Present a TED talk

Making an Introduction

As an introducer, you should assume a secondary supportive role. The person being introduced is the star of the show.

The speaker who is being introduced has a name. Use it. Ask your speaker the exact form of his or her name you should use. Write it down, and get the pronunciation right. Ask the speaker what he or she would like you to say in your introduction. Do as the speaker wishes.

When the introduction has been completed, say, for example, "Mary Peters will talk about . . . (Ladies and gentlemen), Mary Peters." When you have introduced the speaker, step away from the lectern and sit down.

At the end of the presentation, move back to the lectern and stand at the side of the speaker. Say, for example, "Thank you, Mary Peters," and introduce the question-and-answer session if it is to take place immediately following the presentation: "Mary Peters will be happy to answer any questions."

Impromptu Talk

As a scientist you should be able to reply to a question and make spontaneous commentary when called upon unexpectedly. Such impromptu talks require background and a pool of knowledge to draw on.

If you are called to the dais unexpectedly, keep your cool even if someone called on you without warning. If you remain calm despite adverse feelings, you will win the audience on your side.

Walk *slowly*. This gives you a few seconds to think. What should you be thinking about? Your opening sentence—and nothing else.

When you reach the lectern, politely acknowledge the chairman, and look calmly at the audience. Say, for example, "I am delighted to have this opportunity to tell you about . . . " Talk for a few minutes after delivering your opening sentence until you have had time to work out your closing sentence. Try to work out a good closing sentence: For example, you could shortly summarize again what you have talked about. Then, smile for the last time to the audience, nod to the chairman, and leave the dais.

Remember

A good impromptu talk should never be more than three to five minutes long. Be conscious of elapsed time. Usually, five minutes is the outside limit for impromptu remarks.

A strong opening is essential. A strong close is even more important. What comes between should be short and concise. When you have delivered your close, stop and keep silent from then on. Remain polite at all times.

Things to Avoid

Omit afterthoughts. Stop when you come to your first close. Do not open your mouth again no matter what beautiful thoughts float into your mind.

TED Talks

TED (Technology, Entertainment, Design) talks have revolutionized slide-show presentations and delivery styles in science and engineering over the past three decades. Although they require a tremendous amount of preparation in composing and revising a script, repeated practice, and much fine-tuning to produce a presentation of 18 minutes or less, TED presentations are not only highly innovative and engaging but are also very memorable. I encourage you to watch them on YouTube for delivery and design ideas.

30.12 RESOURCES

http://www.toastmasters.org/
Although you may become a member, this site also contains a number of free resources that are useful, from articles about public speaking to videos on speaking tips and advice to overcoming your fear of giving a presentation.

http://web.mit.edu/urop/resources/speaking.html
Presents helpful tips that provide you with advice on how to deliver your message clearly and strongly.

30.13 CHECKLIST FOR AN ORAL PRESENTATION

Use the following checklist to ensure that you have addressed all suggestions in preparing for your talk:

- ☐ 1. Did you practice your talk—a lot?
- ☐ 2. Are your slides informative?
- ☐ 3. Is your talk within the time limit?
- ☐ 4. Did you structure your talk to maximize the attention span of your audience?
- ☐ 5. Are you aware of any appalling sounds or habits you show when presenting?
- ☐ 6. Did you find out who your audience will be?
- ☐ 7. Are your slides logically organized?
- ☐ 8. Do you have an overview slide?
- ☐ 9. Do you have a summary slide?
- ☐ 10. Did you prepare notes?
- ☐ 11. Did you write down:
  - ☐ a. Your opening statement?
  - ☐ b. Important transitions?
  - ☐ c. Concluding remarks?
- ☐ 12. When preparing your slides, did you concentrate on the main points in each portion of your talk?

☐ 13. Do all figures and tables have a title and a legend?
☐ 14. Are visuals and text aligned well in each slide?
☐ 15. Is each slide logically organized and uncluttered?
☐ 16. Did you use visuals where possible rather than text?
☐ 17. Is text used sparingly but informatively?
☐ 18. Is the font large enough?
☐ 19. Are slides/figures/tables kept simple?
☐ 20. Are exhibits attractive? Is color used well?
☐ 21. Did you proofread your text?
☐ 22. Have you familiarized yourself with the setup?
☐ 23. Did you ensure that there will be no ugly compatibility problems between computers or versions of computer programs?
☐ 24. Did you pack a (laser) pointer?

SUMMARY

ORAL PRESENTATION GUIDELINES

- Prepare your talk well ahead of time.
- Get to know your audience.
- Practice, practice, practice.
- Organize your slides.
- As an overview for your presentation:
  - Tell the audience what you are going to tell them.
  - Tell them.
  - Tell them what you have told them.
- Know how to use visual aids.
- Prepare visual aids well ahead of your talk.
- Keep exhibits simple.
- Think graphically.
- Make exhibits look attractive.
- Communicate one main idea per slide.
- Use conservative colors and high contrast between background and writing or figures.
- Follow the 5×5 rule.
- Make lettering large enough.
- Use textual highlights where appropriate.
- Give poster figures and tables a title.
- Choose graphs rather than tables.
- Follow the 3×5 rule for tables : maximum 3 columns and 5 rows.
- Construct well-balanced figure and table slides.
- Stick to the time limit.
- Structure your talk to maximize the attention span of your audience.
- Identify the purpose/question and the take-home message of the study.

- Pick out the most important figure for your core slide.
- Plot your talk by choosing supporting slides.
- Prepare notes.
- Write out your opening sentence in full. Deliver it firmly and accurately.
- Add excitement to your voice, and pause for emphasis.
- Check the set-up ahead of time.
- Make sure you can be heard by the entire audience.
- Speak neither too fast nor too slow.
- Avoid appalling sounds, and pay attention to pitch.
- Explain everything on a slide.
- Use smoothers and transitions.
- Maintain eye contact.
- Face the audience.
- Stay within the presenter's triangle.
- Use gestures.
- Be conscious of body movement.
- Make sure that you are in charge.
- Stay calm and polite.

SPEECH GUIDELINES
- Know how to make a proper introduction.
- Know how to give an impromptu talk.
- Become familiar with TED talks.

PROBLEMS

PROBLEM 30-1 Slide for oral presentation
The following slide is text heavy. Follow the 5 x 5 rule, and reduce the amount of text to maximum five words per bullet point.

Value of Clinic
Patients emphasized the value of:
• The welcome environment of the clinic
• Being able to communicate clearly in their native language
• The comprehensiveness of care offered during a visit
• Being seen by professional and caring clinical teams
• Feeling respected as human beings

PROBLEM 30-2 Slide for oral presentation
Evaluate the following slide. How could it be improved?

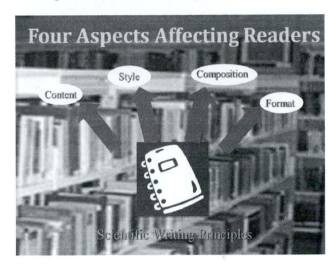

PROBLEM 30-3 Slide for oral presentation
The following slide is text heavy. Reduce the amount of text per bullet point. Suggest what the presenter could do to highlight on the slide what part of the talk he or she is currently presenting.

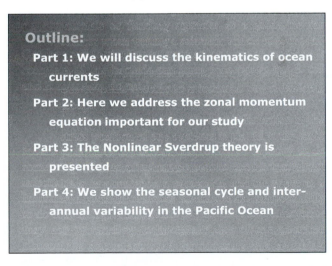

(With permission from Jaclyn Brown, modified)

PROBLEM 30-4 Slide for oral presentation
Assess the following slide. Explain why this is not a good slide.

TABLE 2
List of Proteins Renatured by the Sparse Matrix Approach and Their Optimum Renaturation Conditions

Protein	Molecular mass (Da)	Structure	Standard assay buffer	Optimum renaturation buffer system	Max activity recovered (percentage of initial value)
BAP	80,000 (29)	Homodimer Zn^{2+}, Mg^{2+} cofactor	100 mM NaCl, 5 mM $MgCl_2$, 100 mM Tris, pH 9.5	0.2 M Na acetate, 0.1 M Tris–HCl, pH 8.5, 30% (w/v) PEG 4000	138
HRP	40,000 (30)	Monomer, heme group, Ca^{2+}, carbohydrate	100 mM CH_3COONa, pH 4.2	0.2 M Mg acetate, 0.1 M Na cacodylate, pH 6, 30% (v/v) 2-methyl-2,4-pentanediol	33
β-gal	540,000 (31)	Tetramer, with independent active sites (32)	Z-buffer (see Materials and Methods)	30% (v/v) PEG 1500	51
Lysozyme	14,388 (33)	Monomer	0.1 M K phosphate buffer, pH 7.0	0.2 M Mg acetate, 0.1 M Na cacodylate, pH 6, 30% (v/v) 2-methyl-2,4-pentanediol	333
Sperm bindin	24,000 (23)	Unknown	Seawater	0.2 M Na citrate, 0.1 M Na Hepes, pH 7.5, 30% (v/v) 2-methyl-2,4-pentanediol	100
Recombinant bindin	24,000 (34)	Unknown	Seawater	0.1 M Na Hepes, pH 7.5, 1.6 M Na, K phosphate	100
Trypsin	23,800 (35)	Monomer (α) dimer (β)	10 mM Tris, pH 8.0	0.2 M Mg chloride, 0.1 M Na Hepes, pH 7.5, 30% (w/v) PEG 400	11
Acetylcholinesterase	260,000 (36)	Aggregates, monomer	0.2 M Na phosphate buffer, pH 7.0	0.1 M Na Hepes, pH 7.5, 0.8 M Na phosphate, 0.8 M K phosphate	9

(With permission from Elsevier)

PROBLEM 30-5 Slide for oral presentation
Assess the following slide. Explain why this is not a good slide.

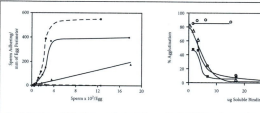

Fig. 4. Species specificity of sperm adhesion in *S. franciscanus* and *S. purpuratus* gametes. The number of adherent sperm was scored as a function of sperm concentration for all possible combinations of *S. franciscanus* and *S. purpuratus* gametes. ●, *S. purpuratus* sperm × *S. purpuratus* eggs; ◆, *S. purpuratus* sperm × *S. franciscanus* eggs; ▲, *S. franciscanus* sperm × *S. purpuratus* eggs; ■, *S. franciscanus* sperm × *S. franciscanus* eggs. Significant numbers of *S. franciscanus* sperm adhere to the surface of *S. purpuratus* eggs. The number of adherent sperm is normalized to account for the larger surface area of *S. franciscanus* eggs.

Fig. 2. Inhibition of egg agglutination by soluble bindin. *S. purpuratus* soluble sperm bindin inhibits egg agglutination non-species specifically. (○) Soluble *Stronglyocentrotus purpuratus* sperm bindin added to *S. purpuratus* eggs in the presence of particulate *S. purpuratus* bindin; (Δ) soluable *S. purpuratus* sperm bindin added to *S. purpuratus* eggs in the presence of particlate *S. franciscanus* bindin; (■) soluable *S. purpuratus* sperm bindin added to *S. franciscanus* eggs in the presence of particulate *S. franciscanus* bindin; (◊) soluable synthetic peptide corresponding to residues 69-130 of *S. purpuratus* bindin added to *S. purpuratus* eggs in the presence of particulate *S. purpuratus* bindin (see text).

(With permission from Elsevier)

PROBLEM 30-6 Oral presentation statements

Explain why the following statements are not good choices for an oral presentation:

1. "Thank you for listening to my talk. I hope it was not too confusing or boring."
2. "On this slide, please focus only on part D and ignore parts A, B, and C."
3. "First, I will provide an overview, then tell you about the methods, show some specific results, and last, I will summarize my talk."
4. "Western blot analysis. Our Western blot analysis showed that . . . Sequencing. When we sequenced the insert . . ."
5. "This finding is in agreement with that of a previously published result reported by Lopez et al. in 2001 where it was shown that emergence of seedlings is temperature and humidity dependent."
6. "It was determined that frogs can hibernate under water for up to six months."

Job Applications

Job Applications and Interviews

To enter into a scientific career, you have to be familiar with documents necessary for job applications. This chapter outlines the most important of these for the scientific and technical fields.

THIS CHAPTER DISCUSSES:

- Curricula vitae and résumés
- Cover letters for job applications
- Research statements
- Teaching statements
- The most common interview questions
- Resources for job applications
- Letters of recommendation

31.1   OVERALL

Many students seek careers in science hoping to improve people's lives and to find knowledge from which all people can benefit. In fact, a background in science and engineering is becoming fundamental to an increasing number of careers as science and technology have become more central to modern society. No longer are scientists needed just in research and development. Scientists have become critically important in other fields as well, such as in business, entrepreneurship, law, politics, security, teaching, and more. Scientists and engineers are involved in establishing and running corporations, making laws, setting policies, and communicating with the public. In addition, they consult, fundraise, invest, and lobby. In short, their careers reach far across industries and institutions and cross national boundaries. With the right leadership abilities and communication skills, there is almost no limit to where scientists are needed and can go (see also http://myidp.sciencecareers.org/).

One such essential communication skill is the ability to apply successfully for jobs. To enter into a successful scientific career, or to change jobs and even careers, you need to be familiar with the necessary documents to take this step. These next sections outline the most important documents required for job applications in the scientific and technical fields.

31.2   CURRICULA VITAE (CV) AND RÉSUMÉS

➤ Keep updating your CV and résumé

The CV or résumé is one of the first items that a potential employer sees. A *curriculum vitae*, commonly referred to as a CV, is a summary of your educational and academic backgrounds and is used primarily in academic and medical settings. A résumé is used for seeking employment in the private sector and contains a summary or listing of your relevant job experience and education. Both of these documents should be updated and maintained on an ongoing basis. As you gain more and more experience, earn additional degrees, get further training, and hold different jobs, information in your CV will have to be adjusted. It is therefore important to keep revising your CV. Realize also that you will not be writing just one draft of your CV/résumé (and other application material). You will need to work on these documents repeatedly over time.

➤ Format your CV or résumé strategically

The use of boldface, italics, spacing (margins and line spacing), and fonts can make a big difference in ease of reading and in finding information for anyone on a search committee. Therefore, do not put everything in the same type. Rather, format the document so that categories are clearly distinguishable and your CV or résumé stands out from others. For example, use boldface and/or italics for headers and subheadings and insert sufficient white space so your document does not look crowded.

You may also wish to use visual highlights for a few select style elements, such as light lines, but do so sparingly. Avoid fancy bullets and other ornaments. CVs or résumés are not meant to be creatively designed, but rather serve as fact sheets. For your CV or résumé, use a common font such as Times, Palatino, or Arial in 12 point.

Be sure to include important key words that appear in the job posting. Sometimes CVs or résumés are read by machines that look specifically for these terms. This is another reason why your CV or résumé should be redone for every job application. Be sure that you also have cleaned up or locked down your social media sites. More often than not, potential employers will look at these to garner more information about you.

International applicants often include the type of personal information on a CV that would not be included on an American CV or résumé. When applying for a job position within the United States, provide only your full name along with your address and email. Do not include personal information such as birthday, marital status, or pictures. Also, do not include a title page.

➤ Make your CV or résumé well-presented and flawless

Typically, CVs have no length restrictions, but it is essential that your CV is clear, concise, and honest. Your CV should also be well presented and flawless. People who review your application will carefully look at your CV or résumé as well as at the accompanying cover letter. They will not only evaluate your experience and skill level but may also note the layout, spacing, grammar, style consistency, and spelling of your text. Therefore, it is essential that the document is well prepared and free of errors. When you are working on your CV or résumé, you may not be able to see errors immediately. To that end, it is good to let some time pass before you read over it a final time. Even better, have others proofread your documents before sending them off.

Ensure that your CV is addressed to the organization to which you are currently applying—not the one to which you applied previously! Check that the address on the letter matches the one on the envelope.

CVs

➤ Tailor your CV specifically to the position and to the organization

➤ Highlight your main qualities on the first page

Customize your CV for each job application (as well as for each grant or fellowship application). Most potential employers will only glance at a CV; therefore, you need to highlight your main qualities on the first page (or two). To make your CV as effective as possible, think about what skills and qualities your potential employer may wish to invest in and why. Then, organize and present your information based on the interests of the employer. It has to be apparent immediately that you are the person best matched to the position in terms of skills, training, experience, and interest. If your CV is tailored to the position and company, you stand a much better chance of getting an interview and being hired.

Do your homework and obtain pertinent information on the employer and the position to which you plan to apply. Note that some institutions and organizations have established formats for CVs. As well, before applying for a position, be sure to check their websites for templates and instructions.

CVs are usually very comprehensive and elaborate on professional history, including every previous employment, academic credential, publication, contribution, or significant achievement. Yet CVs differ widely, depending on personal preference, where you are applying, and your experience. Your CV should start with basic details including your contact information. After your contact details, include your degrees and training in order of your personal preference.

The following list provides an example of the categories you may include in your CV and approximately in what order. Note, though, that you should tailor the order within categories to the job. For example, a teaching position will require you to highlight your teaching skills by moving them closer to the top of the CV. For a research position, on the other hand, you need to highlight your research experience, grant proposals, and publications.

Name & address (no personal information)
Education, with degrees and dates (most recent dates first)
Clinical certifications, with dates
Employment history, brief description and dates (most recent dates first)
Honors and awards (predoctoral and postdoctoral)
Grant funding
Leadership and service
Teaching experience
Laboratory skills
Publications—name boldfaced, first authors easily identified
Invited presentations and seminars
Professional qualifications
Certifications and accreditations
Computer skills
Language proficiency
Unique technical abilities
Professional memberships

In a CV for an academic job, the list of peer-reviewed publications is very important. List only those that are published or accepted for publication and not those submitted or in preparation. Often, the applicant's name is shown in bold in the list of publications to make the name immediately visible. List also any currently funded grants (amount, type, length, and agency) that you can bring with you. Such active grants may be extremely important for securing a new position.

The next example shows a sample CV with appropriate format. Note that educational qualifications and work experience are listed in reverse chronological order and account for the job seeker's entire career history.

Example 31-1   Sample CV

JANET MILLER

Address
Telephone and e-mail

——— EDUCATION ———

PhD Chemistry; *University of Utah*

> Area of Specialization: biological chemistry   2006

B.S. Chemistry, *cum laude*; *Washington University of California*   2001

——— CAREER/ACADEMIC APPOINTMENT ———

Associate Professor Yale University, Department of Biological Chemistry,   2008–present

Assistant Professor Brown University, Department of Chemistry,   2004–2008

——— RESEARCH EXPERIENCE ———

Postdoctoral Fellow: University of Pennsylvania; laboratory of Dr. Enzo Russo,   2002–2004

——— PROFESSIONAL HONORS AND AWARDS ———

Fellowship from Institute for Chemical Studies, London	2011
MacDougal Award	2010
DPP Young Investigator Prize	2009
Overseas Research Studentship Award	2001–2003

——— GRANT FUNDING ———

Current Grants

Agency:	NIH
ID#:	3 U02 03GM0566490-01S6
Title:	"Structural determination of antibiotic binding sites in the bacterial ribosome."
P.I.:	Janet Miller, PhD
Percent effort:	20%
Direct costs per year:	$xxx
Total costs for project period:	$xxx,xxx
Project period:	July 1, 2012–June 3, 2017

. . . .

Past Grants

Agency:	MacDougal Award
P.I.:	Janet Miller, PhD
Percent effort:	50%
Total costs for project period:	$xxx,xxx
Project period:	2010–2013

. . . .

─────────── **INVITED SPEAKING ENGAGEMENTS** ───────────

Synthesis of Y   VIIth International Conference on Y.
Woods Hole, Massachusetts   2011

. . . .

─────────── **COURSES** ───────────

Introductory Biology, University A   2012
Writing in the Sciences, University B   2011

. . . .

─────────── **PROFESSIONAL SERVICES** ───────────

Peer Review Groups/Grant Study Sections:

Microbiology Research Committee
to review   2011
K applications, NIH
Alfred P. Sloan Foundation, NIH   2008–present

─────────── **JOURNAL SERVICE** ───────────

Editorial Board:

Microbiology Reports   2011–present
PLoS ONE   2009–present

Journals:

American Journal of Microbiology, Bulletin of bacteriology, Journal of Parasitology, Journal of Virology, Nature, PLoS Biology   2008–present

. . . .

Publishers:

Springer

Academic Editor:

PLoS Medicine

Professional Service for Professional Organizations:

Advised US Department of Health and Human
Service with regard to vaccination policies   2009

. . . .

Meeting Planning/Participation:

Organizer of Bacteriology Workshop at the
University of Carabobo, Venezuela   2011
Scientific Advisory Board Member of Workshop
*Mitigating the spread of MRSA: Lessons learned
from past outbreaks*, Arizona State University,
Tempe   2010

Yale University Service:

Departmental Committees:

Search Committee. Infectious Disease
Ecologist Faculty position   2011
Microbiology Thesis Committee
member   2010–present
Microbiology PhD, and MPH student orientation   2010

Public Service:

. . . .

──────── PUBLICATIONS ────────

BOOKS

Miller, J. H. (2018). *Chemical synthesis—the new era*, Oxford
University Press.

Articles and Book Chapters

Schulz, P.A., **Miller, J. H.,** Hartmann, E. (2008). Title.
Journal of Chemistry 126: 1–8.

Miller, J. H., Knight, C., Cove, D. (2007). Title. *Science*
261: 92–99.

. . . .

──────── ABSTRACTS ────────

Miller, J. H., and Kabe, C. G. (2008). Title. *Protein Society
abstracts.*

. . . .

──────── REFERENCES ────────

Available on request

_____

Note that before you provide the name and contact information of
your references, you should make your references aware of you putting their
name forward on a job application. You should also inform any references
about the position to which you are applying.

Résumés

➤ Tailor your résumé to the position and to the organization

➤ Make your résumé goal specific and concise

Like the CV, your résumé should be tailor-made to the position for which you intend to apply. It should be job oriented, goal specific, and very concise. In contrast to a CV, however, a résumé contains only experiences directly relevant to a particular position. Thus, typically, résumés for industry are much shorter than those for academia, ranging between two and four pages, depending on your past experience and level. Unlike a CV, a résumé puts less emphasis on academic achievement and more on professional or work experience. Thus, listing topics such as technical skills will be more important than listing talks, abstracts, and memberships in honor societies.

Many résumés in today's job market are created, posted, submitted, and searched electronically. This brings with it not only advantages but also some pitfalls. Generally, it is easier and quicker to update and send out résumés electronically. However, beware that if you mass mail your résumé, it may no longer be tailored to a specific job position, which can have a negative effect on your chances of securing employment.

➤ Highlight your skills and professional experiences in your résumé

Skills you may have acquired should be highlighted, for example, in a section entitled "Technical Expertise and Skills." You may be surprised what you can include in such a section based on what you may have learned from previous positions and trainings. However, be careful not to overstate your skills. Aside from skills, highlight work experiences. Résumés may also contain broader experiences outside of work, such as military service or specific programs and courses you may have taken (AutoCAD, continuing education programs), if they are relevant to the position you are seeking.

If it is apparent that you have looked for ways to expand your education and training to set yourself apart from your peers, you will become more employable. You will also be more attractive to potential employers if you can show consistent and sustained engagement in their particular field of work.

Common mistakes in composing a résumé include:

- Not putting your best, most important, and relevant experiences near the top
- Mentioning specific courses or workshops simply because you particularly enjoyed them rather than for their relevance to the job

➤ Distinguish between chronological and functional résumés

➤ Tailor the order within categories to the job

Three different kinds of résumés exist.

Chronological—In a chronological résumé, your education, work history, and accomplishments are formatted in reverse chronological order. This style of résumé is considered the easiest to read and write and is your safest choice if you are not sure what style to choose for your job application.

This style is particularly effective if you

- have a steady work history that is consistent with your employment goal.
- continuously increased your level of responsibility, promotions, and accomplishments.
- want to mention any prestigious employer or institution.

The chronological style might not be the right choice for you if you

- have gaps in employment.
- change jobs frequently.
- lack work experience.
- are transitioning to a new career.

In these cases, another résumé style may be more suitable.

Functional—In a functional résumé, no dates are listed. Instead, the résumé focuses more on your skill set than on your history. This style is very effective when the potential employer needs to know about your expertise in a specific field.

Use a functional résumé

- to stress your business's management skills.
- to point out specific abilities.
- if you are just launching your career and don't have a history to draw on.
- if you are an executive who has only held one or two jobs for long periods.
- to reenter the job market after a long absence.
- if your work history does not show professional growth.

If you use a functional résumé, you can emphasize and showcase the skills and expertise relevant to the targeted position. You should tailor the order within categories to the job. However, by not providing dates, you also may raise the suspicion of your potential employer as the omission of dates may alert them to look for gaps in your employment record and question your professional growth.

Combination—A combination résumé incorporates characteristics of both the chronological and functional résumés. It allows you to format your information according to your specific situation. The combination résumé usually lists your functional skills and qualifications with brief explanations, followed by reverse chronological listings of current and previous employment, training, publications, and other skills. You can also highlight specific expertise and accomplishments that match the needs of your potential client in these chronological lists.

The combination résumé is best if you

- are looking to change careers and want to highlight the skills that would best match your new career path.

- like the position and believe you will be successful but do not have a strong skill set or much experience in the field.

This format seems to diffuse most suspicions that employers may have while allowing you to highlight your skill sets and past accomplishments that you bring to the position. Such combination résumés are becoming more and more common in the scientific fields, particularly in private industry.

Overall, know where you are in your career and training before you decide on the best résumé format. Evaluate the position to which you are planning to apply and see what skills and experiences are required. Compare the position's requirements to your skill sets. Based on this comparison, decide which résumé format will be best suited for your application. Ensure that your résumé is targeted toward the position you want and that it shows not just your day-to-day work but also your accomplishments. In addition, make sure that your résumé looks neat and professional and does not contain any spelling or grammatical errors.

Following are two sample résumés, one chronological and the other a combination résumé.

 Example 31-2

RÉSUMÉ

Peter Jones

Contact Address   Address; Telephone; Fax; email

PROFESSIONAL EXPERIENCE

Jul 2016–to present   Sr. Scientist I/Sr. Investigator II;
                                                  Bayer Pharmaceuticals

- Led a research group specialized in biologically active molecule synthesis and custom synthesis
- Managed a process lab for preparation and processing of X
- Investigated quality and stability of drug products for preclinical and clinical trials

Nov 2013–Jul 2016   Chemist, Abbott Laboratories

- Responsible for designing and conducting complex, multistep synthesis
- Synthesized carcinogenic substances for chemical toxicology research

Nov 2010–Nov 2013 Postdoctoral Fellow, ABC University

- Studied biology of carcinogens
- Synthesis of carcinogenic compounds for cancer research

EDUCATION

- 2010: *PhD in Biochemistry,* The University of Chicago
- 2005: *BS in Chemistry,* Summa Cum Laude, Concordia College, Minnesota

AWARDS

2015   Nominee for Concordia College Alumni Achievement Award

2016   Outstanding Achievement Award, Abbott Global Medical Affairs

PROFESSIONAL ACTIVITIES

Grant Support

- Ongoing Grants
- Complete Grants

SELECTED PUBLICATIONS

 Example 31-3

RÉSUMÉ

John Ips, PhD
Job Announcement Number: 13-758303

11 Street name
SomePlace, CT 06415
Tel (xxx) Fax (xxx)
Email address

SUMMARY

- Highly motivated microbiologist with extensive experience in bacterial genetics
- Highly developed leadership and interpersonal skills
- Broad experience negotiating and supervising contracts with CROs and academic laboratories
- Excellent skills in NMR, HPLC, IR, UV
- Excellent team player with good verbal and written communication skills

EDUCATION AND TRAINING

July 2014 – Dec. 2016:   **Postdoctoral Fellow**
University of Munich, Germany

2014:   **PhD, Biochemistry** (Magna cum Laude) Humboldt University, Berlin, Germany

PROFESSIONAL EXPERIENCE

2018   *Pfizer, Groton, Connecticut*
Director, Biomedical Research & Development

Main focus: Development of rapid diagnostics devices

- Responsible for strategic planning and development of biomedical programs
- Laboratory management and accreditation

2016–2018   *Rib-X Pharmaceuticals Inc., New Haven, Connecticut*
Director, Ribosomal Targeting and New Technologies

Main focus: Drug Discovery and Development

- Lead the Discovery Biology team and a Target Exploration team
- Effectively guided development of assays needed to validate mechanism of action and functional target location for discovery programs
- Key member of leadership team that planned and directed the research strategies and activities

PUBLICATIONS

Selected articles:

- Joseph, E., McConnell, T., DeRey, J., Lance, L., **Ips, J.A.** A new family of linezolids. *Journal of Bacteriology* 52: 3550–3557 (2018).
- Lance, L., Daese, P., DeRey, J., **Ips, J.A.** *In vitro* activities of oxazolidinones. *Journal of C.* 2: 1653–1662 (2016).
- **Ips, J.A.**, Kano, Z.F., Wan, D. Crystal structure of linezolid. *Journal of Medicinal Chemistry* 335: 3–13 (2015).
- . . .

MEETINGS AND CONGRESSES

Selected presentations:

- 2018   Antibiotic Development, San Diego, CA
- 2016   Oxazolidinones, Washington, D.C.
- . . .

PROFESSIONAL SERVICES

2019   Scientific Editor of Journal Z
2016–2019   Reviewer for the following journals: *Journal of Bacteriology, Journal of Z, . . .*

PROFESSIONAL MEMBERSHIPS

- American Society of Microbiology
- American Society for Biochemistry and Molecular Biology

FOREIGN LANGUAGES

- Fluent—Spanish and German

Resources for résumés

Good online resources to create your résumé include:

https://www.visualcv.com/
http://www.krop.com/online-portfolio-templates

31.3   COVER LETTERS

➤ Tailor the cover letter specifically to the position

When you are applying for a job, you will not only have to send a CV or résumé but also a cover letter. Your cover letter creates a professional impression from the outset. Like the CV, it should be flawless as it can make or break your application. In your cover letter, highlight the most relevant parts of your CV that can show your potential employer what you can do for them. Wherever possible, address your cover letter to a specific person, even if that means you have to call the organization. Check (and double-check) the letter for spelling and grammatical errors. Have others check it as well. If you are sending a hard copy, print it onto good-quality paper.

The cover letter should be tailored specifically to the position and should be 1 to 1½ pages long. After mentioning the purpose of the letter, including the reference number of the job posting, introduce yourself, highlight your accomplishments as relevant to the posting, and state your research goal. Then, state why you believe your expertise would add to the department to which you are applying. The following is an outline for the general organization of a cover letter:

Opening paragraph
   State the purpose of your letter and position reference number
   Mention how you heard about the job
Middle paragraph(s)
   Highlight your past accomplishments
   Describe your research goals
   Explain why you believe you are a good fit for the position
Closing paragraph
   Mention any enclosures (CV, publication samples, Teaching
      Statement, etc.)
   Make positive closing remarks

Example 31-4 shows a sample cover letter for an academic position, while Example 31-5 presents an example of a cover letter for a position in private industry. Both follow the outline given previously.

Example 31-4   Cover letter for an academic position

Date

Dear Prof. Moore:

Purpose and name of position

I am writing to apply for the Assistant Professor position in your Department as posted in the October 10 issue of *Science* (posting number XXX). I am convinced that my solid training in X-ray crystallography, my strong track record in cutting-edge research, and my long-standing passion for teaching make me well-suited for the position outlined in your advertisement.

Accomplishments and goals

In 2010, I completed my graduate studies at the University of California, San Francisco with a PhD degree in biochemistry. Subsequently, I spent two years in Dr. Hannes Kari's group at the ABC University School of Medicine as a postdoctoral fellow, focusing mainly on structure function studies of the ribosomal protein Y. During this period, I was part of a team actively involved in solving several crystal structures of protein complexes involving this protein for several different organisms (see enclosed publications). My current research is focused on . . .

Reason for fit

The research and mission of your institute highly complements my own goals and objectives, and my expertise would add to your department's strength in molecular and structural mechanisms of biological processes. It would be an honor to join the University of Tennessee as an assistant professor, and I believe that my skills and talents will be a valuable asset for the University. In addition to leading my own laboratory, I am very enthusiastic about teaching, in which I have been actively engaged for the past four years. Classes that I have taught and which may be of particular interest to your department include . . .

Enclosures and ending on positive note

Please find enclosed my *curriculum vitae* and Research and Teaching Statements, as well as three of my recent publications. The contact information for my references is included in my *curriculum vitae*, and letters of recommendation will be arriving under separate cover. Thank you for your consideration. I look forward to hearing from you.

Sincerely yours,

Johanna Miller, PhD

Enclosed: CV, Research Statement, Teaching Statement, Publications (3)

Example 31-5   **Cover letter for private industry**

Re: Your posting for Senior Director of Biochemistry

May 15, 2013

Dear Dr. Pederson:

With great interest I have learned about the Senior Director of Biochemistry position listed on your website under posting number XXX, and I am excited to apply for the same.

As will also be apparent from my résumé, my 20+ years of comprehensive research and leadership experience encompassing the biochemical sector affords me an excellent perspective to understand the needs of the posted position. In 2002, I joined XXX Pharmaceuticals where I stayed until December 2011 as the Assistant Director of

Chemical Biology, overseeing a team of four research scientists. During my time as the Director of the discovery biology department, one novel antibiotic molecule was progressed from discovery into the clinic.

In early 2011, I was promoted to Associate Director for Biomedical Research and Development, and in July 2012 accepted a position of Director of Biochemistry at AAA LLC, a startup company that aims to develop a novel approach to improve dramatically the sensitivity of quantifying and identifying microbial pathogens in food, clinical, and/or environmental samples.

Overall, my career path has provided me the opportunity to be personally involved in all challenges associated with promoting, funding, and directing research. At the same time, I also gained solid expertise on turning basic science and early drug discovery efforts into compounds and devices for clinical development.

I would greatly look forward to working as a Senior Director of Biochemistry in your corporation, a position that has always been of the highest interest to me. I am confident that my experience, expertise, and enthusiasm for this position, and the field of health sciences generally, will be a great asset in stimulating, planning, directing, and evaluating of research projects and programs.

Sincerely,

Thomas Wang

31.4   ACCOMPANYING DOCUMENTS

Aside from a cover letter and CV or résumé, you may be asked to send a Research Statement, a Teaching Statement, and letters of recommendation. These documents together will tell your potential employer about your education, skills, and expertise.

Research and Teaching Statements are components of many academic job applications for the life sciences and social sciences. A Research Statement summarizes your research achievements and proposes future studies—with the emphasis primarily on the latter. A Teaching Statement summarizes your teaching achievements, describes your teaching philosophy, and proposes planned courses or other educational components, generally related to your research. Here, too, potential employers are mainly interested in seeing what type of teacher you will be going forward. These statements are usually requested for postdoctoral and faculty positions and are announced in academic job postings.

Even if you are not requested to write a Research or Teaching Statement, consider doing so, as writing them will help you focus your professional goals and improve your interview performance. For your potential employer, these statements, along with your CV, cover letter, and letters of recommendation, are important indications of your job readiness, your areas of specialty, your potential to get grants, your academic ability and research needs, as well as your compatibility with the department or school to which you are applying.

Each scientific field has different expectations for Research and Teaching Statements. It is therefore a good idea to obtain examples of statements. Such examples may be hard to come by, as many researchers consider them confidential information, especially in a competitive field. Try searching online for examples, but also do not hesitate to ask your advisor or other people in your department.

As with all other types of written communication, revise repeatedly, and have someone else read and comment on your statement(s) before you apply. These statements are very important, as they can make or break a job application. They should be written with great care and revised several times before you send them out.

31.5   RESEARCH STATEMENTS

> ➤ State your research achievements, current aims, and future goals

> ➤ Tailor your Research Statement to a particular job posting and institution?

A Research Statement summarizes your research achievements to date, indicates your current aims, and states your future goals. It also describes how your research contributes to the field. The Research Statement should convince potential employers that you are knowledgeable and expert enough to carry out the proposed research. It should also show that your research will be different, important, and innovative. Like the CV and cover letter, your research statement should be tailored to a particular job posting and institution.

> ➤ For a Research Statement, use the following general components:

- (Abstract)
- Background
  - ○ Current research
  - ○ Research agenda
- Relevance

The amount of detail and length of Research Statements vary among disciplines. Usually, Research Statements include an overview of your research, background information to show you are on top of your field, key issues remaining in the field, what directions you intend to take to contribute to the field, and why you are motivated to pursue these studies. Your statement may also include a brief abstract, although its inclusion is not required, and mention how you will incorporate graduate and undergraduate students in your research projects. As a beginning young investigator, do not over- or understate your involvement in a past project. Be honest about your contributions. If you led a project, say so. Similarly, if you were involved as a team member, describe the work as such.

On average, Research Statements are about two to five pages long, and content is logically divided using headings and subheadings. Bullets and figures can also be found in these statements. The following list provides an outline for the different sections of a Research Statement:

Abstract (optional)	sometimes useful, but not required; provide an overview of your research and proposed plans; include background, overall objective, focus of work, approach, and significance.
Background	give context to your research efforts and describe relevant past research important in your future professional plans; keep the "big picture" in mind.
Current research	describe key findings and their importance as well as current promising lines of inquiry.
Research agenda (3- to 5-year)	state your short- and long-term goals, approaches, and expected outcomes; provide your top two or three research questions/specific aims/hypotheses; proposed plans should be specific, credible, and realistic.
Relevance	indicate your studies' relevance to your field, potential employer, and to society.

The following example shows a Research Statement that has been constructed based on the outline in Example 31-5.

Example 31-6

Research Statement

John Smith

Previous Research

Research focus

I have a long-standing interest in the molecular determinants for axon outgrowth in the central nervous system. While my PhD thesis and earlier postdoctoral work focused on cell motility and cell-matrix adhesion, my years as an Associate Research Scientist centered on the mechanism by which CNS myelin inhibits axon outgrowth.

Previous research

During my graduate studies in the Lineberger Cancer Center at the University of North Carolina, Chapel Hill, I studied the regulation of Rho family GTPases in response to extracellular matrix and growth factor signals. As a postdoctoral fellow at Yale University, I studied the molecular determinants for axon outgrowth in the central nervous system (CNS).

Current research findings

In my current postdoctoral work at Harvard University, I have focused on determining the mechanism by which CNS myelin inhibit axon outgrowth. . . . We have elucidated NgR-ligand interactions at the molecular level, and this will allow us to develop peptide antagonists to disrupt this receptor-ligand interaction and thereby promote axon regeneration after injury.

Future Research

Specific aims and expected outcome	**I. Identify mutations in NgR that increase risk for schizophrenia.** These specific NgR mutants fail to transduce myelin signals and function as dominant negatives for endogenous NgR (Budel et al., submitted). I aim to determine the underlying mechanism for NgR signaling in humans, particularly in the case of schizophrenia. Our findings may aid in the development of NgR antagonists, thereby allowing axon regeneration after injury (Li et al., 2004).

II. Determine NgR interactions and effects on other cellular pathways . . .

Overall objective and significance	Overall, my ultimate goal is to determine the roles of these mechanisms in distinct neuropsychiatric disorders and work toward the application of these findings to the recovery of a damaged neural system . . . I take a strong experimental and collaborative approach to addressing these questions . . .
Collaborations	Both of the above projects are collaborative, which I find the most productive (and enjoyable!) manner in which to further scientific understanding. In my current position at Harvard, I also initiated a collaboration with a geneticist to investigate . . . I would bring this approach to ABC University and use it to address questions related to . . .
Relevance	As a young faculty member, I will successfully synergize my neuroanatomical, developmental, molecular, cellular, biochemical, and functional neuroscience background with new multidisciplinary approaches to tackle novel mechanisms in the formation of circuits of distinct subpopulations neurons.
Cited literature	**Literature**

(Betty Liu, research statement, modified)

An alternate possible outline is the following:

Overview/Abstract
 Background/context
 Overall objective
 Focus of work/specific aims
 Overall approach
 Significance

Individual research projects
 List individual projects/aims and for each:
 Rationale
 Research aim and approach
 Expected outcome
 Relevance to institution/department

Overall significance of research

An example of a Research Statement that is composed based on the alternate outline is shown next. The individual components for this Research Statement are indicated.

Example 31-7

Research Statement

Jane Smith

Overview

Background	Air pollution changes our planet's climate, but different types of air pollution have different effects. Sunlight is absorbed by industrial air pollution and by the smoke of burning wood. In turn, this absorption decreases cloud formation and heats the atmosphere (Koren et al., 2004).
Overall objective and experimental approach	My research aims to advance our knowledge of cloud formation in response to pollutants based on mathematical modeling. In particular, I am interested in gaining insight into general effects of atmospheric pollutants on precipitation and climate change. The study of cloud fraction in response to pollutants may provide important insights to explain how weather is affected and why our planet has warmed markedly in the past hundred years.
Specific focus	
Significance	

Individual Research Projects

Collaboration and funding	*1 Cumulus cloud formation and atmospheric pollutants*
	    Collaborators: names
	    Funding: name

Rationale	Clouds are sensitive to land and water surface properties. Satellite images have shown that cumulus cloud formation over the Amazon is affected by heavy air pollution. The Amazon region is generally stable meteorologically as is its cloud formation. Thus, the region provides an ideal model to investigate the effect of smoke from wood burning on the formation of clouds (Koren et al., 2004) . . .
Specific aim and expected outcome	My research aims to determine . . . Findings will allow us to gain insight into . . .

Collaboration and funding	*2 Atmospheric pollutant effect in evaluating climate forcing*
	    Collaborators: names
	    Funding: name

Rationale	Atmospheric pollution causes cloud formation with more numerous but smaller droplets, leading to less precipitation and longer cloud lifetime. Atmospheric pollutants also absorbed incoming solar radiation, however, which can reduce the cloud cover.
Specific aim and expected outcome	My research applies these ideas to a combination of different models and data assimilations to understand . . . The energetics will provide a diagnostic tool for assessing . . .

Overall Significance of Research

Significance	Mathematical modeling will advance our understanding of the effects of atmospheric pollutants on precipitation and climate change. Through the study of cloud formation and climate forcing, this interdisciplinary research will have important impacts on . . .

Discuss and coordinate future research themes and strategies with your mentor. Be sure to represent yourself as an independent researcher, with different goals and achievements than your dissertation advisor. Include anything else that might set you apart from your peers (e.g., publications in top journals, important funding partners or collaborators, or breakthrough studies in a particular area of your interests). To be competitive in research today, you generally need a network. If applicable, indicate that you have established collaborations with various researchers across diverse disciplines.

It may also be helpful to point out potential collaborations that could be established with faculty at the department/university to which you are applying. However, do not claim such collaborations unless collaborators have agreed to them. In addition, and if applicable, mention funding organizations likely to support your research agenda and alternative projects showing the breadth of your interests. For candidates who have more seniority, future goals are an expanded Specific Aims page as required for grant application.

➤ Consider the goals of the institution to which you are applying

To create a strong and compelling Research Statement, consider the goals and facilities of the school you are applying to. Overly ambitious proposals, lack of a clear direction, and unclear significance of study usually result in a weak statement. Write as clearly and precisely as you can, adhering to the basic writing rules and guidelines presented in earlier chapters (see Chapters 1–6). Job committees typically look for the following:

- A clear research focus and direction
- Your independence as an investigator
- Potential for funding
- Summary of accomplishments
- Your fit with the unit
- Resources you will need to be successful in your new job
- Your communication skills; how well you present your research

31.6 TEACHING STATEMENTS

➤ Describe your teaching philosophy and approach

➤ For a teaching statement, use the following general components:

- Teaching philosophy
- Teaching experience

- Teaching goals
- Relevance

➤ Tailor your Teaching Statement to a particular job posting and institution

Aside from a Research Statement, a Teaching Statement is also often requested for job applications of academic positions. The Teaching Statement is similar to the Research Statement in form. It is a one- to two-page essay written in the first person that describes your teaching philosophy and approach. The Teaching Statement should cover your teaching experience as well as your teaching goals. It is typically connected to your research in that you will be expected to teach in the field of your expertise or in closely related fields. Your training and research experience will thus be relevant to your teaching. In addition to your teaching experience and teaching goals, consider including a brief statement on how your teaching would add to the department or school to which you are applying.

Usually, teaching institutions have a list of one to three courses that a candidate must be able to teach. Your Teaching Statement should clearly state that you have taught or have the competency to teach these course. In addition, list areas outside of your research where you would feel comfortable in teaching a formal course and indicate your willingness to learn about new topics. If relevant, state how you would go about teaching medical students—from clinical presentation to pathophysiology, genetics, and biochemistry, to treatment and long-term prognosis. The next two examples provide sample Teaching Statements.

Example 31-8   Teaching Statement

TEACHING STATEMENT

Name

Teaching goal | As a teacher, my primary goal is to develop a student's enthusiasm for, and knowledge and questioning of, ecological understanding.

Classes

Since arriving at UGA, I have developed or codeveloped three of the required classes for our undergraduate and graduate programs in ecology. General Ecology, where the class size is greater than 100 and includes students from across a range of scientific majors, has been the most challenging, but I feel I've made significant strides, which are reflected in my teaching evaluations. I have introduced a focus on data interpretation, hypothesis testing, and concepts into formal lectures by interactively working through case studies taken from primary research papers. I take students outside to develop observational skills . . .

In undergraduate seminar classes, both at freshman (Duke) and senior (UGA) levels, I've focused on both the contemporary topics within the field and on topics of societal relevance. For the former, this might involve discussion of niche and neutral theory and for the latter, evaluation of "Policy Forum" articles from *Science* . . .

In the undergraduate seminar classes, I place a lot of emphasis on professional skill development. As well as developing their writing, formal debate, and discussion, students are expected to select a research question and develop it into a class presentation. At the graduate level, I take this emphasis on skills development further. For example, an essential element for the required Ecosystems Class that I codeveloped with an aquatic ecologist at UGA is proposal preparation and evaluation. All students are expected to generate a proposal following NSF guidelines for ecosystem research and then go a few steps further by peer reviewing others within small groups, responding to peer review . . . Discussion is supported by informal lecturing to provide the background knowledge to facilitate effective discussion of the primary material.

Teaching experience

Advising

The part of teaching I find most rewarding is in the one-to-one development of individuals. In my current position, this role takes a number of guises: I am the academic advisor for a cohort of undergraduate ecology majors, was nominated this year to be a Mentor for freshman students, advise three PhD students . . .

Graduate Where I am the main advisor, my primary objective is to develop a student so that he/she can achieve and begin effectively his/her next career step. At the PhD level, this involves generating scientists capable of independent research. I aim to produce well-rounded PhD students who are conversant in topics across ecology, are independent researchers, effective presenters and proposal writers, experienced in teaching, and also in the process of publishing and reviewing manuscripts. To achieve this, students are required to participate in regular lab meetings, meet with me formally at a regular time every two weeks, present at local and national meetings . . . I also work with them to submit proposals. My approach to postdoctoral advising is similar but with more emphasis on preparation for a faculty position, and so it includes providing opportunities such as involvement as a coinstructor on undergraduate classes.

Specific teaching goals

Undergraduate The requirement for all undergraduate science students in my laboratory is an independent-research project. I usually identify a question for a student to work

Specific
teaching goals

on and then take them through all the steps, from literature review to the writing-up of the experimental results as papers. Students that have conducted such research with me have all entered graduate programs and published their results either as lead or coauthors.

Future teaching
goals

At ABC University, I would be keen to work with both undergraduate and graduate research students interested in soil ecology . . . I would welcome the opportunity to develop further graduate-level classes, such as the ecosystem ecology class I currently teach and codeveloped, and to offer advanced seminars on more specialized topics such as the philosophy of ecosystem ecology . . . I would certainly be interested, if desired, in developing classes that contribute to the undergraduate program as well as in facilitating independent research at this level.

(Mark Bradford, teaching statement, modified)

Example 31-9   **Teaching Statement**

Teaching
philosophy

One of the most rewarding aspects of an academic position is the opportunity to teach and interact with students, supporting their curiosity and helping them along in their career paths. My goals as a teacher are to help the students learn about the importance of science and in particular, ecology.

Overall
teaching goal

The education plan of my work has several components, some of which are interconnected with my research, others falling outside my research topic. Specifically, the plan includes a comprehensive program designed to encourage students of different levels, from undergraduates to PhD applicants, to consider careers in biomedical sciences. The central goal in teaching biomedical sciences to my students is to have them acquire critical assets, including knowledge, skills, and interest.

Teaching
experience

My formal teaching career began as a teaching assistant at X University while I was an undergraduate and graduate student. I was a teaching assistant for a variety of courses including . . . In addition to teaching the standard undergraduate courses as a postdoctoral fellow, I gave a number of lectures on selected topics in evolutionary biology and scientific writing . . .

Undergraduate education

I believe that students learn best by doing rather than by passively listening or watching. Thus, I am a strong proponent of interactive learning, asking students for suggestions, encouraging questions, and promoting discussions during lectures. Moreover, I have students directly apply taught skills to practical examples that I provide inside as well as outside class. To motivate students, I use clear, understandable, and exciting examples from a broad range of applications. For instance . . .

Aside from fostering critical thinking in my students, my additional goal as an educator is to promote good communications skills, as these are ever more important in today's interdisciplinary, and often international, research community. To that end, for the past three years, I have offered a special class on scientific communication once a year in my role as a postdoctoral fellow at X University. I plan to introduce this class to your department on my acceptance as well. In this class, students are presented with basic writing principles, which they learn to apply to writing for publication. Many practical examples set the stage for . . .

Specific teaching goal | Finally, I believe that a good teacher should always look for ways to improve teaching skills. For this reason, I plan to refine my teaching strategies by . . .

Graduate level education

At the graduate level, I will supply more specialized theoretical biology courses overlapping my research. These courses cover . . . In addition, I am planning to offer specific communication courses on grant writing and oral presentation to graduate students. I have developed a manual for use in such classes based on . . .

Aside from teaching graduate classes, I am also enthusiastic in supervising graduate student research. My objective is to have my graduate students participate in weekly group meetings during which updates for research projects are presented and critiqued. Students will have the opportunity to consult with all members of the group and will be exposed to different expertise in addition to receiving advice about their research projects.

Outside the weekly group meetings, graduate students will participate in seminar series . . . In addition, I will encourage my graduate students to attend scientific meetings and prepare presentations for these. As well, I will encourage them to participate actively in writing publications and proposals through drafting content and revising documents.

Specific teaching goal | My supervision and mentorship will take place in the form of individual meetings, e-mails, and phone conversations, as well as group meetings. I will provide advice about scientific questions to be addressed, about research approaches, as well as on various forms of communication in the field. In addition, I will provide career advice ranging from how to respond to sensitive emails to securing a strong CV for the application to a faculty position.

Specific teaching goal	I plan to encourage them not only to work on their research project but also to develop their own ideas of research to pursue as independent investigators. In this regard, I will ask students to present quarterly research plans to the group and . . . Furthermore, I will require them to gain teaching experience through presenting a lecture or two in a course attended by both undergraduate and graduate students . . .
Summary	In summary, I am committed to becoming an effective teacher with the goal of preparing my students with a solid foundation of knowledge, skills, and enthusiasm they will need to succeed professionally.

➤ Be prepared to put a Teaching Portfolio together

A Teaching Statement may be tied to a Teaching Portfolio. Such a portfolio is a collection of materials that illustrate your teaching strengths and accomplishments. This portfolio may or may not be required in the initial job application package. If it is not required initially, you may be asked to provide one later in the screening process. Be prepared for this request. It takes time to put a good portfolio together.

The contents and presentation of teaching portfolios vary widely from individual to individual and from field to field. A teaching portfolio may include:

- Teaching Statement
- List of courses taught
- List of teaching awards and certificates
- Sample course material including a sample syllabus for a course
- Teaching evaluations by students and faculty members
- List of professional development in teaching
- Graded papers
- Teaching video

31.7   THE HIRING PROCESS AND INTERVIEW QUESTIONS

➤ Realize the importance of networking

Networking—that is, making contact and creating a group of acquaintances to form mutually beneficial relationships—has become an important aspect in most scientific fields. Many students overlook this social aspect of professional training. Networking may not always come easy or natural for everyone, but nothing will be gained if you do not at least try to put yourself out there.

Networking can be done in a variety of ways. It is not limited to a setting and time in your life. View it as an ongoing process. You have to work on establishing as well as on keeping relationships. In academia, you are constantly networking—in the classroom, with instructors, during office hours, at the cafeteria, and more. Networking may happen not only in person, but also online through platforms such as LinkedIn, through an alumni database at your Career Services office, at conferences/meetings, etc. Your networks may come in very handy when you are looking for a job or applying to one. You may be able to tap into your networks to find job offers or to get advice on how others managed to get a job in your field of expertise.

Take advantage of following up with people you have met through a thank you note or an email message, or by asking if you can meet again for coffee/lunch to informally interview the person about their job or to receive mentoring advice. Consider getting business cards if you currently hold a position of employment and will be networking at conferences or large gatherings. Exchange contact information. Networking can lead to mentoring relationships, research collaborations, and possibly to future jobs.

➤ Be patient during the hiring process

After you have submitted your job application, the response time can vary widely. In some cases, particularly in industry, you may hear back within a few days. Other times, you may not receive an acknowledgement or response to your application until a few weeks later. The latter is especially true in academia, where the hiring process is notoriously slow. To ensure your application is actually delivered to the hiring manager, consider writing to the manager directly in addition to sending the standard online application. Sending your application materials directly to the hiring manager will usually also get you a quick acknowledgement of receipt.

If you have not heard anything two to four weeks after sending your application, consider writing to the manager or responsible human resource specialist once to inquire if your application has been received. The hiring committee often takes time to collect, review, and then select applicants for an interview. Do not pester the hiring manager by requesting updates every week. Let the process of hiring take its course. You will be contacted if they want you to come to an interview. Frequent emails often do you more harm than good because they let you appear desperate or pushy—and neither appearance leaves a positive impression on anyone hiring. Realize also that often you do not get a response at all. It is typical that only candidates who have been selected for interviews get a notification, while others do not.

➤ Prepare for the interview

If you are invited for an interview, inquire how the meeting(s) will be structured. For most academic positions, you likely have to give a seminar about your work (see Chapter 30 for advice on oral presentations) and will have individual interviews that can span multiple days. If you are seeking a job outside of academia—and these days, most graduates will end up with a position outside of academia—you will likely go through a half a day to a day of meetings and interviews. Most often, you will meet individually with diverse people at the organization for about half an hour each. Some interviews may also happen in small group settings. In industry and especially in academia, you may also be asked to give a seminar about your work, which may include past, present, and future research.

Expect that the interview process may include breakfast, lunch, and/or dinner. For all these occasions, ensure that you have appropriate, professional/business attire, which typically includes suit and tie for men and business dress or pant suits for women. If meals are included, ensure that you have impeccable table manners.

Prepare yourself for the interview process by doing your homework. Prepare slides for a talk, a portfolio of your work (e.g., reprints of key publications), and questions you would like to ask. Moreover, do some research on the organization to which you are applying. Find out about its mission, structure, president, finances, products, news, and so forth. In addition, ask your principle investigator or peers to give you a mock interview. For additional help with potential interview questions, consider going to the career center at your college/university to get an assessment of your strengths and weaknesses or you can take the GallupPoll Strength Quest.

➤ Understand the most common interview questions

For many commonly asked questions you can prepare an answer ahead of time. The 15 most common interview questions are listed along with guidelines on how best to handle them.

1. **Tell me about yourself.**
   This is usually one of the first questions or requests you will receive in an interview, and one that you can easily rehearse beforehand. It allows the interviewer to get to know you a bit more. The best reply is to pitch (briefly) why you are the right person for the position. Do not tell the interviewer your entire employment or personal history. Rather, describe your expertise and important achievements or a few key specific interests or experiences the interviewer should know about, and explain how this expertise or interest relates to the particular job. Display confidence (but not to the point of arrogance) and show enthusiasm.

2. **How did you find out about the position?**
   This question is another chance to show your enthusiasm for the position and organization. If you found out about the position through a friend or professional contact, mention that person and share why you are excited about this opportunity. If you discovered the posting through a job board, explain what caught your eye about the position. If you heard about the company elsewhere, share that.

3. **How familiar are you with our organization?**
   Here, the interviewer wants to see whether you are informed about the organization to which you are applying and whether you care about it. Answer this question by showing that you have a thorough understanding of the organization's goals, mission, structure, services, finances, and so on. You also should know the names of the top two to three key people. Then, explain why you are personally drawn to this job. Feel free to give an example or two if appropriate.

4. **Why are you interested in this job?**

   Organizations want to hire people who are enthusiastic about a job. You should be able to explain why you are passionate about the position. State why the role and the company are a great fit for you. Describe also how your skills match the position to show that you can do the work, deliver great results, and will fit in with the team and culture.

5. **What do you consider your greatest professional strengths?**

   For this question, it is important to list your true strengths (not those the interviewer may want to hear) that are relevant and specific to the position. Give an example or two of how you demonstrated these traits in a professional setting. If you do your homework beforehand, you will be better prepared to answer this question. Study the job posting, particularly the specifications and requirements, to know what is needed for the job.

6. **What are your weaknesses?**

   This question allows the interviewer to find any red flags as well as to assess your honesty and your knowledge about yourself. Mention something that is a challenge for you but state that you are taking steps to improve. Do not state that you are perfect and have no faults, or that you have an irreconcilable problem ("I can never make it into the office before 10 a.m."). Rather, pick a weakness that you can turn it into a positive. (For example, you only know basic applications in Excel but are taking a class to improve your skills in that respect.)

7. **Tell me about your greatest professional achievement.**

   Interviewers like to hear success stories about your past employment. Consider using the STAR method to answer this question: **S**et up the situation, explain the **T**ask, lay out the **A**ction of what you did in detail, and describe the **R**esults you achieved. Try to pick a success story that is relevant to the position for which you are applying.

8. **What big challenge have you faced at work in the past, and how have you dealt with it?**

   This question is asked to let the interviewer find out how you respond to conflict, solve a problem, and use communication skills in challenging situations. Here, too, use the STAR method. Focus on how you managed a situation professionally and, ideally, with a happy ending.

9. **Where do you see yourself in five years?**

   The interviewer wants to see if you have thought about a goal and if the position aligns or leads you to this goal. Consider where this position could take you when you answer this question. If you are not quite clear about your ultimate goal, state that you view the position as an important step in determining this goal.

10. **Where else are you interviewing?**

    This question will let the interviewer determine whether you are serious about the profession, how popular you are, and what the competition is. It is best not to answer that "This is the only company to which

I am applying" as this answer will give the interviewer more power, especially when it comes to negotiations. It is better to state that you are actively looking at other, similar positions for which comparable characteristics are required. If you have other interviews lined up, you can mention this, but then also state that you are most excited about this position. If you are not interviewing elsewhere, state that you are still fairly early in your job search, have applied to other positions, and are generally attracted to positions requiring your specific skills.

Consider carefully which details you disclose when you answer this question. It is best to keep information on names for other interviews and offers confidential. Feel free to state that you prefer to do so. Some companies may actually even require you to keep their information confidential. Occasionally, hiring managers will get very pushy about names and position offers. Remember that you are under no obligation to disclose any of this. However, how you handle this situation will show your ability to navigate sensitive questions and your use of diplomacy. Such skills are often very valued.

11. **Why are you considering leaving your current job?**
    This is not an easy question to answer, but it is frequently asked. Above all, stay positive. Do not make any negative comments about your previous job, colleagues, or employer. It will be viewed as unprofessional. Instead, express that you are looking for new challenges and opportunities to grow. If you lost your previous job, say so; explain if there was restructuring or if the organization ran out of funding.

12. **How did you handle a decision at your previous job with which you disagreed?**
    Disagreements happen. It is important for hiring managers to find out how you handle such situations. Are you respectful, do you act professionally toward others at your work, and are you mindful of the hierarchy? It is okay to point out an opposing view, but you always have to know when to let go.

13. **What are your salary expectations?**
    Although you should avoid talking about salary during your interview, you have to be prepared to answer this question should it come up. So do your homework. Educate yourself about the current salary ranges for this type of position. If the interviewer asks the question, give a wide salary range and mention that the salary will not be an issue. You could state your previous pay to help the interviewer know the scale for negotiation, but it is best if the employer tells you the range first. Finally, if asked, you could respond with another question: "In what range do you usually pay people with my background?"

14. **How soon can you start?**
    This question does not mean you got the job. The interviewer might just like to have this information to plan for the position. If you have to discuss a transition with your current employer, say so. If you can start right away, feel free to mention that.

15. **Do you have any questions for me?**

Aside from being questioned by the hiring manager, an interview also affords you the opportunity to ask questions in order to find out more about your potential future employer. Do not start your interview with your questions. Rather, wait until the opportunity arises naturally or until the interviewer gives you the chance. The next section discusses this aspect of the job interview in more detail.

➤ Think about your own questions ahead of time

Asking questions shows you are interested and also gives you the chance to garner enough information to decide whether the job is the right fit for you. The following are 10 good questions for interviewees to ask.

1. **What is a typical day or week like in the position?**

The answer to this question will give you a good idea about what the position entails and how much time you will be expected to devote to certain tasks. The question will also show your interest in the position itself rather than just on the pay and benefits. One note of caution: ask this question only if you are new to the position and do not have prior experience in the field.

2. **What are the biggest challenges for someone in this position?**

This question allows you to be realistic about the job. Are you prepared to take on these challenges? The question also shows that you are realistic but also want to succeed.

3. **Is there room for professional development?**

The answer to this question will allow you to gauge the hierarchy and upward movement within the organization. Are there clear guidelines and responsibilities for different levels, or is the position a static one? This question shows that you are ambitious and in for the long haul.

4. **How does your organization train employees for this position? Is there any mentorship?**

Asking these questions will help you find out if the organization cares about making their employees successful. Ideally, you do not want to be expected to pick up skills and knowledge simply by observing others. The question also shows that you are not just looking to do the bare minimum but rather to achieve in the role.

5. **How do you measure success in this position? What makes a person in this position outstanding?**

The response will give you insight into what the manager values and about the metrics that will be applied. This question will also signal that you care not just about being average but about being great.

6. **Can you give me an overview of the organizational structure of the office?**

The actual hierarchy of a place may not become clear until you receive a detailed organizational layout of the office or company structure.

This will let you put the posted position into perspective and also help you gauge internal movement and turnover.

7. **How would you describe the culture here?**
Finding out about the overall environment of an office will let you determine if the position is the right fit for you.

8. **Could you describe your management style?**
The answer to this question may provide some insight into what your manager expects and how he or she treats people. The answer may also raise a red flag about a difficult manager or unrealistic expectations.

9. **Are there any reservations you have about my fit for the position?**
This question will allow you to address any doubts the hiring manager may have about you. The reply will also let you determine if these doubts are grounded and if the position might be a bad fit.

10. **What is the time frame for filling the position?**
End your interview with this question so you will be clear about the next steps and the projected time frame.

If you are applying for an academic position, be prepared to answer the following questions as well:

1. What project(s) will you tackle first?
2. How much funding will you need in the next three to five years?
3. How many people are you planning to hire?
4. How much lab space will you need?
5. What access to equipment would you need?
6. What grant proposals are you planning to write in the next five years?
7. What funding are you bringing to the department?
8. How successful have you been in previous proposal applications?
9. How many papers do you expect to publish per year?
10. What classes could you teach/have you taught previously?

31.8   RESOURCES

Resources for General Career Path
http://myidp.sciencecareers.org/

Resources for Composing Research Statements
http://search.sciencemag.org/?searchTerm=writing%20a%20research%20plan&order=tfidf&limit=textFields&pageSize=10&&

https://postdocs.cornell.edu/research-statement

http://theprofessorisin.com/2012/08/30/dr-karens-rules-of- the-research-statement/

http://www.vpul.upenn.edu/careerservices/writtenmaterials/researchstatements.php

Resources for Composing Teaching Statements

http://studentaffairs.duke.edu/career/graduate-students/academic
-career-preparation/teaching-statement

http://chronicle.com/article/How-to-Write-a-Statement-of/45133

http://www.acs.org/content/acs/en/education/students/graduate/
six-tips-for-writing-an-effective-teaching-statement.html

https://cft.vanderbilt.edu/guides-sub-pages/teaching-statements

Resources for Interview Preparation

https://www.glassdoor.com/index.htm

https://www.best-job-interview.com

https://www.uptowork.com

https://www.reddit.com

Printed Resources

Feibelman, Peter. 2011. *A PhD Is Not Enough!* (2nd ed.). New York: Basic Books.

Heiberger, Mary Morris, and Julia Miller Vick. 2008. *The Academic Job Search Handbook* (4th ed.). Philadelphia: University of Pennsylvania Press.

Reis, Richard. 1997. *Tomorrow's Professor: Preparing for Academic Career in Science and Engineering.* New York: IEEE Press.

31.9   LETTERS OF RECOMMENDATION

Requesting a Letter of Recommendation

Along with your CV, cover letter, and Research and Teaching Statements, you will probably need one or more letters of recommendation to secure a good position. When you have to ask someone for a letter of recommendation, explain exactly why the letter is needed and how important it is to you. Always offer to provide information that makes the writing task easier (CV, list of accomplishments, publication list, the due date, means of transmission [mail or email], correct address of recipient). Also, provide the recommender with a brief statement explaining *why* this opportunity is important to you and *how* it fits into your overall research or teaching plan. If the writer cannot or will not provide you with a letter, accept this decline gracefully.

Plan your request. Ask someone who is familiar with your work and knows you well enough to include details about you as a person on the letter. The writer should ideally write well, have experience composing letters of recommendation, and have the highest or most relevant job title. As a general rule, request your letter at least a month or two in advance. Also, send with your request your CV and other documentation of work or service relevant to the position to which you plan to apply so that your letter writer has some material with which to work. If the recommender agrees to writing the letter in general but has a busy schedule, consider offering to write a first draft yourself, which you can subsequently pass on to the recommender.

Writing a Letter of Recommendation

➤ In the letter of recommendation, highlight only positive qualities

There may be times when you are asked to write a letter of recommendation. Writing such a letter can be a privilege or a task. Agree to write a letter only for someone you know well enough and only if you can honestly write a supportive letter. You should have a good understanding of the person's academic or professional history and goals. This letter will be important for the individual who has asked you to provide a recommendation. If you are approached, ask the person for a copy of his or her résumé or CV and a list of accomplishments to give you guidelines to use in composing the letter. This information is especially important if you are not sure what to say. Also, ask for information on the job position to tailor the recommendation to that position. Describe the person truthfully. Highlight only positive qualities. If a point may be perceived as negative by a future employer, include it only if you can transform this point to become the person's strength. For example, "CJ is not a native English speaker but has tried hard to improve his/her language skills and has succeeded." The more personalized the letter of recommendation, the more effective and convincing it will be.

If possible, use formal letterhead and include your title on the signature line to reinforce your credibility.

Content

Letters of recommendation not only confirm a candidate's abilities and experiences, they also can provide insight into the candidate's overall character and work ethic. Above all, letters of recommendation build credibility.

The content of a letter of recommendation should be geared toward the particular job position to which the person is applying and should address any specific questions from the selection committee or hiring manager. In writing letters of recommendation, it is extremely important to impress on the reader that you know this person very well. Avoid a long list of general praises. Instead, comment on the unique personality/traits that the individual exhibits.

➤ Make a letter of recommendation three to four paragraphs long

A good letter of recommendation can take a substantial time to write. It should be three to four paragraphs in length and contain some or all of the following bullets.

For a student

- Academic performance
- Honors and awards
- Initiative, dedication, integrity, reliability, etc.
- Willingness to follow school policy
- Ability to work with others
- Ability to work independently

For a researcher
- Previous position
- Summary of job responsibilities
- Strengths, skills, and talents
- Initiative, dedication, integrity, reliability, etc.
- Ability to work with a team
- Ability to work independently

Format

The following are some general guidelines for the format of a letter of recommendation.

First Paragraph	Introduce yourself as the recommender, and state the purpose of the letter. State how long you have known the person and in what capacity.
Second–Third Paragraph	Start by describing the person in general terms, and then mention specific traits of the person. Give specific examples, and make it relevant to the position being pursued. Include two or three outstanding attributes. Address specific qualities in order of importance.
Fourth Paragraph	Express your specific recommendation and confidence in the individual. Offer to answer any questions and to provide further information if needed.

You may also want to provide a phone number or email address so employers can follow up if they have questions or want more information.

➤ Pay particular attention to the first sentence of the letter of recommendation

The wording of the first sentence will communicate your overall opinion of the individual (often unconsciously). The following are some examples of opening sentences and their respective strengths:

Strength of opening sentence

This letter pertains to . . .   weakest

I am pleased to recommend . . .

It is a pleasure to write this letter of recommendation for . . .

It is a genuine pleasure and honor for me to recommend . . .   strongest

Example 31-10   **Sample letter of recommendation**

Letterhead

Name and address

Date

Dear Dr. Smith:

It is with the greatest pleasure that I write this letter of recommendation for Maria Miller. Maria has been working as a postdoctoral fellow in my laboratory at the ABC University since July 2007.

I have been the laboratory head and chair of the department of epidemiology since 2001. Without hesitation, I can say that the expertise, efficiency, and professionalism shown by Maria in her research were surely among the best. These qualities are reflected in her successful publication record in top scientific journals. Her postdoctoral years at ABC University have been very productive. She has published excellent articles on various worldwide epidemics such as TB, HIV, influenza, and rubella. Maria also produced the first model on WNV transmission. Her research is often methodologically innovative and has important implications for public policy.

Maria developed several new techniques in my laboratory and was sought out by her labmates to help them when their assays did not work. She was always able to ask insightful questions in Departmental Seminars covering a very wide range of topics and was never afraid of making a mistake, plunging right into difficult experiments. Moreover, Maria is gifted with superb written and oral communication skills. She kept a well-organized lab book, wrote up her research promptly and clearly, and produced great figures and posters. She also presented outstanding departmental and conference talks.

While working in my laboratory, Maria advanced her knowledge wherever possible by attending lectures whenever her schedule permitted her to do so. For example, she participated in courses at the School of Medicine on virology and microbiology and collaborated with a number of individuals within these fields. Maria was usually the first in my laboratory every morning, and we all enjoyed the collegiality and enthusiasm that she brought to my laboratory and department. She was always cheerful and ready to help, no matter the task.

In short, I recommend Maria without hesitation and have no doubt that this talented young investigator will succeed in the field of epidemiology. I am confident that her research will result in important contributions to your department. Please do not hesitate to contact me if I can provide any additional information that may be helpful.

Sincerely,

31.10 CHECKLIST FOR THE JOB APPLICATION

Use the following checklist to "dissect" the Research and Teaching Statements systematically:

Overall
☐ 1. Did you tailor your CV or résumé specifically to the position and to the organization?
☐ 2. Did you format your CV or résumé strategically?
☐ 3. Did you highlight your main qualities on the first page?
☐ 4. Is your CV well-presented and flawless?
☐ 5. Did you tailor the cover letter specifically to the position?

Research Statement
☐ 1. Did you state your research achievements, current aims, and future goals?
☐ 2. Does your Research Statement follow one of the two following outlines?
    ☐ a. Background
        Current research
        Research agenda
        Relevance
    ☐ b. Overview/abstract
        Individual research projects
        Overall significance of research

Teaching Statement
☐ 1. Does your Teaching Statement describe your teaching philosophy and approach?
☐ 2. Does your Teaching Statement contain the following?
    ☐ Teaching philosophy
    ☐ Teaching experience
    ☐ Teaching goals
    ☐ Sample course outline or syllabus (if a specific course is required)
    ☐ Relevance

Letter of Recommendation

☐ 1. Did you highlight only positive qualities in the letter of recommendation?
☐ 2. Did you pay particular attention to the first sentence?
☐ 3. Is your letter of recommendation three to four paragraphs long?

Overall Style and Composition

Revise for style and composition based on the basic writing rules and guidelines of the book:

☐ a. Are paragraphs consistent? (Chapter 6, Section 6.2)
☐ b. Are paragraphs cohesive? (Chapter 6, Section 6.3)
☐ c. Are key terms consistent? (Chapter 6, Section 6.3)
☐ d. Are key terms linked? (Chapter 6, Section 6.3)
☐ e. Are transitions used, and do they make sense? (Chapter 6, Section 6.3)
☐ f. Is the action in the verbs? Are nominalizations avoided? (Chapter 4, Section 4.6)
☐ g. Did you vary sentence length and use one idea per sentence? (Chapter 4, Section 4.5)
☐ h. Are lists parallel? (Chapter 4, Section 4.9)
☐ i. Are comparisons written correctly? (Chapter 4, Sections 4.9 and 4.10)
☐ j. Have noun clusters been resolved? (Chapter 4, Section 4.7)
☐ k. Has word location been considered? (Verb follows subject immediately? Old, short information is at the beginning of the sentence? New, long information is at the end of the sentence? (Chapter 3, Section 3.1)
☐ l. Have grammar and technical style been considered (person, voice, tense, pronouns, prepositions, articles)? (Chapter 4, Sections 4.1–4.4)
☐ m. Has tense been used correctly? (Chapter 4, Section 4.4)
☐ n. Are words and phrases precise? (Chapter 2, Sections 2.2 and 2.3)
☐ o. Are nontechnical words and phrases simple? (Chapter 2, Section 2.2)
☐ p. Have unnecessary terms (redundancies, jargon) been reduced? (Chapter 2, Section 2.4)
☐ q. Have spelling and punctuation been checked? (Chapter 4, Section 4.11)

SUMMARY

> JOB APPLICATION GUIDELINES
> - Keep updating your CV or résumé.
> - Format your CV or résumé strategically.
> - Make your CV or résumé well-presented and flawless.
> - Tailor your CV specifically to the position and to the organization.
> - Highlight your main qualities on the first page.
> - Tailor your résumé to the position and to the organization.
> - Make your résumé goal specific and concise.

- Highlight your skills and professional experiences in your résumé.
- Distinguish between chronological and functional résumés.
- Tailor the order within categories to the job.
- Tailor the cover letter specifically to the position.

RESEARCH STATEMENT GUIDELINES

- State your research achievements, current aims, and future
- goals.
- Tailor your Research Statement to a particular job posting and institution.
- For a Research Statement, use the following general components:
  - Abstract
  - Background
    - Current research
    - Research agenda
  - Relevance
- Consider the goals of the institution to which you are applying.

TEACHING STATEMENT GUIDELINES

- Describe your teaching philosophy and approach.
- For a Teaching Statement, use the following general components:
  - Teaching philosophy
  - Teaching experience
  - Teaching goals
  - Relevance
- Tailor your Teaching Statement to a particular job posting and institution.
- Be prepared to put a Teaching Portfolio together.

HIRING AND INTERVIEW GUIDELINES

- Realize the importance of networking.
- Be patient during the hiring process.
- Prepare for the interview.
- Understand the most common interview questions.
- Think about your own questions ahead of time.

LETTER OF RECOMMENDATION GUIDELINES

- In the letter of recommendation, highlight only positive qualities.
- Make a letter of recommendation three to four paragraphs long.
- Pay particular attention to the first sentence of the letter of recommendation.

Appendix A

COMMONLY CONFUSED AND MISUSED WORDS

The following list explains the meaning and use of the most commonly misused words scientific editors encounter. Strunk and White, Fowler, and Perelman have similar lists.

ABILITY, CAPABILITY, CAPACITY

Ability	The talent or skill to accomplish a specific thing. Ability can be measured; capacity cannot. (Some microorganisms have the *ability* to fix nitrogen.)
Capability	The maximum, practical power or ability to do something; often relates to volume and quantities. (A tiger has the capability to carry prey of up to 550 kg.)
Capacity	The maximum level or ability at which something can be held or contained, particularly in terms of volume. (The *capacity* of the beaker was 500 ml.)

ACCEPT, EXCEPT

Accept	*Accept* means to answer affirmatively, to receive something offered with gladness, or to regard something as right or true. (The scientists *accepted* his new theory.)
Except	*Except* is generally construed as a preposition meaning with the exclusion of or other than. It can also be a verb meaning to leave out. (Of the bacteria described, all are gram negative *except Streptococci*.)

ACCURATE, PRECISE, REPRODUCIBLE

Accurate	*Accurate* means errorless or exact. It is often used in the sense of providing a correct reading or measurement. (The readings obtained with the spectrophotometer were *accurate*.)
Precise	*Precise* means to conform strictly to rule or proper form. (The value 5.26 is more *precise* than the value 5.3.)

Reproducible   *Reproducible* means something can be copied or repeated with the same results. (The experiment described by William et al. was *reproducible*.)

ADAPT, ADOPT

Adapt   *Adapt* is a verb and means adjust and make suitable to. (Most organisms *adapt* easily to minor changes in the environment.)

Adopt   *Adopt* is also a verb and means to accept or make one's own. (We *adopted* a new protocol for DNA isolation.)

ADMINISTER, ADMINISTRATE

Administer   *Administration* is the noun of *administer*, which may lead to this common confusion between *administrate* and *administer*. (The drug was *administered* orally.)

Administrate   *Administrate* means to manage or organize.

ADVICE, ADVISE

Advice   *Advice* is a noun meaning suggestion or recommendation. (To measure *dP/dt*, we followed the *advice* of J. R. Boyd.)

Advise   *Advise* is a verb and means to suggest or to give advice. (Dr. Boyd *advised* us to follow his protocol.)

AFFECT, EFFECT

Affect   *Affect* is a verb and means to act on or influence. (The addition of Kl-3 to MZ1 cells *affected* their growth rate [i.e., it could have increased or decreased or induced].)

Affect can also be a noun with a specialized meaning in medicine and psychology: an emotion. (People can experience a positive or negative *affect* as a result of their thoughts.)

Effect   As a noun, it means a result or resultant condition. (We examined the *effect* of Kl-3 on MZ1 cells.)

As a verb, it means to cause or bring about. (The addition of Kl-3 to MZ1 cells *effected* their growth rate [i.e., it had caused or brought about]. She was able to *effect* a change in his attitude.)

AGGRAVATE, IRRITATE

Aggravate   *Aggravate* means to make something worse. (Bright light can *aggravate* a migraine.)

Irritate   (pathologically) *Irritate* means to inflame, disturb, or cause pain or discomfort in a body part (Certain chemicals can *irritate* the skin.).

ALLUDE, ELUDE, REFER, REFERENCE

Allude   *Allude* means to mention indirectly. (The authors of "Basic Medical Microbiology" only *allude* to brain abscesses in Chapter 10.)

Elude   *Elude* means to escape from or to escape the understanding or grasp of something. (The cause of the brain abscesses *eluded* us.)

Refer   *Refer* means to pertain or to direct to a source. (*Refer* to Barett et al. for more detailed listings. Questions *referring* to brain abscesses should be directed to a specialist in the field.)

Reference   *Reference* can be used as a noun, meaning a note in a publication referring the reader to another source. (Do not forget to include *references* in the paper.) The term can also be used as a verb, meaning refer to. (They *referenced* his work.)

ALTERNATELY, ALTERNATIVELY

Alternately   *Alternate* means every other one in a series. (Students *alternately* attended lectures on molecular biology and biochemistry every Saturday.)

Alternatively   A choice between two or more mutually exclusive possibilities. (Students can major in biology or, *alternatively*, in chemistry.)

AMONG, BETWEEN

Among   *Among* means in a group of or the entire number of. It is used to express the relation of one thing to a group. (We discovered one black sheep *among* the white ones.)

Between   Use *between* with two items or more than two items that are considered as distinct individuals. (We found no marked differences *between* our results and those reported in *Nature*.)

AMOUNT, CONCENTRATION, CONTENT, LEVEL, NUMBER

Amount   Quantity that can be measured but not counted. (The total *amount* of yeast extract required for the medium was 12 g.)

Concentration   The density of a solution or the amount of a specified substance in the unit amount of another substance. (The *concentration* of protein in the blood was 2.6 mg/ml.)

Content	A portion of a specified substance. (Soybeans have a high protein *content*.)
Level	Relative position or rank on a scale, often used as a general term for amount, concentration, or content. (Heart rate was at normal *level*. Protein *levels* [i.e., concentration] remained stable.)
Number	A quantity that can be counted. (The *number* of proteins in the aggregates varied.)

ANYBODY, ANY BODY, ANYMORE, ANY MORE, ANYONE, ANY ONE

Anybody	Refers to an unspecified person. Often used in place of everyone. (*Anybody* can get sick.)
Any body	A noun phrase referring to an arbitrary corpse or human form. (We found the head, but we did not see *any body*.) *This rule also applies to everybody, nobody, and somebody.*
Anymore	An adverb denoting time. (Doctors in the Western world don't prescribe chloramphenicol *anymore*.)
Any more	Used with a noun or as an indefinite pronoun. (We don't need *any more* measurements.)
Anyone	Refers to any person. Used like anybody and everyone. (*Anyone* can get sick.)
Any one	The two-word form is used to mean whatever one person or thing of a group. (I would like *any one* of these apples.)

ASSAY, ESSAY

Assay	A test to discover the quality of something. (In this *assay*, a Pt electrode was used.)
Essay	A piece of writing, not poetry or a short story. (In this *essay*, she discussed her work in physical chemistry.)

AS, LIKE

As	*As* is a conjunction and is used before phrases and clauses. (The experiment was performed as described by Peters (3).)
Like	*Like* is a preposition with the meaning of "in the same way as." (He was *like* a son to me.) Note the differences in the uses of *like* and *as*: *Let me speak to you as a father* (= I am your father and I am speaking to you in that character). *Let me speak to you like a father* (= I am not your father but I am speaking to you as your father might).

ASSUME, PRESUME

Assume   *Assume* means to take for granted or to suppose. It is usually associated with a hypothesis in scientific writing. (*Assuming* this hypothesis is correct, food should be supplemented with folic acid and vitamin B.)

Presume   *Presume* means to believe without justification. (Scientists *presume* a common ancestor for the two species.)

ASSURE, ENSURE, INSURE

Assure   *Assure* is to state positively and to give confidence and is used with reference to a person. (Let me *assure* you that we did not forget to add any enzyme.)

Ensure   *Ensure* means to make certain. (When you ligate DNA, you have to *ensure* that you do not forget to add the enzyme.)

Insure   *Insure* also means to make certain and is interchangeable with *ensure*. In American English, *insure* is widely used in the commercial sense of "to guarantee financially against risk." (We insured our car.)

BECAUSE, SINCE

See SINCE, BECAUSE

BUT, AND, BECAUSE

But, and,   Can be used to start a sentence as long as the sentence is
and *because*   complete. (Some bacteria can form endospores. *But E. coli* and *E. sakazakii* do not.)

CAN, MAY

Can   Means to be able to, to have the ability or capacity to do something. (Tetracycline *can* be used to treat urinary tract infections.)

May   Indicates a certain measure of likelihood or possibility to do something. It also refers to permission. (Tetracycline *may* act by binding to a certain site of the bacterial ribosome.)

CAN'T, DIDN'T, HAVEN'T

These contractions cannot be used in scientific writing at all. Write them out instead: *cannot, did not, have not.*

COMPLIMENT, COMPLEMENT

Compliment   *Compliment* means praise. (John *complimented* Jean on her new dress.)

Complement   To *complement* means to mutually complete each other. (The protein bindin and its receptor *complement* each other.)

COMPRISE, COMPOSE, CONSTITUTE

Comprise   The conservative definition of comprise is to include or to contain. (The United States *comprises* many different states.) Avoid the phrase *is comprised of.*

Compose   *Compose* means to make up or to create something. Frequently used in the passive voice. (Water *is composed of* hydrogen and oxygen.)

Constitute   *Constitute* means to make a whole out of its parts, to equal or amount to. (Many different organisms *constitute* a habitat.)

CONSERVATIVE, CONSERVED

Conservative   *Conservative* implies a medical treatment that is limited or treatment with well-established procedures. (Due to complications, we selected *conservative* treatment for this case.)

Conserved   *Conserved* means to keep constant through physical or evolutionary changes. (DNA sequences may be *conserved* between species.)

CONTINUAL, CONTINUOUS

Continual   Repeatedly, occurring at repeated intervals, possibly with interruptions. (Measurements were hampered by *continual* interruptions.)

Continuous   Without interruption, unbroken continuity. (Our experiments were *continuously* interrupted means the experiments were started and then interrupted, and this interruption never stopped. The light spectrum is *continuous*.)

CONTRARY TO, ON THE CONTRARY, ON THE OTHER HAND, IN CONTRAST

Contrary to   Preposition meaning in opposition to. (*Contrary to* our expectations, addition of Mg^{++} did not alter our results.)

On the   *On the contrary* is used when one says a statement is not true.
contrary   It is a subjective statement that indicates opposition and is usually used only in spoken English. ("It's exciting!" "*On the contrary*, it's boring!")

On the other hand	Use *on the other hand* when adding a new and different fact to a statement. Rarely used in scientific English. (It's cold, but *on the other hand,* it's not raining.)
In contrast	*In contrast* is also used for two different facts that are both true, but it points out the surprising difference between them. It is an objective statement for a marked difference and can be used in scientific writing. (pH values increased for procaryotes. *In contrast,* no pH difference was observed in eucaryotes.)

DATUM, DATA

Datum	*Datum* is singular and means one result.
Data	*Data* is the plural form of *datum*. It should be used with the plural form of the verb. (Our *data* indicate that many experiments have to be repeated.) The same applies for *criterion, criteria* and *medium, media*.

DESCRIBE, REPORT

Describe	Patients, persons, or cases are *described*. (We *describe* a patient with gynecomastia induced by omeprazole.)
Report	Cases and diseases are *reported*. (We *report* a case of omeprazole-induced gynecomastia.)

DIE OF, DIE FROM

die of	*Die of* is used for diseases and internal causes of death. (He *died of* cancer. She *died of* hunger.)
die from	*Die from* applies to external causes of death. (She died from her injuries. He died from drinking a poisonous substance.)

DIFFERENT FROM, DIFFERENT THAN

different from	Correct expression. (The binding site of amoxicillin is different from that of penicillin.)
different than	Incorrect.

DOSE, DOSAGE

Dose	A *dose* is a specific amount administered at one time or the total quantity administered. (Patients received an initial *dose* of 10 mg.)
Dosage	*Dosage* is the rate of administering a dose. (The dosage he received was 500 mg every 6 hours.)

ENHANCE, INCREASE, AUGMENT

Increase	A general term that means to become or to make greater. (Although we *increased* the amount of nutrients, the number of bacteria decreased.)
Augment	Means to make something already developed greater. (Addition of A greatly *augmented* the effect of B on C.)
Enhance	Means to make greater in value or effectiveness. In scientific writing, terms such as *increase* or *decrease* are preferred over *enhance* because they are much more precise.

ETC./SO ON, SO FORTH, AND THE LIKE

Etc.	Can only be used when the contents of a noninclusive list are obvious to the reader. It is an imprecise expression and should generally be avoided in scientific writing. Instead, use *such as* or *including* at the start of the list, and put nothing at the end of the list.
So on, so forth, the like	Avoid *and so on, and so forth,* and *and the like.*

EXAMINE, EVALUATE

Examine	Patients are *examined.*
Evaluate	Conditions and diseases are *evaluated.*

FARTHER, FURTHER

Farther	Refers to a physical distance. (You need to drive much *farther* than you think.)
Further	Refers to quantity or time. (The binding of X to Z needs to be examined *further.*)

FEWER, LESS

Fewer	Use *fewer* for items that can be counted (*fewer* cells, *fewer* patients).
Less	Use *less* for items that cannot be counted (*less* medication, *less* water). Exceptions include time and money (*less* than two years ago; Jack has less money than John).

FOLLOW, OBSERVE

Follow	A case is *followed.*
Observe	A patient is *observed.*
	To "follow up" on either approaches jargon, as does the term "follow-up study." However, in scientific writing, the use of the term *follow-up* is increasingly common, as is "follow-up study."

IMPLY, INFER

Imply	To *imply* means to hint at or suggest something. (These results *imply* that there is a key–lock mechanism between the enzyme and its receptor.)
Infer	To *infer* is to conclude or to deduce. (Looking at the interaction between the enzyme and its receptor, we can *infer* that a key–lock mechanism is used.)

INCIDENCE, PREVALENCE

Incidence	Means occurrence or frequency of occurrence. (The *incidence* of macular degeneration is high in the elderly.)
Prevalence	Means total number of cases of a disease in a given population at a specific time. (The *prevalence* of macular degeneration in the Western world in the year 2000 was 1 in 1,000 individuals.)

INCLUDE, CONSIST OF

Include	Means partial; to be made up of, at least in part; to contain. *Include* often implies partial listing. (Various antibiotics, *including* streptomycin and tetracycline, bind to the bacterial ribosome.)
Consist of	Means to be made up of, to be composed of. Usually used when a full list is provided. (Macrolites *consist of.* . . .)

INFECT, INFEST

Infect	Endoparasites *infect* or produce an infection.
Infest	Ectoparasites *infest* or produce an infestation.

INTERVAL, PERIOD

Interval	The amount of time between two specified instants, events, or states. (Optical density was measured at 10-min *intervals*.)
Period	An interval of time characterized by certain conditions and events. (Pterodactylus lived during the Jurassic *period*.)

IRREGARDLESS, REGARDLESS

Regardless	Correct.
Irregardless	Does not exist in English.

IT'S, ITS

It's	Is a contraction of *it is* or *it has*. Preferably spell it out in scientific writing.

Its   Means belonging to. (Bindin is an adhesive protein found at the tip of the sperm cell. *Its* structure is unknown.)

LATER, LATTER, FORMER, LAST, LATEST

Later   Refers to time. (The fertilization membrane was not observed until *later* in the assay.)

Latter   Means the second of two items. (*Hpa*II and *Msp*I both cut the recognition sequence CCGG, the *latter* also when the sequence is methylated.)

Former   Means the first of two items mentioned. (*Hpa*II and *Msp*I both cut the recognition sequence CCGG, the *former* only when the sequence is not methylated.)

Last   *Last* refers to the last item of a series. (The samples were pooled, incubated at room temperature for 10 min, and centrifuged at $300 \times g$ for 20 min. For the *last* step, 10 ml of 80% ethanol was added.)

Latest   *Latest* refers to the most recent item in a chronological series. (The *latest* book on molecular biology was published about one month ago.)

LAY, LIE

Lay   To place or put something. (*Lay* is a transitive verb; that is, it needs an object.) Past tense is *laid;* past participle is *laid.* (Birds *lay* eggs.)

Lie   To rest on a surface. (Lie is an intransitive verb; that is, it cannot take a direct object.) Past tense is *lay,* past participle is *lain.* (The tree line ends where the lake *lies.*)

LOCATE, LOCALIZE

Locate   To determine or specify the position of something. (Next, we *located* the vagus nerve.)

Localize   To confine or to restrict to a particular place. (YT-31 was *localized* in the mitochondria.)

MEDIUM, MEDIA

Medium   Singular noun. Needs a singular verb. (LB *medium* was prepared at room temperature.)

Media   Plural form of *medium.* Needs a plural verb. (For the refined analysis, we used six different types of *media.*)

MILLIMOLE, MILLIMOLAR, MILLIMOLAL

Millimole   (mmol); an amount, not a concentration. (A 1 *millimolar* solution contains 1 *millimole* of a solute in 1 liter of solution [or, a 1 *mM* solution contains 1 *mmol/liter* of solution].)

Millimolar	(mM); a concentration, not an amount. (A 0.5 *millimolal* solution contains 0.5 *mmol* of a solute in 100 g of solvent. The final volume may be more or less than 1 liter.)
Millimolal	A concentration, not an amount.

MUCUS, MUCOUS

Mucus	The noun. (*Mucus* is a viscous substance secreted by the *mucous* membranes.)
Mucous	The adjective meaning containing, producing, or secreting mucus.

MUTANT, MUTATION

Mutant	Refers to a strain of organism, population, allele, or gene that carry one or more mutations. (A *mutant* has no genetic locus, only a phenotype.)
Mutation	A *mutation* is an alteration in the primary sequence of DNA. (A *mutation* can be mapped, but a *mutant* cannot.)

NECESSITATE, REQUIRE

Necessitate	*Necessitate* means to make necessary or unavoidable. (The treatment may *necessitate* certain procedures.)
Require	*Require* means to need something. (She *required* medical attention.)

NEGATIVE, NORMAL

Negative	Tests for microorganisms and reactions may be *negative* or positive.
Normal	Observations, results, or findings are *normal* or *abnormal*. (The patient's behavior was *normal*.)

OPTIMAL, OPTIMUM

Optimal	Adjective meaning most favorable; never used as a noun. (The *optimal* [or *optimum*] temperature for X was 28°C.)
Optimum	Noun meaning the most favorable condition for growth and reproduction; often used as an adjective. (X reached its *optimum* 30 hours after induction.)

OVER, MORE THAN

Over	*Over* can be unclear. (The explosion was observed *over* 3 hours.)
More than	Use *more than* when you refer to numbers. (*More than* 500 people were treated.)

PARAMETER, VARIABLE, CONSTANT

Parameter	Means a constant (of an equation) that varies in other settings of the same general form. Parameters are a set of measurable factors, such as temperature, that define a system and determine its behavior. (Changing the *parameters* of the system will result in a different outcome.)
Variable	A *variable* is a quantity that can change in a given system. (In the equation $y = ax + b$, x is the *variable*.)
Constant	A *constant* is a quantity that is fixed. (π is a *constant*.)

PERCENT, PERCENTAGE

Percent	Means per hundred. Written as % together with number in scientific writing. (The yield of the gene product was less than 5%.)
Percentage	Is a general part or a portion of the whole. Usually with a singular verb. (A small *percentage* of the crop was spoiled.)

PERTINENT, RELEVANT

Pertinent	Means having logical, precise relevance to the matter at hand. (These assignments are *pertinent* to understand the class material.)
Relevant	What relates to the matter or subject at hand. (He performed experiments *relevant* to his research.)

PRINCIPAL, PRINCIPLE

Principal	*Principal* can be either a noun meaning a person or thing of importance or an adjective meaning main or dominant. (The *principal* reason for not observing any gas formation was low temperature.)
Principle	*Principle* can only be a noun and means law or general truth. (This manual presents many writing *principles*.)

QUOTATION, QUOTE

Quotation	*Quotation* is a noun meaning a passage quoted. (Indicate a *quotation* with quotation marks.)
Quote	*Quote* is the verb and means to repeat or to cite. (Sentences copied from other sources should be *quoted*.)

QUANTIFY, QUANTITATE

Quantify   Often interchanged and a matter of personal preference. Both are used with the meaning to determine or measure the quantity of something.

Quantitate   *Quantitate* is preferred if one wants to emphasize that something was measured *precisely.*

RATIONAL, RATIONALE

Rational   An adjective meaning able to reason. (This is not a very *rational* thing to do.)

Rationale   A noun meaning basis or fundamental reason. (The *rationale* behind our theory is that cancer incidences are ever increasing.)

REGIME, REGIMEN

Regime   A *regime* is a form of government.

Regimen   *Regimen* means a regulated system or plan intended to achieve a beneficial effect. (He followed a strict *regimen* to lose weight.)

REMARKABLE, MARKED

Remarkable   Is often incorrectly used to indicate a change that is notable but not significant. The correct word is *marked.* (There was a *marked* increase in binding.)

Marked   See *remarkable.*

REPRESENT, BE

Represent   Means to stand for or to symbolize. (Each data point *represents* the average of five measurements.)

Be   Equal, constitute. (Mercury *is* a toxic substance.)

SINCE, BECAUSE

Since   Use this word only in its temporal sense and not as a substitute for *because.* (We have had nothing but trouble *since* we moved here.)

Because   If you want to indicate causality, use *because.* (The reaction rate dropped *because* temperature dropped.)

SYMPTOMS, SIGNS

Symptoms	*Symptoms* apply to people. (The patient displayed no *symptoms* of the disease.)
Signs	*Signs* apply to animals. (A *sign* that the dog was sick was that she did not eat anymore.)

THAN, THEN

Than	*Than* is a conjugation that introduces an unequal comparison. (Birds are closer related to dinosaurs *than* to mammals.)
Then	Means next in time, space, or order. (The samples were centrifuged at 100 *x g. Then,* the pellet was resuspended.)

TOXICITY, TOXIC

Toxicity	*Toxicity* is the degree to which a substance is poisonous or toxic. (We determined the *toxicity* of compound X.)
Toxic	*Toxic* means poisonous. (Mercury is *toxic* for most organisms.)

VARYING, VARIOUS

Varying	*Varying* means changing. (Precise measurements could not be taken because of *varying* levels of humidity.)
Various	*Various* means different. (The crystals were of *various* sizes.)

VIA, USING

Via	*Via* means by way of. (The students went to the lab *via* the ice cream parlor.)
Using	*Using* means to put into service and by means of. (We isolated the plasmid DNA *using* a commercially available kit.)

WHICH, THAT

	Sometimes the words can be used interchangeably. More often, they cannot.
Which	Use *which* with commas for nondefining (nonessential) phrases or clauses. (Dogs, *which* were treated, recovered.)
That	Use *that* without commas for essential phrases or clauses. A phrase or clause introduced by *that* cannot be omitted without changing the meaning of the sentence. Such essential material should not be set off with commas. (Dogs *that* were treated with antidote recovered.)

WHILE, WHEREAS, ALTHOUGH

While	*While* indicates time and temporal relationship. It means "at the same time that." It is often incorrectly used instead of *although* or *whereas*. (Experiments were performed *while* patients were sleeping.)
Whereas	*Whereas,* often the word the writer intended, means "when in fact" and "in view of the fact that." (*Whereas* X increased, Y decreased.)
Although	*Although* is a conjunction meaning despite the fact that or even though. (*Although* the association between breast cancer risk and variant alleles varied slightly, the *p*-values were not statistically significant.)

USE, UTILIZE

Use	Generally, *use* is the preferred term.
Utilize	*Utilize* means to find a new, profitable, or practical use for something. Viewed as an unnecessary and pretentious substitute for use.

Answer Key

Suggested Answers to Problems

Chapter 2

Problem 2-1 **Precise Words**

1. All OVE mutants showed [**threefold; 30% increased**] iP concentrations. POINT: "Enhanced" is imprecise as well as the wrong word choice. "Increased" is the correct quantitative term for concentration. How much was the increase? Give a quantitative value.

3. [**10%, 12%, 6**] of exoplanets orbit multiple stars. POINT: "Some" is imprecise. How much is some? Use a quantitative term.

5. (Last sentence in an Introduction) The present paper reports on [**XXX**] experiments that were performed to clarify this surprising effect. POINT: What were the experiments, and how were they done? The way this introduction stands, "continuing" does not define anything. "Continuing" is imprecise. Experiments need to be defined; otherwise, there is no point to this sentence.

7. To provide proof of concept for our hypothesis, we studied [**the human rotavirus VP7 serotype 1**] in CV-1 cells. POINT: "A virus" is imprecise. What virus? Name it. The same applies to "host cells": it is more precise to name the host cells.

9. The first transition is [**3%, twofold**] lower in energy than the second transition state.

11. The band showing vibrational splitting of 192/cm in Ne [**displaying**] the most intense peak at 444 nm can be identified [**using**] the A $\rightarrow$ X transition of the dimer Ag_2. POINT: "With" is the one of the vaguest terms in English. Because "with" can mean so many things, it is clearer to use a precise term whenever possible.

Problem 2-2 **Simple Words**

1. support
3. are due to/ are caused by different algorithms.
5. fit; agree
7. For example, . . .

Problem 2-3 **Commonly Confused/Misused Words**

1. **like, as:**
   Plasmids were isolated _**as**_ described by Beates (17).
   Carbon dioxide, neon, helium, methane, krypton, and hydrogen are gaseous components of dry air, _**like**_ argon.

3. **varying, various:**
   Varying water levels in a pond are often the result of climate conditions. Each student received _various_ concentrations of NaCl solution for the experiment. Electrodes can be of _various_ sizes.

5. **that, which:**
   Fish _ that_ live in caves show many adaptations to living in darkness. The value of the standard electrode potential is zero, _which_ forms the basis for calculating cell potentials. It is still a challenge to produce layered black phosphorus nanosheets, _which_ have shown promising applications in electronics.

7. **represents, is:**
   25 mg of ketamine _is_ an overdose of anesthetic for mice. (Note: This sentence can be started with a number because the number refers to an actual scientific value.) The Schrödinger equation _is_ a fundamental equation in quantum mechanics.

9. **can, may:**
   It _ may_ appear that Table 1 contains an essentially complete summary of patterns that occur in electrochemical systems. Huge numbers of species _may_ be at risk of extinction from climate change.

Problem 2-4 **Redundancies and Jargon**

essential	like
many	because
despite	because, due to
can, may	close to, near
about, concerning	to
if	because
most	note that

Problem 2-5 **Redundancies and Jargon**

1. The doubling rate appeared to be **short.**
3. **After 2 hr,** we ended the incubation of CO_2 on an Ag(110) surface.
5. Often, jewel weed **grows close** to poison ivy.
7. Transduction efficiencies *in vivo* were much higher than **those** *in vitro*.
9. Upon heat activation, filament size increased, and the number of buds decreased. Both **of these changes** were only seen for cytokinin mutants.

Problem 2-6 **Redundancies and Jargon**

1. **[OMIT]** Collagen synthesis returned to normal 3 days postinjury.
3. **Many** HIV patients also develop tuberculosis.
5. The data in Table **1 are consistent** with Brokl's (1999) model.
7. In **many** cases, degradation leads to topsoil loss and a reduction in soil fertility.

Problem 2-7 **Abbreviations**

Too many abbreviations make it hard for the reader to follow the paragraph. Abbreviations should be limited to standard abbreviations and to four to five nonstandard ones per 10-page manuscript. Nonstandard abbreviations should only be used if the term occurs more than 10 times in the paper or several times in quick succession.

Problem 2-8 **Mixed Word Choice**

1. A typical scientist spends many long hours, even on the weekend, in **the** laboratory.
3. A graph displaying this data is shown in the slide **after next/in the next slide**.
5. We observed a change in cluster size after **10 min**.
7. Absorbance was measured at **various** time points.
9. The hiring of new faculty is traditionally overseen by the **chair** of a department.
11. Thiophene was discovered as a contaminant of benzene.

Problem 2-9 **Mixed Word Choice**

Sulfonamides were among the first **synthetic** agents used successfully to treat diseases. **Because of** their broad antibacterial activity, these drugs were **formerly** used almost exclusively in the treatment of **many different** diseases. **Fortunately**, other drugs have supplanted the sulfonamides as antimicrobial agents because all pathogenic bacteria are capable of developing resistance to the sulfonamides. Sulfonamides prevent the synthesis of folic acid, **which** is a coenzyme important in amino acid metabolism. Although sulfonamides are for the most part readily tolerated (OMIT), they do have some side **effects**.

Chapter 3

Problem 3-1 **Sentence Interpretation**

The end position in a sentence is more emphasized than the beginning position, and the independent clause is more stressed than any dependent clause. Thus, statement 2 is the one most likely to result in the paper being accepted. Statement 1 is the one least likely to result in acceptance.

sentence	news in main clause	news in end position
1)	− negative	− negative
2)	+ positive	+ positive
3)	+ positive	− negative (dependent clause)
4)	− negative	+ positive (dependent clause)

Problem 3-2 **Word Placement and Flow**

(a) Rainwater often picks up carbon dioxide, resulting in a weak solution of carbonic acid. When such rainwater trickles into the ground in areas with a high limestone content, the carbonic acid slowly dissolves the limestone, forming a cave. The cave grows underground as more and more limestone is dissolved. When a cave's ceiling gets eroded and collapses, a sinkhole forms.

Problem 3-3 **Word Placement and Flow**

Fleas often carry organisms that cause diseases. An example of an organism that is transmitted by fleas to humans is the plague *bacillus*. These *bacilli* migrate from the bite site to the lymph nodes. Enlarged lymph nodes are called "buboes," giving rise to the name "bubonic plague."

Problem 3-5 **Subject–Verb–Object Placement**
1. To date, more than 1,000 exoplanets **have been confirmed, some of them with orbits of just a few hours, others with orbits of more than 1,000 years**.
3. Aside from protein X, **protein Y has been found** to be able to bind RNA. **Protein Y has a sequence very similar to a DNA-binding kinase.**
5. Earth's **primordial atmosphere was blown off** several times after catastrophic impacts with other space bodies. The **early atmosphere consisted of** high levels of helium and neon, which are now only present in high quantities in the innermost mantle and core of the Earth.

Chapter 4

Problem 4-1 **First Person**
1. **We assessed** the limitations of the models used.
3. **We/I recommend** that a triple regimen of antibiotics is given to infants with HIV.
5. **We observed** no delay in feeding. (In a Materials and Methods section, authors may prefer third person or passive voice for this example.)

Problem 4-2 **Active and Passive Voice**
1. **We obtained** the following results: . . .
3. **The team analyzed** the collected specimens and videotape recorded during 25 dives.
5. **We recovered** several cytosine deaminases after 2 days.
7. **We saw** no colonies on blood agar.

Problem 4-3 **Active versus Passive Voice**
1. Deep Water Coral, also known as cold water coral, are most often stony corals belonging to the phylum Cnidaria. One of the most common cold water corals is *Lophelia pertusa*. **Lophelia pertusa are affected negatively by fishing methods such as deep water trawling**. These methods tend to break corals apart and destroy reefs. *To optimize word location, use the passive voice.*

Problem 4-4 **Use of Tense**
1. In our study, tree size **increased** with reduction of pesticides.
3. Females of this species **have** a unique projection from the dorsal part of the thorax.
5. Table 3 **shows** that in our study polychaetes **were** most abundant at debts of 10 to 16 m.

7. Sea urchins **spawn** synchronously in the spring, and seasonal and lunar cycles as well as the presence of phytoplankton **increase** their spawning behavior.

Problem 4-5 **Sentence Length**
Example 2
In the past 540 million years, there have been at least five, and as many as 20, distinct mass extinction events on Earth. In all cases, extinction was preceded by major climate changes, quickly changing atmospheric CO_2, and a drop in sea levels. All of these changes led to up to 97% of all species being extinct and a recovery period of millions of years. This recovery re-sulted in an entirely new set of species on the planet.

(Average of 19.5 words per sentence)

Problem 4-6 **Active Verbs**
1. Reaction products **were measured** by a mass spectrometer. (**We measured** . . .)
3. Mitochondria **were removed** by HPLC. (**We removed** . . .)
5. Amyloid plaques **were measured** twice for each brain. (**We measured** . . .)
7. Our results showed that the vaccine **protected** the dogs.
9. To determine whether cells **migrate**, we dissected mouse brains.
11. We must understand the central puzzle of how O=O bonds **form**.

Problem 4-7 **Noun Clusters**
1. The results are compared using a **plane-wave basis set of high quality**.
3. The **callus tissue of the spinach** culture produced no shoots in 1999.
5. With this unconventional technique, we could easily define the transition temperatures **for the liquid crystal of aromatic hydrocarbons**.
7. Previous work suggested that **fish oil diets rich in vitamin A** protect mice against certain mosquitoes.

Problem 4-8 **Pronouns**
1. **Bacterial cells become avirulent** when they are cured of viruses.
3. Patients often suffer from infections of *Corynebacterium* and from toxins released by **these bacteria**.
5. An action potential triggers calcium channels to open at the synapse. **This opening/chain of events** causes docked synaptic vesicles to fuse to the membrane and release their neurotransmitter content.
7. Another costly and labor-intensive approach is to perform biochemical analysis of the fats in yeast. **This analysis** has been done for some yeast mutants (47). **The method/This analysis** involves thin layer chroma-tography, coupled to gas chromatography.
9. Wind or solar power generation combined with pumped hydroelectric storage is being developed. **This combination** may contribute to the adoption of renewable energy in isolated networks.

Problem 4-9 **Parallelism**
1. The pathogenesis observed in other cells, such as circulating mono-cytes, may differ from **that in** endothelial cells.

3. Dengue hemorrhagic fever can occur in individuals with antibodies from previous dengue virus infections of different serotypes, and **(can) result in** severe bleeding, shock, and death.

5. The pellet was washed with 50 µl of ethanol, recentrifuged, **and then dissolved** in 10 µl of buffer K.

7. The internal pressure must not only depend **on** volume but also on the rate of filling.

9. Biomass constitutes 93% of the total direct heat production from renewables, geothermal 5%, and **solar heating 2%.**

Problem 4-10 Comparisons

1. The kinetics of the protein G accumulation were similar to **those of** endogenous proteins (19).

3. The dendrogram showed that EV109 is closer related to HEV-C **than to other strains**.

5. **Mutant A showed the highest peak**.

7. The different propagation behavior of electrochemical fronts compared to **fronts in** reaction-diffusion systems is due to a different range of spatial coupling.

9. The Austro-Tai populations were found to be a unity not only in culture but also in genetic structure, **unlike** other groups in East Asia.

Problem 4-11 Common Errors

1. Altogether, we measured **XYZ** for 5 days.

3. **Data for Fig. 7 were obtained** during the oxidation of hydrogen at the Pt electrode. OR The effect of oxidation of hydrogen at the Pt electrode can be seen in Fig. 7.

5. The liver, gall bladder, and spleen of each patient **were** examined.

7. Having determined that an anisotropy exists, **we rescanned** the full data set from 1 January 2004 to 31 August 2007.

Problem 4-12 Punctuation

1. Moreover, their results can be explained by the fact that the DNA repair defect of the H2A tail deletion mutant is mainly kinetic.

3. As can be seen in Fig. 3, the amount of amplified product increases in parallel fashion to the amount of inclusions observed.

5. The goal was to select the best, most efficient deaminases possible.

7. Thus, this system of pathways is termed "pioneer metabolism."

9. The book, *Handbook of Scientific Writing*, is very helpful.

Chapter 5

Problem 5-1 Prepositions

1. In connection **with**
2. Compared **to/with**
3. In contrast **to**
4. Search **for/of**
5. Correlated **with**
6. A comparison **of** A **with** B . . . /A comparison **between** A and B . . .

7. Similar **to**
8. To look forward **to**
9. Results shown **in** Fig. 3 . . .
10. Through the decrease **of/in**
11. Analogous **to**
12. Implicit **in**
13. Theorize **about/that**
14. Different **from**
15. Attempt (n.) **at**   attempt (v.) **to**
16. **With/In** respect **to**

Problem 5-2 **Articles**

1. Sea urchin fertilization is **a/the** model system in developmental biology. *Here,* **a** *would be used if there is more than one such model system in developmental biology, whereas* **the** *would be used if the author considers it the only one or most important one.*
3. Back to nature.
5. Inorganic chemistry proves to be a very interesting subject.
7. *Title:* Role of Physical Activity on **the** Severity of Diabetes
9. The theoretical studies cited earlier depict a variety of different patterns in oscillating media.
11. *Figure 4:* Effect of temperature on ctx expression.
13. **The** average rainfall in **the** Sahara Desert is less than 25 mm per year.
15. At the end of its "life," **an** mRNA molecule is degraded.

Problem 5-3 **Verb Forms**

1. The analysis was limited to **determining** the change in the population number.
3. Impurities in our samples were barriers to **obtaining** chemically pure products using standard isolation techniques.
5. The reviewers recommend **adding** more information about the bear habitat.

Problem 5-4 **Verb Forms**

1. The conservation of protein A sequences among the diverse species **suggests** that the protein plays a key role in this metabolism.
3. *S. multiplicata* females did not **choose** hetero-specific mates regardless of water level.
5. Microspheres of like charge were **bound** together with nanoparticles of opposite charge to form clusters of two to nine colloids.
7. One third of the mice **was/were** infected with *P. chabaudi*.

Problem 5-5 **Adjectives and Adverbs**

1. X behaved **differently** than expected.
3. Telomeres, DNA sequences at the ends of chromosomes that keep our chromosomes **intact**, shorten with every cell division and are therefore an indication of aging.
5. The resolution of telescopes is restricted by the diffraction of light, which cannot **selectively** focus on anything smaller than a plane wave.

7. The impact of vaccinating domestic animals on rabies occurrence was evaluated by comparing **ecological** field studies with **empirically** based models.

Problem 5-6 **Mixed ESL Errors**

1. **It** is not clear if the difference in our results was due to the temperature difference in our measurement.
3. The environment surrounding invasive breast tumors **exhibits** significant changes.
5. Although the international community recognizes the importance of species diversity in **the** South American rainforests, deforestation continues.
7. Here, we present a model for vaccine development against Ebola virus. However, in support **of** the model, additional, larger studies of Ebola virus pathogenicity and virulence will be required.
9. **The** European cave hyena was found throughout Europe and Asia in the Pleistocene, and often competed with Neanderthal man and Cro-Magnon for **the** occupation of caves.

Problem 5-7 **Mixed ESL Errors**

(a) Polio, a highly contagious disease, has been considered largely eradicated in the developed world. The disease **causes** paralysis in 0.1–1% of patients. Only a few hundred cases are reported **annually** now, compared to 350,000 cases in 1988. Recently, live polio virus was discovered in Israel (Roberts, 2013). Aside from routine childhood vaccinations, **optimal** vaccine policies are now needed to prevent transmission within the population and forego an outbreak. Statistical and mathematical models parameterized by surveillance data and survey studies may be able to provide recommendations for the needed shift in policies.

Chapter 6

Problem 6-1 **Paragraph Organization**

The fungus-like organism *Phytophthora infestans* causes late blight of potatoes and tomatoes. The organism reproduces asexually through sporangia, small sack-like structures containing zoospores, which grow from infected tissues. **Infection by the pathogen involves two phases: an early biotrophic phase in which *P. infestans* infects the plant tissue and develops specialized feeding systems, and a necrotrophic phase during which disease symptoms appear and host tissue is degraded** (Whisson, 2007).

The third sentence, which functions as another topic sentence here, is about the two phases of infection. These phases are subsequently described in parallel form.

Problem 6-3 **Paragraph Consistency**

1′ **Salmon use different methods to find their way during their homeward journey in the fall.** 1 Salmon use geomagnetic imprinting to return to their fresh water birthplace to spawn. 2 As described by Stabell et al. (15), **adults also use olfactory cues to guide them** back to the stream of their birth. 3 It is unclear whether salmon also use cues other than geomagnetic and chemical imprinting to orient themselves.

The topic sentence states the message of the paragraph, "the different methods salmon use to find their way during their homeward journey in the fall." The topic is the subject in every sentence, resulting in a consistent point of view throughout the paragraph. Sentence 2 is parallel to sentence 1.

Problem 6-5 **Paragraph Consistency**

1 Both GAD-positive cell bodies and processes were found in the **thalamic reticular nucleus and ventral lateral posterior nucleus. 2** Almost all of the neurons in the thalamic reticular nucleus appeared to contain GAD-immunoreactivity. **3 However**, only small round cells in the ventral lateral posterior nucleus were GAD positive.

In the revision, consistent order has been kept for thalamic reticular nucleus and ventral lateral posterior nucleus, reversing their order in the first sentence. In addition, a transition at the beginning of sentence 3 indicates the logical relationship between sentences 2 and 3 more clearly.

Problem 6-7 **Key Terms**

A model system to study the underlying mechanism that connects Type II diabetes, <u>life span</u>, and **<u>fat storage</u>** is the <u>nematode *C. elegans*</u>. **Nematodes** are a great system to study the connection of diabetes, **<u>life span</u>**, and **<u>fat storage</u>** because they have a well-conserved insulin signaling pathway that affects <u>life span</u> and <u>fat storage</u>, suggesting that there is an underlying molecular mechanism that connects insulin signaling, **<u>life span</u>**, and **<u>fat storage</u>**. In addition to the known insulin signaling pathway, the entire genome of the <u>nematode</u> has been sequenced, and many molecular and genetic tools are available that will allow a comprehensive identification of the genes that affect insulin-like signaling, **<u>life span</u>**, and **<u>fat storage</u>** in *C. elegans*.

Key terms are now repeated exactly throughout the paragraph.

Problem 6-9 **Transitions**

1 Coral rely on two main energy sources. **2 First,** they capture planktonic organisms with their tentacles. **3 Second,** they rely on nutrients provided by photosynthetic zooxanthellae, single cell algae that live symbiotically within the coral tissues. **4** Coral bleaching occurs when coral lose their zooxanthellae endosymbionts or zooxanthellae lose their photosynthetic pigments (Glynn, 1996), or both, due to changes in water level, water temperature, UV radiation, sedimentation, pathogens, pollution, and nutrient abundance. **5** Bleaching is reversible if stressors are temporary and not too severe. **6 In contrast,** continuous stress and zooxanthellae depletion leads to coral death.

Problem 6-11 **Paragraph Construction**

A possible way to write this paragraph is the following:
Cancer Cells

Cancer cells, which are malignant tumor cells, differ from normal cells in three ways: (1) they dedifferentiate—for example, ciliated cells in the bronchi lose their cilia; (2) they can metastasize, meaning they can travel to other parts of the body; and (3) they divide rapidly, growing new tumors, because they do not stick to each other as firmly as normal cells do.

Problem 6-13 **Paragraph Construction**

Original

1 Chikungunya, a viral disease spread by mosquitoes, causes severe joint inflammation and pain, high fevers, headache, and rash. **2** The disease is seldom fatal. **3 However,** public health officials are concerned about the spread of the mosquito *Aedes albopictus*. **4** Many viral diseases are transmitted by *Aedes albopictus*, which bites not only during dawn and dusk but also during daytime. **5** It is considered an invasive species and was first imported into the US in tires from Asia in the 1980s (Moore & Mitchell, 1997). **6** Many other nontropical countries have since also been invaded by this vector due to international trade and travel. **7** So far, Italy is the only European country that has had an outbreak (Rezza, 2007). **8** Scientists worry about additional and bigger outbreaks in the developed world.

Arrows connecting key terms show that sentences 2 and 3, as well as sentences 5 and 6, are not linked. In addition, the link between sentences 4 and 5 is not clear, and that between sentences 3 and 4 comes rather late in the sentence, as does the link between sentences 7 and 8.

Revised version

1 Chikungunya, a viral disease spread by mosquitoes, causes severe joint inflammation and pain, high fevers, headache, and rash. **2** The disease is seldom fatal. **3 However, public health officials are** concerned about the **transmission of the disease through** the mosquito *Aedes albopictus*. **4** *Aedes albopictus,* **which bites not only during dawn and dusk but also during daytime, transmits many viral diseases. 5** It is considered an invasive species and was first imported into the US in tires from Asia in the 1980s (Moore & Mitchell, 1997). **6 This vector has since also invaded many other nontropical countries due to international trade and travel. 7 Although the only European country that has had an outbreak is Italy (Rezza, 2007), scientists worry about additional and bigger outbreaks in the developed world.**

In the revised version of the paragraph, the links between all the sentences are clearly established and occur—for the most part—early in the sentences. This way, the reader can recognize a relationship between sentences immediately.

Problem 6-15 **Condensing**

The paragraph can be condensed by only stating the most important information, omitting all other irrelevant results.

Our results indicate that only diamond covered with chemically bound hydrogen displays conductivity at temperatures between 5 and 25°C.

(20 words)

Chapter 7

Problem 7-1 **Outline**

- General overview
  - ○ Global warming has been a major concern over the past decades
  - ○ Temperature rise has been accelerated by humans
  - ○ Ocean levels have risen 15–20 cm in the past century
- Causes for ocean level rise
  - ○ As our climate warms, ocean temperatures rise
  - ○ Water expands at higher temperatures, leading to sea level rise
  - ○ As temperature rises, glaciers and ice sheets melt
  - ○ Polar ice melt increases temperature rise on Earth whereas ice reflects sunlight, oceans absorb light and heat
- Effects of ocean level rise
  - ○ By 2100, ocean levels are expected to rise another 13–94 cm (Haug, 1998)
- Previous Pleistocene interglacials
  - ○ Glacial-interglacial cycles are affected by variations in the Earth's orbit
  - ○ In previous interglacial periods, polar ice was at least partially melted and climate conditions were similar to current conditions on Earth
  - ○ Past climate conditions and sea levels are disputed (3–8)
  - ○ It has been suggested that in some interglacial periods, ocean levels were more than 20 m higher than now (Haug, 1998)
  - ○ Microfossil and isotopic data from marine sediments of the Cariaco Basin support the interpretation that global sea level was 10 to 20 m higher than today during marine isotope stage 11 (Poore & Dowsett, 2001)

Problem 7-3 **Constructing a Paragraph**

The most common active ingredient in insect repellents is DEET (N,N-diethyl-meta-toluamide). DEET was developed by the U.S. military in the 1940s and is easily absorbed into the skin. Although DEET is considered a safe product if used correctly, there have been a few reports of adverse side effects in humans (Osimitz & Grothaus, 1995; Sudakin & Trevathan, 2003). Heavy and frequent dermal exposure can lead to skin irritation and in rare cases, death, especially in young children (Osimitz & Grothaus, 1995; Sudakin & Trevathan, 2003). In addition, neurological damage and death can result from ingestion (Osimitz & Grothaus, 1995; Osimitz & Murphy, 1997).

DEET alternatives include eucalyptus oil containing cineol, 1:5 parts diluted garlic juice, soybean oil, neem oil, marigolds, Avon skin-so-soft bath oil mixed 1:1 with rubbing alcohol, and the juice and extract of Thai lemon grass. True citronella has not been proven to be a very effective mosquito repellent (Revay, 2013).

Chapter 8

Problem 8-1 References

Corrected References are in bold:

Bollag RJ, Waldmann AS, Liskay RM (1989) Homologous recombination in mammalian cells. Annu Rev Genet 23:199–225

Capecchi MR (1989a) Altering the genome by homologous recombination. Science 244:1288–1292

Capecchi MR (1989b) The new mouse genetics: altering the genome by gene targeting. Trends Genet 5:70–76

Offringa R, de Groot JA, Haagsman HJ, Does MP, van den Elzen PJM, Hooykaas PJJ (1990) Extrachromosomal recombination and gene targeting in plant cells after *Agrobacterium*-mediated transformation. EMBO J 9:3077–3084

Offringa R, Franke-van Dijk MEI, de Groot MJA, van den Elzen PJM, Hooykaas PJJ (1993) Non-reciprocal homologous recombination between *Agrobacterium*-transferred DNA and a plant chromosomal locus. Proc Natl Acad Sci USA 90:7346–7350

Paszkowski J, Baur M, Bogucki A, Potrykus I (1988) Gene targeting in plants. EMBO J 7:4021–4026

Puchta H, Hohn B (1996) From centimorgans to base pairs: homologous recombination in plants. Trends Plant Sci 1:340–348

Puchta H, Dujon B, Hohn B (1996) Two different but related mechanisms are used in plants for the repair of genomic double strand breaks by homologous recombination. Proc Natl Acad Sci USA 93:5055–5060

Roth D, Wilson J (1988) Illegitimate recombination in mammalian cells. In: Kucherlapati R, Smith GR (eds) Genetic recombination. American Society for Microbiology, Washington DC, pp 621–653

Schaefer DG, Bisztray G, Zrÿd JP (1994) Genetic transformation of the moss *Physcomitrella patens*. Biotech Agric For 29:349–364

Struhl K (1983) The new yeast genetics. Nature 305:391–397

Problem 8-3 Citations

The carbon ($\delta^{13}C$), nitrogen ($\delta^{15}N$), and sulfur ($\delta^{34}S$) isotope ratios of humans, other animals, and microbes are strongly correlated with the isotope ratios of their dietary inputs (1–5). There are limited differences ($\leq 1\%$) between heterotrophic organisms and their diet in either the $\delta^{13}C$ or $\delta^{34}S$ values **(reference)**. Hydrogen ($\delta^{2}H$) and oxygen ($\delta^{18}O$) isotope ratios of organic matter, however, are more useful, because $\delta^{2}H$ and $\delta^{18}O$ values of precipitation and tap waters vary along geographic gradients **(reference)**. Although differences in the $\delta^{2}H$ and $\delta^{18}O$ values of scalp hair have been noted in humans **(reference)**, less is known about diet–organism patterns of $\delta^{2}H$ and $\delta^{18}O$ values. Four potential sources can be important: dietary organic molecules, dietary waters, drinking waters, and atmospheric diatomic oxygen. Hobson *et al.*

provided evidence that δ^2H values of drinking water were incorporated into different proteinaceous tissues of quail (**reference**). Other research showed that the δ^2H values of bird feathers and butterfly wings (both are largely keratin) and water in the region in which the tissue was produced are highly correlated (14, 15). Kreuzer-Martin *et al.* showed that ~70% of the oxygen and ~30% of the hydrogen atoms in microbial spores (~50% proteinaceous) were derived from the water in the growth medium, whereas the remainder was derived from the organic compounds supplied as substrate (**reference**).

(With permission from National Academy of Sciences, USA)

Problem 8-5 **Paraphrasing**

Three possible versions of paraphrased citations are shown following:

. . . McCaffrey predicts on average 1–3 Mw ≥ 9 earthquakes per century that can occur at any subduction zone (McCaffrey, 2008). He also believes that the past half century with five M9 events reflects temporal clustering and not the long-term average.

. . . according to McCaffrey (McCaffrey, 2008), one to three magnitude 9 or larger size earthquakes occur 100 years at subduction zones, and the increased frequency of M9 earthquake occurrences in the past half century is a higher than long-term average.

. . . an average number of one to three magnitude 9 or larger size earthquakes occurs every century at subduction zones, although sometimes, such as in the past 100 years, M9 earthquakes occur with a higher than normal frequency, however (McCaffrey, 2008).

Problem 8-7 **Paraphrasing**

Version A would be considered plagiarized. Here, sentences have simply been rearranged, and a few words have been substituted. Overall, Version A is too close to the original. In contrast, Version B is paraphrased. It does not only differ very much from the original, it also quotes the original source.

Chapter 9

Problem 9-1 **Figure, Table, or Text?**

1. Table or figure/map depending if actual numbers are important or trends are meant to be displayed
3. Table or graphs. Do not include the photo of the vivarium as it does not directly affect your experimental outcome.
5. Both—a schematic will help to visualize what is being described in the text.

Problem 9-3 **Figure**

The electron micrograph is of great quality, and the viruses are well visible with great resolution. A scale bar would strengthen the picture and provide relevant context for the reader.

Problem 9-5 **Figure**

- Different scales for A and B are confusing
- The heavy *x*-axis bar is distracting
- The key in the middle is confusing

- There are too many tick marks on the *y*-axis
- It is not immediately obvious why the numbers above the bars are there.

As individual values above the graph are apparently important, these data would best be presented in a table.

Chapter 10

Problem 10-1 **Statistics**

$n = 15$; Mean = 14.2; Variance = 3.2; Standard deviation = 1.8

Problem 10-2 **Normal Distribution**

a)   34.15%;   c)   81.85%

Problem 10-3 **Normal Distribution**

0.3% of the time (8 mg/ml –5 mg/ml = 3 mg/ml is three standard deviations or 99.7% of the time the brewing process will run)

Problem 10-4 **Normal Distribution**

a)   2.3%

Problem 10-5 **Reporting Statistics**

a)   No correction needed

Problem 10-6 **Reporting Statistics**

a)   The oxygen production was significantly increased (20%) when the leave segment was 5 cm from the light source instead of 20 cm from the light source ($p = 0.05$).

Problem 10-7 **Statistical Test**

Chi-square test for independence.

Chapter 11

Problem 11-1 **Elements of the Introduction**

Despite good signals for the unknown and question, this Introduction leaves the reader hanging and wondering what exactly was done and what the results are. The reason for its apparent incompleteness is that it does not state the question precisely nor mention any experimental approach. This Introduction also provides a meaningless overview sentence of the discussion to come. Overview sentences such as these should be avoided in research papers as they do not add anything to introductions of investigative papers.

The carotenoid astaxanthin is a red pigment that occurs in specific algae, fish, crustaceans, and in some bird plumages (McGraw and Hardy, 2006). Astaxanthin is an antioxidant and commonly is used as a natural food supplement, food color, and anti-cancer agent among other disease preventative measures.

Like many carotenoids, astaxanthin is a colorful, fat/oil-soluble pigment, provising a reddish and pink coloration (2). While in certain bird species all adult members

Known

display carotenoid-containing feathers rich in color, many gulls and terns, which normally have white feathers, display an abnormal pink tinge (or flush) in various degrees across their populations (McGraw and Hardy, 2006). It has been suggested that this pink tinge arises during feather growth when these birds ingest abnormally high quantities of astaxanthin, which often occurs close to salmon farms (McGraw and Hardy, 2006). However, the exact relationship between astaxanthin and plumage is **not fully understood. Here we examine this relationship** in more detail and discuss its implication.

Unknown

Meaningless overview sentence

Question/ Purpose—not stated precisely

Experimental Approach missing

Problem 11-3 **Elements of the Introduction**

This Introduction is descriptive and should contain background, discovery statement, description of findings, and implication. The Introduction consists largely of background information. The discovery statement is very general and reads like an overview or table of contents. Only a partial description of findings is provided ("We also identified . . ."). No implication is given. Thus, the importance of the findings remains unclear, resulting in a very weak introduction.

Ehrlichiae are obligatory intracellular bacteria that infect leukocytes and platelets of a wide variety of mammals and are transmitted by ticks (1). Dogs can be infected by *E. canis, E. chaffeensis, E. ewingii,* and *Anaplasma phagocytophilum* (*Ehrlichia equi*) (2). Infection with any of these species can cause a severe disease with indistinguishable hematological and clinical anomalies (3).

   E. canis was the first species described in dogs (3-5). It is distributed worldwide, particularly in tropical and sub-tropical regions, and is the causative agent of classical canine monocytic ehrlichiosis, which presents three well-characterized clinical phases. Dogs treated during the acute phase of the disease normally recover rapidly. However, since this stage of the infection can evolve with moderate or imperceptible clinical signs, infected dogs can develop the subclinical phase and some of them reach the chronic phase (6).

E. *chaffeensis* is the etiological agent of human monocytic ehrlichiosis. This potentially mortal disease was reported for the first time in the year 1987 in the USA. *E. canis* was first thought to be the causative agent due to a cross reaction of sera from patients with an antigen preparation from this species (7). In 1991, the bacterium was isolated and characterized at the molecular level. It was then established that it was a distinct species of *Ehrlichia* (8). The recognized natural reservoir for this bacterium is the white-tailed deer (9).

Known	Ehrlichiae with tropism for monocytes, lymphocytes, neutrophils, and platelets from dogs diagnosed by BCS have been reported since 1982 (10). The presence of these rickettsiae in monocytes and platelets has been also confirmed using transmission electron microscopy (11,12). In this study, we describe primary cultures of monocytes from the blood of a dog with canine monocytic ehrlichiosis. We also identified *E. canis* and *E. chaffeensis* using nested PCR with DNA samples extracted from the primary cultures and from dogs with natural and experimental infections
General discovery statement	
Partial description and experimental approach	

Implication missing

(Veterinary Clinical Pathology 37(3). pp. 258–265, 2008)

Chapter 12

Problem 12-1 **Scientific Style and Format in the Materials and Methods Section**

1. b. Tissue samples were collected upon organ removal as reported by Chasse et al. (17). *If details are described in a reference, they do not need to be repeated in the text.*

3. b. For our study, we selected only healthy males between 60 and 80 years. *The overview sentence is not needed—it only adds bulk.*

5. a. The Stress Index Short Form (SI/SF), which is a 36-item questionnaire, was used to assess stress after natural disasters (26). *Verb tense is used correctly here. The statement "which is a 36-item questionnaire" is a statement of general validity and should be present tense.*

Problem 12-2 **Applying Writing Rules**

1. The analyses were performed on an Agilent series 1100 HPLC instrument (Agilent, Waldbronn, Germany). *The additional information is usually not needed and can be omitted.*

3. **The ability to adapt** to changes in light was determined by . . .

5. After centrifugation, 10× buffer **(define the exact content)** was added, and the samples were incubated for 2 min on ice.

Problem 12-3 **Evaluation of a Materials and Methods Subsection**

The purpose of the tests described in this passage is not clear because no such statement has been made. In addition, the last sentence, which explains what types of tests are included in the Wisconsin Card Sorting Test, comes late in the passage, leaving the reader wondering what the test is about throughout the paragraph.

A revised version is shown following. Note that in the revised version, the overall purpose for the test in this particular study is stated (to assess cognitive flexibility), and the background information on the WCST has been moved to the beginning of the paragraph where it helps to provide context.

	Wisconsin Card Sorting Test (WCST)
Purpose of test in study is stated, and background information has been placed at the beginning of the passage to provide context	**Cognitive flexibility was assessed using the Wisconsin Card Sorting Test (WCST)** *(Berg, 1948), in which test subjects have to correctly identify, implement, and remember sorting rules.* To complete this task in our study, participants sorted cards according to color, shape, or number stimuli depicted on the card. Sorting the cards by color was initially verbally reinforced. After a participant responded correctly for 10 consecutive cards in that color category, the participant continued sorting by form and numbers without verbal stimuli or reinforcement. To evaluate our results for cognitive efficiency, the ratio of correct responses to errors was computed, and the numbers of trials, errors, perseverative responses, and perseverative errors were analyzed.

Problem 12-5 **Evaluation of a Materials and Methods Section**

The first 4 paragraphs of this Materials and Methods section have been well constructed. They contain or repeat key terms and transitions. In addition, word location has been considered, and the topic sentences are easily identified. Transitions and relationships between the paragraphs (except for the last one) could be made clearer through the use of transitions.

	Research Question/Purpose: To construct and test a safe, live, attenuated, oral vaccine candidate, IEM108, immune to CTXΦ infection
Experiment Construction of IEM108	**Construction of the candidate IEM108.** The 1.15-kb XbaI fragment containing the upstream regulatory and coding regions of ctxB was recovered from pBR (a pUC19-derived plasmid carrying ctxB and rstR, constructed in our laboratory before) and cloned into the XbaI site of pXXB106, containing *E. coli*-derived thyA (30), resulting in two new constructs, pUTBL1-5 and pUTBL1-6. ctxB and thyA have the same transcriptional direction in pUTBL1-5 and opposite directions in pUTBL1-6 (Fig. 1).
Experiment Construction of IEM108 Organization: chronological	The rstR gene and its upstream sequence were amplified from El Tor strain Bin-43 with primers PrstR1 (CC-GAATTCACTCACCTTGTATTCG) and PrstR2 (CG-GAATTCTCGACATCAAATGGCATG). The amplified fragment was then cloned into the EcoRI site of pUTBL1-5, yielding new construct pUTBL2. Subsequently, an 0.8-kb PvuI fragment of the bla gene in pUTBL2 was deleted to generate pUTBL3. pUTBL3 was then electroporated into IEM101-T to construct IEM108.

Serum vibriocidal antibody assay. Serum vibriocidal antibody titers were measured in a microassay using 96-well plates. The immunized rabbit sera were inactivated at 56°C for 30 min and diluted 1:5 with PBS before use. The prediluted rabbit sera were added into the first well and then serially diluted threefold in PBS. PBS was added to the last well as a negative control. The plates were incubated for 30 min at 37°C with 25 µl of a solution containing 102 CFU of *V. cholerae* Bin-43/ml of culture and 20% guinea pig serum as a complement source in PBS. One hundred fifty microliters of 0.01% 2,3,5-trihenyltetrazolium chloride in LB broth was added to each well, and the plates were further incubated for 4 to 6 h at 37°C until the negative-control wells showed a color change. The reciprocal vibriocidal titer is defined as the highest dilution of serum that completely inhibits growth of Bin-43, i.e., no color change

Measurement of serum vibriocidal antibodies

Rabbit immunization. Eight adult New Zealand White rabbits (2 to 2.5 kg) were divided into naïve, IEM101, and IEM108 groups. The naïve group consisted of two rabbits that were not immunized. Each immunization group had three rabbits. After fasting for 24 h, the rabbits in both immunization groups were anesthetized with ether. After the abdominal skin was sterilized with an iodine tincture and alcohol, the abdominal cavity was opened by vertical incision (under sterile conditions).

General exploration was performed to find the ileocecal region. This region was ligated to the inner wall of the abdomen. Then, 10^9 CFU of vaccine strain IEM101 or IEM108 were injected into the proximal ileum. Finally, the abdominal cavity was closed. The ligature that tied the ileocecal region to the abdominal wall was removed 2 h later, and the rabbits were given water and feed for 28 days. One rabbit of the IEM101 group died after the operation, probably because of heavy anesthesia. Serum samples were collected from the immunized rabbits prior to the immunization and on days 6, 10, 14, 21, and 28 after the vaccination. The serum titers for the anti-CT antibody and vibriocidal antibody were measured as described above.

Consistent Order

Naïve, IEM101, and IEM108 tests

Experiment

Parallel form

Rabbit ileal loop assay and protection model. To evaluate the protection efficacy *in vivo*, the immunized rabbits were challenged with pure CT and four virulent *V. cholerae* strains (395, 119, Wujiang-2, and Bin-43) of different serotypes and biotypes (Table 1) 28 days after the single-dose immunization. Rabbits were anesthetized and their abdomens were opened as described above. Their intestines were tied into 4- to 5-cm-long loops, and then 10^5 to 10^8 CFU of challenge strains or 1, 2, 3, or 4 µg of pure CT were injected into each loop. Normal saline was used as negative control. At 16 to 18 h post challenge, the rabbits were sacrificed and the accumulated fluid from each loop was collected and measured. The ratio of the volume of accumulated fluid (milliliters) to the length of the loop (centimeters) was calculated for each loop in the challenged rabbits.

Evaluation of protection efficacy

(With permisison from American Society for Microbiology)

Chapter 13

Problem 13-1 Scientific Style and Format of a Results Section
1. A 10% change in the length of daylight did not **significantly alter** the expression level of X studied here. **(Chapter 4, Principle 17)**
3. Comparison of the cytokinin concentrations after heat induction showed that production of cytokinin was **increased 10-fold** in ST25 mutants. **(Chapter 2, Principle 2)**
5. The amount of mutated DNA in A and B strains is much less (0 to 25 mutations for each) **than in C and D strains**, exhibiting the highest conservation as described in previous research (34). **(Chapter 4, Principle 20; and missing reference)**
7. Three of the molecules, CB4, CB6, and CB10, **inhibited enzyme activity** more than 50%. **(Chapter 4, Principle 17)**

Problem 13-3 Evaluation and Revision of a Results Section
The original paragraphs contain many experimental details that should not be placed into the Results section. These details need to be removed as shown:

   To evaluate inhibitory effects of the isolated molecules, **inhibition assays were performed. Reaction products were analyzed by 12% PAGE and Western plot analysis. We found that six out of 15 isolated molecules inhibited the kinase reaction at 10 μM markedly, and two of the molecules, A3 and A7, exhibited more than 50% inhibition of the enzyme activity. A 10-fold dilution series of these latter samples was prepared to determine the minimal inhibitory concentration. Molecule A3 exhibited 30% and molecule A7 45% inhibition of the kinase reaction at 1 μM. Buffer A did not interfere with the enzyme.**

Problem 13-5 Evaluation of a Results Section
All the parts of a Results paragraph are there except for an overall interpretation of the results. This section would benefit from such an interpretation.

✓ Background and purpose	Because aggrecan-reactive nets are strongly expressed in the barrel cortex, and the timing of the expression of these PNs coincides with the critical period, we asked if altering whisker sensory input would alter the expression of PNs
✓ Approach	and aggrecan. In our initial study, mice had their whiskers trimmed from the right whisker pad every other day from birth through postnatal day (P) 30. Nissl staining of the barrel cortex illustrated that the development of the barrels was not altered by this manipulation (Fig. 4A,D). Our studies revealed decreased Cat-315 expression in the barrel cortex contralateral to the manipulated whisker pad (Fig. 4B,C) and normal Cat-315 expression in the ipsilateral barrel cortex (Fig. 4E,F) (n = 11) compared to controls (n = 10). The number of Cat-315 positive perineuronal nets
✓ Results	decreased 22% as a result of sensory deprivation.

In addition, the repeated measures ANOVA showed a significant interaction between the hemisphere (left vs. right) and treatment (trimmed vs. not trimmed) ($p < 0.05$) (Fig. 4G). The t-test also showed a significant decrease in Cat-315 expression only in the contralateral barrel cortex of trimmed animals compared to both the left and right barrel cortices of the control animals ($p = 0.0015$, 0.0034, respectively). The total number of cells in the barrel cortex of trimmed animals did not differ from that of the controls. ***Thus, our results indicate that . . .***

✓ More results

Interpretation of results missing

(With permission from The Journal of Neuroscience)

Chapter 14

Problem 14-1 Basic Writing Rules in the Discussion

1. *The way the statement is written is very unprofessional. It is better to present contradicting data in a respectful way:* Therefore, Ebb's data differs from ours because of variations in . . . OR: Ebb did not take X into consideration when evaluating the obtained data. OR: Our results do not agree with those of previous studies (Ebb, 2007).

3. *Authors should never assume that a reader can interpret a figure or table as well as the authors. Authors need to present their data and explain it clearly and not just simply point readers to a figure, letting them figure out the meaning of it themselves.* Counter ions influence X by . . . (Figure 3B).

5. *The statement sounds almost apologetic. The author does not seem to be assertive. Avoid such statements. Rephrase:* The theoretical model presented here provides a new tool to delineate the complex ABC system.

Problem 14-3 Conclusion

Version A is a better concluding paragraph. The end is signaled clearly, the key findings are well summarized and interpreted, and the significance is indicated. Version B is simply listing limitations of the study, making for a very weak ending.

Problem 14-5 Conclusion

Signal of the conclusion	Conclusions
Overall key findings	Habitat heterogeneity, as estimated by an advanced land cover classification, provides a stronger prediction of butterfly species richness in Canada than any previously measured factor. At large spatial scales, virtually all spatial variability (.90%) in butterfly richness patterns is explained by habitat
Specific key findings	heterogeneity with secondary but significant contributions from climate (especially PET) and topography. Patterns of species turnover across the best sampled southern region of
Interpretation of specific key findings	Canada are strongly related to differences in habitat composition, supporting species turnover as the mechanism through

	which land cover diversity may influence butterfly richness.
Specific key findings	Differences in climate are unrelated to butterfly community similarity at this scale, suggesting that the influences of energy
Interpretation of specific key findings	on richness may be indirect or limited to within-habitat diversity. These results have significant conservation implications and indicate that the role of habitat heterogeneity may be
Statement of significance	considerably more important in determining large-scale species-richness patterns than previously assumed.

(With permission from the National Academy of Sciences, U.S.A.)

Chapter 15

Problem 15-1 **Evaluating an Abstract**

The Abstract contains all the important parts and corresponds to the title. All parts are clearly signaled.

Structural basis for the interaction of chloramphenicol, clindamycin, and macrolides with the peptidyl transferase center in eubacteria

Background	Ribosomes, the site of protein synthesis, are a major target for natural and synthetic antibiotics. Detailed knowledge of antibiotic binding sites is the key to understand the mechanisms of drug action. Conversely, drugs are excellent tools for studying the ribosome function. **To elucidate** the structural basis of ribosome-antibiotic interactions, **we determined** the high-resolution
Question/ purpose	
Experimental approach	X-ray structures of the 50S ribosomal subunit of the eubacterium *Deinococcus radiodurans* complexed with the clinically relevant antibiotics chloramphenicol, clindamycin, and the three macrolides: erythromycin, clarithromycin, and roxithromycin. **We found** that antibiotic binding sites are composed
Results	exclusively of segments of 23S rRNA at the peptidyl transferase cavity and do not involve any interaction of the drugs with ribosomal proteins. **Here, we report** the details of antibiotic interactions with the components of their binding sites.
Conclusion/ answer	
Significance	Our results also show the importance of Mg ions for the binding of some drugs. This structural analysis **should facilitate** rational drug design.

(With permission from Macmillan Publishers Ltd.)

Problem 15-3 **Evaluating an Abstract**

In this Abstract, not all required parts are present. The experimental approach, the results, and the conclusion are missing. The abstract is an investigative abstract (background) but contains an overview sentence at the end, which should be omitted or, better yet, replaced. A suggested addition incorporating the missing portions is shown in bold.

Background

Question/
purpose

The cyanobacterial circadian pacemaker is an enzymatic oscillator, which orchestrates the metabolism of the bacteria to fit the day and night alternations of this planet. Interactions among KaiA, KaiB and KaiC, the three components of this oscillator, result in many oscillatory properties in vitro, including an overall phosphorylation level of KaiC, an apparent segregation between synchronized phosphorylation and dephosphorylation reactions, and the size and the composition of this oscillator. To explain these properties, we propose here a molecular mechanism for this pacemaker within the framework of a cyclic catalysis scheme. . . .

Experimental
approach
and results/
description and
conclusion/
implication is
missing

Added
experimental
approach

Added results/
description and
conclusion/
implication

. . .as determined mathematically. The mechanism includes the regulation of KaiC's enzymatic activity by KaiA and KaiB. In this system, high phosphorylation states exhibit high affinities for aggregates of KaiA dimers and KaiB dimers, leading to the formation of their higher oligomeric structures. This model deepens our understanding of the molecular mechanism by which a biological clock operates.

(With permission from Jimin Wang)

Chapter 16

Problem 16-1 **Clear Title**

1. Variation of fossil density **in** Triassic sedimentary deposits—*"with" is imprecise*

3. Classification of Fowl Adenovirus Serotypes **through/by** genome mapping and sequence analysis of the hexon gene—*"with" is imprecise*

5. Temperature dependence of **carbon dioxide sequestration by fir trees**—*the noun cluster "fir tree carbon dioxide sequestration" is unclear*

7. Drug-susceptibility assay for the diagnosis of TB **through** microscopic-observation—*the noun cluster "Microscopic observation drug susceptibility assay" is unclear*

Problem 16-2 **Complete Title**

1. Differences **in terpenes** of old-world and new-world species of hazelnuts

3. Widespread increase of bat mortality rates **in the eastern United States due to White-nose syndrome**

Problem 16-3 **Title Length**

1. **Adaptive, cued seed dispersal in the cactus** *Mammillaria pectinifera*
3. **Effect** of negative mood on persistence in problem solving
5. **Optimizing flu virus vaccination strategies: Cost-effectiveness analysis**
7. **Temperature and light requirements** for seed germination and seedling growth of *Sequoiadendron giganteum*

Problem 16-4 **Creating a Title**

The best title is "4. Milk consumption prevents antioxidant effect of blueberries." It includes all important facts and is clear, complete, and succinct.

Problem 16-5 **Running Title**

1. Transcoronary transplantation after myocardial infarction
3. Inflammatory mechanisms in pulmonary disease
5. Optical manipulation of electron spin in quantum dot
7. Decline and extinction of Brazilian tree frog
9. Rare structural variants in schizophrenia
11. Carbon and hydrogen in alkyne hydrogenation
   OR: Catalytic hydrogenation in microreactors

Chapter 19

Problem 19-1 **Review Introduction**

The introduction of the review paper contains all necessary elements and is complete.

Background Unknown/ problem Topic statement Overview	Mitochondrial genomes differ greatly in size, structural organization, and expression both within and between the kingdoms of eukaryotic organisms. The mitochondrial genomes of higher plants are much larger (200–2,400 kb) and more complex than those of animals (14–42 kb), fungi (18–176 kb), and plastids (120–200 kb) (Refs. 1–4). Although there has been less molecular analysis of the plant mitochondrial genome structure in comparison with the equivalent animal or fungal genomes, the use of a variety of approaches—such as pulsed-field gel electrophoresis (PFGE), moving pictures (movies) during electrophoresis, restriction digestion by rare-cutting enzymes, two-dimensional gel electrophoresis (2DE), and electron microscopy (EM)—has led to substantial recent progress. Here, the implication of these new studies on the understanding of *in vivo* organization and replication of plant mitochondrial genomes is assessed.

(With permission from Elsevier)

Problem 19-3 **Abstract**

The Abstract is that of a review article. Sentence 1 provides a short background. Sentence 2 gives an overview of the problem and together with sentence 3 describes the topic of the review. Sentence 4 provides an overview of the content.

Background	In the last few decades, Africanized honey bees have been spreading throughout South and Central America into the
Problem	southern states of the US. During this spread, they have Africanized European bees largely through crossbreeding during mating flights, which almost always leads to the more aggres-
Topic	sive Africanized bees. Various practices have been applied to
Overview of content	counteract this trend. This review analyzes different practices used to counteract this trend and recommends the optimal approach to ensure pure European honey bees.

Problem 19-5 **Conclusion**

The Conclusion section nicely summarizes what the review presents. It starts with a general, brief overview of the topic, followed by a recap list of subsections covered. Finally, the conclusion section narrows down to a model that can be derived from comparison of sequences and the analysis of findings.

Problem 19-7 **Main Analysis Section**

a) Outline for sea urchin species specificity:

- Sea urchin species specificity of gametes
  - ○ Intraspecies
  - ○ Interspecies
- Interspecies insemination frequency
- Surface macromolecules
  - ○ For sperm
  - ○ For egg
- Macromolecule interaction
- Bindin deletion mutants
  - ○ Species specificity
  - ○ Comparison of protein sequences among different species
- Lock and key model

Chapter 21

Problem 21-1 **Abstract for an LOI**

Background	As the concentration of CO_2 in our atmosphere increases, so does the amount of CO_2 in the ocean's surface waters. Although the oceans are able to remove some of the increasing carbon from the atmosphere, just how much they are able to
Problem	remove has vital implications in determining the potential se-
Objective	verity of global warming. To understand the global carbon cycle, we propose to measure the CO_2 content of seawater
Strategy	over a 5-year period using high-precision methods for measuring total dissolved inorganic carbon (DIC), total alkalinity, and carbon and oxygen isotopes of CO_2. Results from this study will provide insight into CO_2 absorption by sea water and
Significance	ultimately increase our understanding of global warming.

Problem 21-3 **Abstract for an LOI**

Cardiovirus is a common cause of gastroenteritis. For routine vaccination of Chinese infants, a new cardiovirus vaccine has been recommended. Here, we propose to evaluate the impact and cost-effectiveness of the Chinese cardiovirus vaccine program using a dynamic model of cardiovirus transmission. Findings from our analysis can be used to inform policy makers to prevent rotavirus infection.

Problem 21-5 **Personnel Section**

Example:

The proposed project will be led by [name 1, title]. [Name 1] is an expert in X and will oversee all personnel and research activities A and B. She has worked X years on A and made several seminal discoveries in the field (references). The co-PI [Name 2] has expertise in C and D and will primarily be responsible for C of the proposed work. A postdoctoral fellow, who will be hired, will perform E. . . .

Chapter 22

Problem 22-1 **Proposal Abstract**

Background	Plants are our oldest source of medicines. Yet much of Earth's rich plant life remains unexplored. In recent decades, natural drug discovery has concentrated on tropical plants due to their great diversity. However, there is equally much diversity for the plants of our oceans, and this plant life has remained
Problem	untapped. The overall goal of this proposal is to identify and purify natural chemicals of oceanic plants and to test their
Objective	activity as potential medicines. We will apply new chemical fingerprinting technology for our screens of plant life in the
Strategy	Florida Keys and assess them for potential medicinal use using microbial techniques. Identification of plants with compounds active against important human diseases, and subsequent characterization of such compounds, will lay the foundation for new drug development and lead to novel treatment thera-
Significance	pies and better outcomes for patients.

Problem 22-3 **Proposal Abstract**

Habit reversal therapy (HRT), a behavioral treatment for tics, may be effective in treating Tourette syndrome (TS) (Deckersbach, 2006), an inherited neurological disorder characterized by chronic motor and vocal tics. We propose to compare the efficacy of HRT in reducing tics and improving life satisfaction and psychosocial functioning in comparison with supportive psychotherapy (SP) in 100 outpatients with TS. We will determine if HR has specific, sustainable, tic-reducing effects and if SP is effective in improving life satisfaction and psychosocial functioning. Assessments of response inhibition may help to inform and predict treatment response to this disorder.

(96 words)

Problem 22-5 **Intellectual Merit and Broader Impact Statements**

The two statements are well written. In the Intellectual Merit section, the impact of the proposed work is clearly stated (". . . with potential applications beyond the scope of this proposal . . ."), and in the Broader Impact section, the educational component is clearly highlighted, as is the fact that a female student will be educated—both components of importance to the NSF.

Chapter 23

Problem 23-1 **Background and Significance Section of a Grant**

In this section, the background and need are stated, but the objective/aim is missing.

	Healthy older adults often experience mild decline in some areas of cognition. The most prominent cognitive deficits of normal aging include forgetfulness, vulnerability to distraction and other types of interference, as well as impairment in multitasking and mental flexibility (Albert, 1997; Bimonte, 2003). These cognitive functions are the domain of the most evolved part of the human brain known as the prefrontal cortex, the brain region that is the last to fully mature in children and the first to decline as we age. Indeed, prefrontal cortical cognitive abilities begin to weaken already in middle age and are especially impaired when we are stressed. Loss of these organizational abilities is a particular liability in this Information Age when our demanding lives require that we multitask and navigate through endless interferences. Thus, understanding how the prefrontal cortex changes with age is a top priority for rescuing the memory and attention functions we need to survive in our fast-paced, complex world.
Background	
Need is stated but no aim	

Problem 23-3 **Background and Significance Section of a Grant**

In this section, the background, unknown, and the objective are stated and signaled.

	One potential therapeutic approach for treating chronic HBV infection is through therapeutic vaccination to disrupt the immunological tolerance and to induce an immune response that is capable of controlling the virus. Despite the promise of therapeutic vaccination for treating chronic HBV, progress in this area has been limited (16, 29, 34, 64, 73, 83). A successful therapeutic vaccination strategy must accomplish two goals. First, immunological tolerance must be broken, and virus-specific T cells must be generated. Second, these T cells must efficiently perform their effector functions, including killing target cells and producing the antiviral cytokines such as IFN-γ and TNF-α, which can noncytopathically inhibit virus replication. We will test the hypothesis that recombinant VSV vectors are ideally suited for therapeutic vaccination for chronic HBV infection.
Background	
Need	
Objective/ hypothesis	

(Michael Robek, proposal to federal agency)

Chapter 25

Problem 25-1 **Preliminary Results Section**

This paragraph contains all necessary elements of a well-constructed prelimi-nary results section.

	ELISA for secreted HBMS. Because HBMS is a secreted
Purpose	protein (46), we determined whether the HBMS protein
	produced in VSV-infected cells is secreted into the culture
Experimental approach	media. We detected HBMS in the media of VSV-HBMS-infected cells but not in uninfected or recombinant WT VSV-infected cells by both Western blot (data not shown) and
Results	qualitative ELISA (Table 2). We found by quantitative ELISA that secreted HBMS levels reached 35 ng/ml in the media
Interpretation of results	of VSV-HBMS-infected BHK cells by 24 h post infection. Therefore, the HBMS produced in VSV-infected cells is cor-rectly processed for secretion.

(Michael Robek, proposal to federal agency)

Problem 25-3 **Preliminary Results Section**

In this paragraph, context has not been provided. This omission can be con-fusing for reviewers and should be avoided.

Context missing	Calculations with a General Circulation Model of the atmo-sphere (6) indicate that during idealized permanent El Niño-like conditions, the warming of the Eastern equatorial Pacific re-duces the area covered by stratus clouds, thus decreasing the albedo of the planet. At the same time, the atmospheric concen-
Experimental approach	tration of the powerful greenhouse gas, water vapor, increases. This scenario may have happened during the early Pliocene,
Results	amplifying the warm conditions at that time. Consequently, projections of the effect of increasing concentration of green-
Interpretation/ conclusion	house gases on climate should include a thorough consider-ation of the tropical climate conditions.

(Alexey Federov, proposal to federal agency, modified)

Chapter 26

Problem 26-1 **Approach Section**

Purpose of experimental design	In this model, we will improve upon preliminary studies to evaluate more precisely whether factors that can be assessed within a few days of a patient's admission can be used to determine which patients are most likely to be infected with drug resistant strains. We will thereby determine an optimal empirical treatment regimen for the interim period until their drug sensitivity results are available. Key clinical factors to be incorporated into the model include TB treatment history, hospitalization history, sputum smear results, lack of response to first-line treatment, chest X-ray deterioration, and rapid rifampin, kanamycin, and ciprofloxacin resistance testing results. Parameterizing our model with the data collected, we will calculate the likelihoods that a patient has non-MDR, non-XDR MDR, or XDR TB, and will determine what further studies are required to enhance confidence in the likelihood calculations. The model will also account for the benefits and disadvantages of different treatment regimens in terms of treatment outcomes, side effects, costs, pill burdens, and the probability of acquired or amplified resistance.
Analysis	
Expected outcome	
Significance/ justification	

(Alison Galvani, proposal to private foundation, modified)

Problem 26-3 **Approach Section**

Purpose of experimental design	**I. Laboratory and Field Behavior Study** Objective: To determine the function of butterfly wing patterns by characterizing the detailed behavioral responses of insect and predators, in response to the Monarch butterfly, *Danaus plexippus* L. Laboratory experiments will incorporate information observed in the field.
Experimental design	Approach and Analysis: Field experiments will include a) tethering of live butterflies to vegetation to observe who predates them; b) observation of interactions of insect predators, such as mantids, with Monarch butterflies; and c) observation of interactions of avian predators, such as Blue Jays, with the same butterflies. For our experiments, artificial habitats will have representative irradiance, background color and complexity based on ecological information we collect in the Mojave Desert. Our model insect predator will be *Mantis religiosa*, a common mantid species. Using high-speed video, we will examine whether the width and number of stripes affect the target of mantid attacks.
Analysis	
Expected outcome is missing	
Significance/ alternatives are missing	

Chapter 30

Problem 30-1 **Slide for Oral Presentation**

Capturing the meaning of the slide by using as few words as possible is important. Details can be filled in by the presenter.

Value of Clinic

- Welcoming environment

- Communication in native language

- Comprehensive care

- Professional and caring clinical teams

- Respected as human beings

Problem 30-3 **Slide for Oral Presentation**

Capturing the meaning of the slide by using as few words as possible is important. Details can be filled in by the presenter. Individual sections can be highlighted by an arrow, as shown, or by different colored text.

(With permission from Jaclyn Brown, modified)

Outline:

 1. Discussion of the kinematics

2. The zonalmomentum equation

3. Nonlinear Sverdrup theory

4. The seasonal cycle and inter-annual variability

Problem 30-5 **Slide for Oral Presentation**

Several problems exist for this slide: (a) the figures have been copied directly from a publication; (b) they lack a title and unnecessarily show a figure legend; (c) the writing is too small to read and the copies look fuzzy; (d) the font of the heading is a serif font—it is better to use a sans serif font, such as Arial; (e) the heading is uninformative; and (f) the slide looks crowded.

Problem 30-6 Oral Presentation Statements

1. *This ending is very weak and shows that the presenter is not very confident. It is better to end just with a simple "Thank you."*

3. *This statement is uninformative and meaningless. When you provide an overview, make sure that the information you give is informative.*

5. *The speaker here is using written English instead of spoken English. This sentence sounds very awkward when used in speech.*

Glossary of English Grammar Terms

Active Voice refers to the form of the verb used in relation to what the subject is doing. In English, there are only two voices—passive and active. The active voice of a verb is the form of the verb used when the subject is the doer of the action.

> *Passive Voice:* The manuscript was reviewed by the head of the department.
>
> *Active Voice:* The head of the department reviewed the manuscript.

Adjective a word that modifies a noun or pronoun.

> *Example:* The *young* birds accepted the bait readily.

Adverb a word that modifies a verb, an adjective, or another adverb. Adverbs generally answer one of four questions: how, when, where, or to what extent. Adding the suffix *-ly* to an adjective commonly turns the word into an adverb.

> *Examples:* The reaction was *fast.* (how)
>
> The reaction took place *immediately.* (when)

Appositive a noun, noun phrase, or noun clause that follows a noun or pronoun and renames or describes the noun or pronoun. Appositives are often set off by commas.

> *Example:* <u>S. aureus</u>, *a Gram-positive organism*, can carry many resistance genes

Article a type of adjective that makes a noun specific or indefinite. In English, there are three articles: the definite article *the* and the two indefinite articles *a* and *an*. In writing, an *article* is a brief nonfiction composition such as is commonly found in periodicals or journals.

Auxiliary Verb a verb that is used with a main verb. *Be, do,* and *have* are auxiliary verbs. *Can, may, must,* etc. are modal auxiliary verbs.

Clause a group of words containing a subject and verb that forms part of a sentence.

Complement a word that follows a verb and completes the meaning of the sentence or verbal phrase.

Conjunctions words that join words, phrases, or sentence parts.

> *Examples: and, or, for, but, nor, so, yet, either/or, neither/nor, both/and, whether/or, not/but,* and *not only/but also.*

Dangling Modifiers a phrase or clause that says something different from what is meant because words are left out. The meaning of the sentence, therefore, is left "dangling."
> *Example: Having studied* the protocol of Bowen et al., the mice received a diet rich in vitamin B. (Reads like the mice studied the protocol.)

Dependent/Subordinate Clause depends on the rest of the sentence for its meaning. It is usually introduced by a subordinating element such as a subordinating conjunction, transition word, or relative pronoun. It does not express a complete thought, so it does not stand alone. It must always be attached to a main clause that completes the meaning.

Direct Object a noun or pronoun that receives the action of a verb or shows the result of the action. It answers the question "What?" or "Whom?" after an action verb.
> *Example:* We observed *the seals.*

Gerund a verb ending in *-ing* and used as a noun.
> *Example:* We recommend *administering* the drug in combination with X.

Indirect Object tells *to whom* or *for whom* the action of the verb is done and who is receiving the direct object.
> *Example:* The editor returned the manuscript *to us.*

Infinitive the simple present form of a verb used as either a noun, adjective, or adverb. The verb of the infinitive is normally preceded by the word *to*. When the infinitive follows some verbs as the direct object, the *to* may be dropped.
> *Example:* Dr. Pacheco helped *to write* the paper.
> Dr. Pacheco helped *write* the paper.

Jargon specialized language of a particular trade or group.

Main or Independent Clause a clause that is not introduced by a subordinating term. It does not modify anything, and it can stand alone as a complete sentence.

Noun a word that signifies a person, place, thing, idea, action, condition, or quality.

Object in the active voice, a noun or its equivalent that receives the action of the verb. In the passive voice, a noun or its equivalent that does the action of the verb.

Participle There are two participles in English: the present participle and the past participle.
> The *present participle* is formed by adding *-ing* to the base form of a verb. It is used in
> i) Continuous or progressive verb forms
> *Example:* Our competitors were *using* different conditions.
> ii) As an adjective
> *Example:* under *varying* conditions
> The *past participle* is formed by adding *-ed* to the base form unless it is an irregular verb. It is used
> i) As an adjective
> *Example:* a *known* fact, at the *determined* concentration

     ii) With the auxiliary verb "have" to form the past perfect tense
       Example: We have *observed* ...
     iii) With the verb "be" to form the passive
       Example: The solution was *mixed.*

Passive Voice the form of the verb used when the subject is being acted on rather than doing something.

    Passive Voice: The manuscript was reviewed by the head of the department.

    Active Voice: The head of the department reviewed the manuscript.

Person refers to the form of a word as it relates to the subject. In English, there are three persons:

    First person refers to the speaker. The pronouns *I, me, myself, my, mine, we, us, ourselves, our,* and *ours* are first person.

    Second person refers to the one being spoken to. The pronouns *you, yourself, your,* and *yours* are second person.

    Third person refers to the one being spoken about. The pronouns *he, she, it, him, her, himself, herself, himself, his, her, hers, its, they, them, themselves, their,* and *theirs* are third person.

Phrase a group of words not containing a subject and its verb (e.g., *in this experiment, after the eruption*).

Plural means "more than one." To show that a noun is plural, an *-s* or *-es* is normally added to the word. There are a few irregular plurals such as *men, children, women, oxen,* and a number of words taken directly from foreign languages such as *alumni* (plural of alumnus) or *media* (plural of medium).

    Plural form of pronouns—pronouns that take the place of plural nouns such as *we, you,* and *they.*

    Plural form of verbs—verbs that go with a plural subject.

Possessive Case used for a noun or pronoun to show ownership or association. Nearly all nouns and indefinite pronouns show possession by ending with an apostrophe plus an *s.*

    Example: the organism's progeny

Prepositions words that relate a noun or pronoun (called the object of the preposition) to another word in the sentence. The preposition and the object of the preposition together with any modifiers of the object are known as a ***prepositional phrase.***

    Example: in, of, under, over, for, to, at, with

Prepositional Phrase a phrase beginning with a preposition and ending with a noun or pronoun. The phrase relates the noun or pronoun to the rest of the sentence. The noun or pronoun being related by the preposition is called the ***object of the preposition.***

Pronoun a word takes the place of a noun in a sentence.

    Example: this, that, these, both, either

Redundant means "needlessly repetitive."

Sentence a group of words that expresses a thought. A sentence conveys a statement, question, exclamation, or command and must contain a verb and (usually) a subject. It starts with a capital letter and ends with a period (.), question mark (?), or exclamation mark (!).

Singular the form of a word representing or associated with one person, place, or thing. The term is normally used in contrast to plural. The singular form of a verb goes with a singular subject.

Subject the main noun (or equivalent) in a sentence about which something is said or who does something.

Tense the tense of a verb shows the time when an action or condition occurred (past, present, or future).

Verb the word or words that express action or say something about the condition of the subject.

Brief Glossary of Scientific and Technical Terms

Active verb	Verb that expresses the action in a sentence (Example: measure instead of measurement).
APA style	Writing style format for academic journals and books, particularly in the social sciences, based on the style guide of the American Psychological Association (APA).
Biographical sketch	Brief description of a person's academic and professional accomplishments.
BIOSIS	English-language, bibliographic database in the life sciences and biomedical sciences.
CINAHL	The Cumulative Index to Nursing and Allied Health Literature (CINAHL) is the largest English-language nursing, allied health, biomedicine and healthcare index and database.
Chicago style	Writing style format used largely in the social sciences and humanities based on the style guide for American English published by the University of Chicago Press since 1906.
Conference abstract	Abstract of a presentation prepared for a conference in order to gauge interest in the presentation.
Coherence	Logical relationship of sentences through linking of ideas from one sentence to the next.
Cohesion	Logical connection of sentences created through word location.
Continuity	The logical flow between sentences and/or paragraphs as a whole using all techniques of coherence and cohesion.
Core slide	Slide containing the most important information or figure in an oral presentation.
CSE style	Writing style format created by Council of Scientific Editors (CSE; formerly Council of Biology Editors [CBE]); used largely in the sciences.
Current Contents	Search database that provides access to tables of contents, bibliographic information, and abstracts from the most recently published leading scholarly journals.

Descriptive paper	Research articles describing a new discovery, apparatus, or application.
EndNote	Reference program that helps you manage your references.
ESL	Abbreviation for "English as a second language."
Funnel structure	Standard structure of the Introduction of a research paper, starting broadly with background information, narrowing to specific knowledge on a topic, to something unknown or problematic, and to the research question of the paper and its experimental approach.
Galley proofs	Semifinal form of a document that is laid out like the final form and provides the last chance to make minor corrections before printing.
Google Scholar	Free online database for academic publications.
IMRAD	Format followed in writing a scientific research article; abbreviation stands for the order of the different sections of a scientific paper: Introduction, Materials and Methods, Result, and Discussion.
Indicative abstract	Abstract that provides the reader with a general idea of the contents of the paper; does not include any methods or results; used for review articles and book chapters.
Informative paper	Research article discussing research done in response to a hypothesis or to fill a specific gap.
Instruction for Authors	Set of instructions and guidelines provided by scientific journals or funding agencies as a guideline for authors when writing a scientific article or proposal.
Investigative paper	Typical research article discussing research done in response to a hypothesis or to fill a specific gap.
Jargon	Use of terms specific to a technical or professional group.
Jumping word location	In consecutive sentences, the new information in the stress position of one sentence is placed at the topic position of the next sentence.
Key term	Words or short phrases used to identify important ideas in a sentence, a paragraph, and the paper as a whole; usually used to identify your main points in the topic sentence.
Key words	Important words that identify key ideas and points in a document; used for indexing and searching databases.
LOI	Letter of intent, sometimes also referred to as "letter of introduction."
MEDLINE	Database produced by the National Library of Medicine; broad coverage includes basic biomedical research and clinical sciences.

MLA style	Modern Language Association (MLA) is a commonly accepted writing style format, particularly in the liberal arts and humanities.
NASA	National Aeronautics and Space Administration.
NIH	National Institutes of Health.
Nominalizations	Abstract nouns derived from verbs and adjectives (Example: "measurement" instead of "to measure").
Noun clusters	A cluster of nouns or modifying words strung together, often incomprehensibly.
NSF	National Science Foundation.
Parallel form	Placing related ideas of equal weight into the same grammatical form and style.
Paraphrasing	To express someone else's words, thoughts, or ideas in your own words.
Plagiarism	Failing to indicate the source of information in scholarly scientific work; a form of academic misconduct.
Preproposal	Short, overview version of a proposal; submitted ahead of proposal to gauge interest of a potential funder.
Power positions	Location where most important information in a sentence, paragraph, section, or document is placed. Two key positions exist: first and last, whereby first in a sentence, paragraph, or document is more powerful than last.
Presenter's triangle	Triangular space to the left side of the lectern; considered the best place to present without blocking the view of the audience to slides.
Primary sources of literature	Original, peer-reviewed publication of a scientist's new research and theories.
Proposal	Grant application asking for financial support from a potential sponsoring agency.
PsycINFO	Database that covers psychology and related disciplines.
PubMed	A free database developed by the National Center for Biotechnology Information; it contains more than 22 million citations for biomedical literature from MEDLINE, life science journals, and online books.
Pyramid structure	Standard structure of the Discussion section of a research paper; starts with specific key results and their meaning and broadens to what these mean in the field generally and to society overall.
Reference Manager	Reference program that helps you manage your references.
Review paper	Secondary source representing a balanced summary of a timely subject with reference to the literature; summarizes what has been published or researched in a specific field and/or evaluate methods and results.

RFP   Abbreviation for "request for proposal," which is put out largely by funding agencies.

SciFinder   Free online search database for scientific literature.

SCOPUS   Database that provides broad international journal coverage of the sciences and social sciences.

Secondary sources   Literature source that cites, builds on, discusses, or generalizes primary sources (Example: a review article).

Stress position   End position in a sentence where important, new information should be placed.

Tertiary sources   Literature source that generalizes and analyzes primary and secondary sources while attempting to provide a broad overview of a topic (Example: a textbook).

Topic position   Beginning position in a sentence where old, familiar information should be placed.

Topic sentence   First sentence of a paragraph, which provides an overview of the paragraph as well as important key terms.

Web of Science   ISI Citation Databases that provide Web access to Science Citation Index Expanded for science and engineering journals.

Bibliography

Alley, Michael. *The craft of scientific writing.* 3rd ed. Springer, New York, 1996.

Alley, Michael. *The craft of editing: A guide for managers, scientists, and engineers.* Springer, New York, 2000.

Alley, Michael. *The craft of scientific presentations: Critical steps to succeed and critical errors to avoid.* Springer, New York, 2003.

Altman, Rick. *Why most PowerPoint presentations suck and how you can make them better.* 1st ed. Harvest Books, Pleasanton, CA, 2007.

American Medical Association manual of style. 9th ed. Lippincott Williams & Wilkins, Philadelphia, 1997.

American Medical Association manual of style: A guide for authors and editors. 10th ed. Oxford University Press, New York, 2007.

American National Standards Institute, Inc. *American national standard for the abbreviation of titles of periodicals.* American National Standards Institute, New York, 1969.

American National Standards Institute, Inc. *American national standard for the preparation of scientific papers for written or oral presentation.* American National Standards Institute, New York, 1979.

American National Standards Institute, Inc. *American national standard for writing abstracts.* American National Standards Institute, New York, 1979.

Atkinson, Cliff. *Beyond bullet points: Using Microsoft® Office PowerPoint® 2007 to create presentations that inform, motivate, and inspire.* Microsoft Press, Redmond, WA, 2007.

Booth, V. *Communicating in science: Writing a scientific paper and speaking at scientific meetings.* 2nd ed. Cambridge University Press, Cambridge, UK, 1993.

Browner, Warren S. *Publishing and presenting clinical research.* Lippincott Williams & Wilkins, Philadelphia, 1999.

Browning, Beverly. *Perfect phrases for writing grant proposals.* 1st ed. McGraw-Hill, New York, 2007.

Council of Biology Editors Scientific Illustration Committee. *Illustrating science: Standards for publication.* Council of Biology Editors, Bethesda, MD, 1988.

Covey, Franklin. *Style guide for business and technical communication.* Franklin Covey, Salt Lake City, UT, 1997.

Day, Robert A. *How to write and publish a scientific paper.* 5th ed. Oryx Press, Phoenix, AZ, 1998.

Davis, Martha. *Scientific papers and presentations.* Academic Press, New York, 2002.

Davis, M., and Fry, G. *Scientific papers and presentations.* Academic Press, London, 2004.

Dodd, Janet S. *The ACS style guide: A manual for authors and editors.* 2nd ed. American Chemical Society, Washington, DC, 1997.

Ebel, Hans. F., Bliefert, Claus, and Russey, William E. *The art of scientific writing: From student report to professional publications in chemistry and related fields.* 2nd ed. Wiley-VCH, Weinheim, Germany, 2004.

Feibelman, Peter J. *A PhD is not enough: A guide to survival in science.* Basic Books, New York, 1993.

Foley, Stephen Merriam, and Gordon, Joseph Wayne. *Conventions and choices. A brief book of style and usage.* D. C. Heath, Lexington, MA, 1986.

Fowler, H. W. *A dictionary of modern English usage.* 2nd ed. Oxford University Press, New York, 1965.

Fowler, H. W. *A dictionary of modern English usage.* Wordsworth Editions, London, 1997.

Geever, Jane C. *The Foundation Center's guide to proposal writing.* 5th ed. Foundation Center, New York, 2007.

Gerin, William. *Writing the NIH grant proposal: A step-by-step guide.* Sage, Thousand Oaks, CA, 2006.

Gibaldi, Joseph. *MLA handbook for writers of research papers.* 7th ed. Modern Language Association of America, New York, 2009.

Glatzer, Jenna. *Outwitting writer's block and other problems of the pen.* Lyon's Press, Guilford, CT, 2003.

Gopen, George D. *Expectations: Teaching writing from the reader's perspective.* Pearson Longman, New York, 2004.

Gopen, George D., and Swan, Judith A. The science of scientific writing. *American Scientist* 78 (1990): 550–558. https://www.americanscientist.org/issues/id.877,y.0,no.,content.true,page.1,css.print/issue.aspx

Greenbaum, Sidney. *The Oxford English grammar.* Oxford University Press, New York, 1996.

Gustavi, B. *How to write and illustrate a scientific paper.* Cambridge University Press, Cambridge, UK, 2003.

Hacker, Diana. *Rules for writers.* 5th ed. Bedford/St. Martin's, Boston, 2004.

Hall, Mary S., and Howlett, Susan. *Getting funded: The complete guide to writing grant proposals.* 4th ed. Continuing Education Press, Portland, OR, 2003.

Harris, Dianne. *The complete guide to writing effective and award-winning grants: Step-by-step instruction.* Atlantic Publishing, Ocala, FL, 2008.

Henson, Kenneth. *Grant writing in higher education: A step-by-step guide.* Allyn & Bacon, Boston, 2003.

Huth, E. J. *How to write and publish papers in the medical sciences.* Williams & Wilkins, Philadelphia, 1990.

Huth, E. J. *Writing and publishing in medicine.* 3rd ed. Lippincott Williams & Wilkins, Philadelphia, 1998.

Iles, Robert L., and Volkland, Debra. *Guidebook to better medical writing.* Rev. ed. Iles Publications, Olathe, KS, 2003.

International Committee of Medical Journal Editors. *Uniform requirements for manuscripts submitted to biomedical journals: Sample references.* http://www.nlm.nih.gov/bsd/uniform_requirements.html

International Committee of Medical Journal Editors. *Uniform requirements for manuscripts submitted to biomedical journals: Writing and editing of biomedical publication*. http://www.icmje.org/

Katz, Michael Jay. *From research to manuscript: A guide to scientific writing.* Springer, Dordrecht, Germany, 2006.

Knowles, Cynthia. *The first-time grantwriter's guide to success.* Corwin Press, Thousand Oaks, CA, 2002.

Korner, Ann M. *Guide to publishing a scientific paper.* Bioscript Press, Flagstaff, AZ, 2004.

Lindsay, David. *A guide to scientific writing.* 2nd ed. Longman, London, 1995.

Lynch, Jack. *The English language: A user's guide.* Focus Publishing/R. Pullins, Newburyport, MA, 2008.

Malmfors, B., Grossman, M., and Garnsworth, P. *Writing and presenting scientific papers.* Nottingham University Press, Nottingham, UK, 2004.

Matthews, Janice R., Bowen, John M., and Matthews, Robert W. *Successful scientific writing.* Cambridge University Press, Cambridge, UK, 1996.

McMillan, Victoria E. *Writing papers in the biological sciences.* Bedford Books/St. Martin's Press, Boston, 1997.

Mills, H. R. *Techniques of technical training*, 3rd ed. Macmillan, New York, 1977.

Morgan, Scott, and Whitener, Barrett. *Speaking about science: A manual for creating clear presentations.* Cambridge University Press, Cambridge, UK, 2006.

O'Connor, Maeve. *Writing successfully in science.* Chapman and Hall, London, 1991.

Ogden, Thomas E., and Goldberg, Israel A. *Research proposals: A guide to success.* 3rd ed. Academic Press, New York, 2002.

Peat, Jennifer, and Barton, Belinda. *Medical statistics: A guide to data analysis and critical appraisal.* BMJ Books, London, 2005.

Peat, Jennifer, Elliott, Elizabeth, Baur, Louise, and Keena, Victoria. *Scientific writing: Easy when you know how.* BMJ Books, London, 2002.

Penrose, Ann M., and Katz, Stephen B. *Writing in the sciences: Exploring conventions of scientific discourse.* 2nd ed. Longman, London, 2004.

Perelman, Leslie C., Paradis, James, and Barrett, Edward. *The Mayfield handbook of technical and scientific writing.* Mayfield, Mountain View, CA, 1997.

Rogers, S. M. *Mastering scientific and medical writing: A self-help guide.* Springer, Berlin, 2006.

Rubens, Philip. *Science and technical writing: A manual of style.* Henry Holt, New York, 1992.

Ryckman, W. G. *What do you mean by that? The art of speaking and writing clearly.* Dow Jones-Irwin, Homewood, IL, 1980.

Scheier, Lawrence M., and Dewey, William L. (Eds.). *The complete writing guide to NIH behavioral science grants.* Oxford University Press, New York, 2007.

Scientific style and format: The CBE manual for authors, editors and publishers. 6th ed. Council of Biology Editors, Bethesda, MD, 1994.

Sternberg, Robert. *The psychologist's companion: A guide to scientific writing for students and researchers.* 4th ed. Cambridge University Press, Cambridge, UK, 2003.

Strunk, W., Jr., and White, E. B. *The elements of style*. 3rd ed. Macmillan, New York, 1979.

Style Manual Committee, Council of Biology Editors. *Scientific style and format: The CBE manual for authors, editors, and publishers*. 6th ed. Cambridge University Press, Cambridge, UK, 1994.

Sullivan, K. D., and Eggleston, Merilee. *The McGraw-Hill desk reference for editors, writers, and proofreaders*. McGraw-Hill, New York, 2006.

Taylor, Robert B. *The clinician's guide to medical writing*. Springer, New York, 2004.

Teitel, Martin. *"Thank you for submitting your proposal": A foundation director reveals what happens next*. Emerson & Church, Medfield, MA, 2006.

Thurman, Susan. *The everything grammar and style book*. Adams Media, Avon, MA, 2012.

University of Chicago Press Staff. *Chicago manual of style*. 16th ed. University of Chicago Press, Chicago, 2010.

Venolia, Jan. *Rewrite right! Your guide to perfectly polished prose*. 2nd ed. Ten Speed Press, Berkeley, CA, 2000.

Venolia, Jan. *Write right! A desktop digest of punctuation, grammar, and style*. 4th ed. Ten Speed Press, Berkeley, CA, 2001.

Wason, Sara D. *Webster's new world grant writing handbook*. Wiley, Hoboken, NJ, 2004.

Williams, Joseph M. *Style: Ten lessons in clarity and grace*. Scott, Foresman, Glenview, IL, 1988.

Williams, Joseph M. *Style: Lessons in clarity and grace*. 9th ed. Longman, London, 2006.

Yang, Jen Tsi. *An outline of scientific writing. For researchers of English as a foreign language*. World Scientific Publishing, Singapore, 1995.

Yang, Otto O. *Guide to effective grant writing: How to write a successful NIH grant application*. Springer, New York, 2005.

Young, Petey. *Writing and presenting in English: The Rosetta Stone of science*. Elsevier Science, Boston, 2006.

Zeiger, Mimi. *Essentials of writing biomedical research papers*. 2nd ed. McGraw-Hill, New York, 2000.

Credits and References

Problem 6-16 Reprinted from *Behaviour Research & Therapy* 44(8), Thilo Deckersbach, Scott Rauch, Ulrike Buhlmann, and Sabine Wilhelm, Habit reversal versus supportive psychotherapy in Tourette's disorder: A randomized controlled trial and predictors of treatment response, 1079–1090, Copyright (2008), with permission from Elsevier.

Example 7-2 Adapted from *Future Microbiol*, 2(6): 2007, 571–574, with permission of Future Medicine Ltd.

Example 8-14 *Corrosion Science* 50(3). 2008: K.E. García, C.A. Barrero, A.L. Morales, and J.M. Greneche, Lost iron and iron converted into rust in steels submitted to dry–wet corrosion process, 763–772. Copyright (2008), with permission from Elsevier.

Problem 8-2 *Mol Gen Genet* 261: 1999, 92–99. A specific member of the Cab multigene family can be efficiently targeted and disrupted in the moss Physcomitrella patens. A. H. Hofmann, A. C. Codón, A. Ivascu, V. E. A. Russo, C. Knight, D. Cove, D. G. Schaefer, M. Chakhparonian and J.-P. Zrÿd. Copyright Springer (1999). With kind permission of Springer Science and Business Media.

Problem 8-3 *Proceedings of the National Academy of Sciences*, 105(8). 2008: 2788–2793. From the cover: Hydrogen and oxygen isotope ratios in human hair are related to geography. Ehleringer, J.R., Bowen, G.J., Chesson, L.A., West, A.G., Podlesak, D.W., Cerling, T.E. Copyright (2008) National Academy of Sciences, U.S.A.

Problem 8-4 *Proceedings of the National Academy of Sciences*, 105(5). Abrupt climate change and collapse of deep-sea ecosystems, 1556–1560. Yasuhara, M., Cronin, T.M., deMenocal, P.B., Okahashi, H., and Linsley, B.K. Copyright (2008) National Academy of Sciences, U.S.A.

Problem 8-5 Used with permission of Geological Society of America, from Global frequency of magnitude 9 earthquakes, McCaffrey, R., *Geology* 36(3), 2008; permission conveyed through Copyright Clearance Center, Inc.

Problem 8-7 Jefferson, T.A., Stacey, P.J., and Baird, R.W. (2008). A review of Killer Whale interactions with other marine mammals: Predation to coexistence. *Mammal Review* 21(4), 151–180.

Example 9-4a and 9-4b With permission from the author, Rudolf Lurz.

Example 9-5 With permission from the author, Roland Geerken.

Example 9-17b Reprinted with permission from Hanna Richter, Organization and transcriptional regulation of the polyphenol oxidase (PPO) multigene family of the moss *Physcomitrella patens* (Hedw.) B.S.G. and functional gene knockout of PpPPO1, PhD Thesis, Universität Hamburg, 2009.

Used with permission of Lyceum Books, Inc. Permission conveyed through Copyright Clearance Center, Inc.

Example 13-10 Reprinted from *Developmental Biology* 156(1), Lopez, A. Miraglia, S.J., and Glabe C.G., "Structure/Function Analysis of the Sea Urchin Sperm Adhesive Protein Bindin," 24–33, Copyright (1993), with permission from Elsevier.

Example 13-22 With permission from the author, Moshe Herzberg.

Example 13-24 Reprinted by permission from Macmillan Publishers Ltd: *Nature* 413. Schluenzen, F., Zarivach, R., Harms, J., Bashan, A., Tocilj, A., Albrecht, R., Yonath, A., and Franceschi, F. Structural basis for the interaction of antibiotics with the peptidyl transferase centre in eubacteria, 814–821, Copyright (2001).

Problem 13-4 Reprinted from *PLoS Comput Biol* 3(4): e59. Yu H, Kim PM, Sprecher E, Trifonov V, Gerstein M. The Importance of Bottlenecks in Protein Networks: Correlation with Gene Essentiality and Expression Dynamics. 2007.

Problem 13-5 Reprinted with permission from *The Journal of Neuroscience*. Online by Paulette A. McRae, Mary M. Rocco, Gail Kelly, Joshua C. Brumberg, and Russel T. Matthews. "Sensory Deprivation Alters Aggrecan and Perineuronal Net Expression in the Mouse Barrel Cortex," 27(20), Copyright 2007 by Society for Neuroscience. Reproduced with permission of Society for Neuroscience in the format Textbook via Copyright Clearance Center.

Problem 13-6 With permission from the author, Daniele Grunow.

Example 14-1 With permission from the author, Neeta Connally.

Example 14-2 Reprinted from *Developmental Biology* 156(1), Lopez, A., Miraglia, S.J., and Glabe, C.G. "Structure/Function Analysis of the Sea Urchin Sperm Adhesive Protein Bindin," 24–33, Copyright (1993), with permission from Elsevier.

Example 14-3 Reprinted from *Current Genetics* 29(5), Backert, S., Lurz, R., and Börner, T. "Electron microscopic investigation of mitochondrial DNA from *Chenopodium album*," 427–436. Copyright (1996), with permission from Springer.

Example 14-4 Reprinted from Fitzpatrick, M.C., Hampson, K., Cleaveland, S., Meyers, L.A., Townsend, J.P., & Galvani, A.P. (2012). Potential for Rabies Control through Dog Vaccination in Wildlife-Rich Communities in Tanzania. *PLoS Neglected Tropical Diseases.* 6(8):e1796.

Example 14-6 Reprinted from *Developmental Biology 156*(1), Angelika Lopez, Sheri J. Miraglia, and Charles G. Glabe, Structure/Function Analysis of the Sea Urchin Sperm Adhesive Protein Bindin, 24–33, Copyright (1993), with permission from Elsevier.

Example 14-9 Reprinted by permission from Macmillan Publishers Ltd: *Nature* 413. Schluenzen, F., Zarivach, R., Harms, J., Bashan, A., Tocilj, A, Albrecht, R., Yonath, A. and Franceschi, F. Structural basis for the interaction of antibiotics with the peptidyl transferase centre in eubacteria, 814–821, Copyright (2001).

Example 14-12 Reprinted from *Analytical Biochemistry*, 230(1), Hofmann, A., Tai, M., Wong, W., and Glabe, C.G. "A Sparse Matrix Screen to Establish

Initial Conditions for Protein Renaturation," 8–15, Copyright (1995), with permission from Elsevier.

Problem 14-2 Reprinted from *Acta Tropica* 89/2 Yaw Asare Afrane, Eveline Klinkenberg, Pay Drechsel, Kofi Owusu-Daaku, Rolf Garms, and Thomas Kruppa. "Does irrigated urban agriculture influence the transmission of malaria in the city of Kumasi, Ghana?" 125–134, Copyright 2004, with permission from Elsevier.

Problem 14-4 Reprinted from *PNAS* 102(50), Held, I.M., Delworth, T.L., Lu, J., Findell, K.L., and Knutson, T.R. "Simulation of Sahel drought in the 20th and 21st centuries," 17891–17896; Copyright (2005) National Academy of Sciences, U.S.A.

Problem 14-5 Reprinted from *PNAS* 98 (20). Kerr, J.T., Southwood, T.R.E. and Cihlar, J. "Remotely sensed habitat diversity predicts butterfly species richness and community similarity in Canada," 11369. Copyright (2001) National Academy of Sciences, U.S.A.

Problem 14-6 With permission from the author, Neeta Connally.

Example 15-2 Reprinted from the *New England Journal of Medicine* 336(19). Luzuriaga, K., Bryson, Y., Krogstad, P., Robinson, J., Stechenberg, B., Lamson, M., Cort, S., and Sullivan, J.L. "Combination treatment with zidovudine, didanosine, and Nevirapine in infants with human immunodeficiency virus type 1 infection," 1343–1349. Copyright © 1997 Massachusetts Medical Society. All rights reserved.

Example 15-3a Reprinted, with permission, from *Applied Adhesion Science* 3:22, 2015. Patrícia A. Saliba, Alexandra A. Mansur, Dagoberto B. Santos, and Herman S. Mansur. Fusion-bonded epoxy composite coatings on chemically functionalized API steel surfaces for potential deep-water petroleum exploration. Copyright (2015), Herman Mansur.

Example 15-3b Ghosh, S., Mukhopadhyay, P., and Isaacs, L. 2010. *Journal of Systems Chemistry* 1:6.

Example 15-4a Reprinted from *Science* 2003, 302 (5644). Schiestl, F.P., Peakall, R., Mant, J.G., Ibarra, F., Schulz, C., Franke, S., and Francke, W. The Chemistry of Sexual Deception in an Orchid-Wasp Pollination System, 437. Reprinted with permission from AAAS.

Example 15-4b Reprinted from *Proc Natl Acad Sci U S A.* 104(13). Zhang, R., Li, G., Fan, J., Wu, D.L., Molina, M.J. Intensification of Pacific storm track linked to Asian pollution, 5295. Copyright (2007) National Academy of Sciences, U.S.A.

Example 15-5 Reprinted from *Proc Natl Acad Sci U S A.* 105(48). Sanjay Basu, Gretchen B. Chapman, and Alison P. Galvani. Integrating epidemiology, psychology, and economics to achieve HPV vaccination targets. 19018–19023. Copyright (2008) National Academy of Sciences, U.S.A.

Example 15-7 With permission from the author, Irene Bosch.

Problem 15-1 Reprinted by permission from Macmillan Publishers Ltd: *Nature* 413. Schluenzen, F., Zarivach, R., Harms, J., Bashan, A., Tocilj, A, Albrecht, R., Yonath, A., and Franceschi, F. Structural basis for the interaction of antibiotics with the peptidyl transferase centre in eubacteria, 814–821, Copyright (2001).

Problem 15-2 Reprinted with permission from *PNAS 106*(3). Martinus E. Huigens, Foteini G. Pashalidou, Ming-Hui Qian, Tibor Bukovinszky, Hans M. Smid, Joop J. A. van Loon, Marcel Dicke, and Nina E. Fatouros. Hitch-hiking parasitic wasp learns to exploit butterfly antiaphrodisiac, 820. Copyright (2009) National Academy of Sciences, U.S.A.

Problem 15-3 With permission from the author, Jimin Wang.

Example 19-5 From *Science 5*, 321. 2008. Daniel Rosenfeld, Ulrike Lohmann, Graciela B. Raga, Colin D. O'Dowd, Markku Kulmala, Sandro Fuzzi, Anni Reissell, and Meinrat O. Andreae. Flood or Drought: How Do Aerosols Affect Precipitation? 1309. Reprinted with permission from AAAS.

Example 19-9 Reprinted, with permission, from the *Annual Review of Ecology, Evolution and Systematic*, Volume 38. Aronson, R.B., Thatje, S., Clarke, A., Peck, L.S., Blake, D.B., Wilga, C.D., and Seibel, B.A. Climate Change and Invasibility of the Antarctic Benthos, 129–154. Copyright ©2007 by Annual Reviews www.annualreviews.org

Example 19-10 Reprinted from *Trends in Plant Science 2*(12), Backert, S., Nietsen, B.L., and Boerner, T. The mystery of the rings: structure and replication of mitochondrial genomes from higher plants, 477. Copyright (1997), with permission from Elsevier.

Example 19-11 Reprinted from *Science 5*, 321. 2008. Daniel Rosenfeld, Ulrike Lohmann, Graciela B. Raga, Colin D. O'Dowd, Markku Kulmala, Sandro Fuzzi, Anni Reissell, and Meinrat O. Andreae. Flood or Drought: How Do Aerosols Affect Precipitation? 1309. Reprinted with permission from AAAS.

Example 19-12 Reprinted, with permission, from the *Annual Review of Anthropology*, Volume 41. Haselgrove, C. and Krmnicek, S. The Archeology of Money, 235–250. Copyright ©2012 by Annual Reviews www.annualreviews .org. Permission conveyed through Copyright Clearance Center, Inc.

Example 19-13 Reprinted, with permission, from the *Annual Review of Ecology, Evolution and Systematic*, Volume 38. Aronson, R.B., Thatje, S., Clarke, A., Peck, L.S., Blake, D.B., Wilga, C.D., and Seibel, B.A. Climate Change and Invasibility of the Antarctic Benthos, 129–154. Copyright ©2007 by Annual Reviews www.annualreviews.org.

Problem 19-1 Reprinted from *Trends in Plant Science 2*(12), Backert, S., Nietsen, B.L., and Boerner, T. The mystery of the rings: Structure and replication of mitochondrial genomes from higher plants, 477. Copyright (1997), with permission from Elsevier.

Example 21-3 With permission from Alison Galvani, Yale University.

Problem 21-2 With permission from Mark Bradford, Yale University.

Example 22-4a, 22-4b With permission from Patty Lee, Yale University.

Example 22-5 With permission from Jun Korenaga, Yale University.

Example 22-6 With permission from Michael Robek, Yale University.

Example 22-13 With permission from Hong Tang, Yale University.

Problem 22-5 With permission from Jun Korenaga, Yale University.

Example 23-3 With permission from Richard Phillips, Indiana University.

Example 23-8 With permission from Hong Tang, Yale University.

Example 23-10 With permission from Michael Robek, Yale University.

Example 23-13 With permission from Daniel Goldstein, Yale University.

Problem 23-2 With permission from Patty Lee, Yale University.

Problem 23-3 With permission from Michael Robek, Yale University.

Problem 23-4 With permission from Alexey Federov, Yale University.

Example 24-3 With permission from Alison Galvani, Yale University.

Example 24-4 With permission from Daniel Goldstein, Yale University.

Example 25-3 With permission from Michael Robek, Yale University.

Example 25-10 With permission from Jun Korenaga, Yale University.

Example 25-11b With permission from Michael Robek, Yale University.

Problem 25-1 With permission from Michael Robek, Yale University.

Problem 25-2 With permission from Richard Phillips, Indiana University.

Problem 25-3 With permission from Alexey Federov, Yale University.

Example 26-2 With permission from Mark Bradford, Yale University.

Example 26-3 With permission from Michael Robek, Yale University.

Example 26-4 With permission from Michael Robek, Yale University.

Problem 26-1 With permission from Alison Galvani, Yale University.

Example 29-1 With permission from Diana Chu et al., UCSF.

Example 29-3 With permission from Roland Geerken, GTZ.

Example 29-6 With permission from Roland Geerken, GTZ.

Example 29-8 With permission from Roland Geerken, GTZ.

Example 29-13 With permission from Betty Liu, Yale University and currently Keck Foundation.

Example 29-14 With permission from Roland Geerken, GTZ.

Figure 29-4 With permission from Alexey Federov, Yale University, and Jaclyn Brown, Centre for Australian Weather and Climate Research (CAWCR).

Figure 29-5 With permission from Philip Duffy, Yale University.

Figure 29-6 With permission from Pedro Walfir M. Souza Fihlo, Vale Institute of Technology.

Example 30-6 With permission from Patty Lee, Yale University, and Robert Homer, Yale University.

Example 30-7 With permission from Mark Bradford, Yale University.

Example 30-8 With permission from Betty Liu, Yale University.

Example 30-9 Reprinted from *The Neuroscientist* 7(5), Tadzia Grandpre, Stephen M. Strittmatter. Nogo: A Molecular Determinant of Axonal Growth and Regeneration, 10. Copyright (2001) by SAGE Publications. Reprinted by Permission of SAGE Publications.

Figure 30-2 Figure reprinted from Mills, H.R. (1977). *Techniques of Technical Training,* 3rd ed. Macmillan, London; with permission of MacMillan Press.

Problem 30-3 With permission from Jaclyn Brown, Centre for Australian Weather and Climate Research (CAWCR).

Problem 30-4 Table reprinted from *Analytical Biochemistry 230*(1), A. Hofmann, M. Tai, W. Wong, and C.G. Glabe. A Sparse Matrix Screen to Establish Initial Conditions for Protein Renaturation, 8, Copyright (1995), with permission from Elsevier.

Problem 30-5 Figure reprinted from *Developmental Biology* 156(1), Lopez, A., Miraglia, S.J., and Glabe, C.G. Structure/Function Analysis of the Sea Urchin Sperm Adhesive Protein Bindin, p. 10, Copyright (1993), with permission from Elsevier.

Example 31-6 With permission from Betty Liu, Yale University and currently Keck Foundation.

Example 31-8 With permission from Mark Bradford, Yale University.

REFERENCES FOR EXAMPLES AND PROBLEMS

Problem 5-6, exercise 8 O'Connell, J. F. et al. 1999. *Journal of Human Evolution* 36(5), 461–485; Wood, B. 2000. Shaking the Tree: Readings from Nature in the History of Life, 371.

Problem 5-7a Roberts, L. 2013 *Science* 342 (6159), 679–680.

Problem 5-7b Mahowald, N. et al. 1999. *Journal of Geophysical Research* 104(d13), 15895–15916; Twohy, C. H. et al. 2005. *Atmos. Chem. Phys.* 5, 2289–2297.

Example 6-1 Witham, C. S. 2005. *Journal of Volcanology and Geothermal Research* 141(3–4), 299–326.

Example 6-5 Heukelbach, J. and Feldmeier, H. 2008. *The Lancet infectious diseases*, 8(5), 302–309.

Example 6-19 Jones, B. H. et al. 1996. *Am. J. Physiol.* 270, E192–E196.

Problem 6-1 Whisson, S. C. et al. 2007. *Nature* 450: 115–118.

Problem 6-3 Stabell, O. B. 1984. *Biological Reviews* 59(3), 333–388.

Problem 6-6 Roth, L. et al. 2014. *Science* 343 (6167) 171–174; Griffith, C. A. et al. 2003. *Science*, 300(5619), 628–630; Iess, L. et al. 2014. *Science* 344, 78.

Problem 6-9 Glynn, P. W. 1996. *Global Change Biology* 2(6), 495–509.

Problem 6-13 Moore, C. G. and Mitchell, C. J. 1997. *Emerging infectious diseases* 3, 329–334; Rezza, G. et al. 2007. *The Lancet* 370(9602), 1840–1846.

Problem 6-14 Styer, D. F. 2000. *American Journal of Physics* 68 (12), 1090–1096; Larsen, T. 2003. *Optics Express* 11(20), 2589–2596.

Example 7-1 Dietrich, A. et al. 2015. *Eur Child Adolesc Psychiatry* 24(2):141–51.

Problem 7-1 Haug, G. H. et al. 1998. *Paleoceanography,* 13:427; Poore, R. Z. and Dowsett, H. J. 2001. *Geology* 29(1), 71–74.

Problem 7-2 de Klerk, A. 2009. *Advances in Foscher-Tropsch Synthesis, Catalysts, and Catalysis.* CRC Press.

Problem 7-3 Osimitz, T. G. and Grothaus, R. H. 1995. *Journal of the American Mosquito Control Association* 11(2 Pt 2), 274–278; Osimitz, T. G. and Murphy, J. V. 1997. *Clinical Toxicology*, 35(5), 435–441; Sudakin, D. L. and Trevathan, W. R. 2003. *Clinical Toxicology*, 41(6), 831–839; Revay, E. E. et al. 2013. *Acta tropica*, 125(2), 226–230.

Example 8-13 Albert, M. S. 1997. *Philos. Trans. R. Soc. Lond. B: Biol. Sci.* 352, 1703–1709; Bimonte, H. A. et al. 2003. *Neurobiol. Aging* 24, 37–48.

Problem 8-6 WHO 2014. Monthly report March 2014. http://www.who.int/influenza/human_animal_interface/Influenza_Summary_IRA_HA_interface_24March14.pdf?ua=1 (Accessed April 2014); Herfst, S. et al.

2012. *Science* 336 (6088), 1534–1541; CDC, 2014. http://www.cdc.gov/flu/avianflu/h5n1-virus.htm (accessed April 2014).

Example 11-1A Chan, D. K. and Hudspeth, A. J. 2005. *Nature Neuroscience* 8 (2), 149–155; Brownell, W. E. et al. 1985. *Science* 227 (4683), 194–196; Santos-Sacchi, J. et al. 2006. *Journal of Neuroscience* 26 (15), 3992–3998.

Example 11-1B Müller, U. 2008. *Current Opinion in Cell Biology* 20 (5), 557–566; Chan, D. K. and Hudspeth, A. J. 2005. *Nature Neuroscience* 8 (2), 149–155; Wartzog, D. and Ketten, D. R. (1999). "Marine Mammal Sensory Systems." In J. Reynolds and S. Rommel. *Biology of Marine Mammals.* Smithsonian Institution Press. 132; Brownell, W. E. et al. 1985. *Science* 227 (4683), 194–196; Santos-Sacchi, J. et al. 2006. *Journal of Neuroscience* 26 (15), 3992–3998.

Example 11-9 Vialard, J., Menkes, C., Anderson, D. L. T., and Alonso Balmaseda, M. 2003. *J. Phys. Oceanogr.* 33, 105–121.

Example 11-10 de Jong, P. T. 2006. *N Engl J Med.* 355 (14), 1474–1485; Seddon, J. M. et al. 2006. *American Journal of Ophthamology* 141(1), 201–203; Axer-Siegel, R. et al. 2004. *American Journal of Ophthamology* 137(1), 84–89; Thambyrajah, J. and Townend, J. N. 2000 *European Heart Journal* 21, 967–974; Seddon, J. M. et al. 2006. *Arch Ophthalmol* 124, 995–1001; Snow, K. K. and Seddon, J. M. 1999. *Ophthalmic Epidemiol.* 6(2),125–143.

Problem 11-1 McGraw, K. J. and Hardy, L. S. 2006. *J. Field Ornithol* 77(1), 29–33.

Problem 11-2 Kvenvolden, K. 1995. *Organic Geochemistry* 23 (11–12), 997–1008; Hoffmann, R. 2006. *American Scientist* 94 (1), 16–18.

Example 13-8 Moss-Racusin, C.A. et al. 2012. *PNAS* 109 (41) 16474–16479.

Example 13-14 Riffell, J. A. et al. 2008. *PNAS* 105(9), 3404–3409.

Example 14-7 Schulz, P. A. et al. 2001. *Plant Physiology* 126(3), 1224–1231.

Revised Example 15-6 Iverson, C. and Forsythe, S. (2004). *Food Microbiology* 21(6), 771–777.

Problem 22-3 Deckersbach, T. et al. 2006. *Behaviour Research and Therapy* 44(8), 1079–1090.

Problem 23-1 Albert, M. S. 1997. *Philos. Trans. R. Soc. Lond. B: Biol. Sci.* 352, 1703–1709; Bimonte, H. A. et al. 2003. *Neurobiol. Aging* 24, 37–48.

Example 25-9 Fisher, P. G. and Buffler, B. A. A. 2005. *JAMA* 293, 615; Scott, C. B. et al. 1998. *Int J Radiat Oncol Biol Phys* 40 (51); Sathornsumetee, S. and Rich, J. N. 2006. *Expert Rev Anticancer Ther* 6 (1087); Clarke, M. F. 2004. *Nature* 432 (281); Fan, X. et al. 2007. *Seminars in Cancer Biology* 17 (214); S. Bao et al., *Nature* 444 (756); Dean, M. et al. 2005. *Nat Rev Cancer* 5 (275); Kreuter, J. 2001. *Adv Drug Deliv Rev* 47(65).

Examples 25-14 and 25-15 Riffell, J. A. et al. 2008. *PNAS* 105(9), 3404–3409.

Example 29-10 Basso, D. M. et al. 1996. *Exp Neurol* 139, 244–256; Joshi, M. and Fehlings, M. G. 2002. *J Neurotrauma* 19, 175–190.

Example 31-7 Koren, I. et al. 2004. *Science* 303(5662), 1342–1345.

Index

abbreviations
  avoiding 24–25, 330, 341
  Latin-derived 25
  of journals 165t
  punctuation 72–73
  special 25
Abstract. *See also* conference abstracts
Abstract, for LOI 427–428, 438
Abstract, for posters 567–568
Abstract, for research paper
  abbreviations in, 330, 335, 341
  common problems 337–341
  components 330
  condensing 339–340
  conformity 341
  data in 341
  for descriptive papers 334–335
  extraneous information 330
  format 30–335
  graphical 330, 333, 333f 334f
  indicative 330, 340, 392–393
  informative 330–332
  length 339–340
  overview 329
  problems for 343–345
  question/purpose in 331–332
  rejection 341
  revision checklist 341–342
  signaling of information in 336–337, 336t
  structured 330, 332–333
  summary 343
  tense 335
  types of abstracts 340
  voice 335
  word choice 335
Abstract, for research statements 633

Abstract, for review articles 392–393, 405–406
Abstract and specific aims, for proposal
  broader impact statement 450
  common problems 451–453
  components 441
  condensing 452
  first sentences 441–442
  intellectual merit statement 450
  interdependent aims 453
  lay abstracts 443–448
  length 452
  objective in 442, 443f
  overview 440
  problems 455–458
  reasons for rejection 453
  revision checklist 453–455
  signals 451, 451t
  significance and impact in 449
  specific aims 448–449, 453
  summary 455
  technical abstract 443–448
  tense 450
  timeline of research 443f
  writing, applying basic rules 450–451
abstract nouns 55, 58, 84–85
acceptance of paper 375, 380
  with revision 375–376
acknowledgments 352
active verbs 54–59
adjectives and adverbs 19, 88–89, 93–94
algorithms 204, 207–208
American English 66, 67t 72
American Psychological Association (APA) style 165–167
analysis of variance (ANOVA), 220, 221f, 222

Analyze it 227
APA. *See* American Psychological
    Association
Appendix 264–266
approach and research design section
    alternative strategies in 503–506
    clarity and interpretation 511–512
    closing paragraph 506–508
    common problems 509–512
    components 500–501
    detail, level of 510–511
    feasibility 509–510
    format 501–506
    interrelated aims 510
    length 500
    level of certainty in 506
    level of writing 500
    overview 499
    problems 514–515
    revision checklist 512–513
    signals 508, 508*t*
    subsections 501
    summary in 507–508
    summary 513
    tense 500–501
    tone 500–501
articles, use 84–85
audience
    attention span 597*f*
    introduction and 233
    journal choice and 127–128
    oral presentations 584, 597–598
authorship 128–130

background, in research
        paper 234–236, 249
background and significance section,
    for proposals
    amount of detail 470
    coherence and cohesion 470
    elements of 461–466, 473–475
    emphasis 459–460
    format 459–460
    funnel structure for 461–463
    innovation, difference to 471, 481
    length 459–460

for NIH programs 460
    objectivity 470
    organization 470
    overview 459
    problem, difference to 470–471,
        473–475
    references 460–461
    revision checklist 471–472
    sample sections 466–469
    signals 469, 469*t*
    significance and impact
        statement 464–466
    summary in 466
    summary 472–473
    unknown/problem in 463–464
bar graphs 180f 187, 188*f* 189, 191, 198,
        225*f* 226, 593–595
Basic Precept, see also central principle
        14–15
binomial distribution curve 217–218
Binomial nomenclature 25
biographical sketch 523–524
BIOSIS 148
BioOne 149
block diagram 185–186, 186*f*
body actions and motions
    eye contact 604–605
    facing audience 605
    gestures 606
    in oral presentations 604–606
    presenter's triangle and 605, 605*f*
box plot 188–189, 189*f*
British English 66, 67*t* 72–73
budget
    budget narrative/justification 520–521
    components 518
    direct and indirect costs 517–518
    funding sources 521
    investments of organization 521–522
    justifying costs 518–521
    in LOI 432–433
    preparing 517
    sample 519*t* 520

calculate 269
capitalization 67

Central Principle, see also Basic Precept
    14–15
checklists
    abstract and specific aims section
        revision 453–455
    abstract revision 341–342
    approach and research design section
        revision 512–513
    background and significance section
        revision 471–472
    discussions revision 322–323
    innovation revision 481–482
    introduction revision 254–255
    job applications 654–655
    LOI revision 436–437
    materials and methods section
        revision 272–273
    oral presentations 610–611
    posters 581
    preliminary results section
        revision 495–496
    for pre-submission peer
        review 363–365
    results section revision 297–299
    review article revision 402–403
    statistical analysis 229
    titles revision 354
Chicago Manual of Style, 66
chi-square test, 221, 222*f* 223
Chicago style, references 168
CINAHL. *See* Cumulative Index to
    Nursing and Allied Health
    Literature 148
citations
    internet 167
    placement 153
    text 151
    tone and style 152
    wording 153*t*
Cite Seer 149
clarity and conciseness
    in approach and research design
        section 511–512
    in LOI 424–425
    in résumés 626–627
    in revision of proposals 540

of titles 347–348
coherence and cohesion
    in background and significance
        section 470
    continuity and 108–109
    in introduction 242
    key terms and 107–108, 123
    in paragraphs 103–110
    in review articles 401
    transitions for 108–111, 118–20, 121
    word location and 103–106
collaborative arrangements 524–525
commas 70–72
comparisons. *See also* lists and
    comparisons
    ambiguous 63
    in discussion 309
    faulty 63–65
    incomplete 63–65
    overview 61–63, 79–80
    in review articles 387
common errors 65*f*
competing interest statement 525
complex sentences 21, 54
composition
    from data and ideas 135–139
    manuscript from outline 132–139
conclusions
    in approach and research design
        section 506–508
    in discussions 313–315, 317, 325–326
    introduction stating 239
    from other studies 294–295
    posters 571–572
    preliminary results section 493
    results section avoiding 294–295
    review articles 399–401, 407
condensing
    abstract 339
    abstract and specific aims section 452
    establishing importance 112
    excessive detail 115
    figure legends 116
    first draft 360
    intensifiers and hedges limited 115
    negative writing avoided 114

overview words, phrases,
   sentences 113–114
   paragraphs 110–16, 122–123
   titles 349
conference abstracts 330, 560–562,
   567, 584
confidence interval 218, 284
conflict of interest statement 353
consistent point of view, 102–103, 118,
   121–122
continuation or noncompeting
   proposals 412
continuity 106, 108, 316
correlation analysis 221, 222f 224
Council of Scientific Editors (CSE)
   style 165, 167, 168–169
cover letter, manuscript 367–370
   content 368
   Instructions to Authors on 369
   problems 382
   samples 368–370
   well-prepared 367–370
cover letter, proposal
   LOI 434
   overview 544–545
   samples 544–545
cover letter, job application
   overview 621
   samples 631–633
CSE. *See* Council of Scientific Editors
Cumulative Index to Nursing and Allied
   Health Literature (CINAHL), 148
Current Contents 148
curricula vitae (CVs)
   components 622
   for job applications 621–625
   qualifications highlighted 621–622
   sample 623–625
CVs. *See* curricula vitae

data
   in abstract 341
   composition from 135–39
   interpretation in results
     section 282–283
   in preliminary results

section 486–487
sharing 527
statistical analysis 219–221
statistical information with 283–285
trends 214–15
decision tree 222, 223f
   alternative strategies as 503–506
   websites 226–227
Declaration of Helsinki 5
dependent clauses 39–40
dependent variable 189, 202
descriptive papers
   abstract for 334–335
   introduction for 239–241
   results section in 288, 296–297
   tense in, 290
deviation 216–219
dictionaries 26–30
   apps 30
   biological and medical sciences 27–29
   Dictionary of Contemporary
     English 66
   ESL 30
   online 29–30
   other scientific fields 28–29
dictionaries, list of scientific 26–30
direct and indirect costs 517–518
discovery statement 241
discrepancies 311–312, 317t
Discussion, for research paper
   answer to question in 305,
     306–309, 313
   common problems 318–319
   comparisons with published
     results 305–6, 317tt
   components 305–306
   conclusions 313–314, 325–326
   confidence and authority
     conveyed 314
   continuity 316
   controversial topics 308
   discrepancies, unexpected findings,
     limitations 311–312, 317t
   first paragraph 306–308, 324–325,
     326–327
   first person in 316

Discussion (*continued*)
  format 305–306
  generalization in 305, 312–313
  hypotheses in 312–313
  interpretation in 306–307, 318
  irrelevant information 318–319
  key findings and answers to research
    question 305
  last paragraph 313–315, 326–337
  level of certainty 313–314, 314*f*
  middle paragraphs 309–313
  organization 309
  overview 304
  problems for 324–328
  pyramid structure
    for 305–306, 306*f*
  question/purpose in 307–309
  results combined with 316
  revision checklist 322–323
  sample discussion 319–321
  signals 316, 317*t*
  significance stated 318
  summary 305
  targeting readers 305
  tense 316
  voice in 316
  writing guidelines 315–316
dissemination plan/data sharing 527
distribution curves
  binomial distribution 217
  normal and standardized normal
    distribution 218–219, 219f
  Poisson distribution 217–218, 217*f*
  in statistical analysis 217–219
drafting
  authorship 128–30
  ESL authors 132, 140–141
  first draft 131–132
  IMRAD format 131
  momentum 132
  outlining and composing
    manuscript 133–139
  outside help 141–143
  prewriting 126–128
  problems 143–145
  references and 132–133

writer's block 139–140
  writing process and 125–126

education and outreach 527–529
electronic submission 367, 544
emphasis
  in background and significance
    section 459–460
  commas for 70–71
  competition for 37–39
  figures for 178–179
  in oral presentations 600
  of results 290
  word location and 36–37, 39
EndNote 401
endnotes, see also footnotes 350, 352
endorsement letters 529
English as a second language (ESL)
  authors 17, 43–44
  dictionaries 30
  drafting 140–141
  eliminating language barriers 333
Environment Complete 149
equal variance 221
equations 204–209
ESL. See English as a second language
ESL, grammar problems 82–95
  adjectives and adverbs 19, 88–89,
    93–94
  articles 84–85
  nouns and pronouns 89–90
  prepositions 82–84
  problems 91–95
  references 90–91
  verbs 85–89, 92–93
establishing importance 112–113
ethical conduct in science 5
  guidelines 270t
  for human experimentation 269–270
  for laboratory animals 270–271
  materials and methods section
    and 269–271
  phrasing 269–271
  science communication and 5
evaluation plan 529–530
executive summary 530

expected outcomes 531
experimental approach
    details 293–94
    in research paper 234, 239
    in LOI 430–431

facilities, description of 525–526
family foundations 415–416
feasibility
    in approach and research design
        section 509–510
    in innovation section 481
figure or table 178
figure legends 116, 199–200
    general advice 180
    general guidelines 176–78
    graphs 186–198
    lines and curves, 192
    log graphs 193
    misleading readers 182
    posters 181*t* 573–576
    preparing 181–182
    in print 180
    problems 210–213
    production 181
    results pointing to 277–78
    readability 190–191
    on slides 181t 594
    titles 199
    for trends, relationships, and
        emphasis 80–81
    types of, 182–189
figures 176f
    axis labels and scales 192
    formatting and placement 177, 189
final version, manuscript 366
findings
    key 305
    of others 146, 175
    unexpected 311–312, 317*t*
first draft, general 125, 131–132
first draft, research papers
    condensing 360
    content and content location 357–360
    organization and flow 359
    outline 359

proofreading 360
revision 357–360
style in 359–360
transitions in 359–360
first draft, proposal 421–422
first person 49–50, 76
    in discussions 316
5x5 rule 588–590, 594, 612
flow
    in first draft 363
    oral presentation information 585*f*
    of paragraph 99, 102–106
    of technical sentences 40–44, 45–46
    word location and 40–41, 46–47, 103
fonts
    posters 566–567
    in visual aids 590
footnotes 350, 352
formatting, conference
        abstracts 330, 560
formatting, figures and tables 177–78,
        198, 200–204
    formulas 204–209
    graphs 189–191
formatting, letters of
        recommendation 652
formatting, oral presentations 586–587
formatting, posters 564–567
formatting, proposals
    approach and research design
        section 501–506
    background and significance
        section 459–460
    innovation section 470–480
    LOI 425–427
    overview 417o
    preliminary results section 487–492
formatting, research papers
    Abstract 330–335
    Discussion section 305–306
    IMRAD format 131
    Introduction 234–235
    Materials and Methods
        section 265–266
    Results section 285–288, 299–300
formatting, review articles 388–391

formulas 204–209
foundations
    family 416–417
    private 414–415
    relevance to 537
    reviews 550
funders
    in budget 522
    goals and priorities 410–411
    importance of funding
        section 532–533
    interacting with 422–423
    list 412
    locating 410
    for proposals 410–411, 419–420, 554
    questions of 411
    researching 419–420
    resources 416–417
funnel structure
    in background and significance
        section 461–463
    in introductions, research
        paper 234–235, 235f 236
    in review articles 390–391
    in proposals 461–463
future tense 51, 500–501

Gaussian distribution 219
gerunds 88f
GetCITED 149
goals
    of funders 410–411
    institutional 638
    in LOI 430–431
    in research statements 623
    in résumés 626–627
Google Scholar 149
grammar. See also ESL grammar
    errors 43, 65f 82f 361–362
    ESL problems 82f
    modifiers 74–75
    in oral presentations 604
    references 90–91
    subject-verb correspondence 73–74
    technical style and 48–49
grant writing, skills 410, 422

GraphPad Prism 227
graphical abstract 330, 333, 333f 334f
graphs
    axis labels and scales 192,
        192f 195f 196f
    bar 187, 188f 198f 224f 595
    box plot 188–189, 189f
    as figures 186–198
    formatting 189–191
    line 186, 190f 191f 224f
    log-graphs 193, 193f
    pie chart 188, 188f
    problems 212–213
    readability 190–191
    samples 193–199
    scatter plot 187, 187f
    on slides 593
    statistical analysis
        representation 225f
    x-axis and y-axis, 189–190, 194f 197f

hedges 96, 115–117
hiring process 643–648
hyphens
    avoiding 72
    in noun clusters 60
hypothesis, null 219

impact statement
    for abstract and specific aims
        section 442, 443f
    for background and significance
        section 464–466
    in LOI 433, 439
    proposals, special sections 531–532
impromptu talks 608–609
IMRAD format 131
independent sentences 53, 69–70
independent variable 189, 200, 202
indicative abstract 330, 340, 392–393
infinitives 88f
informative abstract 330
innovation section
    common problems 480
    components 477–478
    emphasis 477

feasibility and credibility 481
format 470–480
length 477, 478
overview 476–477
revision checklist 481–482
sample 479–480
signals 480
significance, difference to 481
summary in 479–480
summary 483
instructions to authors 128
for reference list 163, 167
intensifiers and hedges 115
interdependent aims 453
Internet
DOI 169
online sources 209
references 167–169
interpretation, in citations 155–156
in approach and research design
section 511–512
in discussions 306–307, 318
in preliminary results
section 486–487
in results section 282–283
interview
preparing for 644
questions you may
receive 645–648, 649
questions to ask 648–649
Introduction, Methods, Results, and
Discussion (IMRAD), 131, 388
Introduction for LOIs and
preproposals 428–429
Introduction, for posters 568–569
Introduction, for research paper
background information in 234–235
cohesion and coherence 242
common problems 244–249
components and format 234–235
of descriptive paper 240–241
elements 236–240
experimental approach in 239
funnel structure 234, 234f
length 235, 246–249
overview 233

overview sentences and, 250
problems for 253–261
question/purpose in 238–239
results and conclusions in 239
revision checklist 254–255
sample section 250–253
signals 243t
significance and implication in 240
summary 249
tense 242–243
unknown/problem in 237–238
writing guidelines 241–242
Introduction, for review
article 394–396, 405
irregular verbs 86–87
italics 69

jargon 21–23, 33–34, 268, 330, 335
job applications
accompanying documents 633
checklist 654–655
components and format 620–621
cover letters 631–633
CVs 621–625
highlighting main
qualities 620–621
hiring process 643–644
interview, preparing for 644–645
interview questions to
ask 648–649
interview questions
to expect 645–648, 649
letters of recommendation 650–654
overview 619–620
research statements 634–639
resources 649–650
résumés 626–630
sample cover letter 620–22
sample CV 623–625
sample résumé 628–630
summary 655–656
teaching portfolio 643
teaching statements 639–643
journal abbreviations 165t
journal choice 128
JSTOR 149

key terms
  coherence and cohesion
    and 107–108, 119
  for continuity 108
key words 351–352

lay abstract 443–448
layout
  oral presentations 585–586
  posters 564
leadership 431–432
lectern 601
legalese 8
length
  abstract 339–340
  abstract and specific aims section 452
  background and significance
    section 459–460
  introductions 235, 246–249
  letters of recommendation 650
  oral presentations 585–586
  proposals 413
  research statements 634
  technical sentences 53–55, 77–79
  titles 349–350, 355, 391
letter. *See also* cover letters
letter from the editor 374–376
letter of inquiry (LOI)
  abstract 427–428, 438
  budget 432–433
  clarity and conciseness 424–425
  components and format 425–427
  cover letter 434
  detail, in approach section 430–431
  experimental approach 430–431
  impact statement 433, 439
  introduction and
    background 428–429
  leadership and organization
    in 431–432
  level of writing 425–427
  objective and aims 429–430
  outline 435
  overview 424–425
  to private foundations 414–415
  problems 438–439

  revision checklist 436–437
  statement of need 429
  statistics 429
  strategy and goals 430–431
  significance 433
  summary 437
letters of recommendation 650–652
  content 651
  format 652–653
  importance of first sentence 652
  length 652
  sample 653–654
  writing 651–652
level of certainty
  in approach and research design
    section 506
  in discussion 313–314, 314f
limitations 311–312, 317t
line graphs 188, 192f 226f
lists and comparisons
  coordination of ideas 63
  parallel ideas 61–63, 79–80
  in technical sentences 61–63, 80
log-graphs 195
LOI. See letter of inquiry

main clauses 39–40
Materials and Methods, for posters 569
Materials and Methods, for research
    paper
  appendix 264–266
  common problems 271
  components 261–265
  ethical conduct 269–2,1 270t
  format 265–266
  overview 260
  problems for 274–279
  revision checklist 272–273
  references 261–262
  sample section 271–272
  signals 266
  statistical analysis in 264
  subsections 265–266
  summary 273
  technical details 262–264
  tense 268

voice 266–268
    word choice 268–269
mean 215
media literate 6
    peer-reviewed vs nonpeer-reviewed 6
median 215
MEDLINE 148
meta-analyses review articles 386
methods. *See* materials and methods
        section
misplacement of information 105
MLA style 165–166, 168
modifiers 74–75

National Aeronautics and Space
        Administration (NASA), 412
National Institutes of Health (NIH), 5,
        412, 413*t* 448
    background and significance section
        for programs 460
    research plan for 476
    resubmission to 553
    review 545–549
National Library of
        Medicine (NLM), 165
National Science Foundation (NSF),
        412, 413*t* 449
    format 476–477
    resubmission to 553
    review 549–550
    subsections 528
negotiating agreement 554
nervousness 599–600
NIH. *See* National Institutes of Health
NLM. *See* National Library of Medicine
nomenclature and terminology
    common 25
    scientific 26
nominalization 55
nonparametric tests 221
normal distribution curve 218–219,
        219*f*
notes. *See also* outlining
    endnotes and footnotes 350, 352
    for oral presentations 599
nouns

abstract 55, 58, 84–85
    capitalizing proper 68
    ESL grammar 90
    noun clusters 49, 59–60
    singular and plural 74*t*
NSF. *See* National Science Foundation
null hypothesis 219
numbers *versus* numerals 66

objective
    in abstract and specific aims
        section 442, 443*f*
    LOI 429–430
objects, verb placement and 43–44
oral presentations
    5x5 rule for text slides 588–589
    3x5 rule for tables 594
    attention span of audience 597*f*
    audience 584, 597–598
    body actions and motions 604–606
    checklist 610–611
    colors on slides 588
    components 574–77
    conference talks and
        abstracts 330, 584
    contrast color 591–592
    core slide 598
    dos and don'ts 602–603
    emphasis in 600
    end of 606
    flow of information 585*f*
    format 586–587
    giving talk 600–601
    graphs and tables 593–594
    grammar and sentence structure 604
    illustrations 587–596
    impromptu talks 608–609
    layout 585–586
    length 585–586
    making an introduction 608–609
    nervousness 599–600
    notes 599
    opening sentence 599–600
    overview 583, 586–587
    planning 584, 596–600
    practicing 584

oral presentations (*continued*)
  preparing 583–584, 577, 596–600
  presenter's triangle 605f
  problems 612–615
  question/purpose 598
  questions and answers
    period 607–608
  resources 610
  self-improvement suggestions 600
  set-up 601
  TED talks and 610
  text slides 588–592
  time limit 596–597
  title 592
  transitions in 603–604
  vocabulary and style 603–604
  voice and delivery 600–603
  word choice 600–601, 603–604, 615
organization, description of 526–527
originality, lack of 341, 453
outline
  checking during revision of
    proposal 540–542
  first draft 125f 352
  full sentence 134–135
  LOI 435
  manuscript, composing 133–139
  order and organization 133–135
  research statements 635–636
  review articles 389–391
outside help, for drafting 142f
overview sentences 250

paragraph
  closing, in approach and research
    design section 506–508
  coherence and cohesion 103–110
  condensing 110–114, 122–123
  consistent point of view, 102–103,
    118, 121–122
  construction 96f
  continuity 106
  first, in discussion 306–308, 324–325,
    326–327
  flow of 99, 102–106
  last, in discussion 313–315, 326–327

middle, in discussion 309–313
middle of 100–101
misplacement of information 105
missing links 106
mixed word locations 41, 104
organization 99–103,
  117–118
problems 117–123
sentence position 99
structure 96–99
topic order 101–102
topic sentence 99–100, 101
parallel ideas 61–63, 79–80
paraphrasing 157–162
  references 401
passive voice 50–51, 76–77, 266
  in materials and methods
    section 264–267
past tense 51–52, 77
  for completed actions 316, 335
  in introduction 241–242
  in materials and methods
    section 268
  for observations 450
  for results in preliminary results
    section 493
  in results section 290
peer-review
  checklist 363–365
  formal 372–373
  informal 371
  vs nonpeer-review 6
peripheral information 292–293
person
  consistent use 102
  first 49–50, 76, 316
  in technical sentences 49–50, 76
personal statement 533–534
personnel 518, 534–535
photographs 182–184,
  183f 185f 573–576
phrases
  to omit 114
  overview 113–114
  samples for references 141–142
  titles as 346

transition 111t
unnecessary 22–23
pie chart 188, 188f
PIs. *See* principle investigators
placement. *See also* word location
  figures and tables 177–78
  misplacement of information 105
  object 43–44
  references 157–158
  subject 41–43
  verb 43–44
plagiarism 155–162
planning
  dissemination plan/data sharing, 527
  evaluation plan 529–30
  NIH research 460, 476–477
  oral presentations 584, 596–600
  posters 560, 572
podium 600
point of view 102–103
  materials and methods
    section 267–268
Poisson distribution curve 217–18

posters
  abstract 567–568
  acknowledgments 572–573
  background and color 565–566
  checklist 581
  components 562–564
  conclusions 571–572
  conference abstracts 330, 560–562
  figures, 181t 573–576
  focus 562–564
  format 564–567
  function 559–560
  graphics and blank space 564–565
  illustrations 573
  introduction 568–569
  layout 564, 567
  materials and methods section 569
  overview 559–560
  photographs 573–576
  presenting 578
  references 572–573
  resources for preparation and

    presentation 576–577
  results section 570–571, 570f
  revising 577–578
  samples 563f 578–580
  software and hardware
    resources 576–577
  summary 581–582
  tables on 181t 573–576
  text fonts 566–567
  title 567
  useful links 577
  visual illustrations 564
Power position 99–100, 112
preliminary results section
  common problems 494
  conclusions 490, 493
  content 485–487
  credibility 484–485
  data and interpretation 486–487
  figures and tables in 485
  format 487–492
  length 487
  level of writing 486
  organization 487–491
  overview 484
  problems 497–498
  revision checklist 495–496
  signals 494, 494t
  subsections 485, 487–494
  summary in 487, 491–492
  summary 496–497
  tense 493
  transitions in 494
  word choice 492–493
  writing rules 482–494
prepositions
  ESL grammar 82–84
  problems 91–92
preproposals. *See* letter of inquiry
presenter's triangle 605f
present perfect tense 51–52
present tense 51–52, 77
  for conclusions, in proposals 493
  in descriptive papers 290
  in proposal abstract and specific aims
    section 450–451

present tense (*continued*)
 for general statements 316
 in introductions, 241–242
 in materials and methods section 268
 for true statements in abstract 335
prewriting 126–128
primary sources, 147
principle investigators (PIs), 546
private foundations 414–415
project management 535–536
pronouns 86
proofreading 360
proposals. *See also specific sections*
 addressing reviewers'
   comments 553–554
 administrators 415, 418–419
 agreement negotiation 554
 choosing a sponsoring agency 412
 common weaknesses 410
 continuation or noncompeting 412
 corporations 415–417
 factors that kill a proposal 551–552t
 family foundations and public
   charities, 416–417
 federal agencies 412–414, 413t
 first page, importance 421
 format 417
 funders 410–411, 419–20, 554
 funding categories, common 415
 grant writing 421–422
 guidelines 417–418
 interacting with funder 414, 422–423
 length 413
 letter of inquiry (LOI), 414–415
 navigating and composing 409–410
 overview 409–410
 persuasion and 539
 policy changes 414
 preliminary steps 417–420
 preparation 522
 preproposal 411
 private foundations 414–415
 reasons for rejections 551–554
 renewal 412
 request for proposal (RFP), 411
 resources 420–421

resubmission 552–554
revision and submission 539–556
solicited 411
success factors 411
summary 555
tips for postdoctoral fellows and
   junior faculty members 420
types of, 411–412
unsolicited 411
verbal 434–435
proposals, special sections
 biographical sketch 523–524
 collaborative arrangements 524–525
 competing interest statement 525
 description of facilities 525–526
 description of organization 526–527
 dissemination plan/data sharing, 527
 education and outreach 527–529
 endorsement letters and letters of
   reference 529
 evaluation plan 529–530
 executive summary 530
 expected outcomes 531
 first page 421
 future directions 531
 impact statement 531–532
 importance of funding 532–533
 overview 522–523
 personal statement 533–534
 personnel 534–535
 project management 535–536
 publications list 533
 recognition statement 536
 references 536–537
 relevance to foundation 537
 timeline 443f 537–528
 title page 538
PsycINFO 148
publications list 533
public charities 416–417
PubMed 148
punctuation
 abbreviations 72–73
 commas 70–71, 72
 errors 69–73, 81
 hyphens 60, 72

periods 69
    quotation marks 72
    semicolons 69–71
    simple 69
p-value 222, 223, 223–224, 284
pyramid structure 305–306, 306*f*

question/purpose
    in abstract 332, 337
    in discussion 307–309
    in introduction 238–239
    in oral presentation 598
    of study 134
questions
    to ask in interview 648–649
    to expect in interview 645, 648–649
    of funders 411
    in oral presentations 607–608
    research 305
    results section answering 285
questions and answers period 607–608
quotation marks 72

R 228
range 216
readers. *See also* signaling of
        information
    discussion targeting 305
    errors bothering 7*f*
    figures misleading 182
    keeping in mind 6–9
    understanding reading 6–7
    word location expectations 36
recognition statement 536
redundancies, see also jargon 21–23
    complex sentences and 21
    problems 33–34
    unnecessary phrases 22–23
    unnecessary words 22
reference list 173
    general conventions 165
    Instructions to Authors on 163
    journal's style for 163–164
ReferenceManager 401
References 146
    acknowledgments 352

background and significance
        section 461–462
    collecting 131
    common styles 165–167
    drafting and 132–133
    footnotes and endnotes 350. 352
    form and order 151–152
    for formulas, equations,
        algorithms 208
    for ideas and findings of others 146
    Internet citing 163–169
    journal abbreviations 165*t*
    keeping track of 162
    list 163–164
    literature for materials and methods
        section 261–262
    managing 150–53, 161
    online sources 167
    paraphrasing 157*f*
    placement 153–54, 162
    posters 572–573
    primary, secondary, tertiary 147
    problems 170–175
    proposals, special sections 536–537
    review articles 401
    role 152, 153*t*
    in scientific paper 162
    selecting 147–149
    statistical analysis 226
    styles, common 165*f*
    verifying 150, 155–157
    text citations 151–154
regression analysis 221, 225
rejection
    factors that kill proposals 551–554
    of proposal abstract and specific aims
        section 453
    of proposals 551–554
    of research paper 376
    of research paper abstract 341
    responding to 376
    try again 552
relevance to foundation statement 537
renewal proposals 412
request for proposal (RFP), 411
Research Gate 149

research paper structure 358t
research statements 634–638
  abstract 634
  achievements, aims, goals in 634
  components 634–636
  institutional goals and 638
  for job applications 634–638
  length 634
  outline 635, 636
  resources 649
  samples 635–638
  tailoring 634
resubmission of manuscript 377–381
  problem 382
  response letter sample 378, 379, 380
resubmission of proposal 552–554
  sample 553–554
Results, for posters 570–571, 570f
Results, for research paper
  common problems 291–295
  components 281
  conclusions, speculations,
    comparisons of other
    studies 294–295
  control results 281
  data and results 282–283
  in descriptive papers 288,
    296–297
  discussion, combined with 316
  emphasis 286
  evaluation 301–303
  experimental details 293
  first paragraph 285
  format and organization 285–289
  format for descriptive papers 288
  in introduction 239
  interpretation in 282
  overview 280
  problems for 299–303
  revision checklist 297–299
  sample section 295–2973
  signals 291, 291t
  statistical information 283–285
  summary 299
  tense 290
  word choice 289–290
  writing guidelines 289–290

results and conclusion, research
  paper 239–240
résumés
  chronological and functional
    types 627–628
  combination type 628
  goals in 626–627
  for job applications 626–630
  resources 630
  samples 628–630
review, of proposals
  broader impact and 549–550
  intellectual merit and 549
  NIH 545–549
  NSF 549–550
  by private foundations and
    corporations 550
  process for final version 370
  understanding process 545
review articles
  abstract 392–393, 405–406
  clarity 387
  coherence and cohesion 401
  common problems 402
  comparing and contrasting
    information 387
  conclusion 399–401, 407
  format 388–391
  funnel structure for 390–391
  Instructions to Authors on 388
  integrative 386
  Introduction 394–396, 405
  iterations 391
  letters to the editor 386, 387
  literature/narrative 386
  main analysis section 396–399,
    407
  meta-analyses 386
  organization 396–399
  outline 389–391
  overview 385–386
  perspectives 386
  problems 405–406
  references 401
  revision checklist 402–403
  signals 401
  source material 387–388

subsections 389–391, 396–398
summary 403–404
systematic 386
titles 391–392, 407
topic sentences 401
topic statement 394
types of reviews 396
review process, for manuscripts 370
review process, for proposals
  NIH review 545–546
  NIH notification 547–549
  NIH review criteria 546–547
  NSF review 549
  NSF review criteria 549–550
  overview 545
review by private foundations and
    corporations 550
reviewing a manuscript 362–365
revision, for LOIs and proposals
  abstract and specific aims section
    checklist 453–455
  approach and research design section
    checklist 512–513
  asking for assistance 540
  background and significance section
    checklist 471–472
  before sending out the proposal, 540
  checking outline 540–542
  clarity and conciseness 540
  cover letter for submission 544–545
  editing and 540
  first page perfect 540
  innovation checklist 481–482
  iterations 543
  LOI checklist 436–437
  overview 539
  placement of information 540
  power positions and 542
  preliminary results section
    checklist 495–496
  stages of revision 542–543
  submission 543–545
revision, for posters 577–578
revision, for research paper
  abstract checklist 341–342
  content and organization 357
  criticism 360, 361, 362

discussions checklist 322–323
of first draft 357–360
flow 359
grammar and errors 361–362
introduction checklist 254–255
proofreading 360
results section checklist 297–299
style 360–362
of subsequent drafts 360–362
summary 365, 381
titles checklist 354
revision, for review article 402–403
RFP. *See* request for proposals
running title 350–351, 354

sample size 284, 341, 453
SAS 228
scatter plot 187, 187*f*
science communication
  communication and ethics 5
  examples 4
science writing 10
scientific communication. *See* science
  communication.
scientific method 3
scientific misconduct 5
scientific names/taxonomy 68
scientific writing
  books on 142
  guidelines 12
  mastering 12
  vs science writing 10
ScienceDirect 149
Scitation 148
Sci Finder 149
SCOPUS 148
secondary information 286
secondary sources 147
sentence
  analysis and revision 58–59
  complex 21, 53
  composition 96f
  establishing importance 37, 112
  independent 53, 69, 71
  length 53
  order, consistent 101–103
  overview 113, 250

sentence (*continued*)
  point of view, consistent 104
  position 99
  stress position 37–39
  technical 47f
  topic position 37–39, 104
  topic sentence 99
  transition 108–110
sexism 20
SigmaPlot 228
signaling of information
  in abstract 336–337, 336t
  in abstract and specific aims
    section 451, 451t
  in approach and research design
    section 508, 508t
  in background and significance
    section 469, 469t
  in discussions 316, 317t
  in innovation section 480
  in introduction 243, 243t
  in materials and methods
    section 266
  in preliminary results section 494,
    494t
  in results section 291, 291t
  in review articles 401
significance. *See also* background and
    significance section
  in approach and research design
    section of a proposal 506–508
significance, research paper 240
  statistical 219, 224
significance level 224f
significant 290, 493
SimFit 228
site visits 550–551
slides
  with bar graph 595
  evaluating 613–614
  explaining 602
  with figures 181t 594
  one idea per slide 588
  oral presentations 586,
    598–599
  with schematic 595

  with tables 181t 594
  with title 596
smoothers 603–604
solicited proposals 411
sources
  funding 521
  material 148–150, 387
  online 227
  primary 147
  secondary 147
  tertiary 388
species names 25–26
specific aims. *See* abstract and specific
    aims section
speculation 294–295
spelling
  American and British 66, 67t 72
  numbers and numerals 66
SPSS Statistics 228
standard deviation 218, 219, 229, 284
standardized normal distribution
    curve 220–221, 221f
Stata 228
statement of need
  in Innovation 478
  in LOI 429
statements. *See also* impact statement;
    research statements
  competing interest 525
  conflict of interest 353
  personal statement 533–534
  recognition 536
STATISTICA 228
statistical analysis
  ANOVA 220f
  basics 214f
  basic terminology 215
  binomial 217f
  checklist 228
  chi-square test 221
  correlation analysis 221
  distribution curve 216f
  graphing 223
  in Materials and Methods 260
  nonparametric test 221
  null hypothesis 219

normal 218
Poisson 217
problems 229–230
statistical significance 219f
Student's t-test 220
regression analysis 221
reporting 221–223
resources 226–228
strategy
alternative 503–506
in LOI 430–431
stress position 37–39
structure. *See also* funnel structure
paragraphs 96–99
pyramid 305–306, 306f
sentence 604
tables 200–201
structured abstract 330, 332–333
style
APA 165–167
Chicago Manual of Style 66
Chicago 168
CSE 165, 166, 168
MLA 165–166, 168
in first draft 359–360
journal 163–164
oral presentations 603–604
for reference list 163–164
references 153–154, 163, 165–168
technical 48–49
subjects
subject-verb correspondence 69–70
true 86
verb placement and 43–44
word location 41–43
submission. *See also* resubmission
submission, manuscript
electronic 367, 544
guidelines and help 367
overview 366
pre-submission 362–363
pre-submission peer review 363–365
submission, proposal
cover letter 544–545
electronic 544
packaging 544

requirements before submitting 540
suffixes 17

tables
column and row headings 202
formatting 200f
formulas, equation 204
general guidelines 176–77
placement 177–78
on posters 181t 570, 573–578
problems 210–213
producing 200
proof, algorithms 204, 207–208
results section pointing to 281–282
simple structure 200–201
sizes 203–204
slides 181t 594
table *versus* figure 178–180
3x5 rule 594
titles 202
teaching statements 638–643
components 638–639
resources 650
samples 639–643
technical abstracts 443–448
technical details 262–264
technical style 48–49
TED talks 610
tense. *See also* past tense; present tense
future 51, 53, 500–501
in introduction 241–242
in materials and methods section
268
mixed 52
in preliminary results section 493
present perfect 52
in technical sentences 51–53, 77
tone and 500–501
terminology. *See also* vocabulary and
nomenclature and terminology
tertiary sources 388
text citations
Instructions to Authors on 151, 163,
167, 305
for references 151–153
timeline 537–538

title pages 350–352, 538
titles
  capitalization in 71f
  clarity of 347–348
  completeness of 348, 353
  condensing 349
  figures 199
  length 3495–350, 353, 391
  key words 351
  main topic 346
  overview 346
  posters 567
  problems for 355–356
  review articles 391
  revision checklist 354
  running 350–352, 356
  slides with 596
  strong 347–350
  summary 354
  tables 202
  uniqueness 347
  for visual aids 592–593
tone, and tense in proposals 500–501
topic position 37, 39–41
topic sentences 99–100, 104
topic statement 394
transitions
  for coherence and cohesion 103–105,
    118–120, 122
  in first draft 363
  for logical relationships between
    sentences 108–110
  in oral presentations 603–604
  in preliminary results section 494
  words 110
  words, phrases, sentences 111t
t-tests 220f

unknown/problem, in research
  paper 234, 237–238
unknown/problem, in proposals 463–
  464, 485
unsolicited proposals 411–412

variance 218, 221, 222, 223f
verbal proposals 434–435

verbs
  action 55–59
  analysis and revision 58–59
  buried 55
  endings 86
  in ESL grammar 85–89, 92–93
  gerund and infinitive 88
  irregular 86–87
  location within sentence 43–44
  to omit 113
  plural vs singular 85–86
  subject verb correspondence 73–74
  tense 51–53
  weak 56–57
  word choice 18
  word location 44–45
verbal proposals 434–435
visual aids
  attractive 588
  balance figures and tables 594–596
  colors 588
  5x5 rule 588–590, 594, 612
  fonts 590
  graphics 587–588
  graphs 593
  lettering 590
  one idea per slide 588
  for oral presentations 587–596
  preparation 587
  textual highlights 590
  titles for 592–593
vocabulary, oral presentations 603–
  604. *See also* nomenclature and
  terminology; word choice; words
voice. *See also* tone
  active 50–51, 76–77, 316
  passive 50–51, 76–77, 266–267
  in technical sentences 50–51, 76–77

weak verbs 56–57
Web of Science 148
word choice 15f
  adjectives and adverbs 19
  calculate 269
  clearly/it is clear/obvious 289–390,
    492–493

determine 269
did not 289
frequently 15
handling language sensitively 20
jargon 21f 268, 330, 335
for Materials and Methods
    section 268–269
measure 269
misused words 17–18, 22
oral presentations 603–604, 615
precision 15, 30, 448
in preliminary results
    section 492–493
quantify 269
quantitate 269
redundancies 21–23
Results section 289–290
sexism 20
significant 290, 493
simple words 16–17, 335
smoothers 603–604
special cases 17–20
suffixes 17
transitions 603–604
unnecessary 22–23, 113–115
verbs 18
with 15–16
wordiness 110
word location
    coherence and cohesion and 103–105
    competition for emphasis 37–39

complexity and 38–39
consistent point of view 40–43, 104
flow and 40–44, 45–46, 103
jumping word location 40
mixed in paragraphs 40, 104
problems 45–46
readers' expectations 36
subjects 41–43
verbs 43–44
words
    misused 17–18, 31–32, 657–671
    overview 113–14
    simple 16–17, 29–30, 335
    transition 111t
    unnecessary 22–23
writers 7–10
writer's block 139–140
writing genres 10f
    most common problems 10
    science writing vs scientific writing
        10f
writing process 125f
    audience and journal choice 127
    outlining and composing 133–139
    prewriting 126
    drafting 130
written consent 270

x-axis 189–190, 194f 197f

y-axis 189–190, 194f 197f